The Textbook of Angiogenesis
and Lymphangiogenesis: Methods and Applications

Enrique Zudaire · Frank Cuttitta

Editors

The Textbook
of Angiogenesis
and Lymphangiogenesis:
Methods and Applications

 Springer

Editors
Enrique Zudaire
Radiation Oncology
Angiogenesis Core Facility
National Cancer Institute
National Institutes of Health
Bethesda, MD, USA

Frank Cuttitta
Radiation Oncology
Angiogenesis Core Facility
National Cancer Institute
National Institutes of Health
Bethesda, MD, USA

ISBN 978-94-007-4580-3   ISBN 978-94-007-4581-0 (eBook)
DOI 10.1007/978-94-007-4581-0
Springer Dordrecht Heidelberg New York London

Library of Congress Control Number: 2012952031

Printed on acid-free paper

Springer is part of Springer Science+Business Media (www.springer.com)

Contents

The In Vitro Endothelial Cell Tube Formation Assay in 3D Culture on Gelled Basement Membrane Extract

Irina Arnaoutova, Hynda K. Kleinman, Jay George, and Gabriel Benton

Abstract In 1988, it was first observed that endothelial cells placed in 3D culture on a gelled basement membrane substratum, in vitro, form capillary-like structures with a lumen. Since then, the tube formation assay has been widely used to define angiogenic and anti-angiogenic factors as well as to identify genes and signaling pathways important in angiogenesis. More recently, it has been used to identify and characterize progenitor cells, and to better understand cell-cell organization, functional effects, and relation in 3D co-culture. The assay has been successful because it is quick, reliable, flexible, quantitative, easy to perform, and amenable for high throughput screening. The assay measures endothelial cell adhesion, migration, and tubule formation, and thus, it is more comprehensive than simple migration, invasion, or proliferation assays. Tubule formation requires protein synthesis; however, proliferation is not required. A variety of endothelial cells can be used as well, and modified gene expression may be used to define the function of certain genes in angiogenesis. There have been several reviews and methods papers recently published on this assay which detail the uses and methods.

1 Introduction

Angiogenesis (formation of new blood vessels) – plays an important role in many physiological processes, such as embryonic development, reproduction, tissue repair and wound healing; and also in pathological conditions related to tumor growth,

I. Arnaoutova • J. George • G. Benton, Ph.D. (✉)
Trevigen Inc, 8014 Helgerman Court, Gaithersburg, MD 20877, USA
e-mail: gbenton@trevigen.com

H.K. Kleinman
NIH, NIDCR, Bethesda, MD, USA

E. Zudaire and F. Cuttitta (eds.), *The Textbook of Angiogenesis and Lymphangiogenesis: Methods and Applications*, DOI 10.1007/978-94-007-4581-0_1,
© Springer Science+Business Media Dordrecht 2012

Table 1 Uses of the tube assay

Screen for angiogenic and antiangiogenic factors
Test angiogenesis activity during factor purification
Define signaling pathways in angiogenesis
Identify genes up or down regulated in angiogenesis
Define genes functionally important for angiogenesis
Identify and characterize endothelial progenitor cells
Monitor endothelial progenitor cell number in patient blood during therapy
Co-culture experiments (endothelial cells + smooth muscle cells or tumor cells or progenitors or etc.)

rheumatoid arthritis, diabetic retinopathy, inflammatory diseases, etc. [1]. The lack of maintenance or new vessel formation has pathological consequences and results in ischemic disorders, coronary artery disease, and chronic wounds. Various factors have been identified which regulate the process of vessel formation. Vessel formation involves several steps: endothelial cell sensing of the stimulus, alignment, migration, breakdown of the extracellular matrix, cellular organization, formation of cell-cell adhesions, proliferation, establishment of the underlying matrix, and lumen formation. Many in vitro assays have been developed which assess one or more of these steps [2, 3]. This chapter will focus on the widely used tube assay which involves 3D culture of endothelial cells on a gelled basement membrane matrix [4–6]. In this assay, the cells attach, migrate, and form tubes with a lumen. The assay has been in use for more than 20 years, and it is easy to perform, reliable, and fast, taking less than a day to obtain quantitative results. An important point about the assay is that it is highly flexible and can be used in many different ways (Table 1). The format is scalable; large wells may be used to generate sufficient samples for thorough biochemical analysis, while smaller wells may be used for high throughput screening. A variety of types of endothelial cells can be used, including primary cells, transformed cells, patient cells, cells from genetically modified animals, cells with up or down regulated genes by transfection or silencing, etc. The assay has been utilized to identify angiogenic and anti-angiogenic compounds, factors, and genes and to define angiogenesis pathways; more recently, considerable activity has been concentrated on stem cells for functional identification and effects with endothelial cells and other cell types [7–19]. Co-culture of endothelial cells with other cell types has also recently been more widely used, not only to study the role of factors, but also to study the role of cells in a more complex and physiological environment (see section below).

The basement membrane matrix underlies the endothelium and functions to maintain tissue integrity, cell adhesion, cell function, and vascular permeability [20]. A basement membrane matrix is isolated from a tumor and is sold commercially as Cultrex Basement Membrane Extract (BME), Matrigel, and EHS extract. It is a liquid at 4°C and gels at 37°C within minutes. The matrix is rich in laminin-1, collagen IV, proteoglycans, and growth factors and is highly biologically active. When used in 3D culture, this matrix promotes the differentiation of normal cells and supports growth of malignant cells [21, 22]. It is widely used as a 3D culture substratum for a variety of cell types, including stem cells and in vivo for tumor growth and angiogenesis assays.

2　Co-culture

More recently 3D co-culture on basement membrane matrix with endothelial cells and other cell types has become an important research tool. This approach has the potential to more closely mimic the in vivo environment and thus examine cross-talk among cells. In particular, endothelial progenitor cell activity has been assessed in co-culture. Bone marrow endothelial progenitor cells cultured with mesenchymal stem cells showed enhance tube formation and increased expression of endothelial cell specific genes, demonstrating a cross-talk between these two cell types [23]. Although smooth muscle-like cells cannot form tubes alone on Matrigel, they will form cooperative tube networks with CD34+ endothelial progenitor cells. The progenitor cells could differentiate into both, smooth muscle-like or endothelia-like cells on basement membrane matrix in co-culture but not in monoculture [24]. In defining the ability of early and late endothelial progenitor cells to form vessels in vivo, the in vitro tube formation assay was used. The late progenitor cells isolated from cord blood could form tubes in monoculture, form cooperative tubes in co-culture with HUVECs, integrate into already formed HUVEC tubes, and integrate into vessels in vivo, while the early progenitor cells from the circulation could not form tubes in any of the conditions and had poor integration into vessels in vivo. This study suggests the reliability of the tube assay for predicting in vivo activity [19]. Endothelial cells co-cultured with retinal pigmented epithelial cells showed an increase in metalloproteinase-2 mRNA levels and enhanced ability to form capillary-like structures suggesting that endothelial cell contact with retinal pigmented epithelium may be involved in choroidal neovascularization, a leading cause of vision loss [25]. Co-culture with endothelial cells and basement membrane matrix has also been used in engineering tissues. Hepatocytes co-cultured with endothelial cells, stellate cells and cholangiocytes in 3D culture on basement membrane showed hepatocytes decorating endothelial tubes resembling the in vivo sinusoids, bile canaliculi, and lumen with retention of albumen secretion and drug metabolism activity. When this mixture of cells and basement membrane were injected in vivo, a liver-like structure formed [26]. There have been numerous studies of endothelial cells and tumor cells in 3D culture on basement membrane extract [22]. For example, CD151 is associated with poor prognosis in breast cancer and CD151+ MDA-MB-231 breast cancer cells showed increased migration in the presence of endothelial cells but not in the presence of fibroblasts in 3D co-culture on basement membrane [27] whereas CD151- cells were not affected. These examples of co-culture in 3D on basement membrane begin to suggest the many important uses this approach has.

3 Basic Assay Scheme

The 3D tube formation assay is performed easily within 1 day [5, 6] using standard techniques [21]. Primary endothelial cells, such as HUVECs (human umbilical vein endothelial cells), should have no more than seven passages, and they should be at no more than 80% confluence prior to setting up the assay. The cells must be fed or passaged the day before they are harvested to set up the assay. The basement membrane matrix is thawed overnight at 4°C, and plated into the wells that are kept cold on ice. BME is a liquid at 4°C and will polymerize into a gel at temperatures above 10°C, so all handling should be done on ice. The vials containing BME should be inverted gently to mix prior to pipetting. Plates may be centrifuged at $300 \times g$ for 10 min at 4°C to eliminate air bubbles in the BME. The coated plate should be incubated at 37°C to promote gel formation. The endothelial cells should be trypsinized to harvest, and the trypsin should be neutralized with complete medium as soon as the cells detach from the tissue culture flask. Over-trypsinization can be detrimental to cell health and adversely affect the assay. The cells should be gently pipetted up and down several times with a serological pipet to break up clumps and create a single cell suspension. The cells are then counted and diluted as needed, and they are plated at about 45,000 cells per cm^2 in their corresponding assay medium (or 15,000 cells per well of 96-well plate). Positive controls, such as 2% serum or 35 ng/ml bFGF, should be included in separate wells, as well as a negative control of serum-free medium. If the assay is to test for stimulators of angiogenesis, the growth factor reduced basement membrane matrix is recommended; whereas if an inhibitor of angiogenesis is to be tested, then regular basement membrane matrix can be used. The cells are incubated on top of the gelled matrix for 2–24 h. The shorter time is recommended for transformed endothelial cells, such as SVEC4-10 (SV40 transformed murine endothelial cells), which form tubes rapidly (during 2–3 h) and the longer time is recommend for primary endothelial cells, such as HUVECs, which form tubes more slowly (4–18 h). To our knowledge, both endothelial cell types respond similarly to all of the factors tested to date. For co-culture, another cell can be added either at the start of the assay or after the endothelial tubes have formed. At no more than 24 h after plating the cells, the tubes are photographed and measured, as needed. The length of the tubes, number of branch points and/or area of the structures have been used as quantitative measurements (Figs. 1 and 2).

4 Tips for Reproducible Results

Where possible, using the following materials and techniques will improve the reproducibility of your results and maintain high quality of tube formation assay. The vials of BME can be aliquotted and refrozen in usable amounts; store it at −80°C for maximum stability. To avoid air bubbles when plating, cool the

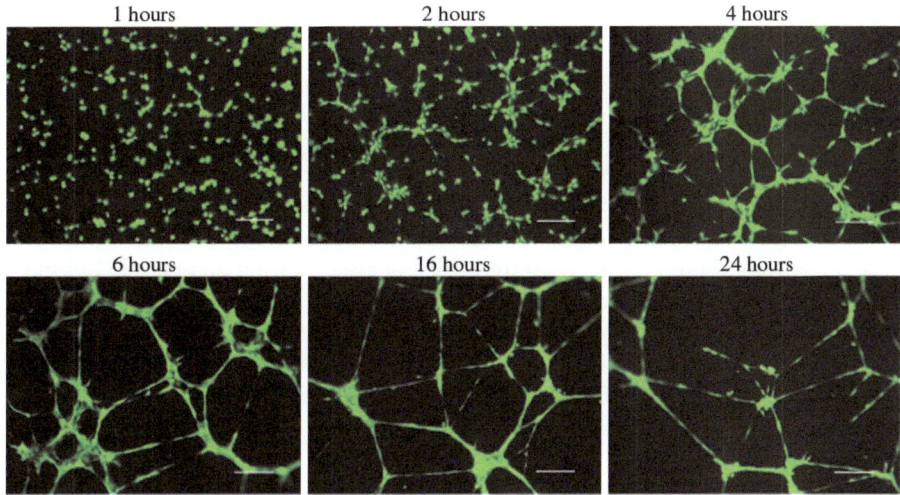

Fig. 1 Dynamics of tube formation by endothelial cells (HUVECs) being plated on reduced growth factor basement membrane extract (RGF BME) in EGM-2 medium. During the first 4 h, cells attach to the matrix, migrate towards each other, and form capillary–like structures which mature by 6–16 h. After 22–24 h, tubes are breaking apart and detaching from the matrix. *Scale bar*, 200 μm

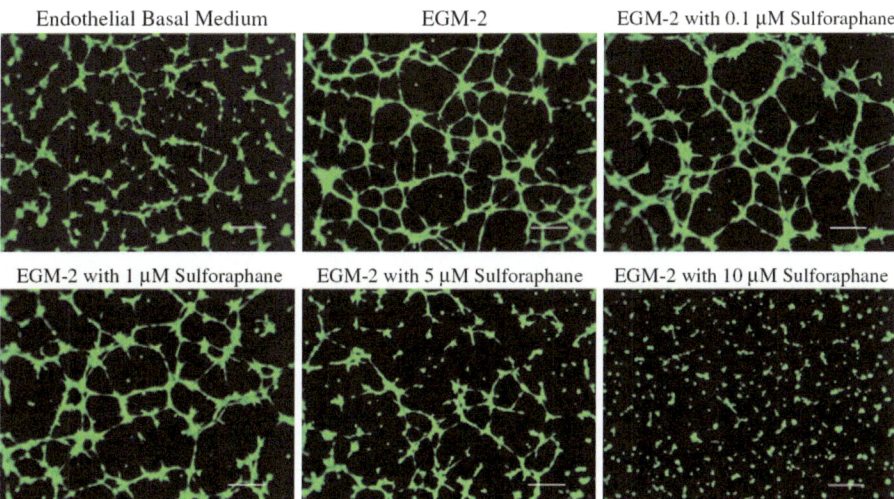

Fig. 2 Effect of angiogenic inhibitor Sulforaphane on the endothelial cell tube formation. HUVECs have been plated on RGF BME and incubated for 5 h in the EGM-2 medium in the absence or presence of Sulforaphane. *Scale bar*, 200 μm

dishes before adding the matrix. Also using a wide bore pipette tip (you may need to cut the pointed end with the sterile scissors) will reduce bubbles in the matrix. Do not try to eject the last bit of matrix as air will come out as well and cause a bubble.

If air bubbles get trapped in the wells, centrifuge the plate at $300 \times g$ for 10 min at 4°C; make sure that centrifuge is pre-cooled to 4°C before placing a plate with basement membrane extract in it. Always use the same amount of matrix per well as varying the stiffness will alter the quality of the tubes formed. We recommend between 150 and 250 ul per cm^2 (or 50 ul per well of 96-well plate). The cells should be at 80% confluence and fed the night before. If using HUVECs or other primary cells, do not exceed passage number 7 as some important properties of the cells change. We found that HUVECs loose their estrogen receptors after passage 6 [28]. The cells should not be used immediately after thawing and culture but rather should be passaged twice before use. Also, use the same lot of positive control protein (s) at the same concentration. We and others have found that VEGF is not useful as a positive control unless used in high amounts. For the best quantitation, we recommend that each test be performed at least in triplicate, better in quadruplicate.

4.1   Protocol (adapted from Arnaoutova et al.[6])

Equipment:

Cell culture incubator (humidified, 5% CO_2)
Biological hood with laminar flow and UV light
Pipettes
Sterile micropipette
37°C water bath
Refrigerated centrifuge with a swing-bucket rotor
Hematocytometer or automatic or cell counter
Inverted phase microscope with 4X and 10X objectives
Inverted phase microscope with fluorescence and 4X and 10X objectives

Materials

Reagents: (sources are given for convenience where appropriate, otherwise any source is acceptable)
Cultrex® Basement Membrane Extract, no phenol red, reduced growth factor (RGF BME; Trevigen, cat. no. 3433-005-01) or Matrigel (BD Bioscience)
Human Umbilical Vein Endothelial Cells (HUVEC; Lonza, cat.no. C2517A)
Trypsin EDTA, 1X
Phosphate-Buffered Saline, 1X
Trypan Blue Solution, 0.4%
Endothelial Cell Growth Medium-2 (EGM-2; Lonza, cat. no. CC-3162)
Endothelial Basal Medium-2 (EBM-2; Lonza, cat. no. CC-3156)
Sulforaphane (Sigma, cat. no. S4441)
Dimethyl Sulfoxide (DMSO)
Dulbecco's Phosphate Buffered Saline, 1X
Calcein AM (Trevigen, cat. no. 4892-010-01)
Cultrex® Cell Staining Solution (Trevigen, cat. no. 3437-100-01)

Methanol
96-well cell culture plates
15 ml conical centrifuge tubes-sterile
Tissue culture flasks, 25 cm^2, filter cap, 50 ml
Disposable sterile plastic pipettes

Reagent Preparation:

Endothelial Cell Growth Medium-2 (EGM-2): Add all supplements and growth
  factors of the EGM-2 SingleQuot Kit to Endothelial Basal Medium-2 (EBM-2)
  and store at 4°C for no longer than 1 month.
bFGF stock solution: Make a 10 ng/µl stock of bFGF in sterile PBS with 0.1%
  bovine serum albumin, aliquot and store at −20°C for no longer than 6 month.
10 mM Sulforaphane: Make a 10 mM stock of Sulforaphane in DMSO solution and
  store at 4°C for no longer than 1 month. For long-term storage, aliquot and store
  at −20°C.
2 mM Calcein AM stock solution: Dissolve lyophilized Calcein AM in DMSO
  solution and keep at 4°C for no longer than 1 month. For long-term storage
  aliquot and store at −20°C.

4.2 Procedure:

Day 0 (day before assay start)

1. Using standard procedures to passage cells, split nearly confluent flasks of
   HUVECs such that cells will be approximately 80% confluent in 24 h. Generally,
   plating 5×10^5– 1×10^6 cells in a 25 cm^2 flask works well.
2. Remove the basement membrane extract from the −20°C or −80°C freezer and
   place on ice in a refrigerator at 4°C.

Day 1 of the assay

3. Place a tube of fully thawed basement membrane extract and labeled 96-well
   plate on ice in a laminar flow hood. Invert a tube with extract a few times to
   mix. Load 50 µl of the basement membrane extract per well of 96-well plate.
4. If air bubbles are trapped inside the extract in the wells, centrifuge the plate at
   4°C for 10 min at 300 g using pre-cooled centrifuge with swing-bucket rotor.
5. Gently transfer 96-well plate to a cell culture incubator and incubate it at 37°C
   for 30 min to allow the basement membrane extract to gel.
6. Warm the PBS, Trypsin-EDTA, Endothelial Cell Growth Medium-2 (EGM-2)
   and Endothelial Basal Medium-2 (EBM-2) in the 37°C water bath.
7. Remove and discard the media from 25 cm^2 flask with HUVECs and rinse cells
   with PBS. Add 1 ml of Trypsin-EDTA to the flask, swirl briefly, and incubate at
   37°C for a few minutes to release the cells. Tap the side of the flask to be sure
   that the cells are detaching.

8. Add 1 ml of growth medium EGM-2; gently pipette solution up and down to make a single cell suspension. Transfer cell suspension in a sterile 15 ml conical tube.

9. Determine cell number and cell viability by mixing 5 µl of cell suspension with 5 µl of Trypan blue and using a hemocytomoter or other cell counter. Typically 1×10^6–1.5×10^6 cells can be harvested from one 25 cm^2 flask at 80–90% confluency.

10. Centrifuge cells at 200 $\times g$ for 3 min in a centrifuge with a swing-bucket rotor.

11. Aspirate supernate and resuspend cell pellet in a basal medium EBM-2 at a concentration of 1.5×10^5 cells per 1 ml. The cells should be gently pipetted up and down a few times to obtain a single cell suspension.

12. Label three sterile conical tubes as follows:

    A. EBM-2 (basal medium without growth factors)
    B. EGM-2 (complete growth medium containing bFGF, VEGF and other growth factors and supplements)
    C. EGM-2 with 10 µM Sulforaphane (inhibitor of angiogenesis)

13. Load 1 ml (1.5×10^5 cells) of a single cell suspension (step 11) in each tube. Centrifuge tubes for 3 min at 200 $\times g$ and aspirate supernate. Leftover cell suspension can either be discarded or seeded into 25 cm^2 cell culture flask in complete growth EGM-2 medium for a future use. Do not disturb cell pellets when aspirating supernate.

14. Add corresponding medium to cell pellets as follows:

    A. 1 ml basal medium EBM-2
    B. 1 ml complete medium EGM-2 and 1.0 µl DMSO
    C. 1 ml complete medium EGM-2 and 1.0 µl of 10 mM Sulforaphane stock solution

15. Carefully resuspend cell pellets in the medium to make single cell suspension at a concentration 1.5×10^5 cells/ml.

16. Gently add 100 µl (15,000 cells) per well of the single cell suspension prepared in steps 14–15 to the corresponding labeled wells of a 96-well plate on top of the gelled basement membrane extract. Be sure that cells are well mixed when adding them to the well. Do not touch the surface of the gel when adding the cells, and add the cells slowly to avoid damage to the gel surface.

17. Incubate 96-well plate at 37°C, 5% CO_2 in the cell culture incubator for a period of 4–16 h, or until desired result is achieved. Examine the plate every hour for tube formation under an inverted microscope with 4X or 10X objective.

18. Once tube formation is observed, there are different options that can be employed to acquire images for quantitation, depending on equipment available:

    18.1 Live tubular network imaging–Photograph the tubular network in the wells using digital camera attached to an inverted microscope with 4X or 10X objective.

18.2 Imaging of fluorescently-labeled live tubes.

    18.2.1 Prepare 6 µM Calcein AM by adding 3 µl of 2 mM Calcein AM stock solution to 1 ml EBM-2 medium.

    18.2.2 Without aspirating the medium, add 50 µl of 6 µM Calcein AM solution per well of 96-well plate.

    18.2.3 Incubate the plate at 37°C, 5% CO_2 for 15–30 min.

    18.2.4 Calcein AM-labeled cells should be observed and photographed using a fluorescent inverted microscope with 485 nm excitation/ 520 nm emission filter.

18.3 Imaging of fixed and stained tubular network.

    18.3.1 Gently aspirate medium from wells and gently rinse wells three times with 100 µl PBS per well.

    18.3.2 Gently aspirate last wash, gently add 100 µl of −20°C- cold methanol per well and incubate the plate for 30 s to a minute (no longer!).

    18.3.3 Gently aspirate methanol and immediately gently rinse wells three times with dH_2O.

    18.3.4 Gently aspirate last wash, gently add 100 µl Cell Staining Solution per well and incubate the plate for 15–30 min at room temperature.

    18.3.5 Rinse wells three times with dH_2O.

    18.3.6 Endothelial cells can be photographed using digital camera attached to an inverted microscope with 4X or 10X objective.

References

1. Folkman J (2003) Angiogenesis and apoptosis. Semin Cancer Biol 13:159–167
2. Auerbach R, Lewis R, Shinners B, Kribai L et al (2003) Angiogenesis assays: a critical overview. Clin Chem 49:32–40
3. Benelli R, Albini A (1999) In vitro models of angiogenesis: the use of Matrigel. Int J Biol Markers 14:243–246
4. Kubota Y, Kleinman HK, Martin GR, Lawley TJ (1988) Role of laminin and basement membrane in the differentiation of human endothelial cells into capillary-like structures. J Cell Biol 107:1589–1597
5. Arnaoutova I, George J, Kleinman HK, Benton G (2009) The endothelial cell tube formation assay on basement membrane turns 20. Angiogenesis 12:267–274
6. Arnaoutova I, Kleinman HK (2010) In vitro angiogenesis: endothelial cell tube formation on a gelled basement membrane extract. Nat Protoc 5:628–635
7. Kinsella JL, Grant DS, Weeks BS, Kleinman HK (1992) Protein kinase C regulates endothelial cell tube formation on basement membrane matrix, Matrigel. Exp Cell Res 199:56–62
8. Grant DS, Kinsella JL, Kibbey MC et al (1995) Matrigel induces thymosin beta4 gene in differentiating endothelial cells. J Cell Sci 108:3685–3694
9. Grove AD, Prabhu VV, Young BL et al (2002) Both protein activation and gene expression are involved in early vascular tube formation in vitro. Clin Cancer Res 8:3019–3026

10. Fukushima K, Murata M, Hachisuga M et al (2008) Gene expression profiles by microarray analysis during matrigel-induced tube formation in a human extravillous trophoblast cell line: comparison with endothelial cells. Placenta 29:898–904

11. Cid MC, Grant DS, Hoffman GS et al (1993) Identification of haptoglobin as an angiogenic factor in sera from patients with systemic vasculitis. J Clin Invest 91:977–985

12. Grant DS, Tashiro K, Sequi-Real B et al (1989) Two different laminin domains mediate the differentiation of human endothelial cells into capillary-like structures. Cell 58:933–943

13. Ades EW, Candal FG, Swerlick RA et al (1992) HMEC-1: establishment of an immortalized human microvascular endothelial cell line. J Invest Dermatol 99:683–690

14. Shen JS, Meng XL, Schiffmann R et al (2007) Establishment and characterization of Fabry disease endothelial cells with an extended lifespan. Mol Genet Metab 92:137–144

15. Grant DS, Lelkes PI, Fukuda K, Kleinman HK (1991) Intracellular mechanisms involved in basement membrane induced blood vessel differentiation in vitro. In Vitro 27:327–335

16. Elkin M, Miao HQ, Nagler A (2000) Halofuginone: a potent inhibitor of critical steps in angiogenesis progression. FASEB J 14:2477–2485

17. Haralabopoulos GC, Grant DS, Kleinman HK (1994) Inhibitors of basement membrane collagen synthesis prevent endothelial cell alignment in Matrigel in vitro and angiogenesis in vivo. Lab Invest 71:575–582

18. Bagley RG, Walter-Yohrling J, Cao X (2003) Endothelial precursor cells as a model of tumor endothelium: characterization and comparison with mature endothelial cells. Cancer Res 63:5866–5873

19. Mukai N, Akahori T, Koaki M (2008) A comparison of the tube forming potentials of early and late endothelial progenitor cells. Exp Cell Res 314:430–440

20. Kleinman HK, Martin GR (2005) Matrigel: basement membrane extracellular matrix with biological activity. Semin Cancer Biol 15:378–386

21. Benton G, George J, Kleinman HK, Arnaoutova I (2009) Advancing science and technology via 3D culture on basement membrane matrix. J Cell Physiol 221:18–25

22. Benton G, Kleinman HK, George J, Arnaoutova I (2011) Multiple uses of basement membrane-like matrix (BME/Matrigel) in vitro and in vivo with tumor cells. Int J Cancer 128:1751–1757

23. AquDirre A, Planell JA, Engel E (2010) Dynamics of bone marrow-derived endothelial progenitor cell/mesenchymal stem cell interaction in co-culture and its implications in angiogenesis. Biochem Biophys Res Commun 400:284–291

24. Guo S, Cheng Y, Ma Y, Yang X (2010) Endothelial progenitor cells derived from CD34+ cells form cooperative vascular networks. Cell Physiol Biochem 26:679–688

25. Dardik R, Livnat T, Nisgav Y et al (2010) Enhancement of angiogeneic potential of endothelial cells by contact with retinal pigmented epithelial cells in a model stimulating pathological conditions. Invest Ophthalmol Vis Sci 51:6188–6195

26. Soto-Gutierrez A, Navaro-Alvarex N, Yagi H et al (2010) Engineering of an hepatic organoid to develop liver assisted devices. Cell Transplant 19:815–822

27. Sadej R, Romanska H, Baldwin G et al (2009) CD151 regulates tumorigenesis by modulating the communication between tumor cells and endothelium. Mol Cancer Res 7:787–798

28. Morales DE, McGowan KA, Grant DS et al (1995) Estrogen promotes angiogenic activity in human umbilical vein endothelial cells in vitro and in a murine model. Circulation 91:755–763

Using Polymers to Build Three Dimensional Coculture Systems for Angiogenesis

Erin B. Lavik

Abstract Combining stem cells with polymers allows one to build new microenvironments and to investigate the role surface molecules, mechanical properties, and soluble factors play on stem cells in a physiologically relevant three dimensional system. In this chapter, we will focus primarily on commercially available polymers and those that can be made very simply with a limited knowledge of chemistry. We will also cover simple techniques to process the polymers to create architectures that can direct cell interactions.

1 Introduction

The concept for combining polymer scaffolds or substrates with cells is as the basis of the field of tissue engineering. It was observed that the introduction of cells alone often could not recapitulate large defects. However, when a polymer scaffold or substrate was combined, some cells have been able to recapitulate tissues [1]. As the field has developed, researchers have come to realize that the polymer scaffolds can be used to recapitulate microenvironments to direct the survival and differentiation of cells including stem cells [2–4].

Polymer scaffolds have become more than a temporary structure. They are being used a tools to engineer the molecular interactions between cells and their environment. They can be used to investigate the molecular cues that influence cell fate and tissue development. By choosing the appropriate polymer and process one can control the mechanical properties, the surface interactions, and the soluble factors that interact with the stem cells. This approach holds tremendous promise for new insight into matrix biology, but to capitalize on this, one must have a basic understanding of the polymers that may act as the foundation for these microenvironments.

E.B. Lavik (✉)
Case Western Reserve University, Cleveland, OH, USA
e-mail: erin.kavuk@case.edu

E. Zudaire and F. Cuttitta (eds.), *The Textbook of Angiogenesis and Lymphangiogenesis: Methods and Applications*, DOI 10.1007/978-94-007-4581-0_2,
© Springer Science+Business Media Dordrecht 2012

In this chapter, we will focus primarily on commercially available polymers and those that can be made very simply with a limited knowledge of chemistry. We will also cover simple techniques to process the polymers to create architectures that can direct cell interactions.

1.1 The Degradable Polyesters

Some of the most commonly used polymers for building scaffolds and designing microenvironments for stem cells are based on degradable polyesters. These polymers are commercially available, and they can be processed into a wide range of structures. Furthermore, by choosing the molecular weight of the polymer, and, in the case of copolymers, the ratio of subunits, one can tailor how stiff the material is, and how fast it degrades.

1.1.1 PLA, PGA, and PLGA

Poly(lactic acid) (PLA), poly(glycolic acid) (PGA), and their copolymer, poly (lactic-*co*-glycolic acid) (PLGA), have been employed in a number of devices approved for use by the FDA [5, 6]. They are sometimes called polyglycolide and polylactide in reference to their monomers. They degrade by hydrolysis wherein water cleaves the ester linkages. While they can degrade by either base hydrolysis (catalyzed by base) or acid hydrolysis, catalyzed by acid, in the body, they degrade by acid hydrolysis to lactic and glycolic acids, which are excreted. The degradation rate is controlled by the molecular weight of the polymer, the crystallinity, and the ratio of glycolic acid to lactic acid subunits. Since lactic acid is more hydrophobic than glycolic acid due to the methyl group, hydrolysis is slower for PLA than PGA of the same molecular weight and degree of crystallinity.

PLA and PLGA in particular are highly processable making them very attractive for the fabrication of complex structures and the degradation can be tailored from a few weeks to years. PGA processing is more limited due to the limited number of solvents in which the polymer will go into solution. Because of their easily tailored degradation, they have been studied extensively for drug delivery purposes. However, the PLGA and the related homopolymers are rather brittle, lack functionalities other than end groups for chemical modification, and exhibit bulk rather than surface degradation which can produce a non-uniform release profile which is less than ideal for certain drugs. These limitations have motivated the development and study of a number of complimentary polymers.

1.1.2 PCL and PHA—More Elastomeric Materials

Polycaprolactone (PCL) is a degradable polyester that is in the FDA approved monocryl suture material sold by Johnson & Johnson. The glass transition

Fig. 1

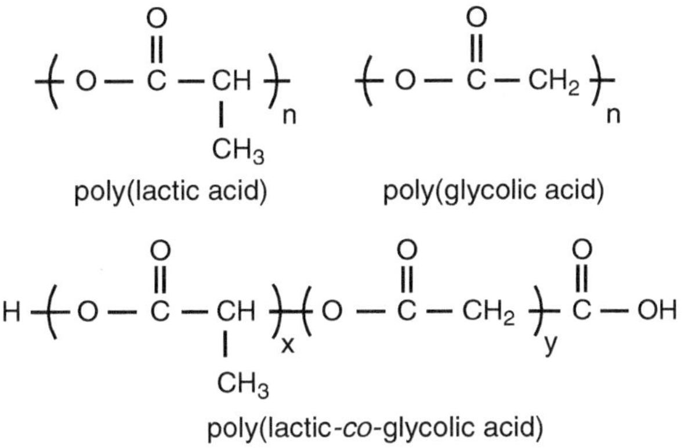

poly(lactic acid)   poly(glycolic acid)

poly(lactic-*co*-glycolic acid)

temperature or T_g of PCL is molecular weight dependent, as is the T_g for polymers generally, but the T_g of PCL is much lower than PLGA (approximately $-60°C$ for PCL as compared to approximately 40–60°C for PLGA [7]). This means that at room temperature and body temperature, PCL is above its glass transition temperature and behaves in a more elastic manner than PLGA. It is often used as a scaffolding material for cells associated with flexible tissues because it tends to be more flexible. Copolymers of caprolactone and lactic acid are also commercially available and allow one to have further control over the stiffness and degradation. The greater degree of lactic acid in the backbone tends to increase the stiffness and decrease the degradation rate.

Poly(hydroxyalkoanoate) (PHA) is an extremely elastic polyester. It is made by bacteria and can be made in plants including corn. It is used as a storage depot by these organisms. The side chains form entanglements that act similarly to cross links. PCL and PHA have been found to be extremely useful when trying to emulate the properties of highly elastic tissues such as arteries and tendon [8].

The modulus or stiffness of a material can play a critical role in angiogenesis [9, 10], so it is helpful to have a range of materials with different stiffnesses for angiogenic applications.

1.2 Hydrogels

Hydrogels are either physically or chemically crosslinked water-soluble polymers that, through crosslinks, become water-swollen network structure. They are generally highly biocompatible due to the enormous amount of water associated with them. Their mechanical properties parallel the properties of soft tissues, and some have been designed to be injected as a liquid which gels in situ [11]. One of the attractions of using hydrogels over the degradable polyesters or other systems is that one can generally visualize the cells within the gels over time, facilitating live imaging. Almost all of the hydrogels described here have some degree of

Fig. 2

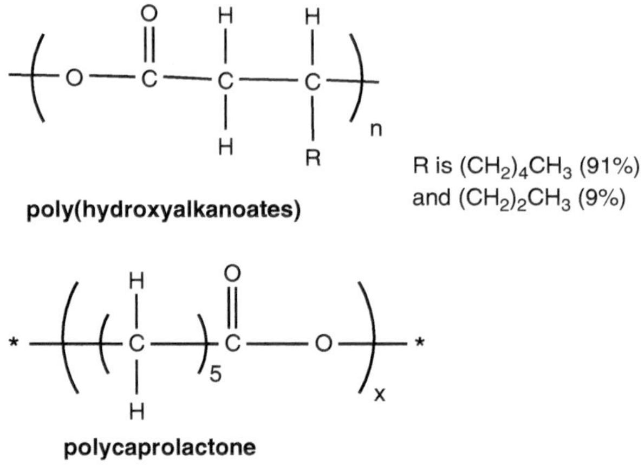

poly(hydroxyalkanoates)

R is $(CH_2)_4CH_3$ (91%)
and $(CH_2)_2CH_3$ (9%)

polycaprolactone

Fig. 3

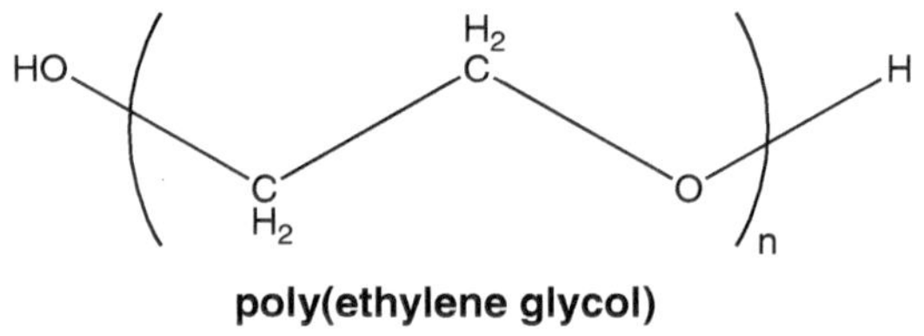

poly(ethylene glycol)

autofluorescence, but by using cell-labels that are excited at long wavelengths, one can mitigate a great deal of the issue.

One of the best known hydrogels used in three-dimensional cell culture is Matrigel®. Matrigel is simple to use, cells can migrate and organize in it [12], and it is commercially available. Another common gel for 3D cell culture is collagen. One can either purchase collagen or do a collagen prep [13]; again, the material is easily obtained and forms a gel readily. The downside of these two gels is that there are strong limits on how much they can be altered. The stiffness can be varied slightly by changing the concentration of the polymer in the gel, but the polymer essentially dictates the chemistry. Synthetic hydrogels or gels based on synthetic and naturally occurring polymers (such as collagen, hyaluronic acid, or other materials) open the door for more variation in the matrix and the potential for directly assessing the role the matrix has on stem cells.

A significant number of synthetic hydrogels used as scaffolds for cells are poly (ethylene glycol) (PEG) -based (Fig. 3). PEG is well tolerated by the body up to a molecular weight of approximately 10,000 g/mole. Above this, the kidneys may have trouble clearing the polymer. It is also soluble both in organics and water making it unique among polymers. In fact, it is highly soluble in water and takes on an extremely hydrated conformation involving an alpha-helical structure with water molecules clustering around the oxygen atoms in the backbone [14]. PEG does not degrade, and hydrogels made from PEG do not degrade unless one uses a degradable linker, peptide, protein, or degradable block such as PLA.

Chemically cross-linked hydrogels may be cross-linked in situ using a chemical initiator [15], or a photoinitiator if the polymer has suitable reactive groups on it such as acrylate groups [2]. The photoinitiated hydrogels use initiators that do not become active until they are exposed to a light of a particular wavelength. The advantage of this over standard initiation schemes is that the hydrogel gelation may be tightly controlled; the un-gelled solution may be injected and then gelled in an controlled manner. These materials also gel quickly, generally in minutes which can be very helpful in encapsulating cells. Photoinitiated hydrogels have been shown to be able to be injected with cells with high cell survival [16, 17] as well as used to deliver growth factors [18].

Physically cross linked hydrogels rely on phase separation of the blocks of block copolymers to gel. The phase separation is generally temperature dependent and reversible. Examples include poly(ethylene)-b-poly(propylene)-b-poly(ethylene) (Pluronics) which has been used in a number of tissue engineering studies [19, 20] and degradable block copolymer systems including ones based on poly(ethylene oxide) and poly(lactic acid) [21, 22].

There are a number of challenges to using hydrogels as the basis for directing stem cells, but the inherent challenges, if addressed, may become the strongest features of using hydrogels to study and control the stem cell microenvironment. Because hydrogels are composed primarily of water soluble polymers, they are very hydrated and typically resist protein adsorption and cell attachment. However, the gels may be chemical modified to include peptides or other moieties to promote cell attachment [16, 23–26]. Therefore, by controlling the concentration of peptides or other groups, one can begin to look at the specific effects of components of the ECM or other molecular structures on stem cell behavior.

1.3 What We Can Model In Vitro with These Systems

Several groups have explored the interactions between endothelial cells and other cell types, the growth factors involved in these interactions, and the role of the matrix molecules in these cocultures [27–30]. These studies have been carried out either in two dimensions in a dish or in vivo or in slice culture. Polymers have the potential to be an excellent intermediate between the reductionist approach of two dimensional culture and the complexities of in vivo or slice systems for investigating interactions [31–34]. Ultimately, these polymer-based approaches may also have therapeutic value in building new vascular systems and several groups look at the constructs with this goal in mind [9, 35–37]. Regardless of one's ultimate goal, being able to select materials with the appropriate properties allows one to move into three dimensions, and there are several systems which are relatively simple that can be employed to take these steps.

Both hydrophobic polymers and hydrogels have been used to grow cells in three dimensions. The former includes polymers such as polystyrene, Dacron, PLA, and PLGA. The latter includes collagen gels, polyacrylamide gels, and PEG gels.

The hydrophobic materials can be processed in a multitude of ways, have a tremendous range of properties, and are capable of being very strong. Hydrogels exhibit properties that are similar to native ECM such as mechanical stiffness but can be more challenging to process and are inherently weaker than many of their hydrophobic counterparts making them challenging at times to handle for cell culture. A few straightforward techniques can help to alleviate some of the challenges.

In tissue culture dishes in two dimensions, endothelial cells are able to organize and form tubes with functional lumen [38]. Endothelial cells are also able to organize and form tubes in collagen. The tube formation assay either in 2D or in collagen in 3D is often used to demonstrate the bioactivity of angiogenic factors [39–42]. Hydrophobic polymers have been used to culture endothelial cells as well and capillary structures have been seen in these materials.

The critical questions become what properties, exactly, do we need and do we want to study with respect to the matrix in three dimensional models. The major properties that have drawn interest with ECM mimics have been the stiffness of the matrix, the presentation (both density and type) of molecules on the matrix, and the pore structure and architecture of the matrix. A number of groups have looked at the role of the elastic modulus on angiogenesis in engineered matrices and found that stiffness does, in fact, impact angiogenesis [9, 10, 43]. There is a balance that needs to be investigated with researcher reporting that very soft matrices with moduli less than 1 kPa may favor angiogenesis when only endothelial cells are cultured [44], but pericytes may require more stiff matrices on the order of 10 kPa or higher [45].

The role of integrins and presentation of peptides for angiogenesis can also be investigated in these model systems under mono- and co-culture conditions [42, 46].

1.4   Moving **In Vivo** *with These Systems: Stable Vessels*

A number of groups have started to investigate the role of different matrices and cell types on angiogenesis and vessel stabilization in vivo using the subcutaneous implantation models. The subcutaneous model is extremely easy to perform, allows one to test several implants or formulations per animal, limiting animal numbers, and can be modified to include a coverslip window to facilitate imaging into the implant over time [47].

The caveat to moving in vivo is that one needs to consider the thickness of the implant. The rule of thumb is that an implant needs to be at least 1 mm thick or thicker to be able to investigate the difference between the host response to the implant itself and the impact of the coculture system one is characterizing. It is therefore very important to normalize the thickness of the implants across groups in these studies.

2   Materials

2.1   Commercially Available Polymers

2.1.1   Degradable Polyesters

PGA, PCL, PLGA, and PLA can be obtained from a wide range of producers including Sigma Aldrich, Polysciences, and Boehringer Ingelheim. PHA has become more challenging to obtain, but a relative of PHA, polyhydroxybuterate (PHB) can be obtained from Sigma Aldrich as well as Polysciences.

2.1.2   Water Soluble Polymers

PEG can be obtained from Sigma Aldrich, Polysciences, and a host of chemical companies. Many of these companies also have chemically modified versions of PEG that may circumvent the need for the chemical reactions outlined in the following sections including acrylated versions and versions that are chemically modified to react with free amines. Our lab uses standard PEG and performs the reactions described below because they are simple and reproducible. Some of the activated PEGs that are commercially available seem to be less reactive in our hands than those we synthesize ourselves.

2.1.3   Other Materials

For PEG/PLL hydrogels: Linear poly-ethylene glycol (PEG), N,N'-Carbonyldii-midazole, Dioxane, Argon, Balloon, Syringe, 2 20 G (1.5 inch) needles, mineral oil bath at 37°C, Spectra-Por dialysis-tubing, 4 l beaker, MilliQ water.

For photopolymerizable PEG/PLL hydrogels: linear PEG 4 K (all OH end groups), methylene chloride (MC) anhydrous, 500 ml round bottom flask, triethylamine (TEA) anhydrous, acryloyl chloride (AC), diethyl ether, dialysis tubing, N,N'-Carbonyldiimidazole (CDI), Dioxane – anhydrous, Argon, Poly-L-lysine MW 1250, Sodium bicarbonate buffer (pH 8.2)

3   Methods

3.1   Fabrication of Hydrogels

3.1.1   Synthesis of PEG/PLL Hydrogels

In our lab, we have developed a PEG/PLL hydrogel. The attraction is that it is a synthetic hydrogel with widely tunable properties that involves extremely simple chemistry. Cells adhere to the continuous network of polylysine. One can also absorb proteins onto the polylysine network or covalently couple molecules through the free amines on the lysines.

Procedure:

Part I: Making the Positive Pressure Device

1. Select a standard oval balloon and a 5 ml syringe.
2. Remove the stopper from the syringe (put it in the regular trash) and place the end of the balloon over the opening in the stopper end of the remaining syringe.
3. Secure the balloon to the syringe by wrapping Teflon tape around the joint.

Part II: Setup of Water Bath and Thermocouple Hot Plate

1. Fill a glass dish 1/3 with mineral oil.
2. Place dish on hot plate and place thermocouple probe just above the bottom of the dish (do not touch the bottom). Add a stirbar.
3. Turn on the stir and heat switch and set temperature set point on the dial to 150°C
4. When the thermocouple LCD stabilizes, press the + key to 37°C (top right number)
5. Wait for the bath temperature to reach 37°C (center main number)

Part III: Reaction (Perform Only in Chemical Hood)

1. Dissolve 7.5 g PEG in 150 ml dioxane @ 37°C.
2. Determine amount of CDI to add. For every hydroxyl there should be an excess of 8 moles of CDI (FW 162.15).
3. Add CDI to the solution (dissolves quickly) and pour into a 250 ml round bottom flask. Place flask just over stirbar in mineral oil bath.
4. Add a stir bar and cover with a rubber septum.
5. Select 2 **sterile** 20 gauge needles. Use new sterile needles for all applications!
6. Insert one needle of onto the balloon/syringe contraption and into the red septum and using the second needle, fill the balloon/flask with argon until the balloon is about the size of the empty space in the flask.
7. Stir for 2 h at 37°C.

Part IV: Dialysis

1. Dialyze in restriction dialysis tubing (size should allow excess CDI to leave and aPEG to remain) in MilliQ water (watching for abundant bubble formation, and leaving the tops of the tubes open and clipped to the sides of the flask until the bubbles cease to form.)
2. Change the water every 2 h for 6 h and then once a day for 72 h.

3. Freeze down the activated PEG solution in several 50 ml centrifuge tubes (never filling past 35 ml) in liquid nitrogen (covered with kimwipes and rubber bands.)
4. Lyophilize for a couple days.
5. Store dessicated and in the freezer.

3.1.2 Synthesis of Photopolymerized Hydrogels

Part I: Acrylation of PEG (Linear or 4-arm)

1. Dissolve 30 g of PEG in 300 ml of anhydrous MC, pour into bottom of 500 ml round bottom flask with a stirbar
2. Quickly add a 9:1 M excess of TEA to PEG-OH groups
3. Add closed off upper flask and flush both flasks with Argon
4. In the upper flask, add a 1:1 M ratio of AC to PEG-OH groups dissolved 1:10 in anhydrous MC
5. Flush flask with Argon
6. Add solution in upper flask to bottom flask but drop-wise due to the exothermic nature of the reaction
7. Stir for 24 h at room temperature
8. Filter out solid precipitate.
9. Precipitate the PEG in 1.4 l of diethyl ether at 4°C
10. Filter to isolate the precipitate
11. Dissolve the precipitate in deionized water and dialyze in water for 72 h
12. Freeze and lyophilize product

Part II: Activation of ACRL-PEG: Activating PEG with N,N′-Carbonyldiimidazole (CDI)

1. Add 7.5 g acrl- PEG and 150 ml anhydrous-dioxane to a 250 ml round bottom flask.
2. Flush with Ar2
3. 8:1 M excess of CDI to PEG OH groups (you can actually do ½ since many are acrylated, but I actually just do an larger excess) to solution and flush with Ar2.
4. Stir for 2 h at 37°C.
5. Dialyze in the restriction dialysis tubing in deionized water for 72 h
6. Freeze and lyophilize
7. Store dessicated at −20°C

Part III: Reacting PLL with Activated ACRL-PEG

1. Dissolve 7 g of PEG with acrylate and activated end-groups in 200 ml
2. 50 mM sodium bicarbonate buffer (pH 8.2)
3. Add drop-wise to 200 mg PLL(1.25 kDa) in sodium bicarbonate buffer.

4. React for 2 h at room temperature stirring constantly.
5. Dialyze in water for 72 h
6. Freeze and lyophilize
7. Store at −20°C

Part IV: Making the Final Gel Product

1. Turn on UV lamp (Black-Ray,UVP B100, 365 nm) – warm up for at least 10 min
2. Prepare a solution of 2.5–5 mg/ml Irgacure 2959 initiator in PBS (protect from light)
3. In a scintillation vial, dissolve polymer initiator solution- vortex until dissolved
4. Place solution under UV light for 4–10 min at 6 mW/cm^2

3.2   Fabrication of Scaffolds Based on Degradable Esters

3.2.1   Fabricating Scaffolds: Salt Leaching

1. Prepare polymer solution. Typically, we use a 5% w/v solution for salt leaching with the solvent being chloroform (1 g polymer/20 ml solvent = 5% solution).
2. Clean plastic vials. The vials are from Cole Parmer #H-08936-00 (phone 708-647-7600). We usually rinse with $CHCl_2$ and then H_2O to remove the residual polymer and salt.
3. Weigh 0.4 g of salt/vial. Add and cap vials.
4. Add 0.24 ml solution to each vial. Cap. Tap to level salt and allow to remain capped for 15–30 min to insure salt is level. (You can use the leveling table with level to make sure your scaffolds are flat.)
5. Uncap vials in hood.
6. Allow CHCl3 to evaporate overnight.
7. Tap out salt-filled scaffolds. They will be very brittle. To get the scaffolds out in one piece, squeeze the containers in one direction, rotate 90° and repeat, then turn over the containers and tap them firmly on the lab bench (on a kimwipe) until they come out.
8. Place the scaffolds in histology cassettes, and place in water (deionized) (2–4 l per 20 scaffolds).
9. Change the water 6X over the course of a day (Approx once per hour).
10. Blot the scaffolds dry.
11. Lyophilize overnight to remove residual water.
12. Store in sealed container with desiccant at −20°C.

3.2.2   Fabrication of Scaffolds: Planar Freezing

Dioxane freezes at 11°C. The degradable polyesters are soluble in liquid dioxane but not solid dioxane. By placing the slide on ice one sets up a temperature gradient between the slide and the surrounding room. The temperature gradient induces the orientation of the dioxane crystals as they form. The solidification expels the polymer. Sublimation of the dioxane using a vacuum system or a lyophilizer leaves behind pores where the dioxane crystallized.

1. Cast 0.4 ml of 5% (w/v) solution of the polymer in dioxane onto a glass slide. I recommend a polymer with a number average molecular weight of 30,000 g/mol or higher. To have the necessary entanglements to stabilize the structure.
2. Place the slide on ice to freeze the dioxane leading to solid–liquid phase separation.
3. After 30–60 s, touch a copper wire which has been immersed in dry ice to the solution to nucleate the freezing process.
4. Once the solution is frozen, place the slides in the freezer at −20 for 1–2 h.
5. Lyophilize the slides. The dioxane is sublimated under vacuum leaving behind a pore structure which is the direct artifact of the solvent crystallization.
6. Remove the scaffolds from the glass. If they are not coming off easily, immerse in water for a few minutes. The slides are very hydrophilic, and the scaffold is very hydrophobic which will help to push the scaffold off of the slide.

3.3 Characterization of Scaffolds

3.3.1 Polymer Characterization

All of the scaffolds here are composed of degradable polymers. While PEG does not degrade, polylysine degrade enzymatically making the gels enzymatically degradable. The degradable polyesters break down by hydrolysis which is acid catalyzed in vivo.

The molecular weight of the polymers drops as they are hydrolyzed which impacts the mechanical properties of the materials. Two of the most basic characterization steps for these materials are determining the architecture of the scaffolds since that directly impacts the ability of the cells to migrate through the material and organize, and the mechanical properties since it is becoming increasingly clear that the mechanical properties have a direct impact on cell behavior, and may influence stem cell behavior and differentiation [48].

3.3.2 Scanning Electron Microscopy (SEM)

Scanning electron microscopy is the simplest way to characterize the architecture of a scaffold whether it be a hydrogel based system or a degradable polyester. The basic principle relies on using a filament to create a stream of electrons that one scans across the surface of a material. Electrons are collected to resolve a three dimensional image. Since polymers are non-conductive, they are usually coated with gold so that the charge does not build up on them.

While there are environmental SEMs in which one can use hydrated materials, the level of hydration is extremely low compared to a hydrogel in normal environments. Therefore, we snap freeze and dry our hydrogels to preserve the architecture of the material.

Preparing a Hydrogel for SEM

Traditional lyophilzation dehydrates and collapses the structure of the hydrogel. However, sublimating the water allows the gel to maintain its swollen architecture.

1. Swell hydrogel in deionized water overnight such that gel reaches equilibrium swelling.
2. Freeze swollen gel at $-20°C$ overnight. While it would seem to make sense to freeze the gel faster using liquid nitrogen, we have found that this disrupts the architecture significantly.
3. Place frozen gel in liquid nitrogen. Keep gel in liquid nitrogen until completely cooled, approximately 1 h.
4. Very quickly, place the gel on the lyophilizer on in a vacuum system (pressure of 100 mTorr or less) before melting occurs. Lyophilize until dry.

3.3.3 Mechanical Behavior

Tensile testing is the most common technique. One needs a very small load cell with most of the polymers used in biomaterials. For hydrogels, parallel plate rheology is the most common means to characterize the system. For the degradable polyesters, tensile testing using an Instron system or other tensile system is often the best choice. We find that a load cell of 1 kN is more than adequate for collecting data with most of the scaffolds described here, but a 100 N load cell would also work with some of the more friable systems. The challenge of using a more sensitive load cell is that if one exceeds the maximum load of the load cell, one can damage the system and the costs of such damage are very high.

3.3.4 Instron Testing

The tensile test is the workhorse of mechanical testing. Instron is a company, and their systems are ubiquitous enough, that tensile testing is sometimes referred to as "Instron testing".

The basic principle is to pull on a sample until it breaks while recording the force and displacement. These can then be converted into stress and strain. The elastic region is the linear, reversible region at low strain, and the plastic region is the irreversible region after the yield stress.

3.3.5 Rheology

Because hydrogels tend to be soft and slippery when hydrated, mechanical testing using an instron can be challenging, even in compression. We have found that the most reliable data can be obtained using a rheometer.

For our hydrogel testing, we use a dynamic frequency sweep test also known as a stress test over a frequency range of 0.01–30 Hz.

3.4 Seeding Cells: General Protocols

The following protocols provide general outlines for approaching cell culture with these materials.

3.4.1 DiI Labeling Cells for In Vitro Experiments

We have found that labeling the cells can allow one to track the cells in the three dimensional scaffolds, particularly the hydrogels.

Preparing DiI Stock Solution

Using CellTracer CM-DiI (C-7001, Molecular Probes) (1 mg solid).
Dissolve 1 mg DiI in 1 ml EtOH.
Aliquot solution in small eppendorf vials (50 μl/vial).
Store in freezer, away from light.

Making DiI Working Solution

Warm DMEM.
Add 50 μl DiI stock to 5 ml warmed DMEM.
Filter w/syringe filter (0.2 μM)

Incubating Cells in DiI Solution

Aspirate medium off 1 T25 dish of cells. Rinse once with PBS.
Add DiI working solution.
Place flask in incubator for 5 min.
Remove flask and place in 4°C fridge for 15 min.
Aspirate DiI solution.
Wash with PBS 2X. Add fresh media. Replace in incubator.

Notes: Protect cells from light at all times. DiI has a relatively short lifetime and high extinction rate. Fixing with formaldehyde should preserve fluorescence of cells.

3.4.2   Coating PEG/PLL Gels with Proteins

Because the gels are charged, proteins adhere to the gels, and one can coat the gels with the protein of interest readily. We have used firbronectin and laminin on the gels with a number of neural cell types.

In the cell culture hood:

1. Prepare PEG/PLL gels with dimensions 1 mm × 5 mm × 10 mm.
2. Sterilize by UV exposure (1 h, 5 mW/cm^2).
3. Prepare protein solution at a concentration of 25 µg/ml.
4. Soak gels for 24 h.
5. Wash three times in sterile PBS.

3.4.3   Cell Encapsulation in the Photopolymerizable PEG/PLL Gels

One of the attractions of the photopolymerizable hydrogels is that cells can be encapsulated in the gels.

In the cell culture hood:

1. Sterilize the ACRL-PEG-PLL macromers and glass scintillation under UV light for 30 min.
2. Dissolve Irgacure 2,959 (I2959) (or another cytocompatible photocrosslinking agent) in culture media at a concentration of 5 mg/ml. and sterilized by 0.22 µm filtration. It can be difficult to dissolve. We often make a stock solution of 100 ml by heating the media and I2959 at 37°C for several days.
3. Suspended the cells in a 10% w/v solution of ACRL-PEG-PLL in photoinitiator solution at a final concentration of 5 × 10^5 cells/ml to 10 million cells/ml.
4. Transfer the cell suspension to a flat bottom glass scintillation vial and place under the UV light (365 nm) for 10 min (6 mW/cm^2).

Remove gels from the scintillation vial and placed into Petri dishes containing 50

3.4.4   Sterilization of Polyester Scaffolds for Seeding

Since PGA, PLA and their copolymers degrade by acid hydrolysis, they will not survive being autoclaved. Commercially available scaffolds arrive unsterilized. While films of these polymers may be successfully sterilized by UV radiation, the UV light will not necessarily penetrate a three dimensional scaffold. Therefore, sterilization by ethylene oxide is the most common technique.

3.4.5   Prewetting of Scaffolds for Seeding

PLA, PLGA, PCL, and PGA are all hydrophobic polymers. Therefore, if one takes a scaffold and immerses it directly in medium, one runs the risk of not having the medium penetrate the three-dimensional porous network. Therefore, it is recommended that the scaffolds be soaked in ethanol to prewet the scaffolds since ethanol wets the polyesters more than water. Soaking the scaffolds in 70% ethanol overnight has also been used in lieu of the sterilization step with ethylene oxide effectively. Essentially, the scaffolds are placed in 70% ethanol at room temperature overnight. Prior to seeding, the scaffolds are washed three times under sterile conditions in PBS and then are seeded with cells as described in the following sections.

3.4.6   Seeding of Scaffolds: Static Seeding

The easiest method for seeding scaffolds is a static method in which a solution of cells in prepared then dripped onto the scaffold in a small dish. After a few hours of incubation to allow for attachment, more medium is added and the scaffold-cell construct is returned to the incubator.

The following protocol was designed for seeding a porous sponge of PLGA, which was 0.75 mm by 1 mm by 6 mm and was approximately 95% porous.

In a cell culture hood:

1. Transfer the scaffolds from 70% ethanol to a 60 mm dish containing 5 ml sterile PBS.
2. Rinse the scaffolds 3X in PBS. Allow scaffolds to remain in PBS while preparing the cell solution.
3. Trypsinize cells and perform a cell count.
4. Centrifuge and resuspend cells at a concentration of 1 million cells/ml.
5. Place each scaffold in a well of a 12-well untreated cell culture dish.
6. Carefully add in a drop-wise fashion 0.12 ml of the cell solution onto each scaffold.
7. Place scaffolds in incubator for 3 h at 37°C and 5% CO_2.
8. After 3 h, very gently add 1.5 ml fresh medium to each well and carefully replace the scaffolds in the incubator for 24 h
9. After 24 h, aspirate the medium and add fresh medium to each well.
10. Subsequently, change medium approximately every 3 days or as often as needed.

3.4.7   Seeding of Scaffolds: Dynamic Seeding

Dynamic seeding and maintenance affords the opportunity for more uniform coverage and better transport of nutrients and gasses throughout the scaffolds. It is the better choice for the seeding of larger scaffolds. The biggest limitation is that few cells adhere so well to a scaffold that they can be seeded dynamically. One will likely have to modify the scaffolds by coating them with an attachment factor or protein to facilitate dynamic seeding. Even if one cannot seed dynamically, one can often culture dynamically once the cells are attached, and this will permit the greatest transport through the scaffold.

While there are a host of ways to seed and maintain scaffolds in a dynamic environment the following protocol using an orbital shaker in the incubator has been found to be very successful for seeding a variety of scaffolds with neural stem cells and is relatively easy. This protocol uses an excess of cells to obtain well-seeded scaffolds very quickly. One can use a lower cell concentration. As long as the cells attach well to the scaffold, which seems to be the case for a variety of neural stem cells on the PLA and PGA scaffolds, one should get good, uniform cell coverage very easily using this method.

In the cell culture hood:

1. Transfer the scaffolds from 70% ethanol to a dish filled with PBS.
2. Rinse 3X in PBS.
3. Allow the scaffolds to soak in PBS while the cell solution is prepared.
4. Trypsinize the neural stem cells, and resuspend in medium.
5. Centrifuge cells. Perform cell count.
6. Resuspend cells to concentration of 1 million cells/ml.
7. Transfer the scaffolds, one to a well in the 12 well dish.
8. Add 2 ml of the cell solution to each of the wells containing a scaffold.
9. Place scaffolds on orbital shaker set to the lowest speed in the incubator (37°C, 5% CO_2).

One day later, aspirate off the old medium and add 2 ml of fresh medium to each well. Replace the scaffolds in the incubator on the orbital shaker.

10. The medium is then changed every few days or as needed, depending on the size of scaffold and number of cells seeded.

4   Examples of Coculture In Vitro

4.1   Methods

4.1.1   Cells Used

In the coculture experiment described, below, endothelial cells were cultured with neural stem cells. The endothelial cells were isolated from the brain and immortalized [49, 50]. They were maintained in brain endothelial cell medium (DMEM, 10% FBS, 10 mM HEPES, 10^{-5} M β − mercaptoethanol, 1% Penicillin/ Streptomycin). Green fluorescent protein (GFP) neural stem cells (NSCs) were isolated from whole brains of postnatal day 1 green fluorescent protein (GFP) mice [51–53]. They were maintained in DMEM/F12 media supplemented with 20 ng/ml epidermal growth factor (EGF), 1 mM L-glutamine, 1% N2 supplement, 1% B27, and 1% penicillin/streptomycin with fungizone. EGF is critical for maintaining NSCs in an undifferentiated state [54]. NSCs were passaged (1:2) approximately every 2 weeks [51].

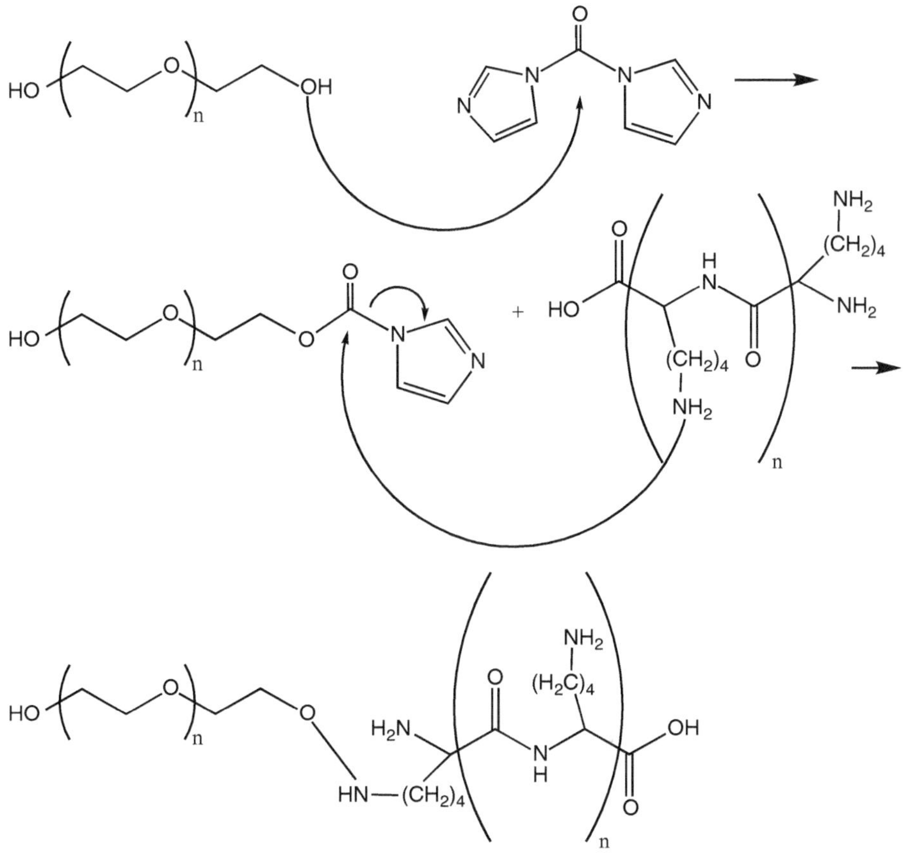

Fig. 4

4.1.2 Hydrogel Preparation

Hydrogel components were sterilized using a 0.22 µm sterile syringe filter prior to crosslinking. The components (activated PEG and polylysine) were mixed in water as described in Sect. 3.1.1 and allowed to gel overnight. Hydrogel disks were cast in scintillation vials and cut using a cork borer to obtain gels that were 1 mm thick and 5 mm in diameter. Gels were sterilized by UV exposure just prior to seeding.

4.1.3 Seeding of Gels

Hydrogel disks 1 mm thick and 5 mm in diameter were seeded with endothelial cells (1 million cells per gel at a concentration of 1×10^6 cells/ml) , NSCs (100,000 cells

Fig. 5

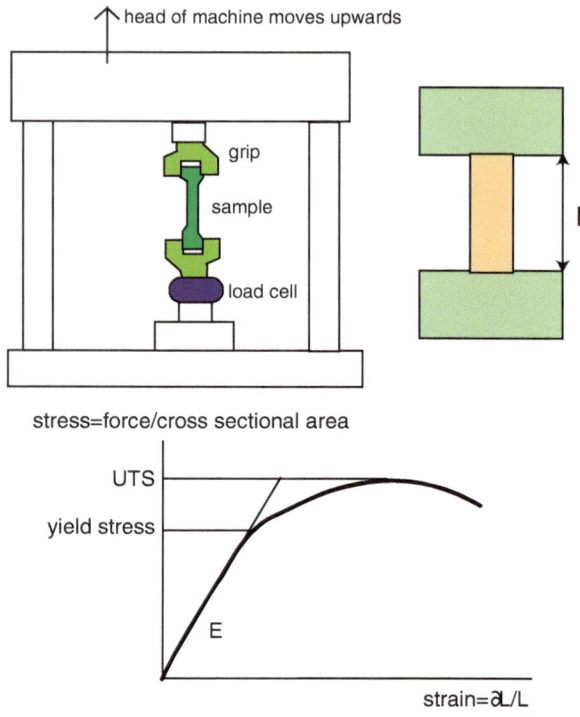

head of machine moves upwards

grip

sample

load cell

L

stress=force/cross sectional area

UTS

yield stress

E

strain=∂L/L

Fig. 6

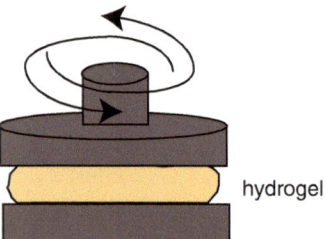

hydrogel

at a concentration of 1×10^5 cells/ml) , or a co-culture of NSCs:endothelial cells at a ratio of 1:10 (1 million BECs and 100,000 NSCs per implant). Seeded gels were maintained in 12 well plates coated with 1% BSA to reduce cell adhesion to the wells.

For in vitro experiments, seeded gels were maintained under static conditions in the endothelial cell media with a media change every 7 days. Hydrogels were fixed in 10% buffered formalin at 3 days and at 1, 2, 4, and 6 weeks for histological analysis.

4.2 Outcomes

In this set of experiments, one of the most exciting findings was that the cells could organize in the gels. NSCs were found to migrate to the endothelial cells. To further augment the coculture system, we started making the hydrogels with macropores. This involves casting the gels around a salt-leached sponge and then degrading the sponge in 1 N NaOH leaving behind a PEG/polylysine hydrogel with macropores in which the endothelial cells can organize. Details of this and the results can be found in [55].

References

1. Langer R, Vacanti JP (1993) Tissue engineering. Science 260:920–926
2. Nguyen KT, West JL (2002) Photopolymerizable hydrogels for tissue engineering applications. Biomaterials 23:4307–4314
3. Chen GP, Ushida T, Tateishi T (2002) Scaffold design for tissue engineering. Macromol Biosci 2:67–77
4. Yang SF, Leong KF, Du ZH, Chua CK (2001) The design of scaffolds for use in tissue engineering. Part 1. Traditional factors. Tissue Eng 7:679–689
5. Lewis DH (1990) Controlled release of bioactive agents from lactide/glycolide polymers. In: Chasin M, Langer R (eds) Biodegradable polymers as drug delivery systems. Marcel Dekker, New York
6. Mooney DJ, Organ G, Vacanti JP, Langer R (1994) Design and fabrication of biodegradable polymer devices to engineer tubular tissues. Cell Transplant 3:203–210
7. Boland ED, Pawlowski KJ, Barnes CP, Simpson DG, Wnek GE, Bowlin GL (2006) Electrospinning of bioresorbable polymers for tissue engineering scaffolds. Polymeric Nanofibers 188–204
8. Zhao K, Deng Y, Chen JC, Chen GQ (2003) Polyhydroxyalkanoate (PHA) scaffolds with good mechanical properties and biocompatibility. Biomaterials 24:1041–1045
9. Kniazeva E, Putnam AJ (2009) Endothelial cell traction and ECM density influence both capillary morphogenesis and maintenance in 3-D. Am J Physiol Cell Physiol 297:C179–C187
10. Krishnan L, Hoying JB, Nguyen H, Song H, Weiss JA (2007) Interaction of angiogenic microvessels with the extracellular matrix. Am J Physiol Heart Circ Physiol 293: H3650–H3658
11. Anseth KS, Metters AT, Bryant SJ, Martens PJ, Elisseeff JH, Bowman CN (2002) In situ forming degradable networks and their application in tissue engineering and drug delivery. J Control Release 78:199–209
12. Valster A, Tran N, Nakada M, Berens M, Chan A, Symons M (2005) Cell migration and invasion assays. Methods 37:208–215
13. Ment L, Stewart W, Scaramuzzino D, Madri J (1997) An in vitro three-dimensional coculture model of cerebral microvascular angiogenesis and differentiation. In Vitro Cell Dev Biol Anim 33:684–691
14. Faraone A, Magazu S, Maisano G, Migliardo P, Tettamanti E, Villari V (1999) The puzzle of poly(ethylene oxide) aggregation in water: experimental findings. J Chem Phys 110:1801–1806
15. Stammen JA, Williams S, Ku DN, Guldberg RE (2001) Mechanical properties of a novel PVA hydrogel in shear and unconfined compression. Biomaterials 22:799–806

16. Mann BK, Gobin AS, Tsai AT, Schmedlen RH, West JL (2001) Smooth muscle cell growth in photopolymerized hydrogels with cell adhesive and proteolytically degradable domains: synthetic ECM analogs for tissue engineering. Biomaterials 22:3045–3051
17. Elisseeff J, Anseth K, McIntosh DSW, Randolph M, Langer R (1999) Transdermal photopolymerization for minimally invasive implantation. Proc Natl Acad Sci USA 96:3104–3107
18. Elisseeff J, McIntosh W, Fu K, Blunk T, Langer R (2001) Controlled-release of IGF-I and TGF-beta 1 in a photopolymerizing hydrogel for cartilage tissue engineering. J Orthop Res 19:1098–1104
19. Liu YC, Chen FG, Liu W, Cui L, Shang QX, Xia WY, Wang J, Cui YM, Yang GH, Liu DL, Wu JJ, Xu R, Buonocore SD, Cao YL (2002) Repairing large porcine full-thickness defects of articular cartilage using autologous chondrocyte-engineered cartilage. Tissue Eng 8:709–721
20. Arevalo-Silva CA, Eavey RD, Cao YL, Vacanti M, Weng YL, Vacanti CA (2000) Internal support of tissue-engineered cartilage. Arch Otolaryngol Head Neck Surg 126:1448–1452
21. Kwon KW, Park MJ, Bae YH, Kim HD, Char K (2002) Gelation behavior of PEO-PLGA-PEO triblock copolymers in water. Polymer 43:3353–3358
22. Peppas NA, Keys KB, Torres-Lugo M, Lowman AM (1999) Poly(ethylene glycol)-containing hydrogels in drug delivery. J Control Release 62:81–87
23. Woerly S, Pinet E, de Robertis L, Van Diep D, Bousmina M (2001) Spinal cord repair with PHPMA hydrogel containing RGD peptides (NeuroGel (TM)). Biomaterials 22:1095–1111
24. Burdick JA, Anseth KS (2002) Photoencapsulation of osteoblasts in injectable RGD-modified PEG hydrogels for bone tissue engineering. Biomaterials 23:4315–4323
25. DeLong SA, Moon JJ, West JL (2005) Covalently immobilized gradients of bFGF on hydrogel scaffolds for directed cell migration. Biomaterials 26:3227–3234
26. Shin H, Zygourakis K, Farach-Carson MC, Yaszemski MJ, Mikos AG (2004) Attachment, proliferation, and migration of marrow stromal osteoblasts cultured on biomimetic hydrogels modified with an osteopontin-derived peptide. Biomaterials 25:895–906
27. Shen Q, Goderie SK, Jin L, Karanth N, Sun Y, Abramova N, Vincent P, Pumiglia K, Temple S (2004) Endothelial cells stimulate self-renewal and expand neurogenesis of neural stem cells. Science 304:1338–1340
28. Ward NL, LaManna JC (2004) The neurovascular unit and its growth factors: coordinated response in the vascular and nervous systems. Neurol Res 26:870–883
29. Li Q, Ford MC, Lavik EB, Madri JA (2006) Modeling the neurovascular niche: VEGF- and BDNF-mediated cross-talk between neural stem cells and endothelial cells: an in vitro study. J Neurosci Res 84:1656–1668
30. Shen Q, Wang Y, Kokovay E, Lin G, Chuang SM, Goderie SK, Roysam B, Temple S (2008) Adult SVZ stem cells lie in a vascular niche: a quantitative analysis of niche cell-cell interactions. Cell Stem Cell 3:289–300
31. Jiang FX, Yurke B, Schloss RS, Firestein BL, Langrana NA (2010) Effect of dynamic stiffness of the substrates on neurite outgrowth by using a DNA-crosslinked hydrogel. Tissue Eng Part A 16(6):1873–1889
32. Seidlits SK, Khaing ZZ, Petersen RR, Nickels JD, Vanscoy JE, Shear JB, Schmidt CE (2010) The effects of hyaluronic acid hydrogels with tunable mechanical properties on neural progenitor cell differentiation. Biomaterials 31(14):3950–3940
33. Bryant SJ, Bender RJ, Durand KL, Anseth KS (2004) Encapsulating Chondrocytes in degrading PEG hydrogels with high modulus: engineering gel structural changes to facilitate cartilaginous tissue production. Biotechnol Bioeng 86:747–755
34. Nuttelman CR, Rice MA, Rydholm AE, Salinas CN, Shah DN, Anseth KS (2008) Macromolecular monomers for the synthesis of hydrogel niches and their application in cell encapsulation and tissue engineering. Prog Polym Sci 33:167–179
35. Peters MC, Polverini PJ, Mooney DJ (2002) Engineering vascular networks in porous polymer matrices. J Biomed Mater Res 60:668–678
36. Borselli C, Oliviero O, Battista S, Ambrosio L, Netti PA (2007) Induction of directional sprouting angiogenesis by matrix gradients. J Biomed Mater Res A 80:297–305

37. Schumann P, Tavassol F, Lindhorst D, Stuehmer C, Bormann KH, Kampmann A, Mulhaupt R, Laschke MW, Menger MD, Gellrich NC, Rucker M (2009) Consequences of seeded cell type on vascularization of tissue engineering constructs in vivo. Microvasc Res 78:180–190

38. Segal MS, Shah R, Afzal A, Perrault CM, Chang K, Schuler A, Beem E, Shaw LC, Li Calzi S, Harrison JK, Tran-Son-Tay R, Grant MB (2006) Nitric oxide cytoskeletal-induced alterations reverse the endothelial progenitor cell migratory defect associated with diabetes. Diabetes 55:102–109

39. Bach TL, Barsigian C, Chalupowicz DG, Busler D, Yaen CH, Grant DS, Martinez J (1998) VE-Cadherin mediates endothelial cell capillary tube formation in fibrin and collagen gels. Exp Cell Res 238:324–334

40. Davis GE, Pintar Allen KA, Salazar R, Maxwell SA (2001) Matrix metalloproteinase-1 and –9 activation by plasmin regulates a novel endothelial cell-mediated mechanism of collagen gel contraction and capillary tube regression in three-dimensional collagen matrices. J Cell Sci 114:917–930

41. Dings RP, van der Schaft DW, Hargittai B, Haseman J, Griffioen AW, Mayo KH (2003) Anti-tumor activity of the novel angiogenesis inhibitor anginex. Cancer Lett 194:55–66

42. Hamada Y, Nokihara K, Okazaki M, Fujitani W, Matsumoto T, Matsuo M, Umakoshi Y, Takahashi J, Matsuura N (2003) Angiogenic activity of osteopontin-derived peptide SVVYGLR. Biochem Biophys Res Commun 310:153–157

43. Sieminski AL, Hebbel RP, Gooch KJ (2004) The relative magnitudes of endothelial force generation and matrix stiffness modulate capillary morphogenesis in vitro. Exp Cell Res 297:574–584

44. Califano JP, Reinhart-King CA (2009) The effects of substrate elasticity on endothelial cell network formation and traction force generation. Conf Proc IEEE Eng Med Biol Soc 2009:3343–3345

45. Lee S, Zeiger A, Maloney JM, Kotecki M, Van Vliet KJ, Herman IM (2010) Pericyte actomyosin-mediated contraction at the cell-material interface can modulate the microvascular niche. J Phys Condens Matter 22:194115

46. Reinhart-King CA (2011) How matrix properties control the self-assembly and maintenance of tissues. Ann Biomed Eng 39:1849–1856

47. Koike N, Fukumura D, Gralla O, Au P, Schechner JS, Jain RK (2004) Tissue engineering: creation of long-lasting blood vessels. Nature 428:138–139

48. Engler AJ, Sen S, Sweeney HL, Discher DE (2006) Matrix elasticity directs stem cell lineage specification. Cell 126:677–689

49. Graesser D, Solowiej A, Bruckner M, Osterweil E, Juedes A, Davis S, Ruddle NH, Engelhardt B, Madri JA (2002) Altered vascular permeability and early onset of experimental autoimmune encephalomyelitis in PECAM-1-deficient mice. J Clin Invest 109:383–392

50. Gratzinger D, Canosa S, Engelhardt B, Madri JA (2003) Platelet endothelial cell adhesion molecule-1 modulates endothelial cell motility through the small G-protein Rho. FASEB J 17:1458–1469

51. Lu B, Kwan T, Kurimoto Y, Shatos M, Lund RD, Young MJ (2002) Transplantation of EGF-responsive neurospheres from GFP transgenic mice into the eyes of rd mice. Brain Res 943:292–300

52. Mizumoto H, Mizumoto K, Shatos MA, Klassen H, Young MJ (2003) Retinal transplantation of neural progenitor cells derived from the brain of GFP transgenic mice. Vision Res 43:1699–1708

53. Sakaguchi DS, Van Hoffelen SJ, Grozdanic SD, Kwon YH, Kardon RH, Young MJ (2005) Neural progenitor cell transplants into the developing and mature central nervous system. Stem Cell Biol Dev Plast 1049:118–134

54. Reynolds BA, Weiss S (1992) Generation of neurons and astrocytes from isolated cells of the adult mammalian central nervous system. Science 255:1707–1710

55. Ford MC, Bertram JP, Hynes SR, Michaud M, Li Q, Young M, Segal SS, Madri JA, Lavik EB (2006) A macroporous hydrogel for the coculture of neural progenitor and endothelial cells to form functional vascular networks in vivo. Proc Natl Acad Sci USA 103:2512–2517

Electron Microscopy in Angiogenesis Research

Ruth M. Hirschberg and Johanna Plendl

Abstract Electron microscopy is a powerful tool for detection and supervision of all key cell differentiation processes of *in vivo* neovascularisation that are also considered important in realistic *in vitro* models of angiogenesis, vasculogenesis and vascular remodelling. It allows detection and supervision of all major vascular differentiation processes for both, *in vivo* and *in vitro* models of neovascularisation, but preparation of samples requires experience and particular diligence in order to preserve the damageable spatial structures. The wealth of available information provided by scanning and transmission electron microscopy approaches may be useful for subsequent, e.g., biochemical or molecular, studies and thus delivers important controls for further experimental designs. In order to preserve the fragile three-dimensional cellular structures for EM, particularly of *in vitro* models, modified processing techniques for both TEM and SEM need to be applied that are emphasised in this chapter. E.g., different pre-embedding and sample taking techniques are provided and illustrated with hands-on photographs from the electron microscopy laboratory. Scanning electron microscopy of vascular micro-corrosion casts – particularly important for studying intussusceptive processes and assessing tumour neovascularisation – is also described.

1 Introduction

Sprouting angiogenesis is a multi-step process involving migration, proliferation and a specific spatial arrangement of endothelial cells that is complemented by surrounding cells such as pericytes and vascular smooth muscle cells (Fig. 1). Another distinct form of microvascular growth and remodelling, i.e. intussusception, is

R.M. Hirschberg • J. Plendl (✉)
Faculty of Veterinary Medicine, Institute for Veterinary Anatomy,
Freie Universität Berlin, Berlin, Germany
e-mail: plendl.johanna@vetmed.fu-berlin.de

E. Zudaire and F. Cuttitta (eds.), *The Textbook of Angiogenesis and Lymphangiogenesis: Methods and Applications*, DOI 10.1007/978-94-007-4581-0_3,
© Springer Science+Business Media Dordrecht 2012

Fig. 1 Low magnification TEM-micrograph of an intestinal microvessel depicting an erythrocyte (Ery) within the lumen and the components of the vessel wall. Note the microvillous-like lumen-side projections (*arrows*) of the endothelial cells (*E*) and the elongated projections of the pericytes (*P*) in close vicinity of the endothelial cells. Original magnification: ×3,150. Courtesy of Monika Sachtleben

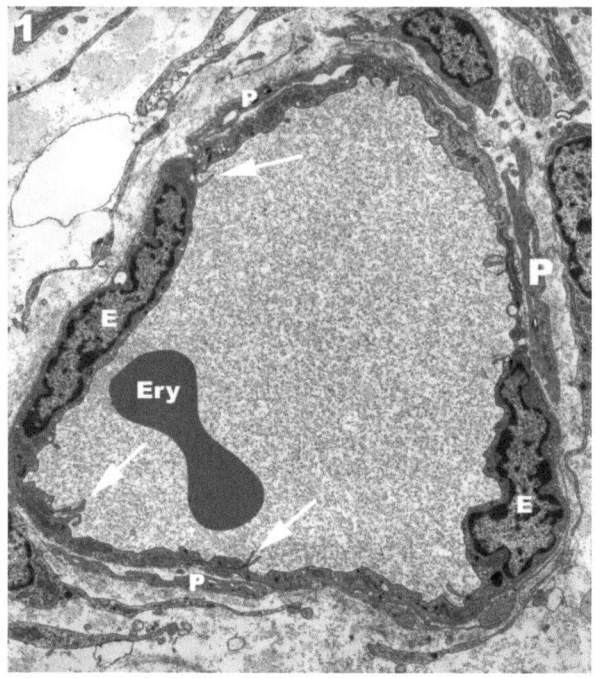

characterised by re-location and re-arrangement of vascular cells rather than cell proliferation (e.g. reviewed in [24, 34, 35, 42]).

Therefore, the morphology of endothelial cells in the different steps of neovascular processes should be of special interest, particularly in *in vitro* models. 'Classical' techniques such as scanning and transmission electron microscopy are infrequently applied, possibly due to the fact that processing cell cultures for electron microscopy (EM) requires particular diligence in order to preserve these damageable spatial structures. Scanning electron microscopy (SEM) is particularly suited to document the three-dimensional arrangement of neovascular cells – an aspect that is often neglected in studies focusing on cellular and molecular control of angiogenesis. Transmission electron microscopy (TEM) is essential for identification of involved cell types and their function-adapted cellular ultra-structure, and for proving lumen-formation in neovascular structures. Thus, EM is a powerful tool for detection and supervision of all key cell differentiation processes of *in vivo* neovascularisation that are also considered important in realistic *in vitro* models of angiogenesis, vasculogenesis and vascular remodelling. These are (Figs. 2, 3, 4, 5, 6, 7, 8, 9, 10, 11): occurrence of function-related cellular junctions; development of specific surface features indicating cellular polarity; production of extracellular matrix material; mechanisms leading to the formation of an internal lumen; specific spatial arrangement of cells within capillary-like networks; intussusception of specific cells or cell groups; detachment of apoptotic cells, i.e. anoïkis [14]. Besides the 'routine' EM procedure, special staining techniques such as ruthenium-red-staining

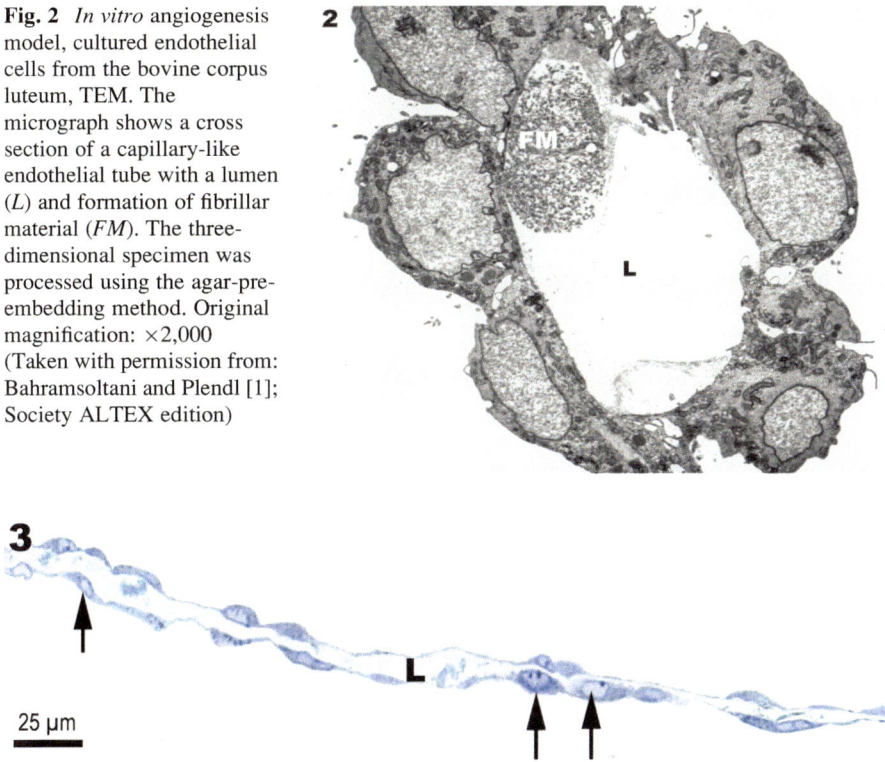

Fig. 2 *In vitro* angiogenesis model, cultured endothelial cells from the bovine corpus luteum, TEM. The micrograph shows a cross section of a capillary-like endothelial tube with a lumen (*L*) and formation of fibrillar material (*FM*). The three-dimensional specimen was processed using the agar-pre-embedding method. Original magnification: ×2,000 (Taken with permission from: Bahramsoltani and Plendl [1]; Society ALTEX edition)

Fig. 3 Capillary-like structure from an in vitro angiogenesis model of human neonatal foreskin endothelial cells, semi-thin section, Richardson stain. The endothelial cells enclose a central lumen (*L*) filled with sparse fibrillar material. The majority of cells protrude to the abluminal side, while individual cells (*arrows*) protrude into the lumen. The three-dimensional specimen was processed using the agar-pre-embedding method (Taken with permission from Lienau et al. [23]) © The American Society for Biochemistry and Molecular Biology

(Fig. 21) or immunogold-labelling allow the detection of specific ultra-structural characteristics of examined cells. Silver-enhanced detection of impregnated cells, or of up-take of gold-labelled substances such as liposomes or other nanoparticles – i.e. prospective carrier substances for therapeutic substances (e.g. [2]), or detection of uptake of vascular contrast media, is an efficient means for complementing respective clinical studies. In addition, SEM of vascular micro-corrosion casts (Figs. 24 and 25) is an elegant technique for three-dimensional assessment of vascular and/or microvascular beds and allows *in vivo* detection of neovascular processes such as sprouting angiogenesis, intussusceptive branching and remodelling, and vascular pruning in health and disease (e.g. [10, 27, 54]). Micro-corrosion casting is therefore frequently employed in *in vivo* models of tumour angiogenesis (e.g. reviewed by: [20, 29]), as well as in respective anti-angiogenic approaches. Micro-corrosion casting of *in vitro* models of angiogenesis may offer an important link between *in vivo* and *in vitro* research approaches [13, 14].

Fig. 4 Specific morphological features in the course of *in vitro* angiogenesis: bovine microvascular endothelial cell culture from the bovine corpus luteum, TEM. (**a**): Development of focal adhaerens junctions (desmosomes, *black arrows*) allow formation of lumenised capillary like tubes. (**b**): Formation of communicative gap junctions (*white arrow*) occurs particularly in the later stages of *in vitro* angiogenesis. Specimen processed using the agar-pre-embedding method (Taken with permission from Hirschberg et al. [14]) © Wiley & Sons

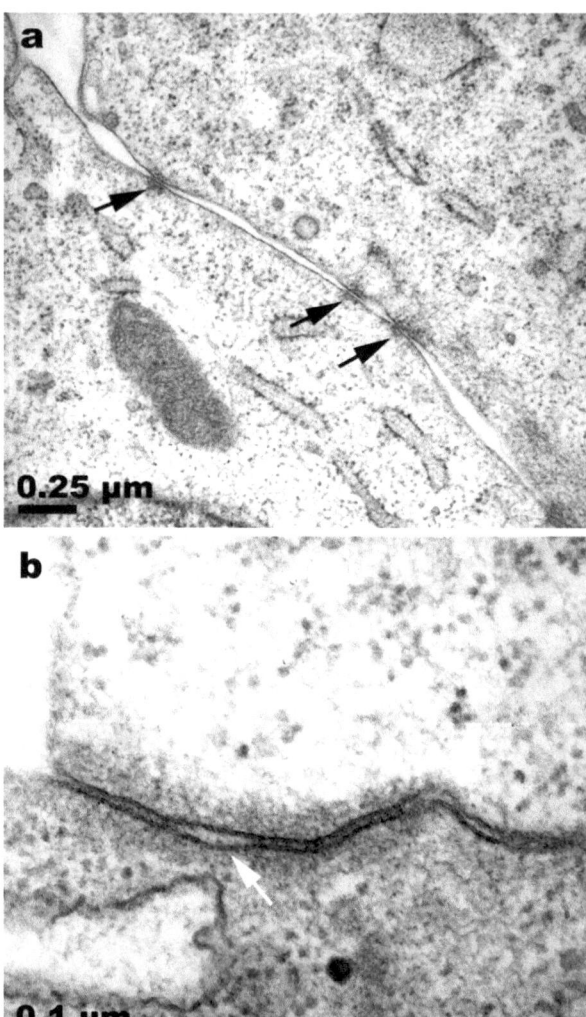

SEM of *in vitro* models allows both, surveillance of multi-cellular formations, even of a whole cellular population within the culture dish (Fig. 7), besides detection of specific structural investigations on cellular and sub-cellular levels, with a comparatively low effort of time and cost. It permits investigation of structural alterations (cell shape, cellular projections and surface modifications) and intercellular interaction, respectively, and thus allows distinguishing of different morphological avatars of angiogenic endothelial cells as well as an overview of cellular grouping and cell behaviour (formation of three-dimensional tubes, detachment of specific cells, intussusceptive figures etc., Figs. 8, 9, 10 and 11) within the entire cell culture. It is therefore particularly well suited to document the

Fig. 5 Specific morphological features in the course of *in vitro* angiogenesis: bovine microvascular endothelial cell culture from the bovine corpus luteum, TEM. Exemplary ultrastructural elements. (**a**): nucleus (*N*), mitochondria (*M*), elements of the cytoskeleton (*white asterisk*) (Courtesy of S. Kaessmeyer); (**b**): nucleus (*N*), Golgi-fields (*arrows*), focally enlarged intercellular space sealed by junctional complexes (*black asterisk*). Specimen processed using the agar-pre-embedding method. Original magnification (**a**): ×10,000, (**b**): ×20,000

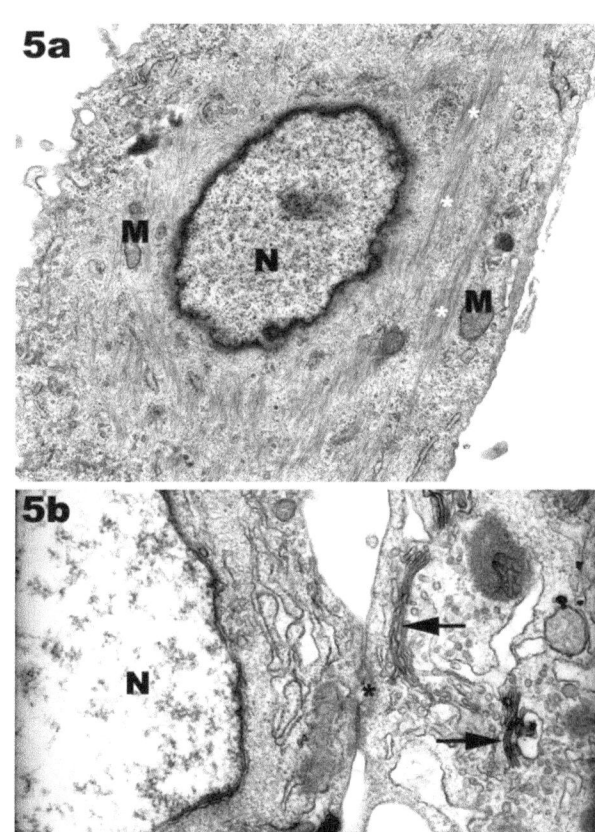

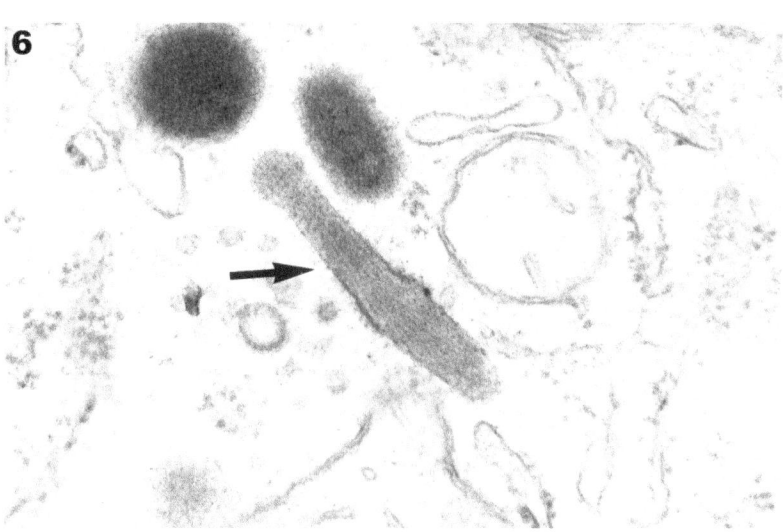

Fig. 6 Endothelial cells in vitro, isolated from the fetal porcine aorta, TEM, cells retrieved and processed as a pellet. The *arrow* indicates a Weibel-Palade-body, typical for endothelial cells. Original magnification: ×9,100 (Taken with permission from Plendl et al. [36]) © Springer

Fig. 7 This overview of a culture with detaching capillary-network-like formations clearly depicts the advantage of SEM-examination: The three-dimensional arrangement of cells within the whole culture dish may be examined, combined with the possibility to display fine surface details on an ultra-structural level. *In vitro* angiogenesis model, bovine microvascular endothelial cells

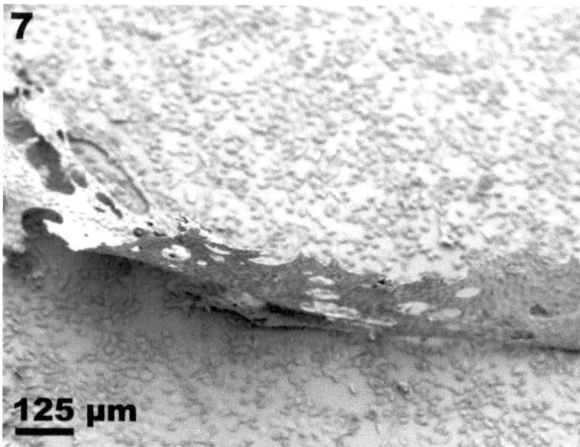

Fig. 8 In vitro angiogenesis model with formation of three-dimensional branching capillary-like tubes, bovine microvascular endothelial cells, SEM. *Asterisks* indicate angiogenic sprouts and the arrow points to a cell group in process of intussusceptive branching remodelling

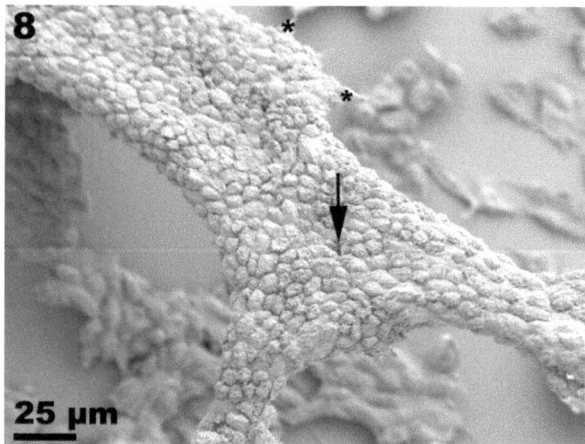

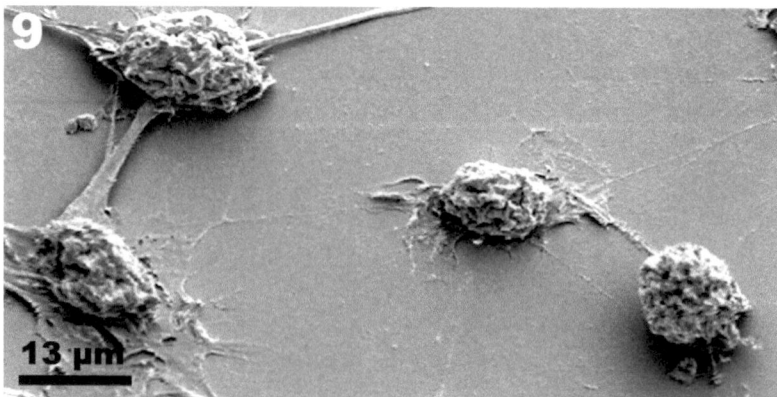

Fig. 9 Surface morphology as revealed by SEM of a migratory endothelial cell with numerous microvilli-like supra-nuclear projections and long invadopodial projections. The cell in the *lower left* displays mild shrinkage artefacts as a result of HMDS-drying. *In vitro* angiogenesis model, bovine microvascular endothelial cells

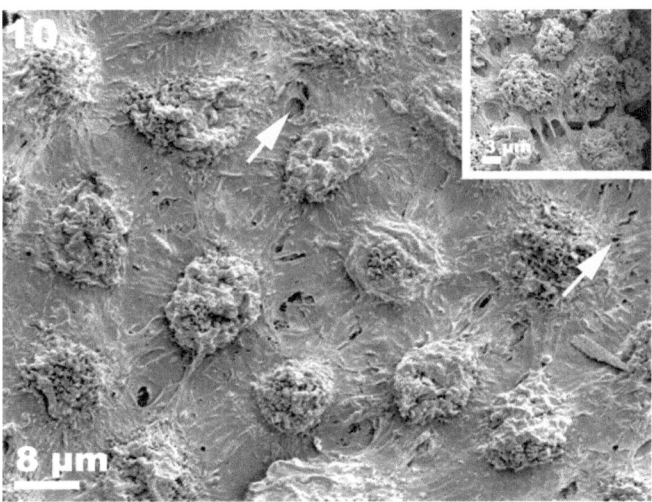

Fig. 10 Surface morphology of endothelial cells within a single-layered, two-dimensional endothelial network: Single cells detach (also see insert) from the cell assembly. In this case it is not easy to discriminate between cells de facto detaching from their respective neighbours and possible shrinking artefacts occurring at cell-to-cell-borders (areas in question indicated by *arrows*). *In vitro* angiogenesis model, bovine microvascular endothelial cells, SEM. Inserted figure taken with permission from [14]. ©Wiley & Sons

Fig. 11 Intussusceptive remodelling of a three-dimensional capillary-like endothelial tube. Intussusceptive cell groups are conspicuous because of their smoother surface structures (*asterisks*). *In vitro* angiogenesis model, bovine microvascular endothelial cells, SEM

three-dimensional arrangement of cells performing the different steps of the angiogenic cascade. Nevertheless, SEM presents static "snap-shots" of the differentiation and/or developmental processes only; therefore consistent sampling in all stages of angiogenesis and/or combination with life cell imaging of the cell cultures is necessary. TEM of *in vitro* models proves lumen-formation and intussusception of capillary-like neovascular formations, and allows detection of specific intercellular contact areas. Thus, combining cell culture with EM may allow a deeper insight into the relationship between molecular control of the angiogenic cascade and the correlated changes in vascular morphology [14].

In conclusion, EM is a cost- but only in part time-efficient instrument for detection and supervision of all major vascular differentiation processes for both, *in vivo* and *in vitro* models of neovascularisation, asking for experience, diligence and attention to detail. The wealth of comparatively quickly available information provided by SEM and TEM approaches may be useful for subsequent, e.g., biochemical or molecular, studies and thus delivers important controls for further experimental designs. In order to preserve the fragile three-dimensional cellular structures for EM, particularly of *in vitro* models, modified processing techniques for both TEM and SEM need to be applied that are emphasised in this chapter.

2   General Equipment for Electron Microscopy

2.1   *Installations, Equipment and Devices*

TEM and/or SEM (Electron microscopes are extremely sensitive to surrounding conditions and require state-of-the-art rooms designed to reduce interference from magnetic fields, background and acoustic vibrations, and barometric pressure and temperature changes.)

Fume hood
Vacuum chamber (if applicable)
Critical point dryer (if applicable)
Sputter coater (for SEM, if applicable)
Phase contrast microscope (if applicable)
Centrifuge (if applicable)
Light microscope
Ultramicrotome (Fig. 17)
Ultrostainer (if applicable)
Knife maker (for glass knives)
Diamond knives
Diamond milling cutter (if applicable)
Heating plate (Fig. 15)

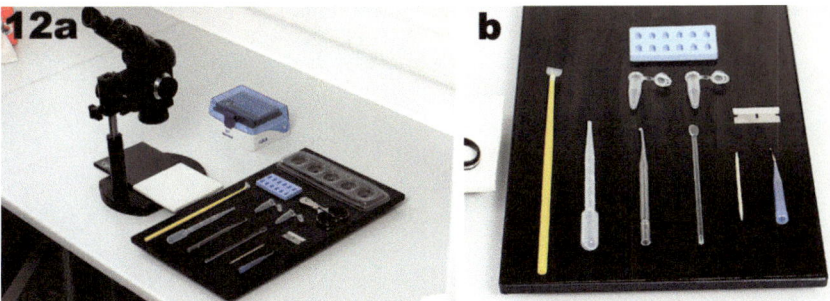

Fig. 12 Laboratory work bench equipment for fixation and (pre-)embedding procedures. (**a**): stereoscopic loupe and *black*-and-*white*-support pad, pipette-tip dispenser; (**b**): *above*: embedding-forms, *middle*: Eppendorf vials, razor blade, *below*: various spatulae and tools with adapted tips for sample taking and positioning, pipette

Incubator

Gas port (for heating and mould-adapting glass- and plastic ware to specific purposes)

Liquid nitrogen container (if applicable)

Three-dimensional adaptable light sources (Fig. 15b)

Binocular loupe (Figs. 12a and 15b)

Band saw (for separation of single wells from multiple-well plates, if applicable)

Chuck and lathe (for removing polystyrene walls of culture vessels from glycid ether-embedded specimens, if applicable)

Soldering iron (for dissection of microcorrosion casts and removing of polystyrene walls of culture vessels, if applicable)

Grid boxes (for TEM specimen-storage, if applicable)

Exsiccator containers and/or special dust-protected drawer containers (for SEM specimen-storage, if applicable)

2.2   Glassware and Supplies

Glass ware for storage of reagents and media

Glass slides and cover slips

Round cover slips (glass and/or plastic, for cell culture, if applicable)

Copper grids (TEM)

Aluminium stubs (SEM)

Pasteur pipettes

Dissection and positioning instruments such as fine forceps, tweezers, spatulas, scissors, razor blades etc. (Fig. 12)

Black-and-white support pads (Figs. 12, 13)

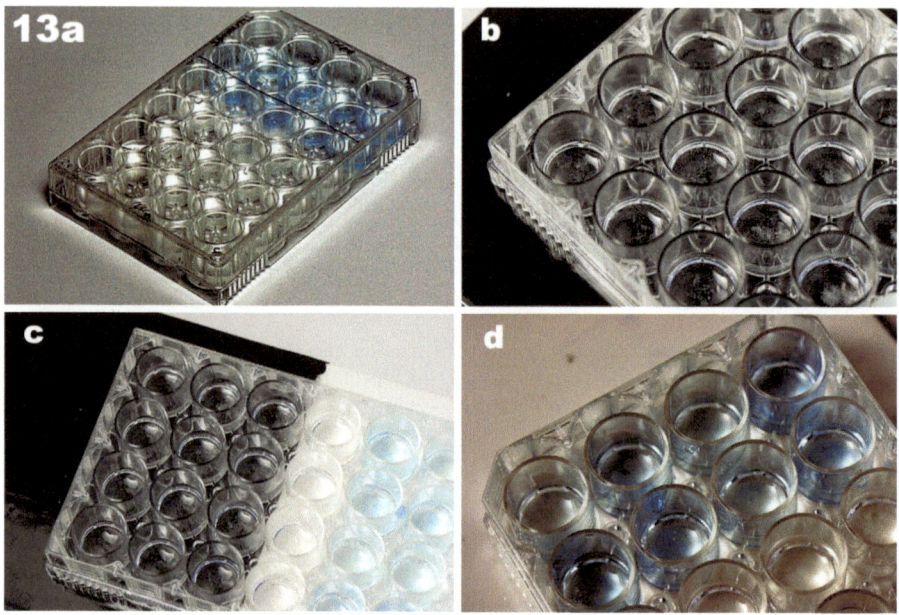

Fig. 13 Advantage of sample-staining (methylene-blue) within the respective culture vessel prior to embedding procedures. (**a**): stained and un-stained cell samples within the multiple-well plate; (**b**): un-stained samples on *black* support pad; (**c**): stained and un-stained samples on *black*-and-*white* support pad; (**d**): stained samples on *white* support-pad. Staining is particularly recommended for samples with very scarce material

2.3   Reagents and Media

2.3.1   Fixation

Glutaraldehyde
Paraformaldehyde
Karnovsky's fixative (a mixture of paraformaldehyde and glutaraldehyde in cacodylate buffer; used often in modified composition: e.g. 2% paraformaldehyde and 2.5% glutaraldehyde in Na-cacodylate buffer, 0.1 M)
Osmium tetroxide

2.3.2   Dehydration

Ethanol, graded 30–100%
Acetone, graded 30–100% (if applicable)

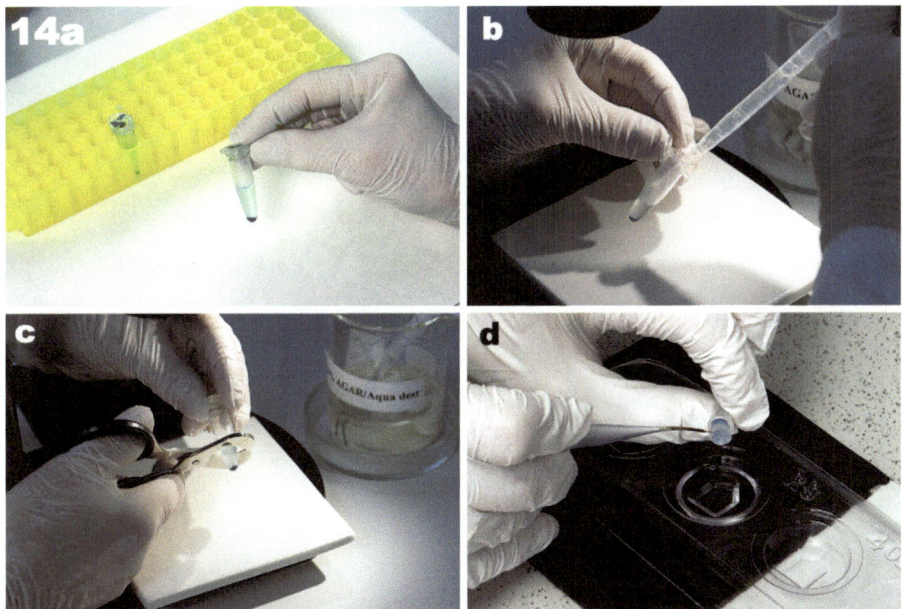

Fig. 14 Retrieval and agar-pre-embedding procedure for cell pellets. (**a**): cells (stained) are sampled as pellets via centrifugation in Eppendorf vials; (**b**): supernatant is carefully removed and replaced by heated agar solution (60°C, 1.5%); (**c**): after cooling, the Eppendorf vial is cut with a scissor; (**d**): removal of hardened agar block with embedded cell pellet from Eppendorf vial, positioning of embedded cell pellet in curing-form filled with liquid agar

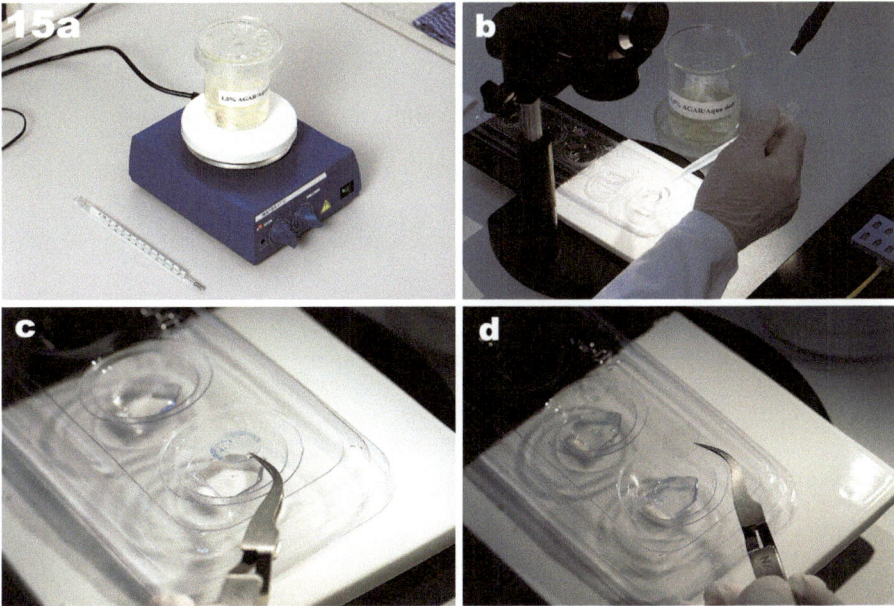

Fig. 15 Agar-pre-embedding procedure for preservation of cells in situ (stained) within the culture vessel. (**a**): preparation of liquid agar (60°C, 1,5%); (**b**): curing form is filled with liquid agar; (**c**): fixed specimen on cover slip is carefully positioned (cell-side down) within agar-filled curing form; (**d**): after cooling, the cover slip is carefully removed from the hardened agar-block with incorporated specimen

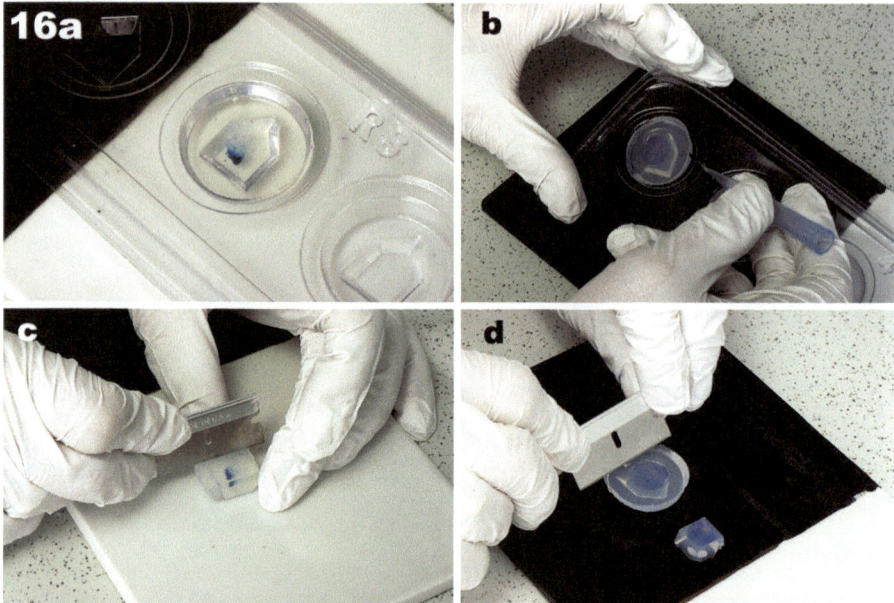

Fig. 16 Taking and trimming of agar-embedded stained specimen. (**a**): agar-embedded specimen within curing form; (**b**): removal of hardened agar-block from curing form; (**c, d**): trimming of hardened agar blocks with razor blade

2.3.3 Embedding and Pre-embedding

Polymerising embedding epoxy resin (e.g. glycid ether/Epon, Araldite, Spurr) or
  acryl resin (e.g. LR White, Lowycryl, Unicryl)
Intermedium (e.g. propylene oxide)
Agar (if applicable)
Bovine serum albumin (if applicable)
Bis-acrylamide, ammonium persulfate, tetramethylethylendiamine (if applicable)

2.3.4 Staining Solutions for Semi-thin Sections

E.g. methylene blue 2%, Giemsa solution, Richardson stain, toluidine blue 0.5%

2.3.5 Buffers

Phosphate-buffered washing solution (PBS)
Na-cacodylate buffer (0.1 M; pH 7.2–7.4)
HEPES-buffer (0.05 M; if applicable)

2.3.6   Post-fixation and Contrasting/Staining for EM

Osmium tetroxide (stock solution 2% in Na-cacodylate buffer)
Potassium ferrocyanide
Uranyl acetate[1] (stock: solution 2% in bi-distilled water)
Lead citrate[2] (stock solution 1–4% in bi-distilled water)
Ruthenium red

2.3.7   SEM

Hexamethyldisilazane (HMDS) or tetramethylsilane
Conductive carbon cement
Osmium tetroxide (if applicable)
Hydrazine hydrate (if applicable)
Thiocarbohydrazide/ruthenium red (if applicable)
Hygroscopic material for exsiccator, preferentially with humidity indicator, e.g.
    blue gel (for SEM specimen's storage, if applicable)

2.3.8   Special Staining Methods (If Applicable)

Ruthenium red (0.05–0.1%)
Two-component silver-enhancement kit for EM[3]
Glycin (0.05 M)
HEPES-buffer (0.05 M)
Na-thiosulfate
Colloidal carbon, particle size 20–50 nm

2.3.9   Corrosion Casting

Pre-polymerised methylmethacrylate kits (e.g. Mercox[4], Batson's No. 17[5]) or
    polyurethane kits (e.g. PU4ii[6])
Monomeric methylmethacrylate (if applicable)

---

[1] e.g. Ultrostain I, Leica Microsystems, Germany.

[2] e.g. Ultrostain II, Leica Microsystems, Germany.

[3] e.g. SPI Supplies, West Chester, PA, USA.

[4] e.g. SPI Supplies, West Chester, PA, USA.

[5] e.g. Polysciences Inc., Warrington PA, USA.

[6] e.g. vasQtec, Zurich, Switzerland.

Pre-treatment solutions: heparin, Ringer solution, sodium chloride (0.9%)
(if applicable)
Injection equipment such as steel and glass needles, syringes, catheters, clamps etc.
Corrosion: potassium hydroxide, sodium hydroxide, enzyme solutions

2.3.10   Other

Culture media (for cell culture material)
Enzyme solutions (for cell harvesting, matrix digestion etc.)

3   Processing *In Vitro* Cell Culture Models for EM

General remarks:
   Generally, tissue sample size for TEM should be as small as possible and not
exceed 1 mm^3. If possible, tissue samples are taken *intra vitam* (under anaesthesia)
or as quickly as possible *post mortem* in order to avoid autolytic alterations. For the
same reason, target tissue and/or samples should be cooled down quickly. Fixation
solution is likewise prepared at low temperatures, and the ratio of sample volume to
fixative volume is at least 1:10 for optimal results of immersion fixation. Vessels
containing fixative should be large enough so that the sample may be moved freely
within the solution. If possible, additional perfusion fixation of test subject or test
tissue is recommended. The respective body part or organ is perfused with fixation
solution (e.g. 'half strength' Karnowsky solution: 0.25% glutaraldehyde, 0.25%
paraformaldehyde in 0.1 M Na-cacodylate buffer, pH 7.2, see Sect. 3.1.1) prior to
sample taking.

In vitro models:
   For *in vitro* models, culture conditions are particularly important regarding
sample taking and embedding processes prior to EM examination. Cells can either
be grown directly on the bottom of glass flasks or multiple-well plates, or on round
cover slips within the well plates. The subsurface of each culture vessel may either
be uncoated or coated (e.g. with collagen or other matrix components). Cells may
be cultured (or co-cultured) in suspension, as colonies, mono-layers, multi-layers,
or – in realistic angiogenesis models – even with formation of three-dimensional
capillary-like networks. All these different culture conditions harbour particular
challenges for TEM and SEM processing.
   Scraping-off of cultured cells from the subsurface or enzymatic proteolysis in
order to loosen cultured cells from respective subsurface coatings, and subsequent
subjecting to pelleting via centrifugation allows relatively uncomplicated access to
cell samples. A disadvantage of this type of sample collection is that sections
through a cell pellet contain only random cell profiles that preclude examination
of cells in their 'natural' orientation and localisation during cell culture – a severe
limitation in realistic three-dimensional neovascular *in vitro* models. Therefore,

different methods for pre-embedding of samples have been developed (q.v.). Mono-layered cultures, or sparse culture material, on the other hand, may prove difficult to see when it comes to orienting blocks for sectioning. To facilitate subsequent handling, specimens can be stained (Fig. 13) with a basic dye such as toluidine blue (0.5%) or methylene blue (2%) after post-fixation in osmium (see below) while they are being dehydrated.

All procedures are carried out wearing protective gloves and goggles under a fume hood. Processed specimens are stored in special dust- and moisture-protected containers, i.e. grid boxes for TEM and exsiccator vessels or special container boxes for SEM.

3.1 TEM

3.1.1 Fixation and Post-fixation of Cells

Culture media is rapidly removed from culture vessels by suction and subsequently substituted by PBS, cacodylate buffer and fixative in the refrigerator (4°C) for 1 to several hours. The cells are then rinsed in cacodylate buffer and post-fixed with osmium tetroxide resp. potassium ferrocyanide. Prior to dehydration (see below), the samples need to be buffer-washed repeatedly for removal of all fixatives.

Generally, low temperature, preferably at 4°C, is recommended for all processing steps. Material from the culture bench needs gentle cooling down from the relatively high culturing temperatures, though; therefore the first washing step with buffer is carried out at room temperature.

Established fixatives for EM:

- glutaraldehyde: 2.5–3.5% if used exclusively; in combination with other fixatives: 1–2%; material may be stored in 1.5% Glutaraldehyde for longer periods; lower concentrations (0.1–0.25%) in combination with formaldehyde can be used for immuno-EM
- formaldehyde: needs to be either freshly prepared from paraformaldehyde or may be stored deep frozen (−20°C) until use
- osmium tetroxide: used for post-fixation (0.5–2%) in cacodylate-buffer; cells may also be fixed by impregnation in osmium vapour for 30–60 min (see Sect. 3.2, SEM)
- glutaraldehyde with formaldehyde: e.g. Karnovsky' fixative: 4% glutaraldehyde and 4% formaldehyde in cacodylate buffer; often employed in modified composition, e.g. 1–2% glutaradehyde with 1–2% formaldehyde in cacodylate buffer
- glutaraldehyde (2.5%) with osmium tetroxide (1%) in cacodylate buffer: recommended for cell cultures; prepare pre-cooled directly before use, fix on ice for 1 h
- osmium-tetroxide (1%) with Ka-ferrocyanide (1.5%) in H_2O: used for post-fixation and contrasting, at least 1 h at room temperature or refrigerated, in the dark

- glutaraldehyde (2,5%) with ruthenium-red (0.05–0.1%): contrasting fixative for depiction of sugar moieties, e.g. in clycocalix etc. (see Sect. 4, special staining procedures)

Preparation of osmium-solutions:
Osmium tetroxide is mostly supplied in solid aggregate state within glass vials. The crystal dissolves only slowly, thus solutions should be prepared 1 day prior to application. The whole procedure is carried out wearing gloves and protection goggles under the fume hood. The vial is thoroughly cleansed and placed onto a fresh sheet of tin foil. Slit the vial slightly, cover vial with tin foil and carefully break the vial. Place vial containing osmium tetroxide into a clean and tightly closable glass container. Prepare 1–4% stock solution with bi-distilled water. Store lightproof until all crystals are dissolved, then prepare aliquots and store at −20°C until use.

Exemplary fixation protocol:

- replace culture media with Na-cacodylate buffer (0.1 M; pH 7.2) at room temperature
- replace buffer with modified Karnovsky' fixative: 2% paraformaldehyde and 2.5% glutaraldehyde in Na-cacodylate buffer, (0.1 M, pH 7.2), keep in the refrigerator (4°C) for 4 h
- rinse 3× in cacodylate buffer
- post-fix (in the dark) with Na-cacodylate-buffered 1% osmium tetroxide and 1.5% potassium ferrocyanide solution for 2 h at 4°C
- wash repeatedly in cacodylate buffer

3.1.2 Sample Taking and Pre-embedding Procedures

After post-fixation in osmium tetroxide and final washing in cacodylate buffer, the cells can then be submitted to different embedding or pre-embedding procedures in order to ascertain the best preservation of cellular structures subject to culture conditions.

a. Retrieval of cell pellets in agar or gelatine (Fig. 14):
   The fixed and washed cells are carefully scraped from the subsurface with a spatula and transferred into a vial containing liquid agar solution (1.5% aqueous agar solution, at approx. 60°C). The cells are then centrifuged at 200 ×g for 5 min. After discarding of the supernatant, fresh agar solution (1.5% aqueous agar solution, at approx. 60°C) is added and left to harden in a refrigerator (4°C) for 30 min (e.g., [15]). Thus, the cells are enclosed as a pellet into a hardened agar block. This method is particularly applicable for subconfluent to confluent cells and may also be achieved employing 1% liquid gelatine solution for centrifugation and liquid 12% gelatine solution that is then solidified for 30 min on ice as 'support' structure of the pellet (e.g., [33]).

b. Pre-embedding of cultured cells *in situ* with liquid agar ([14]; Fig. 15):

Agar- or agarose- pre-embedding – hitherto reported only for embedding of single [57], dispersed [7, 29, 47, 55], pelleted [29] or suspended [25] cultured cells – proves very useful for preservation of the in-situ morphology of delicate contiguous multi-cellular neovascular structures. In our laboratory, use of agar is preferred to pure agarose, because the agar-block resulting after cooling is more stabile than with the gel-like agarose. For cell cultures grown on cover slips (see also procedures e and f), this method may also be applied: cultures fixed on cover slips are placed face-down on liquid agar with a curing form, and after hardening of the agar block the cover slip is gently removed with a fine forceps (Fig. 15c, d). Transfer into polymerising resin requires particular care, though, because the agar-block is lighter than the resin and tends to float on the surface during the polymerisation process. By accurate supervision of the hardening process, the optimal time (after 3–4 h) of appropriate viscosity needs to be observed, and thus a careful re-positioning of the agar-block within the curing form is possible. This procedure is particularly recommended for models of neovascular processes with distinct three-dimensional network formation. A comparable method of gelatine-pre-embedding of cultured cells for preparation of ultra-thin cryosections for immunogold-labelling (q.v.) has been described [33]. Exemplary protocol:

- remove cacodylate washing buffer
- coat cells with liquid agar (1.5% aqueous agar solution, at approx. 60°C)
- refrigerate (4°C) for 30 min
- remove cells incorporated into the hardened agar blocks carefully from culture vessel or cover plate with glass spatula (if necessary, after application of a few drops of buffer solution)
- transfer cell-containing agar-block into cacodylate buffer solution

c. Pre-embedding of cells *in situ* with bovine serum albumin (BSA) or gelatine:

After removal of cacodylate washing buffer, the cells are initially coated with BSA (20%) within the culture vessel and then a few drops of 25% glutaraldehyde solution are added, resulting in a coagulated protein mass incorporating the cultured cells that is then transferred into buffer solution and further processed [14]. A similar procedure may be achieved employing 12% gelatine at 37°C (e.g., [33]). Coating with coagulating BSA or gelatine may result in an inhomogeneous mass that is not easily removable from the respective subsurface in the following steps of the embedding process, but is always applicable for cell pellets (see above).

d. Pre-embedding of cell pellets with BSA and bis-acrylamide [48]:

For cells cultured in suspension or cultures with very scarce material, pre-embedding into solidifying matrices is necessary. Agar (see above), or gelatine, is not translucent, so that difficulties with identifying specimens within the pre-embedded mass may occur. Pre-embedding of cell pellets (attained by centrifugation, see above) in a matrix composed of 4.7%BSA, 4.88% bis-acrylamide, 1.43% paraformaldehyde, 0.24% ammonium persulfate and

2.38% tetramethylethylendiamine (TEMED) results in a translucent and very stabile pre-embedded block. TEMED and ammonium persulfate are used to catalyse the polymerisation of bis-acrylamide. Samples are left to polymerise for 1–1.5 h at room temperature.

e. Processing of cultured cells *in situ* adhering to glass cover slips they were grown on:
    In the subsequent resin embedding process (see below), these specimens are positioned within the curing form with the cover slip facing towards the surface. After polymerisation, the resin block is immersed into liquid nitrogen. From the resulting frozen block, the glass cover slip is then blasted off by mechanical force (e.g., [14]). Removal of glass cover slip is necessary, because the glass cover slip cannot be cut to sample size and thus does not allow sample trimming. Glass cover slips are particularly suited for SEM (see below); therefore this method can be applied when samples from the same multiple-well plate are intended for both, TEM and SEM examination. Otherwise, this method harbours the risk that blasting off of glass plates after deep-freezing may not always be complete, or that removal of the glass is only achieved while concurrently fracturing or destroying the cell samples.

f. Processing of cultured cells *in situ* adhering to plastic cover slips they were grown on:
    Plastic cover slips are suitable for cutting prior to embedding and also for sectioning with microtomes. In the subsequent resin embedding process (see below), these specimens can either be positioned within the curing form with the cover slip facing towards the surface, transferred into buffer solution and then further processed (e.g., [14]). Or the cover slips with adhering fixed cultured cells can also be cut prior to embedding (e.g., [8]; further processing see Sect. 3.1.3). Depending on the respective plastic subsurface employed, it may be necessary to select an appropriate polymerising resin [31]. Fixing and embedding cells cultured on plastic cover slips allows to cut samples to size in all steps of the fixing and embedding procedure, but after hardening the cover slips may in our experience become very hard and 'glass-like' and are therefore likely to damage the diamond knife used for ultra-thin sectioning. Sectioning may thus be achieved only in a plane parallel to the cover slip – a drawback worth considering for some research aims. Like glass cover slips, plastic ones can be mechanically removed after deep-freezing in liquid nitrogen (but with a comparable risk of sample damage, see above).

g. Processing of cultured cells *in situ* within multiple-well plates [50]:
    All washing, fixation, post-fixation, dehydration, contrasting, and embedding (see below) procedures are carried out on cell material within the multiple-well plates. After polymerisation of the embedding resin, the wells are separated by a band saw. The thus acquired 'square dish' is further reduced by removing the thin polystyrene plastic sides: the square is placed in a chuck und the sides are removed by turning on a lathe. Thus, a round cylinder shape of cell incorporated into a round resin block with the bottom of the well still attached is produced. The specimen is immersed into liquid nitrogen for 10 s, and the bottom of the well is then snapped off. While this procedure is very handy because the cell

material can remain within the culture vessel for all EM procedures up to embedding into polymerising resin, removal of the polystyrene multiple-well wall and bottom harbours some risk of sample damage (see above). In our experience with this procedure it is essential to select an appropriate intermedium and polymerising resin, as e.g. propylenoxide and glycid ether may damage plastic multiple-well plates.

For all procedures with a pre-embedding technique, the block (agar, gelatine etc.) containing the incorporated specimens is then trimmed with razor blades for further resin embedding procedure (Fig. 16).

3.1.3   Dehydration and Embedding

Subsequent to all listed pre-embedding procedures, the specimens are dehydrated in an ascending alcohol (or acetone) series: 30 [50] up to 100%, each 5 – max. 30 min (longer residence time in higher concentrations only). The fixed cells are then transferred into and incubated in inter-medium (Fig. 16a) (not necessary when employing acetone for dehydration), a mixture of inter-medium and embedding medium, and finally into the pure polymerising embedding epoxy resin (e.g. glycid ether/Epon, Araldite, Spurr) or acryl resin (e.g. LR White, Lowycryl, Unicryl). These procedures need to be carried out in appropriate – i.e. preferably glass – vessels, as e.g. propylene oxide and glycid ether may damage plastic vessels.

Cells cultured and fixed on plastic cover slips within multiple well plates (see Sect. 3.1.2, item f) may also further be processed within the well plate, up to embedding in resin. The plastic cover slips with adhering cells that have been infiltrated with the liquid resin are then cut in half, and each half is placed cell-side down on a pre-polymerised blank resin block and allowed to polymerise at 70°C for 18 h [8].

Exemplary protocol (all steps at room temperature):

- dehydrate in an ascending ethanol series: 50, 70, 80, 90, 100% ethanol; 15 min each
- transfer into intermedium: 100% propylene oxide; 2 × 15 min
- transfer into mixture of inter-medium and embedding-medium: propylene oxide-glycid ether (1:1), 2 h
- transfer into pure polymerising resin: glycid ether 100%, 4 h
- transfer into gelatine or Beem capsule and polymerise at 60°C for 24–48 h

3.1.4   Semi-thin and Ultra-thin Sectioning, Contrasting

The embedded specimen is then trimmed for sectioning using sharp, de-greased razor blades or a diamond milling cutter. 0.5–2 μm semi-thin sections are cut employing an ultra-microtome (Fig. 17). The sections are then stained (e.g. 2% methylene blue solution, Giemsa solution, or Richardson stain, [41], see below) for light microscopic evaluation. From thus established adequate sample sites, ultra-thin

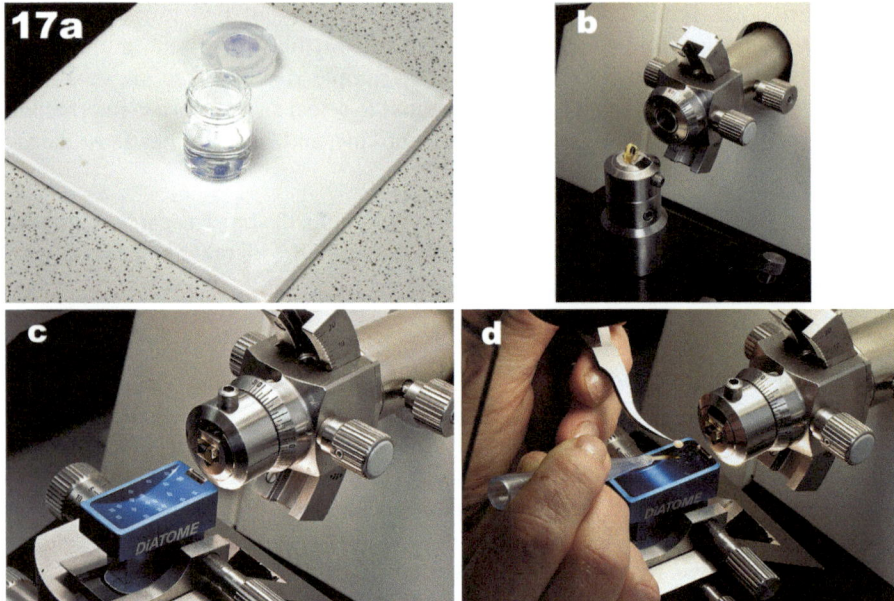

Fig. 17 Further procedure for TEM-samples. (**a**): trimmed agar-block positioned in glass container filled with inter-medium (propylene oxide); (**b**): glycid ether-embedded and trimmed sample, ready for sectioning; (**c**): ultramicrotome – diamond knife and trough containing ultra-thin sections; (**d**): mounting of sections onto copper-grids

sections (50–70 nm) are cut with glass or diamond knife, mounted onto copper grids (or nickel or gold grids for immunogold EM, q.v.) and contrasted with heavy metal salt solutions such as uranyl acetate and lead citrate (usually, pre-fabricated contrasting solutions are employed; for recipes see e.g., [40, 44, 53]).

The diamond knife, and the respective trough of the ultra-microtome, needs to be cleaned regularly in order to avoid contamination of the ultra-thin sections.

Exemplary protocol:

- trim glycid ether-block and cut 1 μm semi-thin sections with the ultramicrotome
- stain semi-this sections for evaluation of further sampling (Richardson stain: [41]; Fig. 2):
  mix stock solution I (1% Azur II in H_2O) and II (1% methylene blue in 1% Na-borate in H_2O) 1:1 and cover dried semi-thin section with staining solution, heat on heating plate (70°C) for 30 s; remove staining solution thoroughly with H_2O; dry section on heating plate
- cut ultra-thin sections (60 nm) and absorb onto copper grids
- contrast ultra-thin sections with aqueous uranyl acetate (2%[7]) and lead citrate[8], 10 min each: apply one strip of parafilm or dental wax onto the bottom of a glass

---

[7] e.g. Ultrostain I, Leica, Germany.

[8] e.g. Ultrostain II, Leica.

dish; pipette same number of drops of uranyl acetate onto strip as number of grids awaiting contrasting; place grids section-side down onto uranyl-acetate drop for 5–10 min; remove single grids and dip repeatedly (approx. 1 min) into bi-distilled water; remove distilled water by applying filter strip to periphery of grid; repeat procedure accordingly with lead citrate.

3.2 SEM

For scanning electron microscopy, cultures grown on (glass) cover slips are preferentially employed, because this stable subsurface enables easy and secure handling of fixed samples in the following procedures. Fixing and drying is carried out on the cover slips within the respective culture vessel.

A so-called "wet SEM" procedure for cultured cells without critical point drying (CPD, see below) and sputter coating has also been described [56]. This technique is particularly recommended for angiogenesis models grown on coated subsurfaces such as matrigel, as these water-enriched gel structures tend to shrink and crack with 'conventional' SEM procedure.

3.2.1 Fixation and Dehydration

The cells are fixed in glutaraldehyde, rinsed in cacodylate buffer impregnated in osmium tetroxide solution (see preparation of osmium solutions, Sect. 3.1.1) and successively dehydrated in an ascending ethanol series.

Exemplary protocol:

- fix cells in 2.5% glutaraldehyde solution for 2 h, at 4°C
- rinse in Na-cacodylate buffer (0.1 M, pH 7.2–7.4), 3 × 10 min, at 4°C
- impregnate in osmium tetroxide solution (1%), for 2 h, at (4°C)
- dehydrate in an ascending ethanol series: 50, 70, 80, 90, 100% ethanol; 15 min each, at room temperature

3.2.2 Drying

Drying of samples can be achieved by employing critical point drying (CPD), air drying from distilled water, or air drying from other organic compounds like hexamethyldisilazane (HMDS), tetramethylsilazane (TMS), acetone or ethanol that are volatile at ambient temperature. CPD renders good results on all kinds of biological specimens but requires the respective equipment. While air drying from distilled water or ethanol is not suitable for most biological specimens (but for vascular corrosion casts, see below), air drying from HMDS is favoured in our laboratory over CPD because of the advantages of ease of handling, low cost and a high rate of success with a justifiably low amount of artificial shrinkage and distortion

Fig. 18 Typical shrinkage and fracture artefacts caused by drying. Cultured granulosa cells from the bovine ovary. SEM

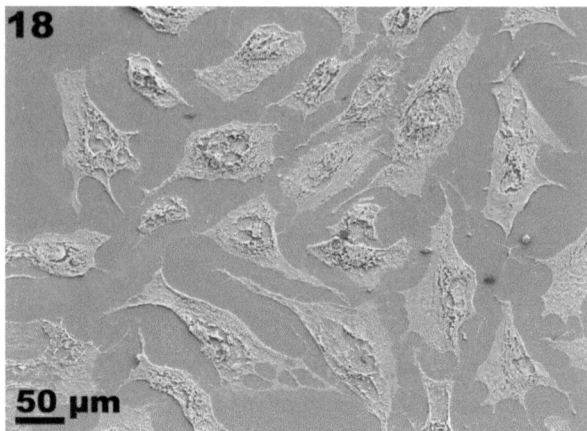

Fig. 19 Drying artefacts may also occur in HMDS-drying, particularly in three-dimensional cultures as shown in this example. *In vitro* angiogenesis model, bovine microvascular endothelial cells, SEM

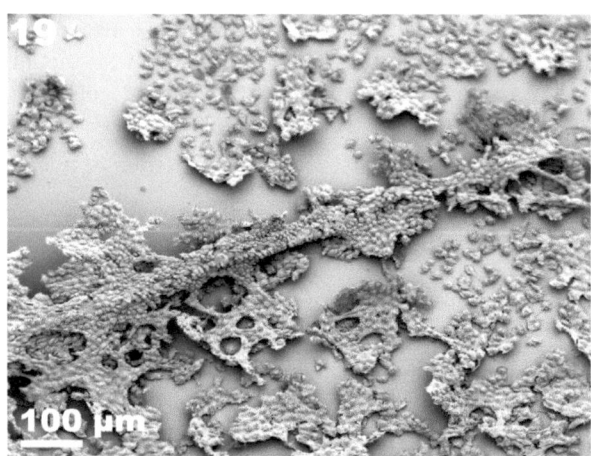

(Figs. 18 and 19). HMDS- or TMS-drying has been repeatedly recommended for mammalian histological specimens when compared to other drying methods (e.g., [39]), and was explicitly suggested for drying endothelial cells [5] as well as vascular specimens [46]. Air drying from acetone apparently renders a much higher amount of shrinkage and distortion [39].

Exemplary protocol:

- after dehydration in an ascending alcohol series, replace ethanol with HMDS (100%)
- place specimen immersed in HMDS under fume hood until completely dried

3.2.3 Mounting, Conductivity and SEM-Examination

Specimens on cover slips are removed from the culture vessel with fine forceps and mounted onto aluminium stubs using a conductive medium such as conductive

Fig. 20 Typical 'electron burn' artefacts after SEM-examination at high magnification (5 kV): the two rectangular darkened areas indicated by the *red oval* and the *white arrows* were caused by the scanning procedure at high magnification. The insert corresponds to the outer of the two darkened areas. Bovine microvascular endothelial cell culture, contacts of cellular projections. Insert figure taken with permission from [14]. © Wiley & Sons

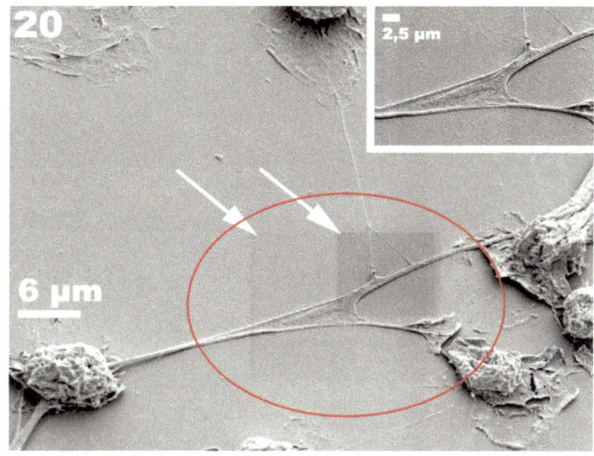

carbon cement, dried-off in a vacuum-chamber for 24 h and subsequently sputter-coated (gold, palladium, chromium or silver in an inert gas, e.g. argon, atmosphere) for conductivity.

Alternatively, or additionally in samples with very uneven and craggy surfaces, specimens may also be rendered conductive by impregnation with osmium tetroxide vapour (see below). Sputter-coating stabilises fragile three-dimensional specimens, but the thin metal film may also obscure very fine surface details of the specimen. Thus, duration of sputter-coating (and thus thickness of metal surface film) needs to be adapted to type of specimens and research aim. Cell culture specimens are mostly sputter-coated for 90–180 s (up to 10 min) with an approximate thickness of 30 nm. Regarding 'troubleshooting' of sputter-coating procedure, also see Sect. 5.2, Sputter-coating of vascular casts.

SEM-examination is usually carried out with low accelerating voltages (5–10 kV). So-called 'burn-artefacts' (Fig. 20) may occur with thinly coated specimens during SEM-examination at higher magnifications.

Storage of processed SEM specimens is best achieved in exsiccator vessels containing hygroscopic substances with added humidity indicator such as blue gel. Alternatively, specimens may be stored in special dust- and moisture-protected drawer-containers.

Impregnation with osmium vapour (e.g., [30]):
Dried samples are placed in a small tightly closed glass container containing osmium tetroxide crystals for 24–48 h. The container is then opened and the samples are left for 15 min or more under the fume hood for sublimation of super-fluous osmium vapour and then placed in a container with vaporised hydrazine hydrate for final removal of superfluous osmium.

Metal coating adds thickness to the specimen and thus possibly obscures fine detail; it may also damage tissue and disrupt delicate structures. An alternative to metal-coating is ligand-mediated osmium binding. Kelley et al. [16] introduced the so-called OTO or OTOTO methods using thiocarbohydracide (T) to enhance osmium (O) binding of soft tissues. A comparable effect is also achieved using ruthenium red (0.05%) as a ligand for osmium binding (ORO/ORORO-method; [4]).

Exemplary protocol OTO/OTOTO [16]:

- fix sample as in routine SEM (Sect. 3.2.1)
- post-fix in 1% osmium tetroxide
- rinse in Na-cacodylate buffer (0.1 M, pH 7.2–7.4), 5–10× (10 min), at 4°C or room temperature
- incubate in an excess of fresh filtered 1% aqueous thiocarbohydrazide for a 10–15 min at room temperature
- rinse 5–10× with distilled water
- incubate in 1% osmium tetroxide for 30 min to 1 h
- rinse 5–10× with distilled water
- (for OTOTO, repeat incubation in thiocarbohydrazide, rinsing, subsequent incubation in osmium tetroxide, and rinsing procedures)
- dehydrate and dry as in routine SEM (Sects. 3.2.1 and 3.2.2)
- mount with conductive medium

(Repeated and thorough rinsing in buffer/distilled water is necessary in order to avoid unwanted brown staining)

3.2.4 Micro-Corrosion Casting of Cultured Cells

Submitting complex three-dimensional cell cultures to corrosion-casting is a new [14] but obvious approach to the problem of preservation of cellular structures for SEM via visualisation of hardened methylmethacrylate negative-patterns (Fig. 21). It is particularly well suited for a direct comparison of *in vitro* cell culture models with microvascular corrosion casts of *in vivo* neovascularisation [13, 14]. The resulting stable high quality casts of the cultured endothelial structures are particularly useful for an overview of cellular development within the whole cell culture dish, because 'conventional' processing for SEM often renders shrinkage and fracture artefacts in the periphery of the culture specimens.

For SEM examination of micro-corrosion casts, cultured cells grown on cover slips are coated with a polymerising methylmethacrylate after suction extraction of the culture medium. After the polymerisation process, the resin covered cover plates need to be removed from the culture vessel. For multiple-well plates, a soldering iron can be carefully applied (caution: the culture cast is also susceptible to heat). The specimens including the cover slips may be subjected to tissue maceration (by enzyme solution, or mild soda lye or caustic potash solution) – in many cases the cover slip will then detach itself from the cast proper when the cultured cells are macerated. For quicker maceration, the cover slip may also be carefully removed from the cast prior to maceration with fine forceps – in case of multilayered or distinctly three-dimensionally arranged cell cultures cells will yield to mild tension and thus may be removed without damage to the cast material.

Fig. 21 Comparison of cell culture (**a**, phase contrast microscopy) and microcorrosion cast of culture (**b**, TEM), bovine microvascular endothelial cells. The microcorrosion cast renders an authentic 'negative' of the surface features of the *in vitro* angiogenesis model including endothelial cell nuclei imprints that are comparative to respective features of in *vivo* vascular corrosion casts (compare with Figs. 23 and 24). *Arrow* = shrinkage artefacts caused by HMDS-drying (Figures taken with permission from Hirschberg et al. [14]; © Wiley & Sons)

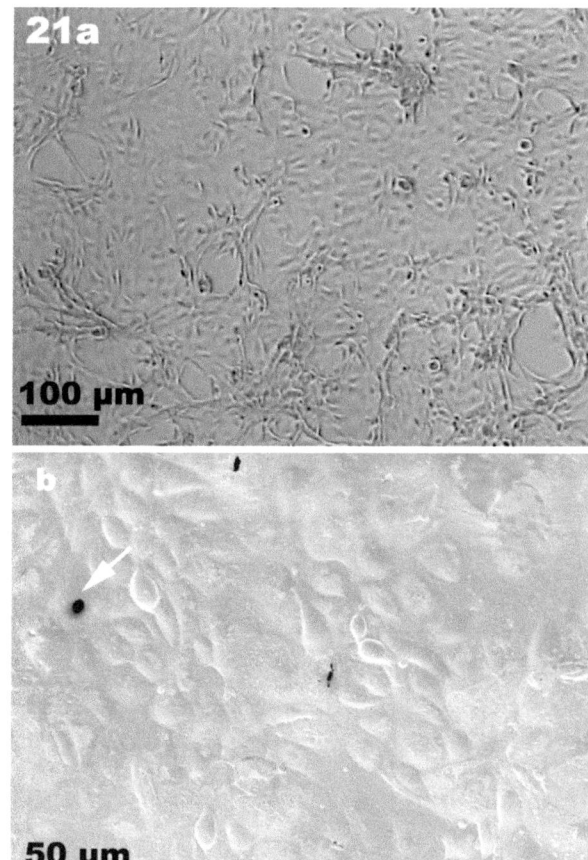

Exemplary protocol:

- prepare polymerising methylmethacrylate with catalyst (see Sect. 5, Microcorrosion casting)
- remove culture medium
- quickly dribble liquid polymerising methylmethacrylate with a pipette onto the cultured cells until the whole surface of the culture vessel is well coated (for enhanced stability of resulting cast apply liberally)
- polymerisation for approx. 24 h at room temperature
- remove cast from culture vessel with soldering iron
- gently remove cover slip from 'cell-side' surface of cast (if possible)
- incubate cast (or cast with cover slip) in enzyme solution for 3 days at 33°C
- rinse cast repeatedly in hand-warm distilled water
- dry cast in HMDS (see above)
- sputter-coat with gold for 2 min (approx. thickness of 30 nm)
- SEM-examination at 5–10 kV

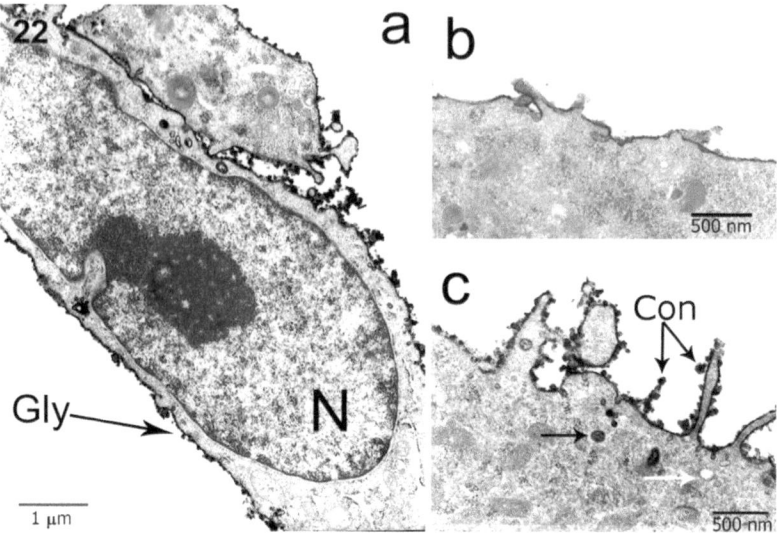

Fig. 22 Murine cardiac microvascular endothelial cells in vitro, TEM, ruthenium red staining; cells retrieved and processed as an agar-pre-embedded pellet. (**a**): The cells are covered with a dense glycocalyx (*Gly*) visualised by ruthenium red staining. (**b, c**): In higher magnification either a thin (**b**) or a thick (**c**) layer with conglomerates (*Con*) of glycocalyx can be observed. The *black arrow* indicates an intracellular vesicle coated with glycocalyx, the *white arrow* points to a vesicle without glycocalyx coat (Taken with permission from: Janczyk et al. [15]; © Oxford University Press)

4 Exemplary Special Staining Methods

Particular ultrastructural details may be distinguished by applying additional staining techniques or by immunocytochemical methods such as immunogold-labelling. Angiogenesis research also includes examination of particle uptake such as liposomes or nanoparticles that in turn may be labelled by gold or silver particles. Uptake of metal-labelled contrast media is another field of applied angiogenesis research. For assessment of ultrastructural localization of nanoparticles with TEM and for further protocols also see [45].

4.1 Pre-treatment with Ruthenium Red for Depiction of Sugar Moieties

To visualise for example the glykocalyx in EM, 0.05% ruthenium red is added to all 'routine' washing, fixing and staining media (e.g., [15]) (Fig. 22).
     Exemplary protocol:

- wash in Na-cacodylate buffer (0.1 M, pH 7.2) with 0.05% ruthenium red for 5 min
- incubate in cacodylate buffer containing 2.5% glutaraldehyde and 0.05% ruthenium red for 60 min

Fig. 23 Bovine ovarial microvascular endothelial cells in vitro, TEM, silver-enhancement of gold-labelled cationic liposomes, cells retrieved and processed via agar-pre-embedding in situ. Within the otherwise un-contrasted specimen, the uptake (*black arrow*) and intracellular distribution (*white arrows*) of the labelled liposomes can be traced. Original magnification ×25,000. Courtesy of Sophie Backhaus

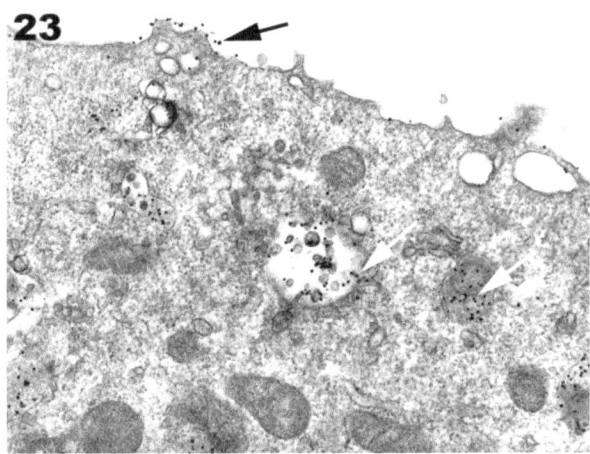

- wash in cacodylate buffer with 0.05% ruthenium red for 5 min
- incubate in Na-cacodylate buffer (0.1 M, pH 7.2)containing 1% osmium tetroxide and 0.05% ruthenium red for 120 min
- wash with pure cacodylate buffer

4.2 Silver-Enhanced Detection of Gold-Labelled Cationic Liposomes

After fixation, liposome-incubated cell cultures are subject to a modified silver-enhancement-technique (according to [49]) (Fig. 23). For facilitating evaluation of gold-labelled structures it is recommended to leave out further contrasting procedures, so that the gold-labelled liposomes are not 'masked' by other metals within the specimen.

Exemplary protocol [10]:

Silver enhancement:

- wash in Na-cacodylate buffer (0.1 M, pH 7.2)
- incubate with glycin (0.05 M in PBS)
- rinse in HEPES-buffer (0.05 M, pH 5.8) containing sucrose (0.2 M)
- wash in distilled water, 3 × 5 min
- incubate with two-component silver enhancement kit for EM (e.g. SPI supplies, USA), 20 min at room temperature, in the dark
- wash in distilled water, 3 × 5 min

Fixation and post-fixation:

- wash in HEPES-buffer (0.05 M, pH 5.8)containing sucrose (0.2 M) for 5 min at room temperature

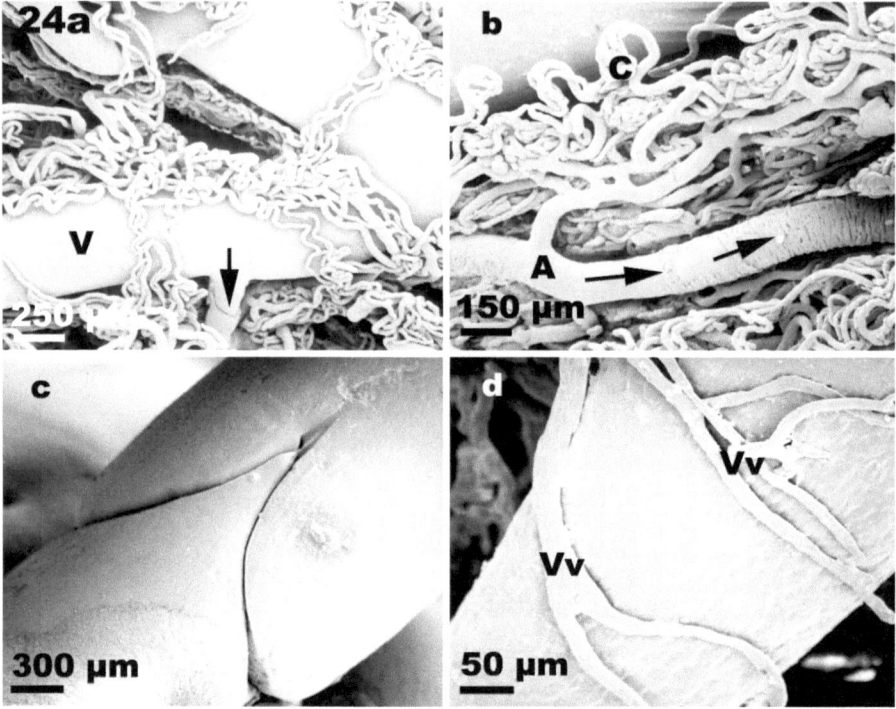

Fig. 24 SEM-examination of microcorrosion casts (bovine pododerma). (**a**): deep pododermal plexus with capillaries draining into veins (*V*), *arrow* indicates cast of bicuspid venous valve; (**b**): deep pododermal artery (*A*) giving rise to glomeriform capillary bed (*C*), *arrows* indicate so-called 'plastic strips' that are interpreted as mummified residues of vascular smooth muscle cells; (**c**): corrosion cast of bigger bicuspid venous valve; (**d**): cast of vasa vasorum (Vv) supplying a pododermal artery (Taken with permission from Hirschberg [11])

- incubate with Na-thiosulfate (0.25 M) in HEPES-buffer (0.05 M, pH 7.4) containing sucrose (0.2 M)
- wash in HEPES-buffer (0.05 M, pH 5.8)for 5 min at room temperature
- wash in 0.05 M Na-cacodylate buffer (0.1 M, pH 7.2) for 5 min at room temperature
- incubate with 0.1% osmium tetroxide in Na-cacodylate buffer (0.1 M, pH 7.2) for 30 min at 4°C
- wash in 0.05 M Na-cacodylate buffer (0.1 M, pH 7.2) for 3 × 10 min
- wash in HEPES-buffer (0.05 M, pH 5.8) containing sucrose (0.2 M) for 10 min

{Contrasting (if applicable, see above)

- contrast with uranyl acetate (2%) for 48 h at 37°C
- wash in HEPES-buffer (0.05 M), 2 × 10 min
- wash in 0.05 M Na-cacodylate buffer (0.1 M, pH 7.2), 2 × 10 min}

Employing the silver-enhancement-technique, ultra-thin sections are mounted onto nickel grids instead of copper grids.

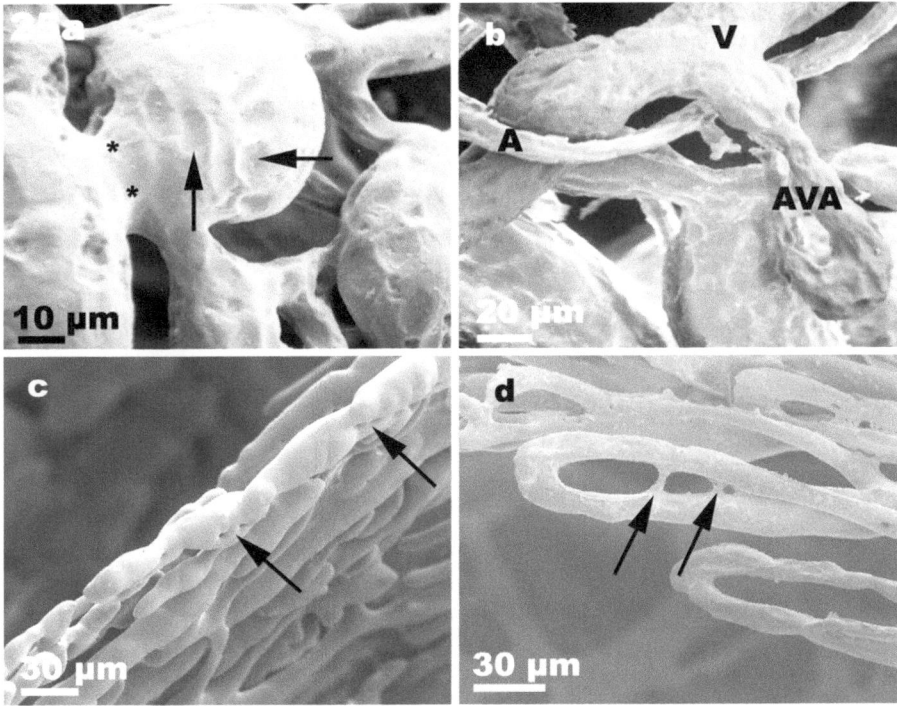

Fig. 25 SEM-examination of microcorrosion casts (bovine pododerma). (**a**): focal dilations within a microvascular network displaying distinct endothelial nuclei imprints (*arrows*) and sphincter-like impressions (*asterisks*); (**b**): arteriovenous shunt (*AVA*) directly connecting an arteriole (*A*) with a venule (*V*), specimens fractured during preparation; also note the distinct and 'vessel-gnomonic' endothelial nuclei imprints; (**c**): small gaps in the cast (*arrows*), typical for the beginning of intussusceptive processes (equivalent of 'pillar formation' dividing vessel lumen); (**d**): intussusceptive remodelling within a thoroughfare channel, the sub-dividing pillars (*arrows*) are already perfused and thus depicted in the cast (Taken with permission from Hirschberg [11]; Figure **d** taken with permission from Hirschberg and Plendl, [13]; © Wiley & Sons)

4.3 Intravascular Application of Colloidal Carbon

Colloidal carbon may be employed as a marker for blood vessels in EM (e.g., [32]). Colloidal carbon is injected into the respective access vessel (e.g. suspension of 10% carbon, particle size 20–50 nm, and 0.9 phenol, stabilised with 4.3% fish glue) until outflow from the egress vessels, then the vessels are clamped. Sample tissue is excised and fixed (e.g. Karnovosky' fixative) for 12–24 h, routinely post-fixed and further processed for TEM, but leaving out the contrasting procedure. Thus, the carbon particles within the microvasculature are easily detectable within the otherwise un-contrasted tissue components.

4.4   Modified Golgi EM Techniques for Detection of Vasculature in Cerebral Tissue

A modified Golgi EM technique, i.e. impregnation with silver nitrate is combined with gold toning and de-impregnation that allows direct detection of sprouting endothelial cells without further contrasting (for further details on procedures see [32]).

5   Microvascular Corrosion Casting

5.1   Corrosion Casting Procedure

Vascular systems can be demonstrated by various injection techniques, and SEM of micro-corrosion casts is facilitated by employing polymerising plastics with high surface-reproducing qualities. Although a widespread technique, micro-corrosion casting is challenging because no 'universal' injection technique has been established that 'guarantees' reproducible results for the respective vascular system in question [12]. An excellent and still relevant overview of applications, techniques and limitations of corrosion casting was established by Lametschwandtner and co-workers [20, 21]. Controversial data exist particularly on effects of pre-treatments, such as whether flushing the respective vascular bed prior to injection of the casting media – e.g. with heparin-, ringer- or physiological sodium chloride solution – is expedient; likewise, whether pre-perfusion with a fixative is recommended or not [21]. In most cases, pre-polymerised methylmethacrylate kits (e.g. Mercox, SPI Supplies, West Chester, PA, USA; Batson's No. 17, Polysciences Inc., Warrington, PA, USA; Tensolcement No. 7, ICI chemicals and Polymers Ltd. Darwen, UK), often diluted with monomeric methylmethacrylate for enhanced capillary filling, are used as casting media for microcorrosion specimens. A newer type of casting media is represented by polyurethane elastomers (e.g. PU4ii; vasQtech, Zurich, Switzerland) that besides SEM of microcorrosion casts (e.g., [19, 22]) may also be employed for 'conventional' histology of paraffin embedded specimens [52]. Comparably, treatment of specimens after injection of the casting media and polymerisation time varies distinctly according to the respective vascular bed in question (e.g., [21]). Maceration in either caustic potash, sodium hydroxide (each at 5–25%) or enzyme solution may be applicable according to size and composition of surrounding tissues (e.g. soft tissue only, or also calcified bones etc.). Intermediary washing steps in warm tap water and/or distilled water increase the quality of the resulting cast. Trimming and sample taking from the vascular casts can be achieved by employing a soldering iron, fine scissors and/or razor blades. Best dissection results (with minimal fracturing) of dense microvascular casts are achieved when the cast is again embedded into a stabilising medium such as gelatine (that needs to be re-macerated afterwards) or carefully deep frozen prior to cutting. Vascular cast samples are then dried and mounted according to routine SEM-procedure (see Sects. 3.2.2 and 3.2.3).

Exemplary protocol:

Preparation of casting media:

- dilute Mercox resin with monomer methylmethacrylate (MMA): 4:1
- add Mercox accelerator (benzoyl peroxide) just prior to injection: 1 (accelerator) : 16 (MMA) : 64 (resin)

Casting procedure:

- cannulate access vessel and rinse with 37°C heparinised Ringer or physiologic saline solution (Heparin: 100 IU/ml) until fluid escapes from egress vessel(s) (optional)
- perfuse vascular bed with fixative (e.g. 'half strength' Karnowsky solution: 0.25% glutaraldehyde, 0.25% paraformaldehyde in 0.1 M Na-cacodylate buffer, pH 7.2) (optinal step)
- inject prepared casting media until it escapes from egress vessel(s)
- clamp first egress vessel(s), then access vessel(s)

For preparation of casting media and the casting procedure itself it is of particular importance to use only containers, syringes, catheters, canulae etc. that are not affected/corroded by methylmethacrylates, i.e. glass or metal. Other plastic material (e.g. polyethylene material) may be affected by methylmethacrylates and should be avoided (or checked in a pre-test) in case of longer injection procedures. Surplus casting media should be disposed of only after complete polymerisation.

The 'pot life' (working time after mixing of resin and catalyst) for Mercox is approx. 2–5 min, the 'cure time' (start of polymerisation) is max. 10 min (depending on prepared resin volume). In case of larger injection volumes, decrease amount of accelerator in order to prolong cure time.

Corrosion procedure:

- immerse either complete cadaver (small animals) or body part into warm water bath (50°C), preferably in a position similar to natural body posture, for 'tempered' polymerisation of the injection media (1 to several hours, according to size)
- immerse specimen in 7.5% (small specimens) – 15% (bigger specimens) NaOH or KOH at 30–60°C, rinse at least once a day with warm distilled water and repeat corrosion cycle as long as necessary to remove all tissue debris from the cast
- rinse carefully in distilled, followed by bi-distilled water

Drying and mounting:

- air-dry specimen either from bi-distilled water or from HMDS (see Sect. 3.2.2, SEM).
- mount and sputter-coat as described in Sect. 3.2.3, SEM

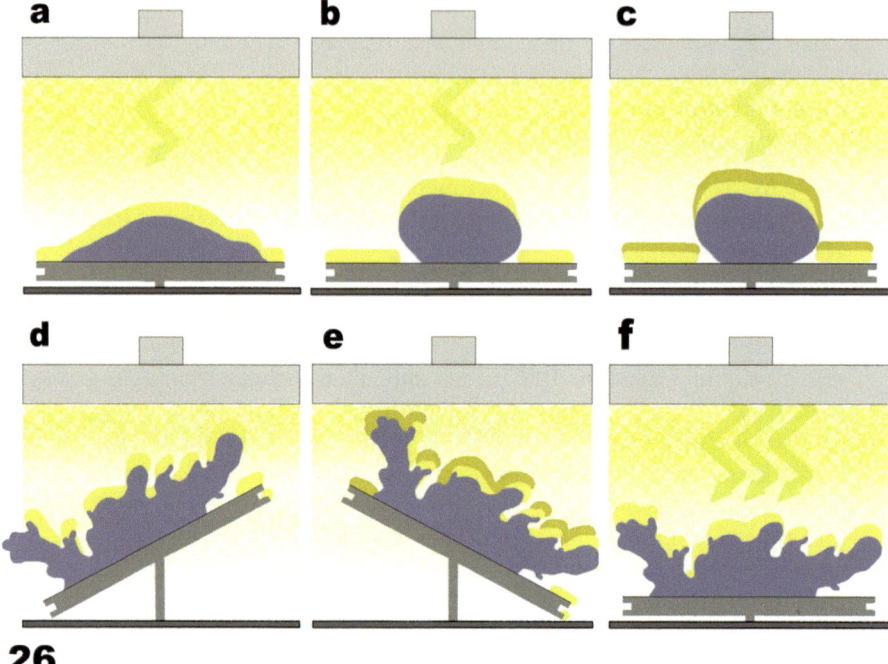

26

Fig. 26 Troubleshooting of sputter-coating process for SEM:(**a**): uncomplicated sputter-coating of small sample with even surface; (**b**): high sample with smaller basis, i.e. surface tapering towards *centre*: side and basal aspects are not covered with metal film (*yellow layer*); (**c**): longer sputtering will not improve coating within 'critical' surface areas and only increases metal film thickness (*darker layer*) on 'non-critical' surface areas, thus possibly obscuring fine surface details; (**d, e**): in large and very uneven samples, such as some microcorrosion casts, sputter coating is improved when specimens are tilted at different angles; (**f**): in difficult samples, running sputter coater at maximal vacuum possible will increase results (Modified according to Protrain [45]; Courtesy of Diemut Starke)

5.2 Sputter-Coating of Vascular Casts

Corrosion casts are routinely sputter-coated with gold for SEM examination (see Sect. 3.2.3) (Fig. 26). However, casts replicating complex three-dimensional angioarchitectures often cannot be satisfactorily sputter coated with conventional sputter coaters, due to 'self-shadowing' effects (i.e. cast elements on the surface of the specimens prevent coating of cast elements in lower levels of the specimen, as they are 'shadowed' behind the superficial cast elements). This may be overcome by applying thicker metal coatings (i.e. longer sputter times. Fig. 26c) – with the risk of stronger charge artefacts during SEM and 'masking' of fine surface structures. Alternatively, sputter-coating at the highest vacuum possible may improve the metal coating procedure (Fig. 26f). It is also recommended to tilt samples during the sputter-coating process at varying angles, particularly in microcorrosion cast specimens that are very large or display very uneven surfaces

(Fig. 26d, e). In such specimens, additional impregnation in osmium tetroxide vapour is recommended (see Sect. 3.2.3). Alternatively, metal coating with equipment allowing rocking in conjunction with rotation ('tumbling') is suggested, as tumbling avoids the effect of 'self-shadowing'. Employing chromium instead of gold renders more stable specimens that allow SEM examinations at higher magnification. 'Tumble coating' with chromium has been recommended for three-dimensional fragile lymphovascular casts (e. g., [4]).

Vascular corrosion casts can also be rendered conductive without sputter coating by osmium impregnation alone (e.g., [30]; see Sect. 3.2.3).

5.3 SEM-Examination of Vascular Cast

Vascular casts are routinely examined at low acceleration voltages (5–10 kV). Qualitative SEM examination of microvascular corrosion casts (Figs. 24 and 25) is achieved by evaluation of characteristic, 'vessel-type-gnomonic' imprints of endothelial cell nuclei, and differentiation of vascular sprouts from incomplete vascular filling 'ends' is also based on occurrence and distribution pattern of these imprints (e.g., [6, 20, 21]). Typical features indicative of intussusceptive angiogenesis (Fig. 24c, d) are also generally recognised (e.g., [24]). Quantitative information of the three-dimensional morphology of vascular networks and neovascular processes can be obtained by taking stereo-paired images obtained by SEM (tilt angles from 6° to 20°) and calculating vectors of for instance intercapillary distances employing the mathematical phenomenon of parallaxis (e.g., [18, 28, 37]). Recently, a simple tool for stereological assessment of digital images has been presented [51].

Acknowledgements The authors particularly wish to thank our experienced EM laboratory staff Monika Sachtleben and Verena Holle (née Eckert-Funke), for their ever-lasting enthusiasm and expert support. Many of our findings and experiences from processing cell culture models of angiogenesis for EM were previously published in the article "Electron microscopy of cultured angiogenic endothelial cells" [14] and are used with permission. The presented agar-pre-embedding technique allowing optimal sample retrieval for processing three-dimensional in vitro-angiogenesis models, in particular, is primarily the result of Monika Sachtleben's "penchant for perfectionism" within her EM laboratory.

Our graphic designer team also supported this chapter: Martin Werner assisted us with the hands-on photos from the EM laboratory, and Diemut Starke supplied the excellent schematic drawing illustrating the trouble-shooting process in sputter-coating for SEM.

Our colleagues and former doctorate students Dr. Mahtab Bahramsoltani (Institute of Veterinary Anatomy, Faculty of Veterinary Medicine, University of Leipzig), Dr. Sabine Kaessmeyer (Institute of Veterinary Anatomy, Faculty of Veterinary Medicine, Freie Universität Berlin), Dr. Jasmin Lienau (Centre for Musculoskeletal Surgery, Charité, Berlin), Dr. Pawel Janczyk (Federal Institute for Risk Assessment, Unit Molecular Diagnostics and Genetics, Department of Biological Safety, Berlin) and Dr. Sophie Backhaus (née Hansen) generously supported this chapter with brilliant EM micrographs.

References

1. Bahramsoltani M, Plendl J (2004) Ein neues in vitro-Modell zur Quantifizierung der Angiogenese in vitro. (A new in vitro model to quantify angiogenesis). ALTEX 21:227–244
2. Bartczak D, Sanchez-Elsner T, Louafi F, Millar TM, Kanara AG (2011) Receptor-mediated interactions between colloidal gold nanoparticles and human umbilical vein endothelial cells. Small 7:388–394
3. Belton CM (1979) Application of ruthenium red ligand binding of osmium: scanning electron microscopy of larval tapeworms. Micron 10:1–4
4. Belz GT, Auchterlonie GJ (1995) An Investigation of the use of chromium, platinum and gold coating for scanning electron microscopy of casts of lymphoid tissues. Micron 26:141–144
5. Braet F, De Zanger R, Wisse E (1997) Drying Cells for SEM, AFM and TEM by hexamethyl-disilazane: a study on hepatic endothelial cells. J Microsc 186:84–87
6. Christofferson RH, Nilsson BO (1988) Microvascular corrosion casting with analysis in the scanning electron microscope. Scanning 19:43–63
7. Dvorak AM, Wiberg L, Monahan-Earley RA, Galli SJ (1990) A simple technique to facilitate the ultrastructural analysis of cells in soft agar culture systems: demonstration of the development in vitro of morphologically mature mast cells and phagocytic macrophages from the bone marrow cells of genetically mast cell-deficient W/Wv or congenic normal mice. Lab Invest 62:774–781
8. Hamill RJ, Vann JM, Proctor RA (1986) Phagocytosis of Stapylococcus aureus by cultured bovine aortic endothelial cells: model for postadherence events in endovascular infections. Infect Immun 54:833–836
9. Hansen S (2009) Aufnahme und Verteilung von Liposomen und Liposom-Konjugaten in Endothelzellen in vitro. (Uptake and distribution of liposomes and liposome-conjugates in endothelial cells in vitro). Dissertation thesis, Institute of Veterinary Anatomy, Faculty of Veterinary Medicine, Freie Universität, Berlin
10. Heinzer S, Kuhn G, Krucker T, Meyer E, Ulmann-Schuler A, Stampanoni M, Gassmann M, Marti HH, Müller R, Volgel J (2008) Novel three-dimensional analysis tool for vascular trees indicates complete micro-networks, not single capillaries, as the angiogenic endpoint in mice overexpressing human VEGF(165) in the brain. Neuroimage 39:1549–1558
11. Hirschberg RM (1999) Die Feinstruktur der Blutgefäße an der gesunden und erkrankten Rinderklaue (The microvasculature of the healthy and diseases bovine claw). Dissertation, Institute of Veterinary Anatomy, Faculty of Veterinary Medicine, Freie Universität, Berlin
12. Hirschberg RM, Muelling CKW, Bragulla H (1999) Microvasculature of the bovine claw demonstrated by improved micro-corrosion-casting technique. Microsc Res Tech 45:184–197
13. Hirschberg RM, Plendl J (2005) Pododermal angiogenesis and angioadaptation in the bovine claw. Microsc Res Tech 66:145–155
14. Hirschberg RM, Sachtleben M, Plendl J (2005) Electron microscopy of cultured angiogenic endothelial cells. Microsc Res Tech 67:248–259
15. Janzcyk P, Hansen S, Bahramsoltani M, Plendl J (2010) The glycocalyx of human, bovine and murine microvascular endothelial cells cultured in vitro. J Electron Microsc 59:291–298
16. Kelley RO, Dekker RA, Bluemink JG (1973) Ligand-mediated osmium binding: its application in coating biological specimens for scanning electron microscopy. J Ultrastruct Res 45:254–258
17. Konerding MA, Miodonski AJ, Lametschwandtner A (1995) Microvascular corrosion casting in the study of tumor vascularity: a review. Scanning Microsc 9:1233–1243
18. Konerding MA, Turhan A, Ravnic DJ, Lin M, Fuchs C, Secomb TW, Tsuda A, Mentzer SJ (2010) Inflammation-induced intussusceptive angiogenesis in murine colitis. Anat Rec 293:849–857
19. Krucker T, Lang A, Meyer EP (2006) New polyurethane-based material for vascular corrosion casting with improved physical and imaging characteristics. Microsc Res Tech 69:138–147

20. Lametschwandtner A, Lametschwandtner U, Weiger T (1984) Scanning electron microscopy of vascular corrosion casts – technique and applications. Scanning Electron Microsc (Pt2):663–695

21. Lametschwandtner A, Lametschwandtner U, Weiger T (1990) Scanning electron microscopy of vascular corrosion casts – technique and applications updated review. Scanning Microsc 4:889–941

22. Lee GS, Filipovic N, Lin M, Gibney BC, Simpson DC, Konerding MA, Tsuda A, Mentzer SJ (2011) Intravascular pillars and pruning in the extraembryonic vessels of chick embryos. Dev Dyn 240:1335–1343

23. Lienau J, Kaletta C, Teifel M, Naujoks K, Bhoola K, Plendl J (2005) Morphology and transfection study of human microvascular endothelial cell angiogenesis: an in vitro three-dimensional model. Biol Chem 386:167–175

24. Makanya AN, Hlushchuk R, Djonov VG (2009) Intussusceptive angiogenesis and its role in vascular morphogenesis, patterning and remodelling. Angiogenesis 12:113–123

25. Mansy SS (2004) Agarose cell block: innovated technique fort he processing of urine cytology for electron microscopy examination. Ultrastruct Pathol 28:15–21

26. McDonald DM, Choyke PL (2003) Imaging of angiogenesis: from microscope to clinic. Nat Med 9:713–725

27. Meyer EP, Ulmann-Schuler A, Staufenbiel M, Krucker T (2008) Altered morphology and 3D architecture of brain vasculature in a mouse model for Alzheimer's disease. Proc Natl Acad Sci USA 105:3587–3592

28. Minnich B, Bartel H, Lametschwandtner A (2001) Quantitative microvascular corrosion casting by 2D- and 3D-morphometry. Ital J Anat Embryol 106:213–220

29. Moskaluk CA, Stoler MH (2002) Agarose mold embedding of cultured cells for tissue microarrays. Diagn Mol Pathol 11:234–238

30. Murakami T, Unehira M, Kawakami H, Kubotsu A (1973) Osmium impregnation of methyl methacrylate vascular casts for scanning electron microscopy. Arch Histol Jpn 36:119–124

31. Murray AG, Schulze H, Blauw E (1991) In situ embedding of cell monolayers cultured on plastic surfaces for electron microscopy. Biotech Histochem 66:269–272

32. Nousek-Goebl NA, Press MF (1986) Golgi-Electron microscopic study of sprouting endothelial cells in the neonatal rat cerebellar cortex. Dev Brain Res 30:67–73

33. Oorschot V, de Wit H, Annaert WG, Klumperman J (2002) A novel flat-embedding method to prepare ultrathin cryosections from cultured cells in their in situ orientation. J Histochem Cytochem 50:1067–1080

34. Patan S (2000) Vasculogenesis and angiogenesis as mechanisms of vascular network formation, growth and remodelling. J Neurooncol 50:1–15

35. Patan S (2004) Vasculogenesis and angiogenesis. Cancer Treat Res 117:3–32

36. Plendl J, Neumüller C, Vollmar A, Auerbach R, Sinowatz F (1996) Isolation and characterization of endothelial cells from different organs of fetal pigs. Anat Embryol 194:445–456

37. Polykandriotis E, Arkudas A, Beier JP, Dragu A, Rath S, Pryymachuk G, Schmidt VJ, Lametschwandtner A, Hordch RE, Kneser U (2011) The impact of VEGF and bFGF on vascular steriomorphology in the context of angiogenic neo-arborisation after vascular induction. J Electron Microsc. doi:10.1093/jmicro/dfr025

38. Protrain (2011) Courses in electron microscopy – Hints and tips. Chapman S. (ed) Internet publication: http://www.emcourses.com/tips.htm (21.07.2011)

39. Reville WJ, Heapes MM, O'Sullivan VR (1994) A survey to assess the ultrastructural preservation of fixed biological samples after air-drying from tetramethylsilane. J Electron Microsc 43:111–115

40. Reynolds ES (1963) The use of lead citrate at high pH as an electron-opaque stain for electron microscopy. J Cell Biol 17:208

41. Richardson KC, Jarett L, Finke EH (1960) Embedding in epoxy resins for ultrathin sectioning in electron microscopy. Stain Technol 35:313–325

42. Riesau W (1997) Mechanisms of angiogenesis. Nature 386:671–674

43. Romeis B (2010) Mikroskopische Technik (Microscopical technique). In: Mulisch M, Welsch U (eds). 18th ed. Spektrum Akademischer Verlag, Heidelberg

44. Sato T (1967) A modified method for lead staining of thin sections. J Electron Microsc 16:133

45. Schrand AM, Schlager JJ, Dai L, Hussain SM (2010) Preparation of cells for assessing ultrastructural localization of nanoparticles with transmission electron microscopy. Nat Protoc 5:744–757

46. Slizova D, Krs O, Pospisilova B (2003) Alternative method of rapid drying vascular specimens for scanning electron microscopy. J Endovasc Ther 10:285–287

47. Taniguchi Y, Tamatani R, Kawarai Y (1994) A reliable method of embedding a small amount of dispersed cells for electron microscopy. J Electron Microsc 43:48–50

48. Taupin P (2008) A simple and direct pre-embedding technique for ultrastructure of scarce biological specimens. Biotech Histochem 83:253–257

49. Thurston G, McLean JW, Rizen M, Baluk P, Haskell A, Murphy TH, Hanahan D, McDonald DM (1998) Cationic liposomes target angiogenic endothelial cells in tumors and chronic inflammation in mice. J Clin Invest 101:1401–1413

50. Townsend LE, Gover JL, Seymour MS (1996) Transmission electron microscopy procedure for endothelial cell culture with angiogenesis. Methods Cell Sci 18:15–18

51. Tschanz SA, Burri PH, Weibel ER (2011) A simple tool for stereological assessment of digital images: the STEPanizer. J Microsc 243:47–59

52. Ulmann-Schuler A, Krucker T, Meyer EP (2007) PU4ii vascular casting combined with histological staining methods. In: Proceedings advances in vascular casting, University of Salzburg, pp 94–97

53. Venable JH, Coggeshall R (1965) A simplified lead citrate stain for use in electron microscopy. J Cell Biol 25:407–408

54. Wei W, Popov V, Walocha JA, Wen J, Bello-Reuss E (2006) Evidence of angiogenesis and microvascular regression in autosomal-dominant polycystic kidney disease kidneys: a corrosion cast study. Kidney Int 70:1261–1268

55. Widehn S, Kindblom LG (1990) Agarose embedding: a new method for the ultrastructural examination of the in-situ morphology of cell cultures. Ultrastruct Pathol 14:81–85

56. Yoshida D, Noha M, Watanabe K, Sugisaki Y, Teramoto A (2001) Novel approach to analysis of in vitro tumor angiogenesis with a variable-pressure scanning electron microscope: suppression by matrix metalloproteinase inhibitor SI-27. Brain Tumor Pathol 18:89–100

57. Yuan LC, Gulyas BJ (1981) An improved method for processing single cells for electron microscopy utilizing agarose. Anat Rec 201:273–281

Methods to Assess Vascular Permeability During Angiogenic Processes

Alexander N. Garcia and Joe G.N. Garcia

Abstract Angiogenesis is a multi-faceted process that is, in part, reliant on dynamic regulation of the endothelium that lines the bodies vasculature. Here we have detailed angiogenic factors VEGF, angiopoietin 1 and 2, sphingosine 1-Phosphate (S1P), and hepatocyte growth factor, all of which act, in part, via alterations of endothelial barrier function. As these alterations in endothelial integrity provide insight into angiogenic responses, we further specify detailed protocols that may be used in its study. These include Transwell Albumin clearance, Evans Blue Albumin Permeability assays, and Trans-endothelial Electrical Resistance.

1 Introduction

The vascular endothelium represents a thin layer of cells lining the innermost portion of the vascular bed within every organ/tissue. Endothelial cell (EC) monolayers serve as a semi-permeable cellular barrier separating blood vessels from surrounding tissues and regulates multiple key biological processes including organ fluid balance and exchange of fluids and cells as well as solute transport between the two compartments. The endothelial barrier is dynamically regulated via cytoskeletal protein-driven conformational alterations in EC morphology in response to stimuli of physiological and pathological origin [1]. For example, an increase in vascular permeability is a necessary feature of the body's defense

A.N. Garcia
Department of Pharmacology, The University of Illinois at Chicago, Chicago, IL 60612, USA

J.G.N. Garcia, M.D. (✉)
Department of Medicine, Institute for Personalized Respiratory Medicine, The University of Illinois at Chicago, Chicago, IL 60612, USA

Earl Bane Professor of Medicine, Pharmacology and Bioengineering, The University of Illinois at Chicago, 1737 West Polk Street, 310 AOB (MC672), Chicago, IL 60612, USA
e-mail: jggarcia@uic.edu

E. Zudaire and F. Cuttitta (eds.), *The Textbook of Angiogenesis and Lymphangiogenesis: Methods and Applications*, DOI 10.1007/978-94-007-4581-0_4,
© Springer Science+Business Media Dordrecht 2012

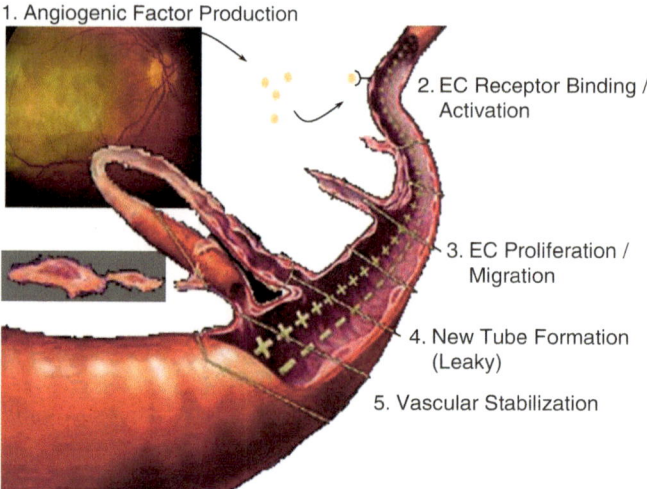

Fig. 1 Generation of new capillaries by sprouting from pre-existing vessels is a complex multi-step process. This process is driven in large part by the dynamic regulation of endothelial barrier function. During the early phases of angiogenesis, the endothelium becomes more permeable in order to facilitate tube formation and elongation. Afterwards, barrier function increases in order to strengthen and stabilize the newly formed vessel and allow the regulation of flow of nutrients

mechanism to provide injured tissues with access to leucocytes, resulting in tissue edema due to fluid extravasation. However, during conditions of intense inflammation, such as observed in sepsis, the opportunity exists for profound vascular permeability resulting in fluid accumulation in multiple organs leading to progressive organ dysfunction or organ failure and increased morbidity and mortality.

As can be seen in Fig. 1, alterations in vascular permeability are also requisite steps in the angiogenic process and ultimately critical for tissue growth and homeostasis due to delivery of nutrients and oxygen, as well as removal of waste products and wound healing. In the embryo, endothelial cells differentiate from angioblasts and then give rise to the early vascular plexus and primitive blood vessels, a process referred to as vasculogenesis. Angiogenesis is defined by the generation of new capillaries by endothelial cells either by sprouting or by splitting from pre-existing vessels. Sprouting angiogenesis involves endothelial detachment from the basement membrane, migration, and subsequent proliferation, tube formation, and, finally, functional maturation of the new vessel (Fig. 1). During the very early steps of angiogenesis, cell-cell contacts weaken in order to allow cells to migrate, invade the underlying tissue and form new tubes that continue to branch, ultimately forming a complex network of vessels with newly strengthened cell-cell adhesions. Angiogenesis characterized by the splitting of existing vessels involves endothelial proliferation within the vessel of origin followed by the generation of a separate lumen and a capillary pillar that then splits from the original vessel and matures into a new vessel. The formation of this network is regulated in large part by the delicate balance of various growth factors in the body with this dynamic vascular regulation of the endothelium representing a key feature of angiogenesis.

A number of angiogenic factors are now recognized as capable of altering endothelial cell barrier properties via receptor-mediated signaling networks involving extracellular matrix proteins and integrin receptors. One such critical factor is vascular endothelial growth factor (VEGF) whose key role in vasculogenesis was apparent in mice lacking the VEGF receptor, Flt-1, which failed to develop fully functional blood vessels. In addition, VEGF has been found to mediate angiogenesis in malignant gliomas and a number of other tumors. These findings have led to the inhibition of VEGF as a promising therapeutic strategy in the management of patients with advanced malignancies. As the precise dynamic regulation of endothelial barrier function is a key feature of angiogenesis, measurements of endothelial permeability can provide valuable insight into the mechanisms of vasculogenesis. As this field of inquiry has developed, it has become clear that modulation of vascular barrier properties by angiogenic agents is a critical aspect of their angiogenic properties. As angiogenic events occur as part of the pathogenesis of a number of cancers and autoimmune diseases, vascular barrier properties have been viewed as a potential target in therapeutic decision-making. Below we detail barrier regulatory aspects of several angiogenic factors which exhibit barrier–modulatory properties, VEGF, angiopoietin 1& 2, HGF and S1P.

2 Angiogenic Factors with Vascular Permeability-Altering Properties

2.1 Vascular Endothelial Growth Factor (VEGF)

VEGF was originally described by the Dvorak lab as vascular permeability factor [2] and is a potent angiogenic factor whose expression is critical for normal vascular development [3, 4]. VEGF functions as a mediator of increased vascular permeability, in both intact vessels [5], in endothelial cell culture [6], and in several organs by directly regulating vascular permeability to water and proteins. Lung over-expression of VEGF induces increased pulmonary vascular permeability resulting in marked pulmonary edema and plasma VEGF levels are significantly elevated in acute lung injury patients. These studies highlight VEGF gene as an attractive barrier-regulatory ALI candidate and molecular target in ALI therapeutic strategies.

The VEGF family of proteins include VEGF-A, VEGF-B, VEGF-C, VEGF-D, the viral homologue VEGF-E, and placental growth factor (PlGF). The primary member of the VEGF family of proteins, VEGF-A, mediates its effects through fms-like tyrosine kinase (Flt-1) and kinase insert domain-containing receptor (KDR; Flk-1 in mice), although studies have shown that activation of Flk-1, but not Flt-1, is required for increased endothelial permeability [7]. Upon binding of

VEGF to its receptor, activation of PLC-γ results in formation of diacylglycerol (DAG) and inositol trisphosphate (IP$_3$), resulting in increased intracellular calcium concentration from both intracellular stores and stores-independent channels [8–10], ultimately resulting in increased paracellular permeability [11].

VEGF gradients driven by hypoxia [12] cause DLL4 expression to become upregulated in tip cells, resulting in activation of NOTCH in stalk cells [13]. This results in downregulation of VEGFR-2 in stalk cells, making these cells less responsive than tip cells [14], which exhibit increased motility and filopodia extensions. Stalk cells then elongate, forming the lumen of the newly formed vessel. The formation of tight junctions and adherens junctions between neighboring endothelial cells results in a firm barrier that offers mechanical support to the vessel. Autocrine VEGF released by endothelial cells maintains vascular homeostasis [15] while paracrine VEGF released by tumors or stromal cells results in increased vessel branching [16].

2.2   Angiopoietin 1 and 2

Additional angiogenic factors with potent barrier-regulatory properties include angiopoietin 1 and angiopoietin 2, growth factors which are critical for normal vascular development. The angiopoietin family represents three ligands, Ang-1, Ang-2, Ang-4 for the family of tyrosine kinases (Tie 1 & 2 receptors) selectively expressed in the vascular endothelium. Whereas VEGF induces EC differentiation and migration, angiopoietin 1 serves to stabilize vascular networks. Angiopoietin 1 and angiopoietin 4 modulate EC permeability by altering the state of adherens junctions and specifically inhibit vascular leakage in response to VEGF or other barrier-disruptive agents, as well as promoting vessel maturation. In contrast, angiopoietin 2 antagonizes angiopoietin 1 and promotes barrier dysregulation by blocking the ability of angiopoietin 1 to activate its receptor, Tie 2. The angiopoietin and Tie signaling system offers a mechanism in which vessels can maintain quiescence while retaining the ability to respond to angiogenic stimuli. No known ligand for Tie-1 has been identified and little is known regarding its function in vessels. Ang-1/Tie 2 act as barrier-stabilizing mediators that oppose VEGF-mediated increase in endothelial permeability in various en vivo and cell culture models [17, 18]. This signaling pathway favors EC survival and maintenance of the endothelial barrier [19]. Genetically engineered mice lacking Ang-1 or Tie-2 are capable of developing a primary vasculature, but die as a result of failure to undergo vessel remodeling [20]. Stimulation of endothelial cells with Ang-2 results in increased vascular permeability, and primes the endothelial monolayer for sprouting angiogenesis [21]. Over-expression of Ang-2 mimics the early embryonic death observed in Tie-2 or Ang-1 knockout models.

2.3 Sphingosine 1-Phosphate/Sphingosine 1-Phosphate Receptor 1

Sphingosine 1-phosphate (S1P) is a biologically active phospholipid found in nanomolar to micromolar concentrations in human and animal serum [22]. Sphingosine kinase, and S1P lyase regulate the levels of S1P in plasma with platelets being the primary storage site of S1P [23]. S1P released from activated platelets acts on a family of heterotrimeric G-protein coupled receptors S1P-1, S1P-2, and S1P-3 found on the surface of the vascular endothelium. Stimulation of these receptors with S1P induces proliferation, calcium mobilization, adhesion molecule expression, and suppression of apoptosis [22, 24, 25]. Furthermore, we have shown S1P to be a potent endothelial cell chemoattractant, and induces endothelial cells to proliferate and differentiate to form capillary like structures that indicate early blood vessels [5, 26]. In murine models of LPS, S1P has been shown to attenuate pulmonary vascular leakage and inflammation [27] through activation of Gi heterotrimeric G-proteins, resulting in downstream activation and translocation of Rac-1. This in turn results in cortical actin assembly and strengthening in adherens junction adhesions between neighboring cells [5].

S1P exerts these barrier protective effects via ligation of sphingosine 1-phosphate receptor 1, S1PR1, a pertussis toxin-sensitive Gi coupled receptor which induces Rac GTPase-dependent substantial increases in cortical actin polymerization critical to endothelial cell barrier enhancement. S1PR1 activation enhances the organization and redistribution of VE-cadherin and β-catenin in junctional complexes in endothelium by phosphorylation of cadherin as well as p120-catenin and inducing the formation of cadherin/catenin/actin complexes. Understanding the role of S1P in enhancing EC barrier function makes it an important molecule for therapeutic applications that reverse the loss of EC barrier integrity. A compelling argument for S1P1 as an a critical barrier-regulatory receptor is that S1PR1 not only its ability to transduce signals which restore barrier integrity but that S1P1 is the target for transactivation by receptors for other potent barrier-protective agonists. These include EPCR (receptor for activated protein C), c-met (receptor for hepatocyte growth factor), CD44 (receptor for high molecular weight hyaluronan), and the ATP receptor.

2.4 Hepatocyte Growth Factor

We have shown that hepatocyte growth factor (HGF), a well-known angiogenic factor that regulates cell growth, motility, and morphogenesis, further functions as a potent endothelial barrier protective agonist via stabilization of the actin cytoskeleton [28]. HGF activates proto-oncogenic receptor tyrosine kinase c-met, ultimately resulting in activation of Rac1 involving the Rac-specific GEF, Tiam1, and inactivation of RhoA small GTPases with increased PAK1 phosphorylation [29]. Depletion of c-met and other downstream effectors using siRNA indicates that these proteins are required for HGF-induced increases in transendothelial electrical resistance [30]. Furthermore, HGF has been shown to be protective in mouse models of LPS-induced acute lung injury [30]. HGF originates from mesenchymal cells to act primarily on epithelial and endothelial cells, but has also been shown to interact with hematopoietic progenitor cells. Activation of its receptor results in stimulation of mitogenesis, cell motility, and matrix invasion. These properties support HGF as a important participant in angiogenesis, tumorogenesis, and tissue regeneration [29]. HGF signals via c-met to recruit CD44v10, a key transactivated receptor for CD44, into caveolin-enriched microdomains (CEMs) or lipid rafts [203]. In experiments using siRNA, both c-met and CD44 were found to be important in HGF-induced increases in EC transendogthelial electrical resistance (TER) [30]. Furthermore, pretreatment of ECs with the CEM-interfering compound methyl-b-cyclodextran also prevented HGF-induced increases in TER [30]. In a mouse model of LPS-induced lung inflammation, HGF was protective against markers of lung inflammation, an effect not noted in CD44 knockout mice [30]. The signaling mechanism involved in HGF-induced EC barrier enhancement is complex, with important roles for c-met, CD44, and CEM formation. HGF produced Rac-dependent increases in the levels of cortical actin, cortactin translocation, and cortical levels of phosphorylated MLC [28]. Further mechanistic studies found that HGF-induced EC barrier enhancement critically involves PI-3-kinaseactivity, distinguishing the mechanism of HGF-induced barrier enhancement from that of S1P [28], with important roles for MAPKs (ERK and p38) and PKC in HGF-induced EC barrier enhancement [28]. Attention to the role of improved cell–cell or cell–matrix adhesion elicited by HGF found that HGF produced increased b-catenin localization to the EC periphery alongside cortical actin and increased association of b-catenin with VE-cadherin [28]. The cell signaling effectors of HGF (PI-3-kinase, ERK, p38, PKC) were found to converge at phosphorylation of glycogen synthase kinase-3b, which regulates the association of b-catenin and cadherin, thereby controlling cell–cell adhesion [28].

3 Methods to Assess Endothelial Cell Permeability and Vascular Barrier Function *In Vitro* and *In Vivo*

3.1 Albumin Clearance Across Endothelial Cell Monolayers (Transwell)

The measurement of monolayer permeability to radiolabeled albumin is a direct measure of protein permeability. Confluent monolayers of endothelium have low albumin permeability, whereas disruption of monolayer integrity leads to formation of intracellular gaps, allowing increased flux of albumin across the barrier(Fig. 5). Serum albumin, radiolabeled on tyrosine residues with ^{125}I, has often been used to assess endothelial permeability [31]. The method for assessing permeability is based on measuring albumin flux across the monolayer of endothelial cells grown on the membrane inserts of a transwell system. This technique utilizes the *Kedem-Katchalsky* derived equation for determining the overall flux of a solute representing the sum of convective and diffusive components:

$$J_s = J_v(1 - \sigma)C_s + PS(\Delta C)$$

where J_s is the solute flux, J_v is the volume flux of fluid; σ is the osmotic reflection coefficient; C_s is the mean concentration of the solute within the pore; P is permeability; S is surface area; and ΔC is the difference in solute concentration across the monolayer. As hydrostatic or osmotic pressure gradients are not imposed on the monolayer, fluid flux for the system is equal to zero and J_s reflects the remaining diffusive component of solute flux, allowing permeability calculation as $P = J_s / S\Delta C$.

Materials:
- Sephadex G-25 column (Pharmacia Inc., Clayton, NC)
- 12 mm Transwell support insert equipped with polyester membrane filter having a pore diameter of 0.4 μm (Corning, Lowell, MA)
- CO_2 incubator
- *Gamma* counter (GMI, Minneapolis, MN)
- Bovine Serum Albumin (Sigma, St. Louis, MO)
- Phosphate-buffered Saline (Invitrogen, Carlsbad, CA)
- Na^{125}I (GE Healthcare Biosciences, Pittsburgh, PA)
- Iodination Beads (Pierce Chemical, Rockford, IL)
- Hanks Balanced Salt Solution with $CaCl_2$ and $MgCl_2$ (Invitrogen, Carlsbad, CA)
- 2% gelatin type B from bovine skin (Sigma, St. Louis, MO)

Albumin Iodination (adopted from original protocol by Du et al. [32]).
- Wash iodination beads with 500 μL of PBS twice.
- Dry beads on filter paper.
- Add beads to 250 μL PBS containing 0.5 mCi Na^{125}I.
- Allow reaction to occur for 5 min.
- Add 50 μg purified albumin dissolved in 250 μL PBS to the reaction mixture.
- Allow reaction to occur for 5 min.
- Stop the reaction by removing the solution from the iodination beads in the reaction vessel.
- Separate radiolabeled proteins from free ^{125}I using a 30 cm × 3.5 mm column packed with Sephadex G-25, which is saturated with unlabeled BSA and equilibrated with saline solution. Steps 9 through 16 (below) assess percentage of contaminant free iodine (which should be <0.3% of the total radioactivity).
- Count total radioactivity of a 250-μL sample in a *Gamma* counter.
- Add TCA to an identical 250-μL sample to a final concentration of 10% in order to precipitate protein.

- Incubate on ice for 30 min.
- Centrifuge for 10 min at 15,000 x g at 4°C.
- Remove supernatant, and add an equal volume of 80% acetone to pellet to wash away residual TCA.
- Vortex and centrifuge again as above. Repeat process for four times.
- Aspirate acetone and dry pellet.
- Measure counts of radioactivity and protein concentration. Compare with initial measurement to obtain purity of iodinated protein solution relative to free [125]I

Measurement of Transmembrane Labeled Albumin Flux [33]
- Add 1.5 mL of medium to lower chamber of 12-well plate.
- Add 0.5 mL medium to the transwell insert.
- Incubate at 37°C for 1 h in CO_2 cell culture incubator.
- Remove medium from transwell insert and add 0.2% gelatin in 0.5 mL PBS
- Incubate at 37°C for 30 min in CO_2 cell culture incubator.
- Remove gelatin and add 0.5 mL of cell suspension containing 1.0×10^5 cells/mL onto the transwell insert membrane.
- Grow cells to a confluent monolayer.
- Aspirate medium and wash cell with HBSS containing Ca^{2+} and Mg^{2+} twice to remove non-adherent cells.
- Add warm HBSS buffer containing 68 nM ^{125}I-albumin and 1.5 µM unlabeled albumin in the upper chamber.
- Remove media from the lower chamber and add 1.5 mL HBSS buffer containing 1.5 µM unlabeled albumin.
- Over the course of 90 min, take 50 µL samples from the lower chamber every 15 min and analyze for *gamma* radioactivity.
- Plot the amount of ^{125}I-albumin in the lower chamber against time of sample collection.
- Fit the data to a linear regression.
- Calculate the slope of the best fitting linear regression to obtain the rate (flux) of ^{125}I-albumin flux across the monolayer, J_s (µg sec^{-1}).
- Calculate the permeability P (cm/s) using the equation [10], $P = J_s / S\Delta C$, where Js is ^{125}I-albumin flux across the monolayer (mg sec^{-1}); S is surface area (cm^2); ΔC is initial concentration of ^{125}I-albumin in the top well in (mg cm^{-3}).

3.2   Evans Blue Albumin

Evans Blue dye exhibits high affinity to albumin [34] and Evans Blue-labeled Albumin (EBA) has been effectively used as an alternative to radioisotope labeling for assessment of albumin extravasation in airways and across EC monolayers [34]. Both Evans Blue and ^{125}I-labeled albumin yield comparable results when albumin flux was measured in lavage fluid from isolated lungs and across endothelial monolayers [34].

Materials:
- Spectrophotometer
- Centrifuge
- 0.22 μm filter (Millipore, Billerica, MA)
- 5–0 silk continuous suture (DemeTech Corporation, Miami, FL)
- Evans Blue Dye (Sigma, St. Louis, MO)
- Bovine Serum Albumin (Sigma, St. Louis, MO)
- Phosphate-buffered Saline (Invitrogen, Carlsbad, CA)
- Ketamine
- Formamide
- Solution 1 (preparation of Evans Blue labeled albumin (EBA)).

Assessment of Evans Blue Clearance:
- Dissolve Evans Blue to a concentration of 5 mg/mL in phosphate-buffered saline (Calcium and Magnesium free).
- Add BSA to a concentration of 4 g/100 mL.
- Mix well by stirring with magnetic bar.
- Let solution stand at room temperature for 30 min.
- Filter solution through sterile 0.22 μM filter. The aliquots of EBA solution can be stored at −80°C.

Protocol: (Adopted from original protocol by Moitra et al. [35])
- Anesthetize mice according to an approved protocol for animal use by the University Animal Care Committee.
- Expose the trachea and right internal jugular vein through neck incision.
- Inject EBA through jugular vein at 30 mg/ kg body weight.
- Close the neck incision with a 5–0 silk continuous suture.
- After 2 h, make a vertical inline incision along the sternum in order to provide access to the heart and lungs.
- Make an incision in the left atrium and inject 6–10 mL of phosphate buffered saline through the right ventricle in order to remove intravascular EBA.
- Remove the heart and lungs *in bloc*.
- Separate lung lobes from the heart and trachea; weigh lung tissue.
- Homogenize the lungs in formamide (1 mL formamide/100 mg lung).
- Incubate for 24 h at 60°C.
- Centrifuge samples at 12,000 x*g* for 30 min.
- Remove supernatant.
- Measure absorbance (OD) at 620 nm and 740 nm against a blank containing 50% formamide in PBS.

Corrected Absorbance (620) = Absorbance(620) − (1.426 x Absorbance (740) + .03)

Generate an Evans Blue standard absorbance curve to obtain the concentration of Evans Blue in the tissue. The Evans Blue concentration read in μg/μL can then be converted to μg/g of wet weight of lungs by multiplying the Evans Blue concentration by the dilution factor of the homogenate (1 mL formamide/100 mg lung).

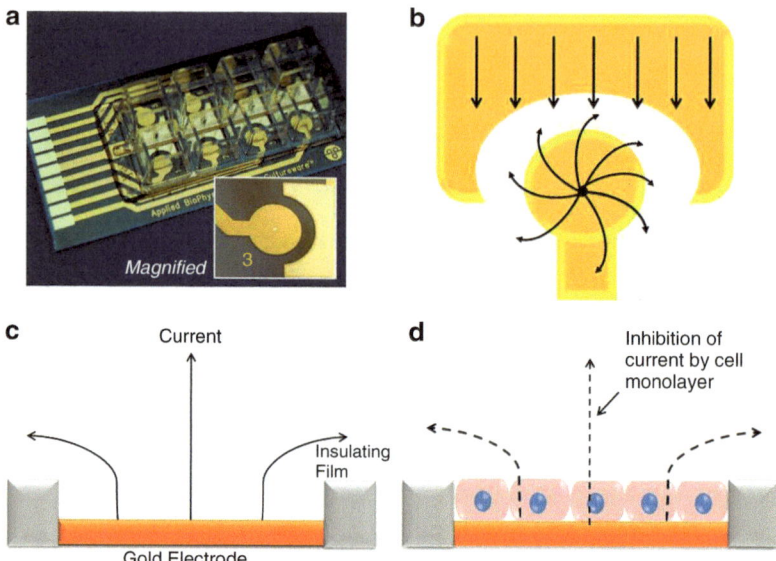

Fig. 2 (**a**) Commercially available 8W1E ECIS cultureware from Applied Biophysics (Troy, NY). (**b**) Schematic diagram of current flow between active and counter electrode. (**c**) In the absence of cells, current flows freely from the active gold electrode. Endothelial cell monolayers grown on the surface (**d**) inhibit this current flow, resulting in increased resistance

3.3   Trans-endothelial Electrical Resistance

Electric Cell-Substrate Impedance Sensing system (ECIS^{TM})[36–38] is a non-invasive biophysical approach to assess cell shape changes in real time. This method monitors time-resolved impedance of a non-invasive current supplied by gold electrodes located on the bottom of the cell culture plate, through the cell culture medium (Figs. 2 and 3). The system software extracts the resistive component of the total impedance from the raw data.

The cells grown on the surface of gold electrodes impede AC current flow between the small active electrode and the large counter-electrode according to monolayer resistance. Disruption of the integrity of the endothelial monolayer, as mediated by pro-inflammatory mediators (e.g., thrombin), results in cell contraction and shape change, thus leading to a decrease in monolayer resistance [36]. In addition, ECIS^{TM} technique can be used to monitor cell attachment, spreading, migration and wound healing. Figure 4 depicts the barrier enhancement properties of the lipid angiogenic factor, sphingosine 1-phosphate, in a dose –dependent manner and the reversal of thrombin –induced barrier dysfunction (Fig. 6).

Investigators have found usually find it convenient to simultaneously analyze the resistance of several control and treated monolayers. To do this, it is necessary to use a specialized slide that has eight individual wells. The measurement system can accommodate up to two 8-electrode arrays for a maximum of 16 individual measurements in any given run.

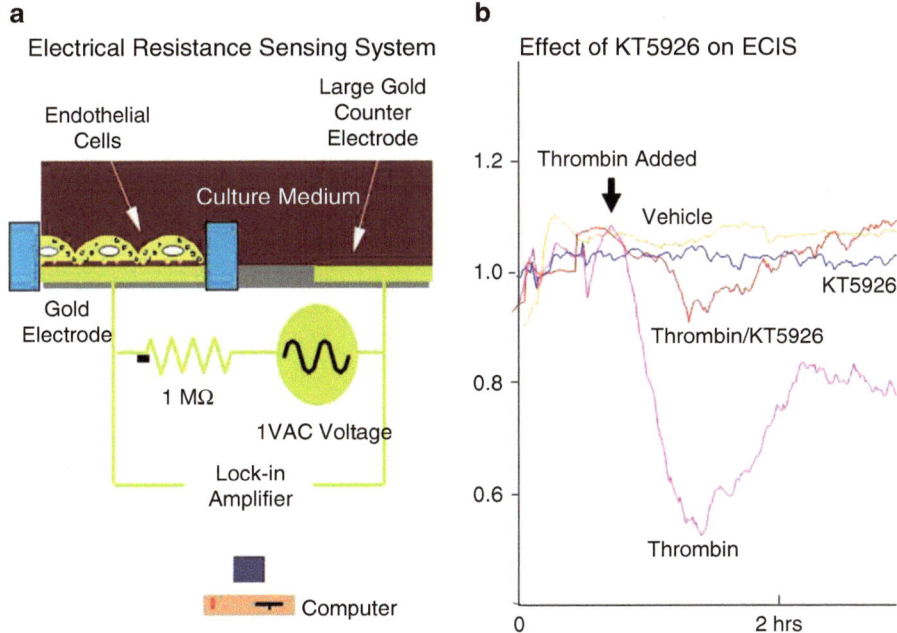

Fig. 3 (a) Schematic of current flow between the active gold electrode and large gold counter in electrode in the presence of an endothelial cell monolayer. Data is presented as a graphical trace of normalized electrical resistance. The addition of the permeability-inducing agonist, thrombin (**b**) induces the formation of intracellular gaps within an endothelial monolayer, allowing for increased current flow between the active and counter electrode which in turns results in decreased normalized electrical resistance. The addition of KT59256, an inhibitor of the central cytoskeletal bioregulatory protein, myosin light chain kinase, prevents this decline in barrier function

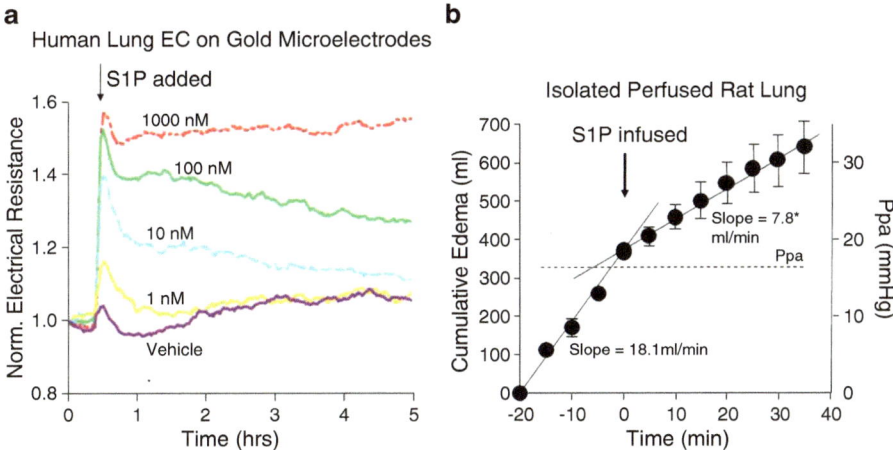

Fig. 4 (a) Dose-dependent effects of S1P on endothelial cells cultured on gold electrodes. Addition of as little as 1nM concentration of S1P results in a significant increases in normalized resistance, further increasing in a dose-dependent manner as high as 1000nM. Note the extended duration of this response. (**b**) Microvascular filtration coefficient ($K_{f,c}$) in the murine lung treated with S1P. The rate of lung weight gain due to formation of pulmonary edema is markedly reduced in lungs following perfusion with S1P

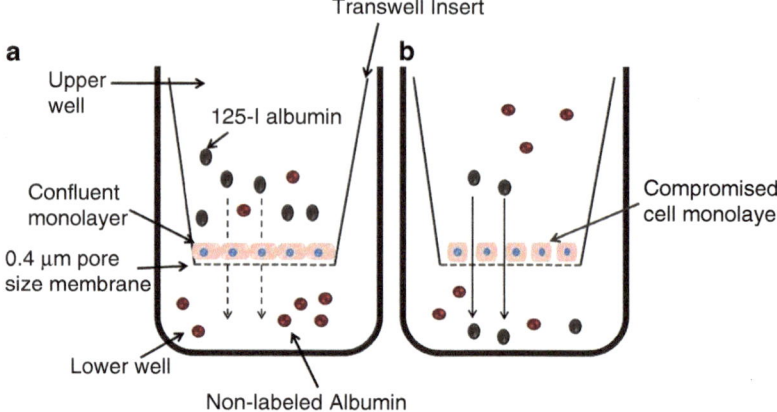

Fig. 5 Homeostatic endothelial cell barrier function (**a**) results in a reduced concentration of ^{125}I-albumin in the lower chamber due to inhibited passage across the monolayer. When permeability of the monolayer increases as a result of the addition of an inflammatory mediator (**b**) paracellular gaps form between cells thereby providing for the passage of ^{125}I-albumin across the monolayer into the lower chamber

Fig. 6 Tracing of normalized electrical resistance in endothelium treated with thrombin, sphingosine 1-phosphate, or both. Treatment with thrombin results in a large decrease in normalized resistance that returns to baseline levels over the course of 3 h. Treatment with S1P, however, results in a large increase of normalized resistance above basal values and reverts the loss of barrier function induced by thrombin

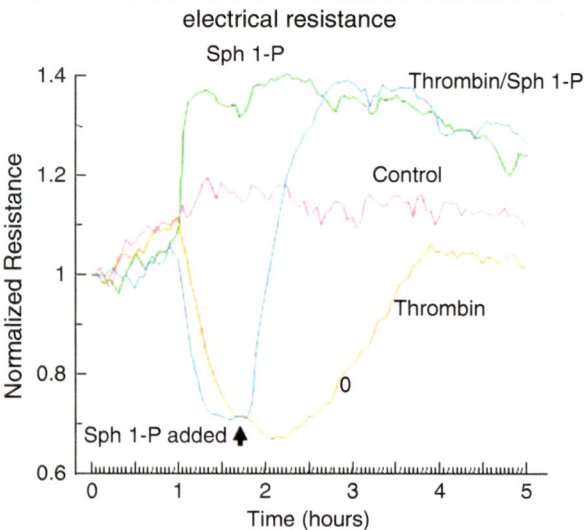

Materials:

- Electric Cell Substrate Impedance Sensing system (Applied Biophysics Inc., http://www.biophysics.com)
- Specialized slides with eight individual wells
- CO_2 incubator
- Computer and printer
- 2% gelatin type B from bovine skin (Sigma, St. Louis, MO)

Protocol: Adopted from original protocol by Garcia et al. [5]

- Coat the surface of each well of the slide with gelatin. Add ~ 200 μL 0.2% gelatin diluted in 1X Phosphate Buffered Saline (PBS) to each well and incubate for 15–20 min at 37°C.
- Remove gelatin by aspiration. Add ~80,000–100,000 primary endothelial cells re-suspended in 400 μL of cell culture medium.
- Grow cells for 3–5 days until a confluent monolayer is established. Control cell confluence by phase contrast microscopy.
- Change media 12 h before experiment.

Data Acquisition

- Using the supplied connectors, connect the arrays of gold electrodes to the ECIS electronics and check electrical connectivity with ECIS software by running "check electrodes" mode. The program will automatically detect if any electrode is not appropriately connected.
- Set up frequency of ac current at 50 Hz, period of data collection, and time-interval between data points. Data points are automatically recorded, displayed as a graph of absolute or normalized impendence (resistance) vs time and saved.
- Record the basal impendence (resistance) for at least 30 min.
- Add mediator (agent) while continuing data acquisition.

Data Analysis

- Upon completion of experiment, export data to *csv* format and import data to Excel, SigmaPlot etc.

4   Conclusion

The complexity of the angiogenic process itself requires a multi-directional approach in order to study its underlying mechanisms. As the process of angiogenesis relies heavily on transient alterations of endothelial barrier function, methods to study permeability are able to provide valuable insights into these processes. Here we present multiple approaches to assess these alterations of permeability directly both *in vivo* and *in vitro*. As angiogenic events are key features required for a variety of diseases, vascular barrier function has lent itself to become a potential target in the treatment of such diseases.

References

1. Dudek SM, Garcia JG (2001) Cytoskeletal regulation of pulmonary vascular permeability. J Appl Physiol 91:1487–1500
2. Dvorak HF (2006) Discovery of vascular permeability factor (VPF). Exp Cell Res 312:522–526

3. Ferrara N, Carver-Moore K, Chen H, Dowd M, Lu L, O'Shea KS, Powell-Braxton L, Hillan KJ, Moore MW (1996) Heterozygous embryonic lethality induced by targeted inactivation of the VEGF gene. Nature 380:439–442

4. Gerwins P, Skoldenberg E, Claesson-Welsh L (2000) Function of fibroblast growth factors and vascular endothelial growth factors and their receptors in angiogenesis. Crit Rev Oncol Hematol 34:185–194

5. Garcia JG, Liu F, Verin AD, Birukova A, Dechert MA, Gerthoffer WT, Bamburg JR, English D (2001) Sphingosine 1-phosphate promotes endothelial cell barrier integrity by Edg-dependent cytoskeletal rearrangement. J Clin Invest 108:689–701

6. Vogel C, Bauer A, Wiesnet M, Preissner KT, Schaper W, Marti HH, Fischer S (2007) Flt-1, but not Flk-1 mediates hyperpermeability through activation of the PI3-K/Akt pathway. J Cell Physiol 212:236–243

7. Esser S, Lampugnani MG, Corada M, Dejana E, Risau W (1998) Vascular endothelial growth factor induces VE-cadherin tyrosine phosphorylation in endothelial cells. J Cell Sci 111(Pt 13):1853–1865

8. Bates DO, Curry FE (1997) Vascular endothelial growth factor increases microvascular permeability via a Ca(2+)-dependent pathway. Am J Physiol 273:H687–H694

9. Jho D, Mehta D, Ahmmed G, Gao XP, Tiruppathi C, Broman M, Malik AB (2005) Angiopoietin-1 opposes VEGF-induced increase in endothelial permeability by inhibiting TRPC1-dependent Ca2 influx. Circ Res 96:1282–1290

10. Pocock TM, Foster RR, Bates DO (2004) Evidence of a role for TRPC channels in VEGF-mediated increased vascular permeability in vivo. Am J Physiol Heart Circ Physiol 286: H1015–H1026

11. Bates DO, Hillman NJ, Williams B, Neal CR, Pocock TM (2002) Regulation of microvascular permeability by vascular endothelial growth factors. J Anat 200:581–597

12. de Oliveira LR, Hamm A, Mazzone M (2011) Growing tumor vessels: more than one way to skin a cat – implications for angiogenesis targeted cancer therapies. Mol Aspects Med 32:71–87

13. Carmeliet P, Jain RK (2011) Molecular mechanisms and clinical applications of angiogenesis. Nature 473:298–307

14. Phng LK, Gerhardt H (2009) Angiogenesis: a team effort coordinated by notch. Dev Cell 16:196–208

15. Lee S, Chen TT, Barber CL, Jordan MC, Murdock J, Desai S, Ferrara N, Nagy A, Roos KP, Iruela-Arispe ML (2007) Autocrine VEGF signaling is required for vascular homeostasis. Cell 130:691–703

16. Stockmann C, Doedens A, Weidemann A, Zhang N, Takeda N, Greenberg JI, Cheresh DA, Johnson RS (2008) Deletion of vascular endothelial growth factor in myeloid cells accelerates tumorigenesis. Nature 456:814–818

17. Thurston G, Rudge JS, Ioffe E, Zhou H, Ross L, Croll SD, Glazer N, Holash J, McDonald DM, Yancopoulos GD (2000) Angiopoietin-1 protects the adult vasculature against plasma leakage. Nat Med 6:460–463

18. Thurston G, Suri C, Smith K, McClain J, Sato TN, Yancopoulos GD, McDonald DM (1999) Leakage-resistant blood vessels in mice transgenically overexpressing angiopoietin-1. Science 286:2511–2514

19. Augustin HG, Koh GY, Thurston G, Alitalo K (2009) Control of vascular morphogenesis and homeostasis through the angiopoietin-Tie system. Nat Rev Mol Cell Biol 10:165–177

20. Sato TN, Tozawa Y, Deutsch U, Wolburg-Buchholz K, Fujiwara Y, Gendron-Maguire M, Gridley T, Wolburg H, Risau W, Qin Y (1995) Distinct roles of the receptor tyrosine kinases Tie-1 and Tie-2 in blood vessel formation. Nature 376:70–74

21. Maisonpierre PC, Suri C, Jones PF, Bartunkova S, Wiegand SJ, Radziejewski C, Compton D, McClain J, Aldrich TH, Papadopoulos N et al (1997) Angiopoietin-2, a natural antagonist for Tie2 that disrupts in vivo angiogenesis. Science 277:55–60

22. An S, Zheng Y, Bleu T (2000) Sphingosine 1-phosphate-induced cell proliferation, survival, and related signaling events mediated by G protein-coupled receptors Edg3 and Edg5. J Biol Chem 275:288–296

23. Yatomi Y, Ohmori T, Rile G, Kazama F, Okamoto H, Sano T, Satoh K, Kume S, Tigyi G, Igarashi Y et al (2000) Sphingosine 1-phosphate as a major bioactive lysophospholipid that is released from platelets and interacts with endothelial cells. Blood 96:3431–3438

24. English D, Garcia JG, Brindley DN (2001) Platelet-released phospholipids link haemostasis and angiogenesis. Cardiovasc Res 49:588–599

25. Lee MJ, Thangada S, Claffey KP, Ancellin N, Liu CH, Kluk M, Volpi M, Sha'afi RI, Hla T (1999) Vascular endothelial cell adherens junction assembly and morphogenesis induced by sphingosine-1-phosphate. Cell 99:301–312

26. English D, Welch Z, Kovala AT, Harvey K, Volpert OV, Brindley DN, Garcia JG (2000) Sphingosine 1-phosphate released from platelets during clotting accounts for the potent endothelial cell chemotactic activity of blood serum and provides a novel link between hemostasis and angiogenesis. FASEB J 14:2255–2265

27. Peng X, Hassoun PM, Sammani S, McVerry BJ, Burne MJ, Rabb H, Pearse D, Tuder RM, Garcia JG (2004) Protective effects of sphingosine 1-phosphate in murine endotoxin-induced inflammatory lung injury. Am J Respir Crit Care Med 169:1245–1251

28. Liu F, Schaphorst KL, Verin AD, Jacobs K, Birukova A, Day RM, Bogatcheva N, Bottaro DP, Garcia JG (2002) Hepatocyte growth factor enhances endothelial cell barrier function and cortical cytoskeletal rearrangement: potential role of glycogen synthase kinase-3beta. FASEB J 16:950–962

29. Birukova AA, Alekseeva E, Mikaelyan A, Birukov KG (2007) HGF attenuates thrombin-induced endothelial permeability by Tiam1-mediated activation of the Rac pathway and by Tiam1/Rac-dependent inhibition of the Rho pathway. FASEB J 21:2776–2786

30. Singleton PA, Salgia R, Moreno-Vinasco L, Moitra J, Sammani S, Mirzapoiazova T, Garcia JG (2007) CD44 regulates hepatocyte growth factor-mediated vascular integrity. Role of c-Met, Tiam1/Rac1, dynamin 2, and cortactin. J Biol Chem 282:30643–30657

31. Garcia JG, Siflinger-Birnboim A, Bizios R, Del Vecchio PJ, Fenton JW 2nd, Malik AB (1986) Thrombin-induced increase in albumin permeability across the endothelium. J Cell Physiol 128:96–104

32. Du T, Alfa MJ (2004) Translocation of Clostridium difficile toxin B across polarized Caco-2 cell monolayers is enhanced by toxin A. Can J Infect Dis 15:83–88

33. John TA, Vogel SM, Tiruppathi C, Malik AB, Minshall RD (2003) Quantitative analysis of albumin uptake and transport in the rat microvessel endothelial monolayer. Am J Physiol Lung Cell Mol Physiol 284:L187–L196

34. Patterson CE, Rhoades RA, Garcia JG (1992) Evans blue dye as a marker of albumin clearance in cultured endothelial monolayer and isolated lung. J Appl Physiol 72:865–873

35. Moitra J, Sammani S, Garcia JG (2007) Re-evaluation of Evans Blue dye as a marker of albumin clearance in murine models of acute lung injury. Transl Res 150:253–265

36. Tiruppathi C, Malik AB, Del Vecchio PJ, Keese CR, Giaever I (1992) Electrical method for detection of endothelial cell shape change in real time: assessment of endothelial barrier function. Proc Natl Acad Sci USA 89:7919–7923

37. Wegener J, Keese CR, Giaever I (2000) Electric cell-substrate impedance sensing (ECIS) as a noninvasive means to monitor the kinetics of cell spreading to artificial surfaces. Exp Cell Res 259:158–166

38. Ellis CA, Tiruppathi C, Sandoval R, Niles WD, Malik AB (1999) Time course of recovery of endothelial cell surface thrombin receptor (PAR-1) expression. Am J Physiol 276:C38–C45

Quantitative Methods to Study Adipose Angiogenesis

Sharon Lim, Jennifer Honek, Ziquan Cao, Takahiro Seki, Yuan Xue, and Yihai Cao

Abstract The adipose tissue is one of the most vascularized tissues in the body. Both deposition and energy expenditure in the adipose tissues are dependent on appropriate vascular structures and functions that in their pathological settings may cause obesity and metabolic disorders. Emerging recent studies demonstrate that regulation of blood vessel functions in the adipose tissue by angiogenesis modulators significantly affects the size and metabolic status of adipose depots, suggesting that the adipose vasculature is an important target for the treatment of obesity and metabolic diseases. Additionally, angiogenesis modulators have been implied for the treatment of obesity- and diabetes-associated clinical disorders such as cancer, cardiovascular disease, ophthalmological disorders, and chronic ulcers. In this book chapter, we describe methodologies developed in our and other laboratories to study structural and functional aspects of the adipose vasculature in relation to the metabolic status of adipose depots.

Keywords Angiogenesis • Endothelial cell • Adipose tissue • Obesity • Diabetes • Adipocyte • White adipose tissue • Brown adipose tissue • Vasculature • Cold • Metabolism • Immunohistochemistry • Whole-mount staining

Sharon Lim and Jennifer Honek contributed equally to this work.

S. Lim • J. Honek • T. Seki • Y. Xue
Department of Microbiology, Tumor and Cell Biology, Karolinska Institute,
171 77 Stockholm, Sweden

Z. Cao
Department of Medicine and Health Sciences, Linköping University, Linköping, Sweden

Y. Cao, M.D., Ph.D. (✉)
Department of Microbiology, Tumor and Cell Biology, Karolinska Institute,
171 77 Stockholm, Sweden

Department of Medicine and Health Sciences, Linköping University, Linköping, Sweden
e-mail: yihai.cao@ki.se

E. Zudaire and F. Cuttitta (eds.), *The Textbook of Angiogenesis and Lymphangiogenesis: Methods and Applications*, DOI 10.1007/978-94-007-4581-0_5,
© Springer Science+Business Media Dordrecht 2012

1   Introduction

According to a recent report by the World Health Organization, the global overweight adult population has reached more than one billion and at least 30% of this population suffers from clinical obesity (BMI >30) [1]. In particular, the ratio of overweight versus healthy lean individuals in developed countries such as USA and a majority of European countries has grown more than 60%. It is predicted that this ratio will continue to increase and lean individuals will represent only a small minority of our society in the future. Obesity is probably the leading threat for human health. Almost all common and lethal human disorders such as cardiovascular disease, type 2 diabetes, cancer, chronic inflammation and ischemic limb diseases are coupled to obesity [2]. Thus, prevention and treatment of obesity are one of the most important steps for improvement of human health.

Adult humans have two types of adipose tissues, i.e., white adipose tissue (WAT) and brown adipose tissue (BAT), both of which can undergo expansion and regression at any given life time. While WAT stores excessive energy molecules as fat, active BAT can sufficiently burn the stored energy molecules by releasing heat. It has been estimated that full activation of the 63 g of BAT in an adult human could burn approximately 4 kg of white fat per year [3]. In small mammals such as rodents, WAT can be converted into a BAT-like phenotype under certain circumstances such as cold [4, 5]. Adipose plasticity has also offered an exciting opportunity for therapeutic intervention. Similar to adipocytes, blood vessels undergo alterations during the expansion and regression of the adipose tissue.

Emerging experimental evidence demonstrates that the vascular system plays an essential and active role in modulation of adipose tissue growth or regression and its metabolic functions [1, 2]. For example, an extremely high density of microvessels in BAT is associated with an active metabolic phenotype by perfusion of oxygenated blood as a fuel. Inhibition of angiogenesis in expanding WATs in animal experimental models leads to a lean phenotype in obese animal models [6, 7]. These preclinical findings suggest that targeting the adipose vascular offers a novel option for possible treatment of obesity and metabolic diseases. To define novel therapeutic targets and to explore molecular mechanisms of the adipose vasculature in modulation of adipose tissue growth and function, it is essential to develop rigorous experimental assay systems that allow us to quantitatively study adipose microvessels under different metabolic conditions [8].

2   Angiogenesis in Adipose Tissues

Angiogenesis is the process of growing new blood vessels from preexisting ones and this process is involved in controlling growth, remodeling and functions of many tissues and organs under physiological and pathological conditions [9]. Angiogenesis is required for embryonic development, tissue growth and tissue repair.

As one of the largest tissues in the human body, the adipose tissue constantly undergoes expansion and regression throughout the lifetime. Like most other tissues in the body, adipose tissue growth critically relies on angiogenesis [6, 10]. Consequently, treatment with anti-angiogenic agents can prevent the expansion and even induce regression of adipose tissue both in genetically obese mice as well as in high fat diet models [2, 6, 7, 10]. The adipose vasculature regulates various processes of adipogenesis including: (1) Supplying oxygen and nutrients; (2) Providing cytokines and growth factors; (3) Supply of bone marrow-derived stem cells, which may differentiate into pre-adipocytes and adipocytes, endothelial cells as well as pericytes; (4) Facilitating infiltration of inflammatory cells; and (5) Removal of metabolic waste products. The intimate interplay between adipocytes and the vasculature also provides an outstanding opportunity of developing therapeutic strategy for the treatment of obesity and metabolic diseases by targeting angiogenesis.

The adipose tissue contains various other cell types including endothelial cells, pericytes, macrophages and mesenchymal cells, which interact very closely with one another. In the adult humans, WAT predominantly accumulates in the subcutaneous regions and in the intra-abdominal area. While WAT is crucial for heat insulation, mechanical protection of inner organs as well as energy storage, an excessive amount of the adipose tissue can cause tremendous damage of the health condition [11]. Another important function of WAT is the endocrine regulation of multiple other tissues and organs by producing a number of hormones, cytokines/adipokines and angiogenic factors such as vascular endothelial growth factor (VEGF), leptin, adiponectin, resistin, interleukin-6 (IL-6), IL-8, hepatocyte growth factor (HGF), angiopoietin (Ang)-1 and Ang-2, fibroblast growth factor (FGF)-2, leptin, estrogen, transforming growth factor (TGF)-α and -β as well as matrix metalloproteinase (MMP)-2 and -9 [10, 12–19]. These WAT-derived factors not only control adipocyte survival, proliferation and differentiation in the local environment, but also have a significant impact on other tissues via the endocrine mechanism [20, 21].

Both WAT and BAT are highly vascularized tissues and blood vessel density in BAT is several-fold higher relative to WAT [5]. This extremely high vessel density correlates with the metabolic activity of BAT [5]. However, vascularity is not the only difference between white fat and brown fat. WAT adipocytes consist of large diameter cells, which are characterized by a spherical morphology. Their large unilocular lipid droplets are surrounded by a thin layer of the cytoplasm. Conversely, adipocytes in BAT consists of smaller cells and multilocular lipid droplets as well as high density cytoplasmic contents. The most characteristic feature of brown adipocytes is the expression of the mitochondrial uncoupling protein 1 (UCP1), which mediates thermogenesis-related energy metabolism [22]. In mice, BAT is primarily located dorsally in the subcutaneous and interscapular region. For many years, it has been assumed that BAT is only present in the newborn human but not in adults. However, several recent studies demonstrate that a significant amount of BAT exists in adult humans [3, 23, 24] and that the BAT depot is concentrated in the fascial plane in the ventral neck and thorax bilaterally [25]. Similar to mice, the

BAT activity in adult humans can be activated by exposing to a cold environment [24]. The number and structure of adipose vasculature are critical determinant to ensure the optimal functions of BAT. Excessive or defective angiogenesis might lead to abnormal function of WAT and BAT, which eventually result in development of various diseases.

3 Immunohistochemistry of Adipose Blood Vessels

3.1 Reagents and Equipments

- 1× PBS
- 4′,6-diamidino-2-phenylindole, dilactate (DAPI, dilactate) (Invitrogen, cat. no. D3571)
- Adobe Photoshop CS3 or later versions (Adobe)
- Alexa Fluor 488 donkey anti-rabbit IgG (H + L) antibody (Invitrogen, cat. no. A-21206)
- Alexa Fluor 555 goat anti-rat IgG (H + L) antibody (Invitrogen, cat. no. A-21434)
- Antigen unmasking solution, Citric Acid Based (Vector Laboratories, cat. no. H-3300)
- BD falcon 50 ml polypropylene conical tubes (BD Biosciences, cat. no. 358206) (cat. no. 70318–04)
- Confocal microscope (e.g., Nikon D-eclipse C1)
- Confocal software (e.g., EZ-C1 3.9 Nikon digital eclipse)
- Costar 6-well cell culture plates (Corning, cat. no. 3516)
- Costar 96-well cell culture plates (Corning, cat. no. 3596)
- Cryotome (Histolab Products AB, cat. no. HM500OM)
- Distilled water (dH₂O)
- Dried fat-free milk (e.g., Semper, Sweden)
- Ethanol 99.7% (Solveco AB, cat. no 200-578-6)
- Forceps (AgnTho's AB, cat. no. 08-060-120)
- Hamster anti-mouse podoplanin (Angiobio, cat. no. 11–033)
- Hematoxylin Solution, Harris modified (Sigma-Aldrich, cat. no. HHS16)
- Hoechst 33342, trihydrochloride, trihydrate, 100 mg (Invitrogen, cat. no. H1399)
- Humidified incubation chamber for slides
- Methanol (Sigma-Aldrich, cat. no. 32213)
- Mice all animal studies should conform to all relevant ethics regulations and must be reviewed and approved by government and institutional animal care and use committees.
- Microscope cover slips (VWR International, cat. no. 631–0135)
- Non-immune goat serum (Vector Laboratories, cat. no. S-1000)
- O.C.T. compound Tissue-Tek (Sakura Tissue-Tek, cat. no 4583)
- PAP pen for immunostaining (Sigma-Aldrich, cat. no. Z672548)
- Paraformaldehyde (PFA) (Sigma-Aldrich, cat. no. 441244)

- Pertex (Histolab Products Ab, cat. no. 00801)
- Proteinase K (Invitrogen, cat. no. 25530–049)
- Rabbit anti-mouse lymphatic vessel endothelial hyaluronan receptor-1 antibody (LYVE-1; AngioBio, cat. no. 11–034)
- Rat anti-mouse CD31 monoclonal antibody, MEC13.3 (BD Pharmingen, cat. no. 553370)
- Rocking board (VWR International, cat. no. 444–0341)
- Scalpel blade (AgnTho's AB, cat. no. 02-040-010)
- Scalpel blade holder (AgnTho's AB, cat. no. 02-030-030)
- Spatula/microspoon (VWR International, cat. no. 231–1354)
- Superfrost Plus microscope slides (Thermo Scientific, cat. no. 4951plus)
- Tetramethylrhodamine-labeled dextran (2,000,000 MW, lysine fixable, Invitrogen, cat. no. D-1818)
- Timer (Fisher Scientific, cat. no. FB70232)
- Triton X-100 Polyoxyethylene octylphenyl ether (Acros Organics, cat. no. 215680010)
- Vectashield mounting medium (Vector Laboratories, cat. no. H-1000)
- Vertical staining jar with glass lid (Electron Microscopy Sciences, cat. no. 70318–04)
- Water bath up to 60°C (e.g., Lauda Aqualine Al5)
- Xylen (Histolab Products Ab, cat. no. 02080)

3.2   Buffers and Solutions

4% Paraformaldehyde (PFA): Add 20 g PFA powder to 1× PBS. Dissolve PFA in water bath or use a magnetic stirrer. Store 4% PFA at 4°C for up to 2 weeks.

0.3% Triton X-100: Add 0.9 ml Triton X-100 to 29.10 ml 1× PBS to a final volume of 30 ml of 0.3% Triton X-100 (PBST) (vol/vol). Vortex and store the solution at room temperature.

3% blocking buffer: Add 1.5 g of dried fat-free milk powder to 50 ml PBST for a final volume of 50 ml of 3% blocking buffer.

3.3   Whole-Mount

To get a general overview of the adipose tissue vasculature and the structural properties of blood vessels, whole-mount immunohistochemistry is the method of choice. For this method, it is recommended to use freshly dissected adipose tissues from different depots, including WAT or BAT. After careful dissection of the tissue, it should be fixed in freshly prepared 4% PFA for about 12 h at 4°C.

On day 2, the PFA-fixed tissues are transferred into petri-dishes containing 1× PBS. Using a scalpel blade, thin slices (approximately 5 mm × 5 mm) are prepared by carefully securing the adipose tissue with a pair of forceps and meanwhile sectioning the tissue by applying slight pressure to the scalpel blade to produce even sections. The tissue should be washed in 1× PBS for 1 h on a rocking board to remove PFA. Incubate the adipose tissues with proteinase K 20 µg ml^{-1} in 10 mM Tris–HCl buffer (pH 7.4) at room temperature for 5 min to digest the tissues. For further permeabilization, incubate the tissues with 100% methanol for 30 min at room temperature. Perform this step in a chemical fume hood due to the toxicity of methanol. After thoroughly washing tissue sections with 1× PBS for three times (last wash is carried out for 1 h on a rocking board), block non-specific binding sites using 3% blocking buffer (1.5 g of dried fat-free milk powder in 50 ml PBST consisting of 0.3% vol/vol Triton X-100 in PBS) for 12–24 h at 4°C on a rocking board.

On day 3, remove blocking buffer and wash the tissue with PBST prior to incubation with one or several primary antibodies (12–24 h at 4°C on a rocking board). On day 4, wash the tissue with PBST for 1.5 h, followed by incubation with 3% blocking buffer at 4°C on a rocking board for 1.5 h. Incubate the adipose tissue samples are with appropriate dilutions of secondary antibodies in 3% blocking buffer for 2 h at room temperature on a rocking board. After incubation with blocking buffer (1:1 dilution with PBST) for 1 h at room temperature, wash the samples at 4°C with PBST for about 12 h on a rocking board. Mount the stained tissue sections are on microscope glass slides in the following day. For mounting, add 1–2 drops of Vectashield mounting medium (Vector Laboratories, cat. no. H-1000) per slide onto the tissue samples and cover the samples with microscope cover slips. Until analysis of the samples with confocal microscopy, the slides can be stored for a few days at 4°C or for several weeks at −20°C. In order to prevent the tissues from drying, it is recommended to seal the cover slips using commercial nail polish.

Use a confocal microscope (e.g. Nikon D-eclipse C1 and EZ-C1 3.90 software), to harvest 3-dimensional images of stained adipose tissue samples can be acquired. It is recommended to obtain confocal images at 10×, 20×, 40× or 60× magnifications at 5 µm × 8–10 layers. The specific choice of magnification, layer thickness and number of layers should be adapted according to desirable information.

To investigate the overall structure of blood vessels in the adipose tissue, several endothelial cell markers are recommended for staining. These include: antibodies targeting CD31 (Fig. 1), CD34 or isolectin (1:200 dilution). All these proteins are expressed on blood vessel endothelial cells. Using one of these markers in a whole-mount setting provides conclusive information on general structures of the vasculature. Furthermore, with these markers, the presence of microvessels and capillaries can be detected that might have been newly formed upon a switch to the angiogenic phenotype.

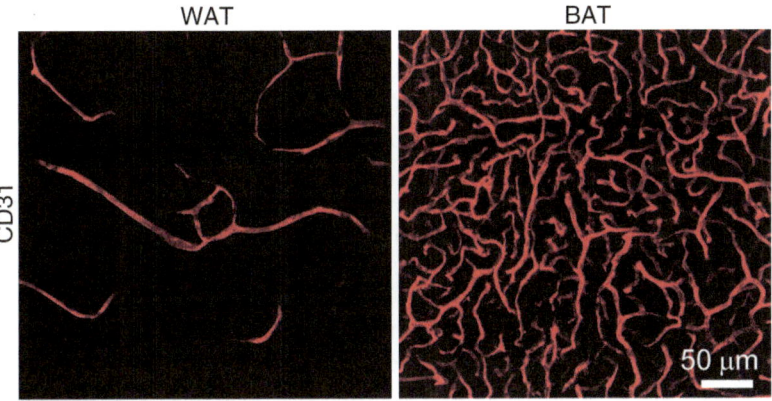

Fig. 1 Adipose tissue vasculature. Whole-mount staining on white (*WAT*) and brown adipose tissue (*BAT*) of C57Bl mice was performed using an anti-CD31 antibody and 3-dimensional images were obtained by confocal microscopy. *Red* signals visualize blood vessels. *Scale bar* = 50 μm

3.4 Paraffin-Embedded Tissues

Embed PFA-fixed adipose tissues into paraffin blocks. Prepare tissue sections in 5 and 3 μm thickness from embedded WAT and BAT, respectively, using a micro-tome device. Transfer thin paraffin membranes onto superfrost microscope slides and incubate the slides for 2 h at 60°C to remove excessive paraffin. After cooling the slides to room temperature, use immediately for further applications or store slides for up to 1–2 years until further use.

De-paraffinization the sections by washing 2 × 5 min with xylen and rehydrate the sections in an ethanol series (99.7%, 90% and 70% ethanol, 2 × 5 min) and in distilled water (dH$_2$O) for 5 min. For antigen retrieval, boil the slides in an unmasking solution in the microwave at approximately 100°C for 10 min. After cooling, wash slides with 1× PBS twice (5 min each). The tissue sections are encircled with a PAP hydrophobic barrier immunohistochemistry pen. In order to block non-specific antigen binding sites, slides are incubated with 1× PBS containing 4% non-immune goat serum. After washing the slides twice in 1× PBS, incubate at 4°C for 12 h with one or several primary antibodies at an appropriate dilution in 1× PBS containing 4% normal goat serum. Wash slides with 1× PBS (5 min each). This step is followed by incubation with a secondary antibody at specified dilutions in 1× PBS containing 4% normal goat serum. Depending on which antibodies are used, incubation is carried out for 30 min to 1 h at room temperature. Secondary antibodies include an Alexa Fluor-555 conjugated goat anti-rat antibody and a Cy3-conjugated goat anti-rabbit antibody.

After washing the tissue with 1× PBS twice for 5 min each, nuclei can be counterstained using 1 μM Hoechst 33342 or Propidium Iodide (PI) for 2 min at room temperature. Slides are washed in PBS and then mounted on the microscope slides with one drop of Vectashield mounting medium per slide. Tissues are covered with microscope cover slips and can be stored at 4°C in the dark for a few days or at −20°C for several weeks. Stained tissue can be examined using a fluorescent microscope equipped with an imaging software. It is recommended to harvest fluorescence images at 10×, 20×, 40× or 60× magnifications.

For certain markers it is desired to use a chromogenic detection method instead of fluorescence-based detection. In these cases, some additional steps need to be carried out. Prior to blocking non-specific binding sites, endogenous peroxidase activity has to be inactivated by incubation with 3% H_2O_2 for 10 min at room temperature. Furthermore, the sections are immersed with one drop of actin and biotin blocking solution per slide, respectively for 10 min at room temperature. Between the actin and the biotin blocking step, tissue slides are washed with 1× PBS twice. The secondary antibodies used in this setting are horseradish peroxidase (HRP)-conjugated antibodies (usually diluted at 1:100). Upon addition of the substrate 3-3′ diaminobenzidine (DAB), a brownish product is formed. The reaction is stopped by immersing the slides in dH_2O. If desired, nuclei can be counterstained by briefly incubating the slides in hematoxylin. Slides are dehydrated in 70%, 90% and 99.7% ethanol (2 × 5 min each) and can then be mounted in Pertex solution and covered with microscope cover slips. The advantage of chromogenic detection is that problems caused by potential autofluorescence of red blood cells or the adipose tissue itself can be avoided. However, this method only allows for detection of one marker at a time while different markers can be co-stained when using fluorochrome-linked secondary antibodies.

For detection of endothelial cells and analysis of blood vessels, endothelial cell specific markers such as CD31 or isolectin (diluted 1:50 and 1:100, respectively) can be used. For quantitative analysis of the ratio of blood vessels per adipocyte, it is recommended to use an anti-isolectin antibody and the chromogenic detection method. To study the BAT-like phenotype, the brown adipose tissue specific Uncoupling protein-1 (UCP-1) serves as a marker. As adipose tissue expansion and obesity are frequently associated with inflammation, it can be of interest to investigate infiltrating inflammatory cells using markers such as F4/80, which stains macrophages.

3.5 Cryosections

The adipose vasculature has been extensively studied [1, 2]. However, the development of lymphatic vessels, lymphangiogenesis in adipose tissue is less known. In this chapter, we will discuss the use of podoplanin and LYVE-1, the two commonly used markers to identify lymphatic vessels [26, 27]. LYVE-1 is a type I integral membrane glycoprotein that has 41% homology with CD44 receptor for hyaluronan. LYVE-1 expression is almost exclusively restricted to lymphatic endothelium in most tissues [28]. However, in the adipose tissue the expression of LYVE-1 is not exclusive to the lymphatic endothelium and some inflammatory cells express this marker (Fig. 2). Another marker used for differentiating blood endothelial cells

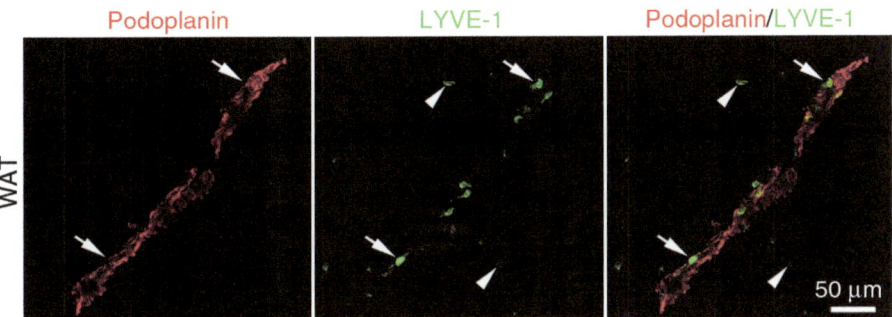

Fig. 2 Lymphatic endothelial cells in adipose tissues. Cryosections of WAT from C57Bl mice were embedded in O.C.T compound, sectioned and stained with podoplanin and LYVE-1 antibodies. *Arrows* show double-positive lymphatic vessels and *arrowheads* show non-lymphatic signals. *Scale bar* = 50 μm

from lymphatic endothelial cells is podoplanin. Podoplanin is exclusively expressed in the lymphatic endothelium but not in the blood endothelium [29]. Podoplanin seems to be a better marker in identifying lymphatic vessels in the adipose tissue.

To perform immunohistochemistry on adipose tissues, embed dissected adipose tissue from mouse in a plastic cryomould and fill the cryomould with Tissue-Tek, an O.C.T compound on dry ice. Upon solidifying, these frozen moulds can be stored in −80°C until they are sectioned. To ensure the ease of cutting the cryosections, the temperature of the cryotome should be lowered to −30°C. This temperature is slightly lower than for cutting other organs due to the high fat content and soft integrity of the adipose tissues. Since the adipose tissue is relatively soft compared to other peripheral organs, it is rather challenging to section adipose tissue to a thickness less than 10–15 μm and yet maintaining tissue structure and integrity. Use a superfrost glass slide to ensure that the tissue sectioned is securely mounted throughout the rest of the staining procedures. Mounting cryosectioned tissues on normal glass slide would risk loss of tissues during the subsequent staining.

Begin the staining procedure by adapting the tissue slides on a flat surface for 30 min at room temperature. Fix the tissue slides by immersing the tissues in 100% acetone in a vertical staining jar/Coplin jar for 10 min. To prevent any loss of tissues on the slides, it is preferable to immerse the tissue slides in vertical staining/ Coplin jar containing acetone or other working buffer instead of pouring acetone or working buffer directly onto the samples in the vertical staining jar which might dislodge the tissues. Next, wash the tissue slides in 1× PBS three times for 3–5 min each. After the last wash, remove excess solution with a paper, wipe around the tissue without touching the tissue. Encircle the tissue with a PAP hydrophobic barrier immunohistochemistry pen to minimize the amount of reagent required. To block non-specific signals, incubate the tissue section with blocking buffer for 30 min at room temperature. Typically, 80 150 μl of blocking buffer/antibodies should be sufficient to completely cover the tissue. Place the slides in a moist incubation chamber. Wash the tissue slides in 1× PBS three times for 3–5 min each. Incubate the slides with primary antibodies diluted in 1× PBS supplemented with non-immune whole serum at 4°C overnight or 1 h at room temperature depending on the antibodies. Dilute primary antibody in 1× PBS supplemented with whole serum and incubate at 4°C overnight in a moist incubation chamber.

On the second day, wash the tissue slides in 1× PBS three times for 3–5 min each. Incubate the slides with secondary antibodies diluted in 1× PBS supplemented with non-immune whole serum for 2 h at room temperature. Wash the tissue slides in 1× PBS three times for 3–5 min each. DAPI can be used to counterstain the tissue sections to visualize cell nuclei. Incubate the tissue sections with DAPI (dilute 1:1,000) in 1× PBS for 1 min followed by washing in 1× PBS. Remove excess PBS using paper and mount the slide using 1–2 drops of Vectashield mounting medium. Cover the sections with cover-slips, and slides can be stored at −20°C until examination with a microscope.

In the staining of lymphatic vessels, primary antibodies, podoplanin and LVYE-1, hamster anti-podoplanin (dilute at 1:200) and rabbit anti-mouse (dilute at 1:200) are diluted in 1× PBS supplemented with goat serum and incubated at 4°C overnight. Secondary antibodies such as Alexa Fluor 568 goat anti-hamster IgG (dilute at 1:400) and Cy2-congugated goat anti-rabbit IgG (1:400) are used to visualize lymphatic vessels.

To study vascular remodeling and maturation, vascular smooth muscle cells (VSMCs) markers such as anti-α-SMA (1:500) can be co-stained with endothelial cell markers, CD31 or CD34. Positive α-SMA VSMCs are usually located on the large arterial blood vessels as seen in *wt* in Fig. 3. However, in FOXC2 transgenic mice, positive α-SMA VSMCs are redistributed onto microvessels demonstrating remodeling of microvessels.

3.6   Analysis of Vascular Functions

Tetramethyrhodamine-labeled dextran can be used to study blood perfusion in adipose tissue. Inject dextran into the tail vein of mice, followed by counterstaining with endothelial cell specific marker, CD31 or CD34 (see immunohistochemistry whole mount). To facilitate the ease of injection, it is recommended to warm the mouse tail with infrared light bulb ~37°C or immerse the mouse tail in a water bath ~37°C for approximately 10–15 min to dilate the tail vein. Fill the syringe with 100 μl of dextran and remove all air bubbles to avoid air embolism. Thereafter, place the mouse in a mouse restrainer leaving the tail free for tail vein dextran injection. Since dextran is rather costly, tail vein injection is preferably performed by an experienced person. Locate the lateral vein of the mouse; it should be visible after the exposure to either infrared light bulb or water bath. Inject 100 μl of dextran (12.5 mg/ml) with moderate speed and pressure into the tail vein. During injection, no bulge or pressure should be felt, any pressure felt reflects a failed injection. It is preferable to make the first injection close to the tip of the tail first as this allows

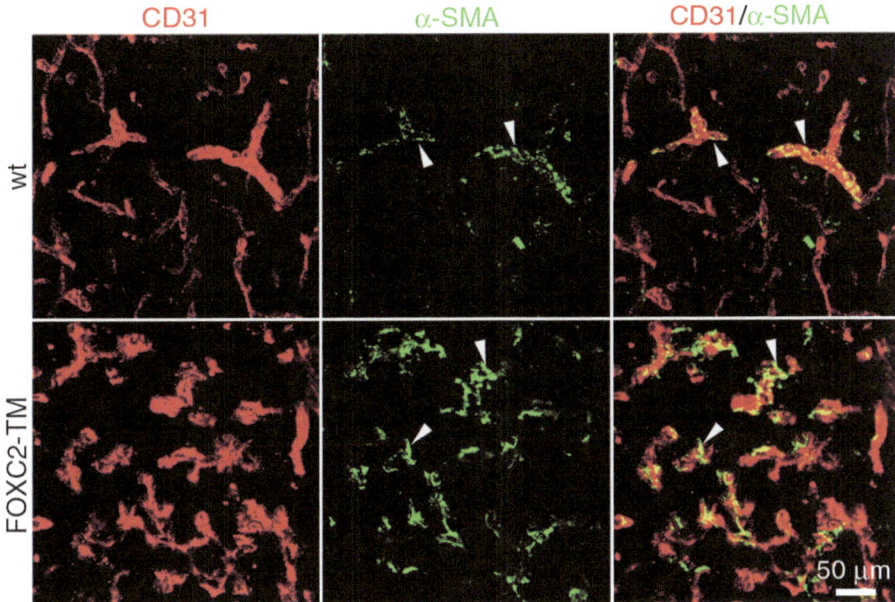

Fig. 3 Alteration of vascular structure in adipose tissues. Cryosections of WAT from *wt* and FOXC2 transgenic mice were co-stained with endothelial cell marker, rat anti-mouse CD31 and vascular smooth muscle cell marker (VSMC), mouse anti-human α-SMA antibodies. *Red* signals represent endothelial cells stained with an Alexa Fluor 555 goat anti-rat IgG antibody and green signals represents VSMC stained with fluorescein-labeled goat anti mouse antibody. *Arrowheads* show double-positive areas. *Scale bar* = 50 μm

subsequent attempts. Exert slight pressure on the tail after withdrawing the needle to stop bleeding. Return the mouse to its homecage and allow dextran to circulate for 15 min before sacrificing the mouse.

Adipose tissues harvested should be fixed either in freshly prepared 4% PFA overnight and proceed on to whole-mount or embedded in Tissue-Tek O.C.T-filled moulds followed by immunohistochemistry on cryosections see immunohisto-chemistry: Whole-mount or cryosections. Since tetramethyrhodamine-labeled dex-tran is red, it is necessary to consider a suitable fluorochrome-linked secondary antibody conjugate for CD31 or CD34 antibodies. In Fig. 4, a goat anti-rat CD31 antibody (1:200) is used to detect endothelial cells and a rabbit anti-goat conjugated Cy5 (1:400) secondary antibody is used for visualization.

3.7   Quantification of Adipose Vasculature

When quantifying the tissue vascularity by vessel area/field, it should be noted that an increase of adipocyte size might misleadingly imply a decrease of the vessel number. However, the ratio of vessels per adipocyte might not have

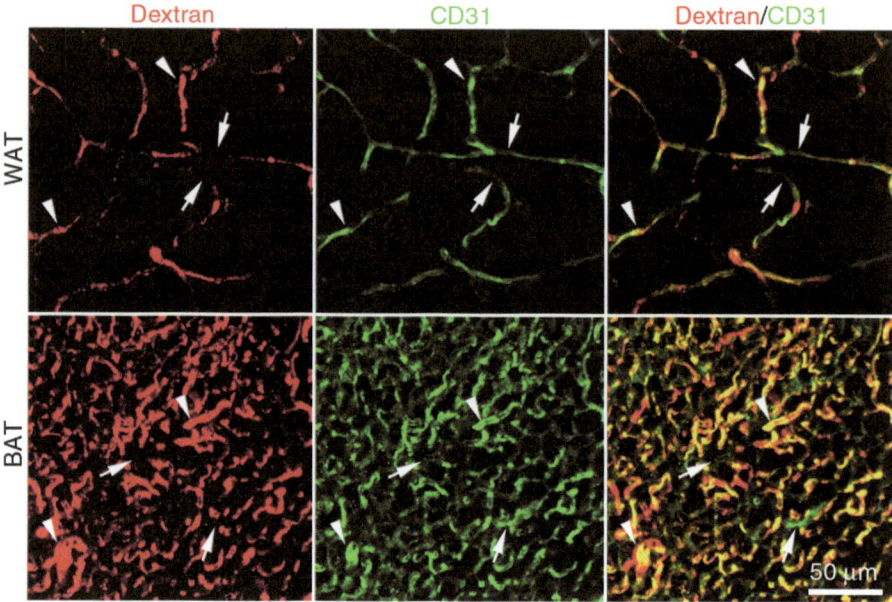

Fig. 4 Vascular function in adipose tissues. Whole-mount staining was performed on *WAT* and *BAT* depots of dextran perfused C57Bl mice. PFA fixed tissues were further co-stained with goat anti-rat CD31 and Cy5-conjugated goat anti-rat antibody. *Red* signals represent dextran perfused vessels and green signals represent endothelial cells. *Arrows* show non-perfused blood vessels and *arrowheads* show perfused vessels. *Scale bar* = 50 μm

been altered. Instead, a larger adipocyte size automatically leads to a decrease of the area per field that can be covered by vessels. If the total size of adipose depot increases, however, the ratio of vessels per adipocyte might still remain the same. To avoid misinterpretations, it is recommended to quantify the relative number of vessels per adipocyte, instead (Fig. 5). Based on 3-dimensional pictures of whole mount stained adipose tissue, total area of blood vessels per field is quantified in ten randomly chosen microscopic fields using Adobe Photoshop CS3 software.

For quantifying vessel number or number of vessels per adipocyte, paraffin embedded sections of adipose tissues fixed in 4% PFA, are used. Adipose tissue sections are stained with hematoxylin and isolectin to examine the morphology. To quantify the number of vessels, a scoring system is used giving a score of 1 to very short vessels, a score of 2 to vessels of an intermediate length and a score of 3 to long vessels. The sum of these scores represents the total number of vessels. Adipocyte cross-sectional areas are measured in cells from five pictures of isolectin staining using Adobe Photoshop for example CS3 software. Quantification of adipose vasculature vessels per adipocyte is calculated using the formula total numbers of blood vessels per field/total number of adipocytes per field.

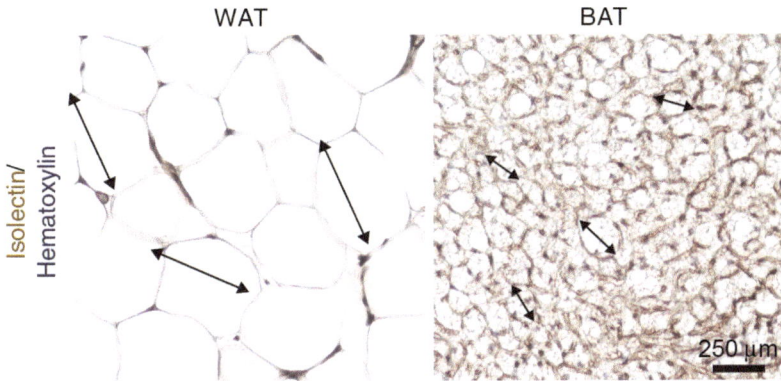

Fig. 5 Quantitative analysis of adipose tissue vasculature. Paraffin sections on white (*WAT*) and brown adipose tissue (*BAT*) from C57Bl mice was stained with an isolectin antibody and chromogenic visualization using DAB was performed. Nuclei were co-stained with Hematoxylin. *Scale bar* = 250 μm

4   Conclusion and Outlook

Our current methods provide the state-of-the-art methodology to study the number and structure of mircovessels in both BAT and WAT. Unlike other tissues, immunohistological staining of adipose tissues may encounter tremendous technical difficulties owing to light density and large lipid contents of the adipose tissue. To develop optimal and functional methods, we have tried different conditions and share the current protocols with other researchers. Although our current methods may not be the best optimal protocols, they ensure an experienced researcher to study the intimate interplay between the adipose vasculature and adipocyte functions. When studying the vasculature function in the adipose tissue, it is crucial to consider the metabolic status and the microenvironment in a particular tissue. Since both adipose tissue and microvessels exhibit tremendous plasticity, immunohistochemical methods may generate different type of results if the same tissue is exposed under different conditions. This is particularly true to the BAT, which can become activated under different physiological and pathological conditions. For example, cold acclimation of mice can lead to marked increase of BAT activity accompanying a high density of blood vessels.

The adipose tissue mass has a significant impact on human heath. Excessive or deficient amounts of adipose tissue in the body may lead to serious pathological consequences of an individual. Emerging evidence from our laboratory and from others demonstrate that microvessels within the adipose tissue play a pivotal role in regulation of adipose mass expansion and shrinkage. Moreover, blood vessels might play dual roles in controlling the adipose tissue expansion and functions, depending on the metabolic status [1]. For example, in the metabolically quiescent WAT, angiogenesis might lead to further expansion of the adipose tissue mass. Inversely, the switch of an angiogenic phenotype in the BAT might facilitate energy expenditure, leading to a lean and healthy phenotype. In the light of the dual functions of the adipose vasculature, therapeutic interventions targeting blood vessels should also be carefully designed to different adipose depots. The same therapeutic agent may have opposing biological consequences depending on the metabolic status of the host.

Another interesting aspect of the adipose vasculature is its relation to development of obesity-related disorders. Virtually, all obesity-related disorders are associated with angiogenesis. As the adipose tissue is one of the largest organs in the body, the absolute amount of the angiogenic factors produced in this tissue would have an immense impact on other vasculatures distributed in various tissues and organs. This is particular true for obese individuals who carry the largest load of adipose tissues relative to other tissues. High expression levels of angiogenic factors may enter the circulation to facilitate development of obesity-associated disorders and complications such as diabetic retinopathy and nephropathy in which excessive angiogenesis exist. Additionally, adipose tissues also provide a rich-source of stem cells that can be differentiated into other cell types under pathological conditions. The adipose mesenchymal cells are defined as pluripotent stem cells that have potentials to differentiate into many other kinds of cells including vascular endothelial cells.

Taken together, to study the structure and functions of the adipose tissue may provide new mechanistic insights on the intimate interactions between the vascular and adipose compartments. The mechanistic understanding of vascular functions in the adipose tissue would help us to develop novel therapeutic strategies for the treatment of obesity and its related disorders by targeting the vasculature. Encouraged by the beneficial outcomes in the clinic of anti-angiogenic drugs for the treatment of cancer and ophthalmological disorders, we strongly believe that angiogenesis modulators would one day produce clinical benefits for obese patients. This current protocol paper would provide an excellent tool to reach this important goal.

Acknowledgements Y.C.'s laboratory is supported through research grants from the Swedish Research Council, the Swedish Cancer Foundation, the Karolinska Institute Foundation, the Karolinska Institute distinguished professor award, the Torsten Soderbergs foundation, the European Union Integrated Project of Metoxia (Project no. 222741) and the European Research Council (ERC) advanced grant ANGIOFAT (Project no 250021).

References

1. Cao Y (2010) Adipose tissue angiogenesis as a therapeutic target for obesity and metabolic diseases. Nat Rev Drug Discov 9:107–115
2. Cao Y (2007) Angiogenesis modulates adipogenesis and obesity. J Clin Invest 117:2362–2368
3. Virtanen KA, Lidell ME, Orava J et al (2009) Functional brown adipose tissue in healthy adults. N Engl J Med 360:1518–1525
4. Feldmann HM, Golozoubova V, Cannon B et al (2009) UCP1 ablation induces obesity and abolishes diet-induced thermogenesis in mice exempt from thermal stress by living at thermoneutrality. Cell Metab 9:203–209

5. Xue Y, Petrovic N, Cao R et al (2009) Hypoxia-independent angiogenesis in adipose tissues during cold acclimation. Cell Metab 9:99–109
6. Rupnick MA, Panigrahy D, Zhang CY et al (2002) Adipose tissue mass can be regulated through the vasculature. Proc Natl Acad Sci USA 99:10730–10735
7. Brakenhielm E, Cao R, Gao B et al (2004) Angiogenesis inhibitor, TNP-470, prevents diet-induced and genetic obesity in mice. Circ Res 94:1579–1588
8. Xue Y, Lim S, Brakenhielm E et al (2010) Adipose angiogenesis: quantitative methods to study microvessel growth, regression and remodeling in vivo. Nat Protoc 5:912–920
9. Folkman J (1995) Angiogenesis in cancer, vascular, rheumatoid and other disease. Nat Med 1:27–31
10. Brakenhielm E, Cao Y (2008) Angiogenesis in adipose tissue. Methods Mol Biol 456:65–81
11. Gersh I, Still MA (1945) Blood vessels in fat tissue relation to problems of gas exchange. J Exp Med 81:219–232
12. Asano A, Kimura K, Saito M (1999) Cold-induced mRNA expression of angiogenic factors in rat brown adipose tissue. J Vet Med Sci 61:403–409
13. Bouloumie A, Drexler HC, Lafontan M et al (1998) Leptin, the product of Ob gene, promotes angiogenesis. Circ Res 83:1059–1066
14. Bouloumie A, Sengenes C, Portolan G et al (2001) Adipocyte produces matrix metallopro-teinases 2 and 9: involvement in adipose differentiation. Diabetes 50:2080–2086
15. Cao R, Brakenhielm E, Wahlestedt C et al (2001) Leptin induces vascular permeability and synergistically stimulates angiogenesis with FGF-2 and VEGF. Proc Natl Acad Sci USA 98:6390–6395
16. Dallabrida SM, Zurakowski D, Shih SC et al (2003) Adipose tissue growth and regression are regulated by angiopoietin-1. Biochem Biophys Res Commun 311:563–571
17. Rehman J, Traktuev D, Li J et al (2004) Secretion of angiogenic and antiapoptotic factors by human adipose stromal cells. Circulation 109:1292–1298
18. Stacker SA, Runting AS, Caesar C et al (2000) The 3 T3-L1 fibroblast to adipocyte conversion is accompanied by increased expression of angiopoietin-1, a ligand for tie2. Growth Factors 18:177–191
19. Voros G, Maquoi E, Demeulemeester D et al (2005) Modulation of angiogenesis during adipose tissue development in murine models of obesity. Endocrinology 146:4545–4554
20. Powell K (2007) Obesity: the two faces of fat. Nature 447:525–527
21. Tang W, Zeve D, Suh JM et al (2008) White fat progenitor cells reside in the adipose vasculature. Science 322:583–586
22. Cioffi F, Senese R, de Lange P et al (2009) Uncoupling proteins: a complex journey to function discovery. Biofactors 35:417–428
23. Cypess AM, Lehman S, Williams G et al (2009) Identification and importance of brown adipose tissue in adult humans. N Engl J Med 360:1509–1517
24. van Marken Lichtenbelt WD, Vanhommerig JW, Smulders NM et al (2009) Cold-activated brown adipose tissue in healthy men. N Engl J Med 360:1500–1508
25. Heaton JM (1972) The distribution of brown adipose tissue in the human. J Anat 112:35–39
26. Bjorndahl M, Cao R, Nissen LJ et al (2005) Insulin-like growth factors 1 and 2 induce lymphangiogenesis in vivo. Proc Natl Acad Sci USA 102:15593–15598
27. Cao R, Bjorndahl MA, Gallego MI et al (2006) Hepatocyte growth factor is a lymphangiogenic factor with an indirect mechanism of action. Blood 107:3531–3536
28. Banerji S, Ni J, Wang SX et al (1999) LYVE-1, a new homologue of the CD44 glycoprotein, is a lymph-specific receptor for hyaluronan. J Cell Biol 144:789–801
29. Breiteneder-Geleff S, Soleiman A, Kowalski H et al (1999) Angiosarcomas express mixed endothelial phenotypes of blood and lymphatic capillaries: podoplanin as a specific marker for lymphatic endothelium. Am J Pathol 154:385–394

Methodologic Approaches to Investigate Vascular Tube Morphogenesis and Maturation Events in 3D Extracellular Matrices *In Vitro* and *In Vivo*

Amber N. Stratman, Dae Joong Kim, Anastasia Sacharidou, Katherine R. Speichinger, and George E. Davis

Abstract *De novo* blood vessel formation is regulated by two major processes, termed vasculogenesis and angiogenesis. A key aspect of this formation is the process of endothelial cell (EC) lumen and tube assembly. Major advances have been made in our understanding of the EC lumen formation process primarily through the development and utilization of *in vitro* models of this process in 3D extracellular matrices. Recent advances include the identification of an EC lumen signaling complex that controls EC tubulogenesis in 3D collagen matrices and determination of growth factor requirements that are required for these events under serum-free defined conditions. Components of the lumen signaling complex include the collagen-binding integrin, $\alpha2\beta1$, the membrane-type 1 matrix metallo-proteinase, MT1-MMP, junction adhesion molecules B and C, Par3, Par6b and the Rho GTPase, Cdc42. This complex of proteins controls EC lumen and tube formation and establishes the signaling conditions necessary to both form and sustain an EC tube network in 3D matrices. This EC tube network induces the recruitment of pericytes which then affect tube and extracellular matrix remodeling events to regulation the maturation of tubes. Methodologies utilized to address

A.N. Stratman • D.J. Kim • A. Sacharidou • K.R. Speichinger
Department of Medical Pharmacology and Physiology, Dalton Cardiovascular Research Center, University of Missouri School of Medicine, MA415 Medical Sciences Building, Columbia 65212, MO, USA

G.E. Davis, M.D., Ph.D. (✉)
Department of Medical Pharmacology and Physiology, Dalton Cardiovascular Research Center, University of Missouri School of Medicine, MA415 Medical Sciences Building, Columbia 65212, MO, USA

Department of Pathology and Anatomical Sciences, Dalton Cardiovascular Research Center, University of Missouri School of Medicine, MA415 Medical Sciences Building, Columbia 65212, MO, USA
e-mail: davisgeo@health.missouri.edu

E. Zudaire and F. Cuttitta (eds.), *The Textbook of Angiogenesis and Lymphangiogenesis: Methods and Applications*, DOI 10.1007/978-94-007-4581-0_6,
© Springer Science+Business Media Dordrecht 2012

these events are presented to illustrate how the cellular and molecular basis for EC tube morphogenesis and stabilization are currently investigated.

1 Introduction

The molecular control of endothelial cell (EC) lumen and tube formation is affected by multiple factors and signals [1–6]. Over the past two decades, considerable progress has been made to uncover the molecular basis for this specific biological response. This progress has primarily occurred through the use of *in vitro* approaches utilizing 3D extracellular matrix environments with human ECs (Figs. 1 and 2) [1, 2, 4]. Also, recent years have seen the development of Zebrafish, mouse, and quail approaches to address questions in vascular tube morphogenesis [7–9]. Much effort has been put forward from our laboratory using both *in vitro* and *in vivo* approaches to elucidate the molecular basis for these events [1, 4, 9–17]. For this purpose, we have developed 3D collagen type I and fibrin microassay systems that mimic vasculogenic and angiogenic events [18]. In these models, human ECs are seeded either as single cells or aggregated clusters in 3D collagen type I matrices using defined serum-free media conditions such that ECs undergo dramatic morphologic changes which lead to interconnecting tube networks (Figs. 1 and 2). A consequence of EC lumen and tube formation is the generation of vascular guidance tunnels which represent proteinase-generated physical spaces within the 3D extracellular matrix [14, 16] (Fig. 3). Vascular guidance tunnels

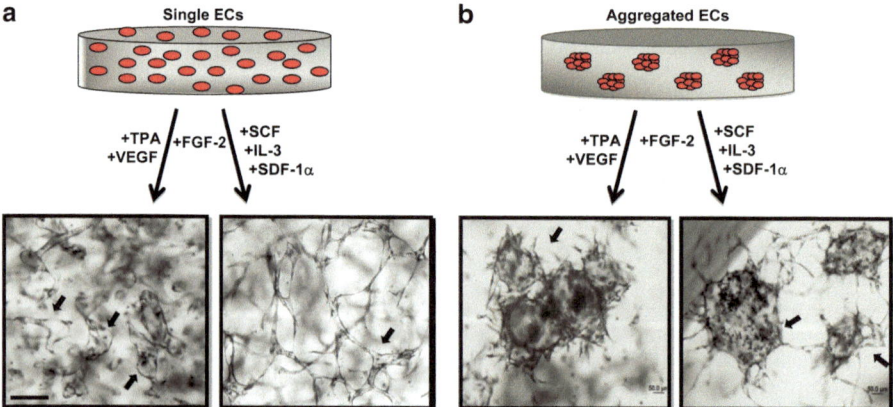

Fig. 1 *Models of EC Vasculogenic Tube Assembly in 3D collagen matrices.* Diagrammatic representation of vasculogenic tube assembly microassay systems that have been developed in our laboratory. (**a**) In the *left panels*, an assay is illustrated where single ECs are suspended in 3D collagen matrices and allowed to undergo morphogenesis in two ways with the indicated varying serum-free defined conditions. In condition #1, phorbol ester (TPA), VEGF and FGF-2 are added into the culture media and cultures are fixed at 24 h. In condition #2, SCF, IL-3 and SDF-1α are added to the collagen matrix and FGF-2 is added into the media and cultures are fixed at 72 h. *Arrows* indicate lumen and tube structures. *Left panels- Bar* equals 100 μm. (**b**) In the *right panels*, assays are performed just like the single EC assay except that ECs are pre-aggregated. *Right panels- Bar* equals 50 μm

Time lapse imaging of EC vasculogenic tube assembly

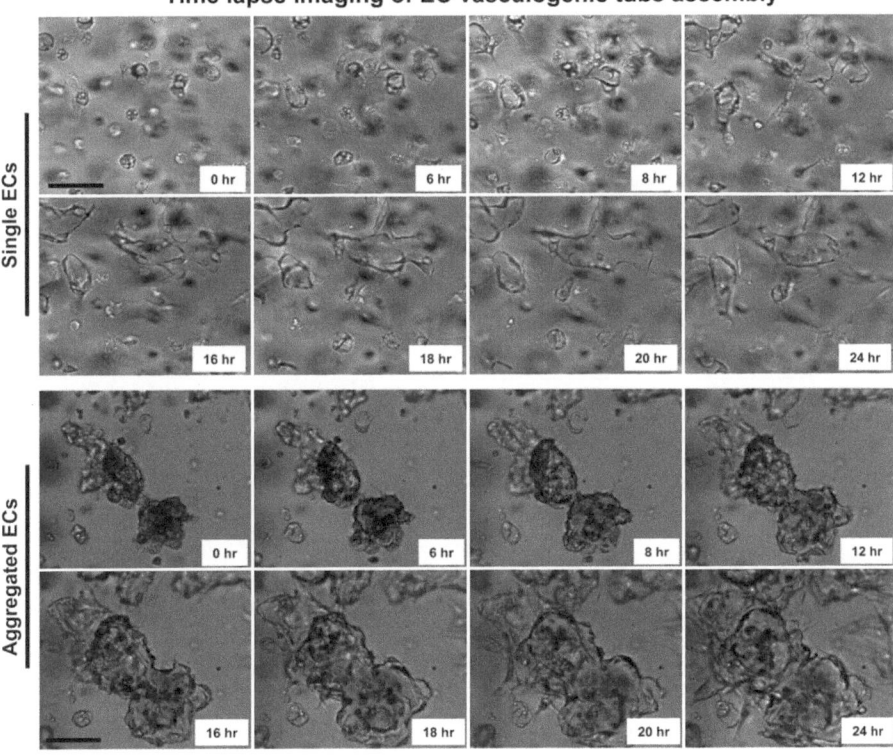

Fig. 2 *Time lapse imaging of EC vasculogenic tube assembly with single versus aggregated ECs in 3D collagen matrices.* Time lapse imaging analysis of EC vasculogenic lumen and tube assembly in 3D collagen matrices over a 24 h period. Images are shown at the indicated times from both single and aggregated EC cultures which demonstrates EC lumen and tube assembly. *Bar* equals 100 μm

allow for EC tube assembly, remodeling and play a key role in allowing mural cells recruited to tubes to migrate along the EC tube abluminal surface networks of EC lined tubes to stimulate vascular basement membrane assembly [19] (Figs. 3 and 4). Also, ECs and mural cells such as pericytes can be co-cultured in 3D matrices such that both cell types co-assemble into EC-lined tubes with associated pericytes [19] (Fig. 4). In this manner, many studies are ongoing elucidating how pericytes influence EC tube morphogenesis, remodeling and maturation events [15, 19–21].

After much work over the years our laboratory and others have generated the molecular tools and approaches to elucidate the genetic and molecular requirements controlling the pathway of EC lumen and tube formation. New vessel formation is regulated by two major processes, vasculogenesis and angiogenesis, whereby endothelial cells (ECs) undergo morphogenic processes including cell motility, lumen formation, sprouting/branching, and remodeling in response to growth factors and cytokine stimuli as well as extracellular matrix [1, 6, 22–24]. Lumenization of vessels is required for establishing blood flow and stabilizing

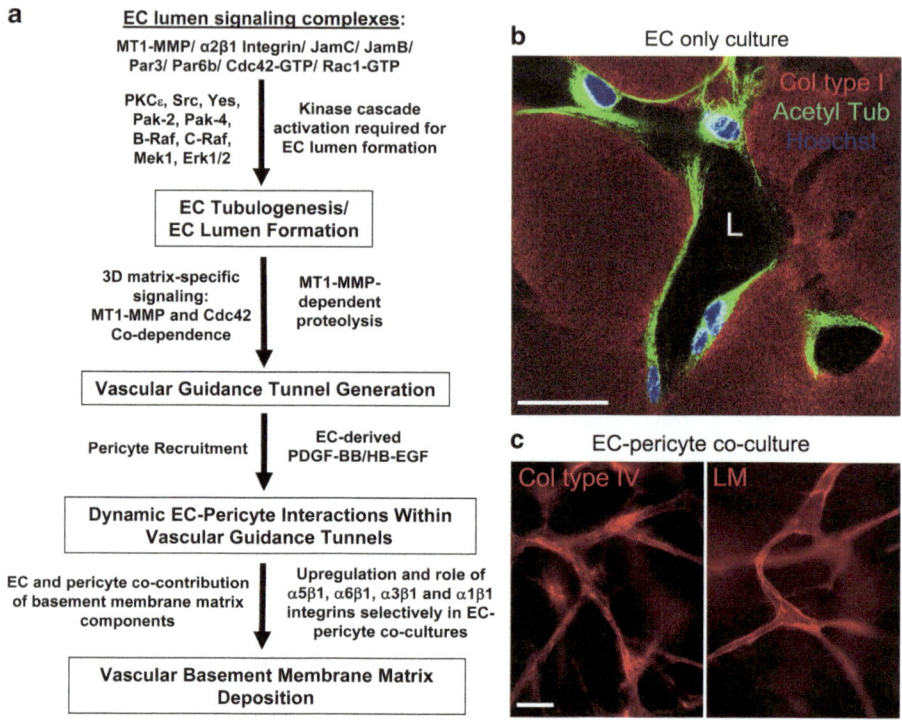

Fig. 3 *Molecular mechanisms controlling EC tubulogenesis and EC-pericyte tube co-assembly and maturation events in 3D collagen matrices.* (**a**) Schematic diagram illustrates molecular and signaling requirements for EC tube formation and vascular guidance tunnel generation. Pericyte recruitment and vascular basement membrane matrix assembly occurs due to EC-pericyte interactions within vascular guidance tunnel matrix spaces. (**b**) Confocal microscopic image of an EC only culture at 72 h. Collagen type I (Col type I) was stained red and acetylated tubulin (Acetyl Tub) was stained *green*, while nuclei are stained *blue* using Hoechst dye. *L* indicates lumen. *Bar* equals 50 μm. (**c**) EC-pericyte co-cultures were established and fixed after 5 days. The gels were immunostained for the indicated basement membrane matrix components, collagen type IV (Col type IV) and laminin (LM). *Bar* equals 50 μm

vessel networks during vascular development and thus represents a critical step during vascular EC tube morphogenesis [3].

Overall, these events are controlled by key molecules including integrins, extracellular matrices, proteinases, Rho GTPases and growth factor/ cytokines that tightly control the signaling cascades that are necessary for EC vasculogenesis and angiogenesis [1, 4, 6]. For example, we have recently reported that combinations of hematopoietic stem cell cytokines, such as stem cell factor (SCF), interleukin-3 (IL-3) and stromal-derived factor-1α (SDF-1α), control vascular tube morphogenesis and sprouting under defined serum-free conditions [21] (Fig. 5). In contrast, VEGF and FGF-2, when added together, but in the absence of hematopoietic cytokines, fail to stimulate these events under the same conditions.

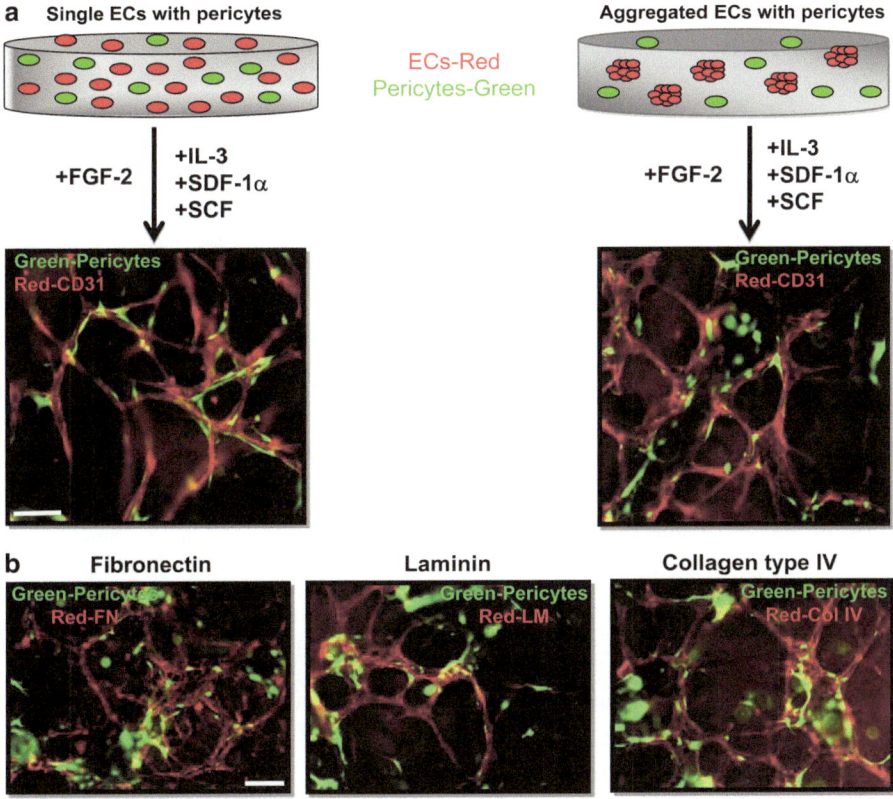

Fig. 4 *Models of EC-pericyte tube co-assembly and vascular basement membrane formation in 3D collagen matrices.* (**a**) Diagrammatic representation of the 3D co-culture microassay systems developed for either single or aggregated tube formation in the presence of pericytes. In either system, endothelial cells and pericytes are seeded together at a 5:1 ratio within collagen type I matrices. EC-pericyte tube morphogenesis occurs in the presence of FGF-2 and the hematopoietic cytokines, SCF, IL-3, and SDF1-α for 5 days. Cultures were fixed and stained for CD31 which is *red*, while the pericytes were GFP-labeled. Pericytes are observed to be associated with the abluminal surface of EC tubes. *Bar* equals 50 μm. (**b**) EC tubes generated from EC-EC aggregates are shown with associated GFP-labeled pericytes and were stained with the indicated basement membrane matrix antibodies (*red*), fibronectin (FN), laminin (LM), and collagen type IV (Col IV). *Bar* equals 50 μm

Sprouting responses strongly occur when the three indicated hematopoietic cytokines, VEGF and FGF-2 are all added together under these defined serum-free conditions [21]. Of great interest is that much of the influence of VEGF and FGF-2 could be attained by priming ECs with these factors overnight and then performing the invasion assay system using the three hematopoietic cytokines (SCF, IL-3, SDF-1α) [21]. Using this EC sprouting assay system, we illustrate that siRNA suppression of MT1-MMP, MT2-MMP, Cdc42 and Rac1 blocks EC sprouting and tube formation while siRNA suppression of RhoA and MT3-MMP does not (Fig. 5). These data

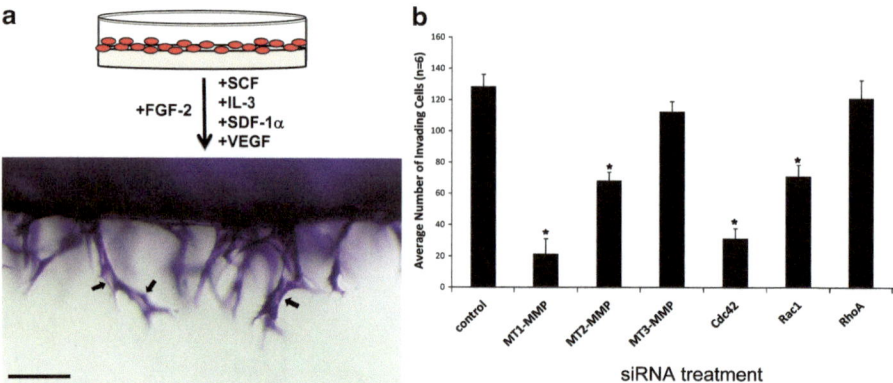

Fig. 5 *An EC angiogenic sprouting model in 3D collagen matrices.* (**a**) Diagrammatic representation of EC angiogenic sprouting in 3D collagen type I matrices during which ECs invade collagen matrices and undergo tube morphogenesis. ECs undergo angiogenic sprouting in the presence of FGF-2 in the media and a mixture of the hematopoietic cytokines SCF, IL-3, and SDF-1α, along with VEGF which are added to the collagen matrix. The image shows a cross-section of a 3D gel assay after 24 h that was stained after fixation with toluidine blue. *Arrows* indicate invading EC sprouts with lumen structures. *Bar* equals 100 μm. (**b**) ECs were treated with the indicated siRNAs, angiogenic sprouting assays were established and after 24 h, the cultures were fixed and quantitated for sprouting. *Asterisk* indicates statistical significance at $p < 0.01$

demonstrate that both membrane metalloproteinase activity, Cdc42 and Rac1 are critical regulators of EC sprouting responses in 3D collagen matrices.

Thus, these new types of approaches are essential to define the specific factors and conditions which control various stages of vascular morphogenesis [21]. Furthermore, using multiple genetic and molecular manipulations such as siRNA suppression, adenoviral and lentiviral transfection, use of chemical inhibitors and blocking antibodies in conjunction with our 3D defined microassay systems, we have been able to show that EC lumen and tube formation is controlled by a signaling complex comprising the membrane-type 1 matrix metalloproteinase (MT1-MMP), α2β1 integrin, the junction adhesion molecules (Jam)B and JamC, the polarity molecules, Par3 and Par6b, and the Rho GTPase, Cdc42 [4, 12–14, 18]. Together, this complex of proteins controls human EC lumen formation (a 3D matrix-specific process) and directly regulates Cdc42 activation in 3D matrices, but not on 2D matrix surfaces [4, 14].

These data reveal that Rho GTPases and, in particular Cdc42 and Rac-1, are critical regulators of EC lumen and tube formation and thus warrant further investigation on how they control these biological processes. Utilizing our *in vitro* EC morphogenic models we have been able to carry out a detailed biochemical analysis of these Rho GTPases during EC tubulogenesis from either single EC or pre-aggregated cultures in 3D collagen matrices. We have been able to show that both Cdc42 and Rac1 are highly activated over time as EC lumen formation progresses (while RhoA activity is suppressed) and, furthermore, they directly activate a signaling cascade that involves p21-activated kinase (Pak)2, Pak4, B-Raf, C-Raf, and Erk1/2 [12, 14]. In addition, we recently demonstrated a critical role for protein kinase C (PKC) epsilon, and Src family kinases, Src and Yes, during these events [13].

Here, we describe methodologies that are routinely utilized by our laboratory to reveal the mechanisms controlling vascular morphogenesis and stabilization. We describe the microassay systems as well as the different techniques that have been used in conjunction with these systems to reveal the cell and molecular basis for human vascular tube morphogenesis and stabilization in 3D extracellular matrices.

2 *In Vitro* Models Mimicking Vasculogenesis and Angiogenesis Using Human Endothelial Cells in 3D Collagen Matrices

Our 3D *in vitro* models closely resemble the processes of embryonic vasculogenesis as well as angiogenic sprouting events during development or postnatal life [2, 4, 25, 26]. In addition, our assays can be performed as co-culture assays whereby endothelial cells are mixed with pericytes to enable us to study endothelial tube remodeling, maturation, and stabilization mechanisms [15, 19–21] (Fig. 4). The assays are fully compatible with important molecular and biochemical techniques to evaluate gene expression and function, signaling cascades, proteomic analyses, and protein-protein interactions controlling these processes [11–16, 18, 27]. They are also compatible with real-time and confocal imaging approaches to visualize these biological events [11, 14–16, 19–21].

2.1 *EC Lumen and Tube Formation from Single Cells and Aggregates Using* In Vitro *3D Collagen Assays*

Human umbilical vein endothelial cells (HUVECs) that are used for any of our *in vitro* assays were purchased from Lonza and were cultured between passages 2 and 6 as previously described [18].

2.1.1 For Single EC Lumen Formation Assays

1. HUVECs are grown to confluency on gelatin coated flasks using 1x Medium 199 (M199) (Invitrogen) supplemented with 20% fetal calf serum, bovine hypothalamic extract at 100 µg/ml and heparin at 50 µg/ml as previously described [28]. Media was changed every 3–4 days.
2. Trypsinize HUVECs and resuspend them at a concentration of 1×10^7 cells/ml in 1× M199 medium. Mix well but gently with a pipette to resolve any cell clumps. Place it on ice.

3. For our 3D microassay systems you can prepare a 3.75 mg/ml collagen type I gel
   solution or a 2.5 mg/ml collagen type I gel solution. All components have to be
   kept on ice at all times so as to prevent collagen from polymerizing. With
   2×10^6 cells, one can make 1 ml of either of these collagen gel concentrations.

   (a) For 1 ml of gel with a concentration of 3.75 mg/ml collagen type I, use the
       following: 525 µl of 7.1 mg/ml type I collagen resuspended in 0.1% acetic
       acid, 58.5 µl of 10× M199, 3.15 µl of 5 N NaOH and 213 µl of 1× M199.
   (b) For 1 ml of gel with a concentration of 2.5 mg/ml collagen type I, use the
       following: 350 µl of 7.1 mg/ml type I collagen resuspended in 0.1% acetic
       acid, 39 µl of 10× M199, 2.1 µl of 5 N NaOH and 409 µl of 1× M199.
       Remember to mildly finger mix the gel solution each time a component is
       added.

4. When the gel is done add 200 µl of HUVECs directly to the gel for a final
   concentration of 2×10^6 cells/ml. Do not pipette the mixture up and down to
   avoid creating air bubbles. In order to mix the gel solution one has to finger mix
   the tube (best performed in a 50 ml conical tube). When everything is sufficiently mixed, add 28 µl per well to a 96 half-area well (clear flat bottom TC-
   treated microplate) (Costar). Every so often, one has to tap the plate gently on
   one side to make sure the gel solution is evenly distributed in every well.
5. When this is done, place the plate in a 37°C incubator with 5% CO_2 for 30 min in
   order for the collagen to polymerize and for the pH to equilibrate.
6. While the plate is polymerizing, the feeding media should be prepared. For every
   well 100 µl of feeding media is required containing 1× M199 with:

   (a) 1:250 dilution RSII (reduced serum supplement) which is prepared as
       previously described [18].
   (b) 1:1,000 dilution FGF-2 (R&D Systems) (stock at 40 µg/ml) (40 ng/ml final
       concentration)
   (c) 1:250 dilution VEGF-165 (R&D Systems) (stock at 10 µg/ml) (40 ng/ml
       final concentration)
   (d) 1:100 dilution Ascorbic Acid (stock at 5 mg/ml) (Sigma) (50 µg/ml final
       concentration)
   (e) 1:20,000 dilution 12-O-tetradecanoyl-phorbol-13-acetate (TPA) (Sigma)
       (stock at 1 mg/ml in absolute ethanol) (50 ng/ml final concentration)

7. Return the plate in the 37°C incubator with 5% CO_2 and allow ECs to undergo
   lumen and tube formation for at least 24–48 h.

2.1.2 For Aggregate EC Assays

1. Before trypsinizing the cells, a 35 mm polystyrene (not tissue culture treated)
   dish (Corning) is coated with 1 ml of 1% BSA in PBS solution.

2. Trypsinize HUVECs and resuspend them in 3 ml of $1\times$ M199 medium with a 1:250 dilution of RSII. Mix well but gently with a pipette to resolve any cell clumps.
3. Place the mixture on the BSA coated dish.
4. Place in the 37°C incubator with 5% CO_2 for 3 h. Gently swirl the cells every hour to promote cell-cell contact.
5. After 3 h, pour the cell aggregates in a 50 ml centrifuge tube without pipetting up and down. Let the aggregates settle for 20 min.
6. Then remove all the media except for 500 µl which contains the aggregated ECs. For our 3D microassay systems with aggregates you can either use 3.75 mg/ml collagen type I gel solution or a 2.5 mg/ml collagen type I gel solution and prepare them as described above.
7. When the gel is done, add 200 µl of aggregated HUVECs directly to the gel for a final concentration of ~2 × 10^6 cells/ml. Do not pipette the mixture up and down to avoid creating air bubbles or breaking up the aggregates. In order to mix the gel solution one has to finger mix the tube mildly. When everything is sufficiently mixed, add 28 µl in every well of a 96 well half-area microplate as described above.

When this is completed, place the plate in a 37°C incubator with 5% CO_2 for 30 min in order for the collagen to polymerize, equilibrate the pH and then feed the cultures with the media conditions as described above.

2.2 EC-pericyte Co-culture Model Used for Evaluating EC Tube Remodeling and Maturation Events Including Vascular Basement Membrane Matrix Formation

HUVECs and pericytes are grown to confluency on gelatin coated flasks as stated before.

1. Trypsinize HUVECs and pericytes and resuspend them at a concentration of 1×10^7 cells/ml in $1\times$ M199 medium. Mix well but gently with a pipette to resolve any cell clumps. Place it on ice.
2. For this 3D microassay system, we use 2.5 mg/ml collagen type I gel solution. All components have to be kept on ice at all times so as to prevent collagen from polymerizing. With 2 × 10^6 ECs, 1 ml of collagen gel can be made. For a 1 ml, 2.5 mg/ml collagen gel solution, use the following: 350 µl of 7.1 mg/ml type I collagen resuspended in 0.1% acetic acid, 39 µl of $10\times$ M199, 2.1 µl of 5 N NaOH and 369 µl of $1\times$ M199.
3. To the collagen gel mix, add the 3 growth factors: 1:250 dilution (200 ng/ml final concentration) of IL-3 (R&D Systems) (IL-3 stock at 50 µg/ml), 1:500 dilution (200 ng/ml final concentration) of SDF-1α (R&D Systems) (SDF-1α stock at

100 μg/ml), and 1:250 (200 ng/ml final concentration) of SCF (R&D Systems) (SCF stock at 50 μg/ml).

4. After these additions and gentle mixing (maintain on ice) add 200 μl of HUVECs directly to the gel for a final concentration of 2×10^6 ECs/ml and 40 μl of pericytes for a final concentration of 0.4×10^6 pericytes/ml. Our typical assays utilize GFP-labeled pericytes as described [19, 20]. When everything is sufficiently mixed, then add 28 μl per well of a 96 half-area well clear flat bottom TC-treated microplate.

5. When done place the plate in a 37°C incubator with 5% CO_2 for 30 min in order for the collagen to polymerize and equilibrate the pH.

6. While the plate is polymerizing, the culture media should be prepared. For every well 100 μl of feeding media is added containing $1\times$ M199 with:

   (a) 1:250 dilution RSII
   (b) 1:1,000 dilution (40 ng/ml FGF-2) (FGF-2 stock is 40 μg/ml)
   (c) 1:100 dilution (50 μg/ml Ascorbic Acid)

Return the plate to the 37°C incubator with 5% CO_2 and allow ECs to undergo lumen and tube formation as well as pericyte recruitment for 3–5 days.

This system can also be utilized as described above without the addition of pericytes (Fig. 1). In this manner, experiments can be performed to assess the role of ECs alone versus EC-pericyte co-cultures to investigate EC tube formation or stabilization events. This approach and system was previously utilized to determine that EC-pericyte interactions were required in order for EC-lined tubes to deposit a vascular basement membrane [19, 20]. EC only cultures failed to deposit a basement membrane matrix on their own. This microassay system can also be utilized with aggregated ECs (Fig. 1). Follow the procedures described above in order to form aggregates and use those instead of the single EC suspension.

2.3 Hematopoietic Stem Cell Cytokine- and VEGF-Induced Sprouting Assay

1. For this 3D microassay system, we utilize 2.5 mg/ml collagen type I gel solution. All components have to be kept on ice at all times so as to prevent collagen from polymerizing. For a 1 ml collagen gel at 2.5 mg/ml of collagen, we use:

   350 μl of 7.1 mg/ml type I collagen resuspended in 0.1% acetic acid,
   39 μl of 10× M199, 2.1 μl of 5 N NaOH and 609 μl of 1× M199.
   Then add four growth factors to this collagen gel mixture:
   IL-3 1:250 dilution (200 ng/ml final concentration), SCF 1:250 dilution (200 ng/ml final concentration), SDF-1α 1:500 dilution (200 ng/ml final concentration), VEGF-165 1:50 dilution (200 ng/ml final concentration) (VEGF stock concentration of 10 μg/ml).

2. Pipet 28 µl in every well of a 96 half-area well clear flat bottom TC-treated microplate.
3. When this is completed, place the plate in a 37°C incubator with 5% CO_2 for 30 min in order for the collagen to polymerize and equilibrate.
4. Trypsinize ECs and resuspend them to 4×10^5 cells/ml. Make feeding media with:

   1:250 dilution RSII
   1:1,000 dilution (40 ng/ml FGF-2 final concentration)
   1:100 dilution (50 µg/ml Ascorbic Acid final concentration)

5. After the gels have polymerized and pH equilibrated, add 100 µl of cells to each well.
6. Incubate plate for 24 h. Cultures are fixed with 3% glutaraldehyde in PBS and then stained with toluidine blue.

2.4 EC Tube Regression Model with ECs Alone or EC-pericyte Co-cultures

This model system is based on several publications from our laboratory demonstrating the ability of EC-lined tubes to undergo tube regression in a manner dependent on matrix metalloproteinase (MMP)-1, MMP-10 and plasmin, which is generated following conversion of plasminogen to plasmin through EC-derived plasminogen activators [15, 29, 30]. Furthermore, we demonstrated that this tube regression response could be abrogated as a result of pericyte addition to the EC cultures [15]. This ability of pericytes to stabilize EC tubes is due to EC-pericyte interactions which resulted in the production of EC-derived tissue inhibitor of metalloproteinase (TIMP)-2 and pericyte-derived TIMP-3 [15]. These TIMPs blocked the activity of MMP-1 and MMP-10 and thus prevented the collagen matrix degradation events that underlie this tube regression response.

1. Make a 1 ml collagen gel solution as described above with ECs alone or EC-pericytes. Finger mix thoroughly. The cultures are set up as described above into 96 well half-area tissue culture plates.
2. 100 µl of feeding media is added containing $1 \times$ M199 with:

   (a) 1:250 dilution RSII
   (b) 1:1,000 dilution (40 ng/ml FGF-2)
   (c) 1:250 dilution (40 ng/ml VEGF-165)
   (d) 1:100 dilution (50 µg/ml Ascorbic Acid)
   (e) 1:20,000 dilution (50 ng/ml 12-O-tetradecanoyl-phorbol-13-acetate-TPA)
   (f) 1:200–1:500 dilution (2–5 µg/ml plasminogen) (plasminogen stock solution at 1 mg/ml) (American Diagnostica)

3. Return the plate in the 37°C incubator with 5% CO_2 and monitor cultures over a 48 h period.

2.5 Time-Lapse Imaging of EC Vasculogenesis

Time lapse analysis of EC lumen and tube formation or angiogenesis in 3D collagen matrices is an important tool (Fig. 2). Using time lapse analysis, we have been able to demonstrate the formation of pinocytic vacuoles that coalesce to participate in lumen formation, the assembly of ECs into multicellular tube networks and the recruitment of pericytes to EC-lined tubes [14–16, 19, 21]. What is quite striking about these studies is how dynamic ECs and pericytes are during tube assembly, remodeling and maturation events [19, 21] (Fig. 2). Our time lapse analysis is performed using an inverted Nikon microscope (Eclipse TE 2000-U) equipped with fluorescence and a digital camera (Photometrics Cool-Snap HQ2) as well as an associated temperature and CO_2 controlled chamber. To analyze the images collected, Metamorph software is used to create stacks which are then compiled and saved as either AVIs or quick-time files.

EC vasculogenic or angiogenic assays are set up as described above. Water should be added to surrounding wells to prevent dehydration of the gels. A thin glass plate is added on top of the plate to increase optical clarity of the images and to reduce condensation on the lid of the plate. The temperature controlled chamber should be set at 37°C with a continuous infusion of 5% CO_2. Before starting the image acquisition the plate is incubated for 1 h in the 37°C cell incubator with 5% CO_2 to allow for proper equilibration.

2.6 Data Analysis

Our 3D in vitro systems allow for a large variety of analyses of the various biological processes being examined. Assays can be stopped at any desired point and can proceed for 2–3 weeks. The assays can be fixed for lumen area analysis or invasion but they can also be fixed for immunohistochemistry or electron microscopic analysis as well.

2.6.1 Normal Fixation of Assays

1. Remove media from each well with a pipet. (Consider saving the conditioned media in order to assess secreted proteins that may influence the responses being analyzed.)
2. Replace media with 150 µl 3% glutaraldehyde in 1× PBS.

3. The fixative should stay on at least for 2 h. Plates could be stored at 4°C and stained later.
4. Remove glutaraldehyde with pipet and replace with 100 µl of 1% Toluidine blue (Sigma) in 30% methanol.
5. Allow the stain to proceed for 30 min.
6. Remove stain, wash plate(s) three times with water.
7. On third wash fill wells up so that a meniscus is visible above each well.
8. Invert plate in a plastic container that has water. Be careful so that there are no bubbles between wells and water and let sit on a stir plate for several hours. Check color to see when destaining is sufficient.
9. Store plates at 4°C.

2.6.2 Fixation of Assays to Be Used for Immunofluorescence Staining

1. Remove media from each well with a pipet.
2. Replace media with 150 µl 2% Paraformaldehyde in PBS.
3. The fixative should stay on overnight for better results. Plates could be stored at 4°C with 2% Paraformaldehyde and stained later.

2.6.3 Culture Lysates for SDS-PAGE Analysis

At any time point one can prepare lysates from the cultures to assess the expression or phosphorylation states of proteins during vasculogenesis or angiogenesis.

1. Remove the media.
2. Pick the desired number of gels using sterile forceps. For every collagen gel collected, add 75 µl of pre-heated, at 100°C, 1.5× Laemmli sample buffer containing 7.5% beta-mercaptoethanol.
3. After the collection of gels is done the mixture is boiled again for 10 min at 100°C.
4. Cool down samples before using for loading on a SDS-PAGE minigel.
5. We typically use 30–40 µl of sample per lane.

2.6.4 Microscopy/Imaging and Statistical Analysis

Visualization and image acquisition of EC vasculogenesis and angiogenesis were performed using an inverted microscope (CKX41; Olympus) as previously described [18]. For every condition we usually acquire 15–20 images to be analyzed. Each picture is then imported into Metamorph software and the lumen area is traced. In the same manner, pericyte recruitment can be assessed as well by counting the number of pericytes found on each tube. The data are imported into a Microsoft Excel file for statistical analysis. Statistical analysis for our 3D in vitro systems is performed using SPSS 11.0 software (SPSS, Inc.) and is typically assessed by using a Paired-Samples t-test with a p value of less than 0.05.

3 Molecular and Genetic Manipulation of Endothelial Cells for Use in 3D *In Vitro* Morphogenesis Systems

Our 3D *in vitro* assays are flexible enough to allow for a wide variety of biochemical or genetic manipulations of the cells used. This ability has allowed our laboratory to assess many molecules, genes, and signaling pathways that are directly involved in EC lumen and tube formation, stabilization, regression and angiogenic sprouting [1, 3, 4, 31].

3.1   Single and Double siRNA Transfection Protocol for HUVECs

Our laboratory has developed and very successfully utilized the siRNA technique in order to study the effects of knocking down one or multiple genes during EC vascular morphogenesis, stabilization or regression [12–15, 30]. Originally our laboratory used Smartpool siRNAs (Dharmacon) and from that work various genes were identified but in the past few years, we have been utilizing single stealth siRNAs (Invitrogen) which may be more potent in the degree of siRNA suppression using our protocol. Our siRNA treatment of ECs utilizes 90–95% confluent monolayers of cells grown in a 25 cm^2 surface area flask. If more or less cells need to be transfected, the protocol below can be proportionally scaled up (e.g. a 75 cm^2 flask). To assess the degree of siRNA gene knockdown, we typically use SDS-PAGE analysis, RT-PCR or real-time PCR.

1. Cells need to be at least 90–95% confluent the day of the siRNA transfection.
2. Add 578 µl of Opti-Mem I and 17.5 µl siPORT™ Amine (Ambion) to a 1.7 ml eppendorf tube.
3. Vortex the contents for 5–10 s and then quick spin at 2,000 rpm for 10 s.
4. Incubate at room temperature for 30 min.
5. Add 3.6 µl of a 20 µM stock siRNA solution for a final concentration of 20 nM per siRNA. If double of multiple genes are needed to be simultaneously knockdown then add 3.6 µl of the additional siRNA into the mixture. So far we have successfully knocked down up to three genes at the same time with high efficiency and no apparent negative effects on the cells transfected.
6. After the siRNAs are added, finger mix gently to mix reagents and briefly quick spin at 2,000 rpm for 10 s.
7. Let the mixture incubate for 20 min at room temperature.
8. While waiting for the 20 min incubation, wash the ECs twice with 3 ml of DMEM without any antibiotics.

9. After the second wash add 3 ml of DMEM that contains 1% fetal calf serum to each flask.
10. Following the 20 min incubation, gently pipette the siRNA mixture onto the cells.
11. Mix all the components by swirling the tissue flask.
12. Incubate the flasks in a 37°C incubator with 5% CO_2 for 4–5 h.
13. When the incubation is over wash the cells twice with 3 ml DMEM without antibiotics and feed the cells with appropriate feeding media that contains NO antibiotics.
14. 48 h later repeat the siRNA transfection protocol.
15. 24 h after the second siRNA transfection the cells are ready to be used in experiments.

3.2   Adenoviral Transfection Protocol

Our laboratory has frequently utilized recombinant adenoviruses to evaluate the functional roles of specific genes during EC lumen and tube formation. Prior work with adenoviruses expressing wild type and mutant versions of Cdc42, Rac1, RhoA, MT1-MMP, JamB, JamC and a variety of kinases have been used during EC lumen and tube formation [11–13, 32] and adenoviruses expressing MMP-1 and MMP-10 have been used to study EC tube regression [30]. For generating recombinant adenoviruses, we have utilized protocols that were previously described by others [33]. The following is a protocol for using recombinant adenoviruses to transfect ECs and then for utilizing the cells in tube formation assays in 3D collagen matrices.

1. You need 100% confluent flasks of cells on the day of transfection.
2. Wash the cells twice with 1× M199 (10 ml for a T-75 tissue flask or 3 ml for a T-25 tissue flask). All media should contain the normal amount of antibiotics.
3. After the washing is done add 7 ml (for a T-75) or 2.5 ml (for a T-25) of 1× M199 with antibiotics to each flask.
4. Do not add any serum.
5. Add the desired volume of your adenovirus and mix.
6. Incubated the cells for 5 h in a 37°C incubator with 5% CO_2.
7. After incubation is done remove the media with the virus and feed your cells with the normal media. No need to wash the cells.
8. Cells will be ready to use 24 h after transfection.

3.3 Lentiviral Transfection Protocol

Lentiviruses are generated using the ViraPower Lentiviral Expression system (Invitrogen) and utilize a lentiviral construct (pLenti6/ V5 TOPO) with a blasticidin resistance gene. For generating lentiviruses follow the manufacturer's specifications.

3.3.1 Transfection with Lentiviruses

Use freshly split cells at about 80–90% confluence.

1. Prepare ahead of time fresh POLYBRENE (Hexadimethrine bromide stored at 4°C)
   Make the stock of POLYBRENE at 6 mg/ml

2. For transfection of a T-75 (75 cm^2 surface area) tissue flask you need:

   3 ml of complete media (feeding media + 10% FCS)
   5–6 ml of lentiviral supernatant
   18 μl of the stock POLYBRENE (final concentration: 12 μg/ml)

   If the surface area of tissue flask to be used is different one needs to scale up or down the ingredients for the lentivirus transfection.

3. Incubate cells with mixture overnight in a 37°C incubator with 5% CO_2.
4. The next morning add 7 ml fresh media, to have a final volume of 15 ml of feeding media (media + 10% FCS) and return the tissue flask back in the incubator.
5. Next day remove media and add 15 ml of fresh feeding media
6. 3 days after the first feeding the media needs to be changed and at this time add blasticidin at a final concentration of 5 μg/ml. Keep the stock of blasticidin at 10 mg/ml in -20°C.
7. Keep feeding the tissue flask twice a week until flask becomes 100% confluent.

   Cells are ready to characterize and can be used for experiments. Make sure the cells are expanded and frozen down for long-term storage in a liquid nitrogen tank.

3.4 siRNA and Adenoviral Transfection Protocol Combined

To combine the siRNA protocol with adenoviral transfection is very straight-forward using our 3D systems. The siRNA protocol precedes the adenoviral transfection protocol. Follow the siRNA transfection protocol as described above. The day after the second siRNA transfection instead of setting up a

3D assay we proceed immediately to the adenoviral transfection protocol. The 3D assay can be set up 24 h after the adenovirus transfection, which is 48 h after the second siRNA transfection. This technique has been successfully used for rescuing the inhibitory siRNA effects of gene knockdowns for the junctional adhesion molecules JamB and JamC such that EC lumen formation is restored [14].

3.5 Use of Blocking Antibodies, Soluble Factors and Chemical Inhibitors

Previously, we have shown that our 3D microassay systems are nicely compatible with the use of blocking antibodies, soluble factors (e.g. recombinant proteins and receptor-Fc chimeras) and chemical inhibitors. These reagents can be added either in the feeding media or within the gel, when the assay is setup. Dose response experiments need to be performed to establish the desired concentration of the reagent to be utilized.

4 Assessment of Rho GTPase Activation During EC Tube and Lumen Formation in 3D Collagen Matrices

Activation of Rho GTPases is a critical regulatory step in many key cellular behaviors [34], and we have shown their fundamental importance in vascular morphogenic events. When activation occurs, Rho GTPases bind to and activate downstream effectors which can affect many aspects of cell behavior, including actin and microtubule cytoskeletal dynamics, transcriptional regulation, cell cycle progression, vesicular trafficking, and membrane polarity [34, 35]. The most well known of the Rho GTPases are Cdc42, Rac1 and RhoA which are involved in EC vascular morphogenic events that involve cell migration, proliferation, differentiation, and cell polarity. Cdc42 and Rac1 have been shown to play key roles in EC lumen and tube formation in 3D collagen matrices while RhoA has been implicated in EC tube collapse mechanisms as well as inhibition of EC lumen formation [1, 4]. Thus, the study of Rho GTPase activation provides very useful information toward elucidating the underlying mechanistic basis for vascular morphogenesis. In order to assay for the activation of these Rho GTPases one needs to use special type of beads which have GST-fused binding domain of human p21 activated kinase protein called PAK. GTP-bound Cdc42 and Rac1 (activated state) will only associated with this binding domain.

In order to assess the activation state of these Rho GTPases we have modified the S-tag pull-down assay. A general schematic is shown that delineates the steps in this

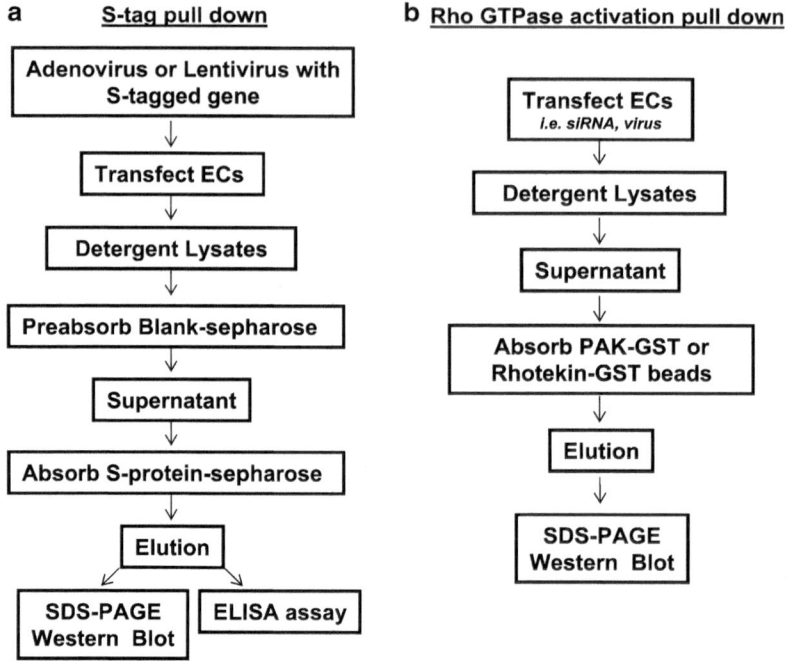

Fig. 6 *Biochemical and signaling analysis of EC lumen and tube formation in 3D collagen matrices.* Schematic representation of the biochemical and signaling analyses used for our 3D EC tube morphogenic microassay systems. (**a**) Schematic representation of the S-tag pull down assay used to study protein complexes which control EC vasculogenic or angiogenic events in 3D collagen type I matrices. (**b**) Schematic representation of the Rho GTPase activation assay used to study activation of Cdc42, Rac-1 and RhoA during EC vasculogenesis and angiogenesis. For this assay, we use special beads coated with PAK-GST or Rhotekin-GST fusion proteins that selectively bind to activated Cdc42/Rac1 or RhoA, respectively

process (Fig. 6). Detergent lysates are prepared from the 3D in vitro assay while ECs are undergoing morphogenesis using the following technique:

4.1 Detergent Lysis Preparation

1. Normal EC vasculogenic assay is set up as previously described [18]. Time course lysates at 0, 4, 8, 16 and 24 h or single time point lysates can be prepared and used for the pull downs.
2. At each time point selected, 20 collagen gels are removed using forceps and are placed in a 1.5 ml eppendorf tube, containing 1 ml cold detergent lysis buffer.
3. The detergent lysis buffer is made up as follows:

   a. 1× TBS, pH 8.0

    b. 1% Triton X-100

    c. 1 mM $CaCl_2$

    d. 1 mM $MgCl_2$

    e. Complete EDTA-free protease inhibitor cocktail tablets

    f. 100 μM GTP-γS

    g. 150 μl of 1 mg/ml high purity collagenase

4. Gels are placed in the above lysis solution and then on a rocker at 4°C until all collagen is dissolved. Then the cell lysis is centrifuged at 14,000 rpm for 20 min at 4°C.

5. 1 ml of the cleared cell lysis is used for the pull down assay.

4.2 Pull-Down Assay

1. 30 μl of PAK-GST or Rhotekin-GST beads are placed in a 1.5 ml eppendorf tube.

2. Beads are washed twice with 200 μl of washing buffer. Washing buffer contains the same reagents as the lysis buffer except it contains 0.1% Triton X-100 instead of 1%.

3. When beads are washed the cleared cell lysate is placed on the beads.

4. Beads are set on a rocker at 4°C for 1 h.

5. When the incubation is done beads are centrifuged for 30 s at 2000 rpm.

6. Supernatant is removed and beads are washed three times with 200 μl washing buffer.

7. When washing is done 120 μl of 1.5× sample buffer containing 7.5% beta-mercaptoethanol is added to the beads. Beads are then boiled for 10 min at 100°C.

8. To examine the GTP bound Cdc42, Rac1 or RhoA we use SDS-PAGE analysis. We normally load 40 μl of each sample to each well.

The Washing buffer is made up as follows:

(a) 1× TBS, pH 8.0

(b) 0.1% Triton X-100

(c) 1 mM $CaCl_2$

(d) 1 mM $MgCl_2$

(e) Complete EDTA-free protease inhibitor cocktail tablets

(f) 100 μM GTP-γS

5   Biochemical Analysis of Signaling Pathways and Complexes Controlling 3D Lumen and Tube Formation in Collagen Matrices

To establish the signaling cascades that control EC lumen and tube formation, pericyte recruitment to EC tubes, angiogenic sprouting or EC tube regression, a detailed molecular approach is required. Our microassay models in 96 well plates are particularly well suited for such an analysis. For all these processes, our studies have shown that key regulator molecules can be identified for each of these pathways. For example, we have recently identified a critical EC lumen signaling complex that controls EC tubulogenesis in 3D collagen matrices [14]. To examine the associations of these molecules with each other and their downstream or upstream effectors we have extensively used the S-tag fusion system [14]. The S-tagged proteins are readily captured from detergents lysates using S-protein agarose. This system allows for high binding specificity and rapid protein purification. Thus, we generated S-tag fusion adenoviruses expressing MT1-MMP, JamB and Cdc42 [12, 14]. The generation of the S-GFP-Cdc42 adenovirus construct was described previously [12]. This S-tag fusion approach is easily applied to other proteins that are required to identify the binding partners and signaling networks that control vascular morphogenic, stabilization and regression events.

5.1   Generation of MT1-MMP-S-tag and JamB-S-tag Vectors

For S-tag fusion constructs, the pIEX-5 vector was utilized (Novagen). Standard PCR, restriction digest and ligation protocols were used to clone each gene into the S-tag vector. The primers used for MT1-MMP wt were:

5′ AGAGATCTGCCACCATGTCTCCCGCCCAAAGACCC and
3′ AGGCGGCCGCGACCTTGTCCAGCAGGGAACG and for JamB wt were:
5′AGAGACTGCCACCATGGCGAGGAGGAG and
3′ AGGCGACAATATAAAGGATTTGTGGCTT

Each S-tag fused gene was then subsequently cloned into pAd-Track-CMV using the following PCR primers:
For the MT1-MMP-S-tag we used:

5′AGAGATCTGCCACCATGTCTCCCGCCCAAAGACCC and
3′ AGAAGCTTTTAGAATATACAGCACTTCCTTTTGGG and for the JamB-
    S-tag we used
5′ AGAGATCTGCCACCATGGCGAGGAGGAG and
3′AGTCTAGATTAGAATATACAGCACTTCCTTTTGGG

Once the S-tag fused genes were cloned into pAd-Track-CMV, standard recombinant adenoviral production protocols were carried out as previously described [33].

5.2 The S-tag Pull-Down Assay

1. Wash T75 HUVECs 2X w/ 10 ml M199. After second wash, add 7 ml M199 and your virus.
2. After 5–6 h, aspirate media and feed w/ 15 ml endothelial cell growth media.
3. Next day, set up EC vasculogenic assay.
4. Add 1 ml lysis buffer to 1.7 ml tubes and keep at 4°C. At each time point, add 150 μl of 1 mg/ml collagenase (high purified) (Sigma) (warm up collagenase in water bath) to your lysis buffer and remove and add 20 gels. Mix well at RT for 5 min. Rock 20–30 min or until collagen is all dissolved at 4°C. As you are ready to perform the pull-down assay, spin for 15–20 min at 4°C. You should save some of starting lysates at this step if you wish. As described above, lysates are pre-cleared by incubation with Sepharose 4B beads.
5. While pre-clearing, prepare 200–300 μl S-agarose bead slurry as in Step 5. After 2 h of pre-clearing, quick spin and take 1 ml (or more, depending on how much beads you are using) of pre-cleared lysates and transfer to S-agarose beads.
6. Rock for 60 min at 4°C.
7. After this time, aspirate supernatant. Wash beads 3X w/ 500 μl of washing buffer. After each wash, make sure to allow beads to settle down. After your final wash, you want to remove washing buffer as much as you can without losing your beads (use P20 or P200 pipette to remove any residual buffer).
8. After third wash, add 125–250 μl (125 μl/ 50 μl beads) of 1× sample buffer with 7.5% beta-mercaptoethanol to your beads (depending on the amount of beads you used). Vortex and boil for 10–15 min. Store the samples at −20°C.
9. Run your Western blots (load 40 μl per lane).

6 Staining Protocols for 3D In Vitro Tube Morphogenesis and Maturation Models

Our 3D microassay systems are highly compatible with general staining protocols (Figs. 3 and 4). The following is a general protocol for staining 3D in vitro gels for basement membrane components, vascular guidance tunnels or cell surface antigens (Figs. 3 and 4). The gels are removed from the 96 well plates the staining protocols are performed in 48 well plates.

1. 2% Paraformaldehyde fixed gels are required. Fixing should be at least for 2 h at in 4°C without any shaking. Remove fixative and add 500 μl of 100 mM Tris-Glycine, pH 7.5 into each well and shake at 4°C for 1 h.
2. Remove the Tris-Glycine and add 500 μl of 1% Triton X-100 in TBS in each well and shake at 4°C for 1 h.
3. Remove this buffer add 500 μl of the appropriate buffer containing 1% Triton X-100 with 1% serum of the species used for the secondary antibody and shake at 4°C for 1 h.

4. When the time is up, remove the blocking buffer and add 350 μl of fresh blocking buffer with the desired primary antibody and incubate at 4°C overnight.
5. Next day remove the primary antibody and wash three times for 10 min each with Tween 20/ Saline solution at room temperature.
6. When the washing is over add 350 μl of fresh blocker and the appropriate secondary antibody coupled to an AlexaFluor reagent (Molecular Probes) and incubate at 4°C for 2 h. This is a light sensitive step so turn off the lights.
7. After 2 h, remove the secondary antibody and wash at least six times with Tween 20/Saline solution over a period of 24–48 h before image acquisition.

For immunostains examining the extracellular deposition of vascular basement membrane proteins, we change the above protocol and do not include Triton X-100 in any of the steps [19]. We have successfully used this latter protocol demonstrating that EC-pericyte interactions control the deposition of vascular basement membranes [19, 20] (Figs. 3 and 4). We have also utilized this protocol to stain the quail chorioallantoic membrane to detect the extracellular deposition of basement membrane matrix *in vivo* [19, 20].

7 Evaluation of Quail Vascular Development In Vivo

7.1 Injection Protocol

Fertilized quail eggs (*Coturnix japonica*) are placed in a 37°C incubator to initiate development (Fig. 7). Injection of the quail eggs could occur at any developmental stage. We generally set up our injections 24 or 48 h after the eggs are placed in the incubator. The injections occurred using a 27 gauge through the apex of the egg. All drugs injected into quails are first tested for their potency and their effects in EC morphogenesis in a 3D assay. From the 3D tests a dose is chosen to be used for quail injections. For each egg, a 50 μl drug mix is prepared.

The embryos and chorioallantoic membrane tissues (CAM) are typically harvested at day 6 of development and fixed in 2% paraformaldehyde for further analysis. Immunostaining of an embryonic day 6 CAM with the EC-specific antibody, QH-1, is shown to demonstrate the pattern of developing vessels in the CAM (Fig. 7). Embryos can also be photographed at this point to visually show the effects of each drug treatment (Fig. 7).

7.2 CAM Staining Protocol

The staining occurs in 48 well plates. A small piece of the CAM is cut and placed in a well and the fluorescence immunostaining protocol is performed as described above. The QH1 monoclonal antibody is used to specifically label quail ECs in the

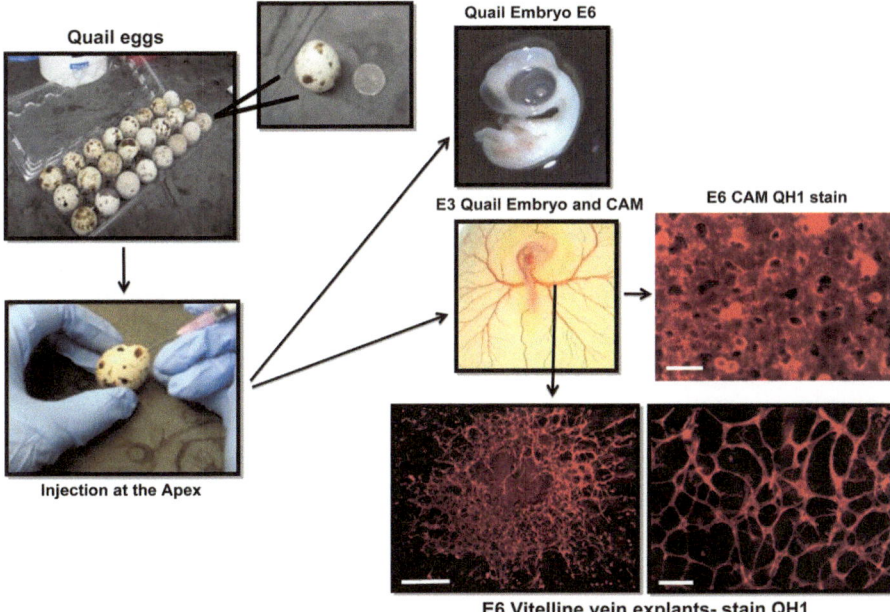

Fig. 7 *Quail vasculogenic tube assembly in vivo and quail vitelline vessel explant assay in vitro.* Image representation of the *in-vivo* quail system and analysis used in our lab for *in-vivo* EC vasculogenesis. The use of quail eggs (*Coturnix japonica*) is very convenient due to their small size. Injections of drugs, peptides, inhibitors and growth factors are performed at the apex of each egg. Quail embryos are usually harvested at embryonic day E6. During embryo harvest we separate the embryo body to be used for photography, paraffin sectioning and H + E staining and the CAM for QH1 staining (EC quail specific marker). *Bar* equals 25 μm. In addition the quail vitelline vein can be harvested from E6 embryos. The vein is then cleared from all blood and sectioned into smaller pieces which are used as explants in 3D collagen assays in the presence of the hematopoietic cytokines IL-3, SDF-1α, SCF. FGF-2 is added to the culture media. Explants were treated overnight with VEGF and FGF-2 to prime the ECs. Quail EC morphogenesis was allowed to occur for 5 days and after fixation, immunostaining was performed using QH1 antibodies to selectively stain quail ECs. *Left* image, *Bar* equals 100 μm. *Right* image, *Bar* equals 50 μm

CAM tissue (Developmental Studies Hybridoma Bank) and can also be used to label ECs in immunohistochemical staining of paraffin sections of quail embryos.

7.3   *Embryonic Quail* Ex-Vivo *Explant Assay to Study Vasculogenesis*

1. Quail embryos are allowed to develop until day 6, at which time the vitelline vein is dissected. Typical explants were 1 cm in length, with a typical vessel that is 200 μm in diameter [21].
2. Sterile vessels are washed in clear culture media (M199) and individually cannulated onto 200 μm glass pipettes. Vessels are secured to the cannula and the mesenchymal tissue surrounding the vessel trimmed.

3. Culture media is then flushed through the vessel to evacuate the blood.
4. Each vessel explant is then cut into 0.5 mm segments, with approximately 5–7 segments generated per vessel.
5. Vessel segments are split into equal numbers and placed into M199 media containing RSII for control conditions or culture media containing RSII, 40 ng/mL VEGF-A, and 40 ng/mL FGF-2 for primed vessel conditions as described [21].
6. After priming for 16 h, vessel segments are placed into 2.5 mg/ml 3D collagen type I matrices containing 200 ng/mL SCF, IL-3, and SDF-1α [21].
7. Cultures are allowed to assemble over 3–5 days, at which time cultures are fixed in 2% paraformaldehyde and immunostained for the quail EC-specific marker QH1 (Fig. 7).

8 Conclusions

Significant advances in our understanding of the molecular basis for EC lumen and tube formation have been elucidated using *in vitro* 3D collagen type I models as well as our new approaches to assess quail vascular tube formation *in vivo*. All of these important advances have been made using new technologies which have allowed us to address questions such as utilizing gene expression approaches through adenoviral and lentiviral gene transfer or siRNA suppression. Signaling pathways controlling these events have been elucidated using biochemical approaches such as evaluating protein kinase cascades (using selective chemical inhibitors, siRNAs or dominant negative mutants), S-tag and Rho GTPase activation pull down experiments, DNA microarray analysis and proteomic analysis. In addition, the use of 3D time-lapse light and fluorescence microscopy has allowed for real-time visualization of EC morphogenic processes including vasculogenesis, angiogenesis, pericyte recruitment, and EC tube regression. Furthermore, these time-lapse experiments have been coupled with the molecular approaches to assess the temporal influence of specific molecules and signaling cascades controlling vascular morphogenesis. It is clear that the *in vitro* approaches described have had a major impact toward our understanding of the cellular and molecular control of vascular tube formation, stabilization and regression. Finally, it is essential that such efforts continue to investigate the molecular details underlying these events and also correlate these important findings in conjunction with *in vivo* approaches to address these issues.

Acknowledgements This work was supported by NIH grants HL59373, HL79460, HL87308, and HL105606 to GED.

References

1. Davis GE, Stratman AN, Sacharidou A et al (2011) Molecular basis for endothelial lumen formation and tubulogenesis during vasculogenesis and angiogenic sprouting. Int Rev Cell Mol Biol 288:101–165
2. Davis GE, Koh W, Stratman AN (2007) Mechanisms controlling human endothelial lumen formation and tube assembly in three-dimensional extracellular matrices. Birth Defects Res C Embryo Today 81:270–285
3. Iruela-Arispe ML, Davis GE (2009) Cellular and molecular mechanisms of vascular lumen formation. Dev Cell 16:222–231
4. Sacharidou A, Stratman AN, Davis GE (2012) Molecular mechanisms controlling vascular lumen formation in three-dimensional extracellular matrices. Cell Tissues Organs 195 (1–2):122–143
5. Xu K, Cleaver O (2011) Tubulogenesis during blood vessel formation. Semin Cell Dev Biol 22 (9):993–1004
6. Senger DR, Davis GE (2011) Angiogenesis. Cold Spring Harb Perspect Biol 3:a005090
7. Kamei M, Saunders WB, Bayless KJ et al (2006) Endothelial tubes assemble from intracellular vacuoles in vivo. Nature 442:453–456
8. Zovein AC, Alfonso Luque A, Turlo KA et al (2010) Beta1 integrin establishes endothelial cell polarity and arteriolar lumen formation via a Par3-dependent mechanism. Dev Cell 18:39–51
9. Xu K, Sacharidou A, Fu S et al (2011) Blood vessel tubulogenesis requires Rasip1 regulation of GTPase signaling. Dev Cell 20:1–14
10. Bayless KJ, Salazar R, Davis GE (2000) RGD-dependent vacuolation and lumen formation observed during endothelial cell morphogenesis in three-dimensional fibrin matrices involves the alpha(v)beta(3) and alpha(5)beta(1) integrins. Am J Pathol 156:1673–1683
11. Bayless KJ, Davis GE (2002) The Cdc42 and Rac1 GTPases are required for capillary lumen formation in three-dimensional extracellular matrices. J Cell Sci 115:1123–1136
12. Koh W, Mahan RD, Davis GE (2008) Cdc42- and Rac1-mediated endothelial lumen formation requires Pak2, Pak4 and Par3, and PKC-dependent signaling. J Cell Sci 121:989–1001
13. Koh W, Sachidanandam K, Stratman AN et al (2009) Formation of endothelial lumens requires a coordinated PKC{epsilon}-, Src-, Pak- and Raf-kinase-dependent signaling cascade downstream of Cdc42 activation. J Cell Sci 122:1812–1822
14. Sacharidou A, Koh W, Stratman AN et al (2010) Endothelial lumen signaling complexes control 3D matrix-specific tubulogenesis through interdependent Cdc42- and MT1-MMP-mediated events. Blood 115(25):5259–5269
15. Saunders WB, Bohnsack BL, Faske JB et al (2006) Coregulation of vascular tube stabilization by endothelial cell TIMP-2 and pericyte TIMP-3. J Cell Biol 175:179–191
16. Stratman AN, Saunders WB, Sacharidou A et al (2009) Endothelial cell lumen and vascular guidance tunnel formation requires MT1-MMP-dependent proteolysis in 3-dimensional collagen matrices. Blood 114:237–247
17. Yuan L, Sacharidou A, Stratman AN et al (2011) RhoJ is an endothelial cell-restricted Rho GTPase that mediates vascular morphogenesis and is regulated by the transcription factor ERG. Blood 118:1145–1153
18. Koh W, Stratman AN, Sacharidou A et al (2008) In vitro three dimensional collagen matrix models of endothelial lumen formation during vasculogenesis and angiogenesis. Methods Enzymol 443:83–101
19. Stratman AN, Malotte KM, Mahan RD et al (2009) Pericyte recruitment during vasculogenic tube assembly stimulates endothelial basement membrane matrix formation. Blood 114:5091–5101
20. Stratman AN, Schwindt AE, Malotte KM et al (2010) Endothelial-derived PDGF-BB and HB-EGF coordinately regulate pericyte recruitment during vasculogenic tube assembly and stabilization. Blood 116:4720–4730
21. Stratman AN, Davis MJ, Davis GE (2011) VEGF and FGF prime vascular tube morphogenesis and sprouting directed by hematopoietic stem cell cytokines. Blood 117:3709–3719

22. Adams RH, Alitalo K (2007) Molecular regulation of angiogenesis and lymphangiogenesis. Nat Rev Mol Cell Biol 8:464–478
23. Holderfield MT, Hughes CC (2008) Crosstalk between vascular endothelial growth factor, notch, and transforming growth factor-beta in vascular morphogenesis. Circ Res 102:637–652
24. Hynes RO (2009) The extracellular matrix: not just pretty fibrils. Science 326:1216–1219
25. Davis GE, Bayless KJ, Mavila A (2002) Molecular basis of endothelial cell morphogenesis in three-dimensional extracellular matrices. Anat Rec 268:252–275
26. Davis GE, Stratman AN, Sacharidou A (2011) Molecular control of vascular tube morphogenesis and stabilization: regulation by extracellular matrix, matrix metalloproteinases and endothelial cell-pericyte interactions. In: Gerecht S (ed) Biophysical regulation of vascular differentiation. Springer, New York, pp 17–47
27. Bell SE, Mavila A, Salazar R et al (2001) Differential gene expression during capillary morphogenesis in 3D collagen matrices: regulated expression of genes involved in basement membrane matrix assembly, cell cycle progression, cellular differentiation and G-protein signaling. J Cell Sci 114:2755–2773
28. Davis GE, Camarillo CW (1996) An alpha 2 beta 1 integrin-dependent pinocytic mechanism involving intracellular vacuole formation and coalescence regulates capillary lumen and tube formation in three-dimensional collagen matrix. Exp Cell Res 224:39–51
29. Davis GE, Pintar Allen KA, Salazar R et al (2001) Matrix metalloproteinase-1 and −9 activation by plasmin regulates a novel endothelial cell-mediated mechanism of collagen gel contraction and capillary tube regression in three-dimensional collagen matrices. J Cell Sci 114:917–930
30. Saunders WB, Bayless KJ, Davis GE (2005) MMP-1 activation by serine proteases and MMP-10 induces human capillary tubular network collapse and regression in 3D collagen matrices. J Cell Sci 118:2325–2340
31. Davis GE, Senger DR (2008) Extracellular matrix mediates a molecular balance between vascular morphogenesis and regression. Curr Opin Hematol 15:197–203
32. Bayless KJ, Davis GE (2004) Microtubule depolymerization rapidly collapses capillary tube networks in vitro and angiogenic vessels in vivo through the small GTPase Rho. J Biol Chem 279:11686–11695
33. He TC, Zhou S, da Costa LT et al (1998) A simplified system for generating recombinant adenoviruses. Proc Natl Acad Sci USA 95:2509–2514
34. Hall A (2005) Rho GTPases and the control of cell behaviour. Biochem Soc Trans 33:891–895
35. Etienne-Manneville S (2004) Cdc42–the centre of polarity. J Cell Sci 117:1291–1300

Preparation and Analysis of Aortic Ring Cultures for the Study of Angiogenesis Ex Vivo

Roberto F. Nicosia, Giovanni Ligresti, and Alfred C. Aplin

Abstract Protocols outlined in this chapter illustrate how to prepare and analyze angiogenic cultures of rat or mouse aorta. Aortic rings embedded in gels of extracellular matrix generate vascular outgrowths that can be visualized and monitored over time with inverted microscopy. The angiogenic response is measured by counting vessels or with image analysis. The expression of angioregulatory genes is evaluated by quantitative real-time RT-PCR, immunocytochemistry, and ELISA. Angiogenesis is modulated by adding growth factors, cytokines or chemical inhibitors to the growth medium. Aortic rings isolated from genetically modified animals or transduced with viral vectors are used to evaluate how gene disruption or overexpression affects the angiogenic response. Aortic ring cultures can be used to investigate molecular mechanisms of angiogenesis and test the efficacy of stimulators and inhibitors of the angiogenic process. As such this assay is an invaluable tool for both basic and applied angiogenesis research.

1 Introduction

Angiogenesis, the formation of new blood vessels, plays an important role in the progression of many diseases, the healing of wounds and the revascularization of ischemic organs [1, 2]. Research in angiogenesis has grown exponentially since the existence of soluble angiogenic factors was first proven in the late 1960s [3]. Major impetus to this field was given by Judah Folkman's hypothesis that tumor growth is

R.F. Nicosia, M.D., Ph.D. (✉)
Pathology and Laboratory Medicine Services, Veterans Administration Puget Sound
Health Care System, 1660 South Columbian Way, Seattle 98108, WA, USA

Department of Pathology, University of Washington, Seattle, WA, USA
e-mail: roberto.nicosia@va.gov

G. Ligresti • A.C. Aplin
Department of Pathology, University of Washington, Seattle, WA, USA

E. Zudaire and F. Cuttitta (eds.), *The Textbook of Angiogenesis and*
Lymphangiogenesis: Methods and Applications, DOI 10.1007/978-94-007-4581-0_7,
© Springer Science+Business Media Dordrecht 2012

angiogenesis-dependent [4]. Following studies, fueled by advances in cell and molecular biology, led to the identification of many molecular regulators of the angiogenic response. The search for clinically effective inhibitors of the angiogenic process came of age with the approval by the FDA of anti-angiogenic drugs for the treatment of cancer and age-related macular degeneration of the retina [5–7].

Many experimental models have been developed to study mechanisms of angiogenesis and test the efficacy of stimulators and inhibitors of the angiogenic process [8–10]. Among these is the aortic ring model. This model is based on the capacity of rat and mouse aortic explants to sprout and form neovessels when embedded in three-dimensional gels of collagen or other biological matrices. Following our original report [11], the aortic ring model has been progressively refined and has now become one of the most commonly used methods for the study of angiogenesis [12]. This paper details procedures currently used in our laboratory to prepare and analyze aortic ring cultures.

2   Basic Principles

The angiogenic response of the aortic rings is a self-limited process triggered by the injury of the dissection procedure. Sprouting of microvessels from the explants is regulated by a cascade of gene activation involving growth factors, inflammatory cytokines and chemokines, matrix metalloproteinases, and extracellular matrix molecules. Angiogenesis in aortic cultures does not require addition of serum or growth factors, provided that the concentration of soluble mediators released by the explants is maintained at stimulatory levels using culture wells of appropriate size [13, 14]. The endogenous mechanisms that regulate angiogenesis are more robust in rat than mouse aorta cultures, probably due to the greater vessel size and growth factor release in the rat model. Spontaneous vessel sprouting can be obtained in mouse cultures using smaller volumes of growth media (see below) [14]. The extent of aortic angiogenesis is influenced by the age and genetic background of the animals [15–17]. It is therefore recommended to consistently use animals of the same age and strain.

3   Advantages

The aortic ring model can be considered an *ex vivo* bridge between the simplified setting of *in vitro* assays and the complex environment of the live animal. As such this model combines advantages of both systems. Like the *in vitro* assays with isolated endothelial cells, aortic cultures can be visualized and monitored on a daily basis under an inverted microscope. Similarly, the growth medium and extracellular matrix in which vessels grow can be modified to test the effect of soluble and solid phase regulators of the angiogenic response. Unlike the endothelial cell strains used for in vitro assays, however, the native endothelium of the aortic outgrowths has not been modified by repeated passages in culture and behaves like endothelial cells do *in vivo*. Microvessels formed in aortic cultures are lined by endothelial cells with well defined luminal and abluminal polarity, and enveloped by pericytes [13, 14]. As typically observed during angiogenesis in vivo, microvessels formed in aortic cultures are associated with macrophages and fibroblasts which promote the angiogenic response by paracrine and juxtacrine mechanisms [14, 18, 19]. Proteins released by the aortic explants and its outgrowth accumulate in the conditioned medium and can be analyzed by standard immunochemical assays including ELISA and Western blotting without the confounding effects of serum.

4 Disadvantages

The main disadvantage of the aortic ring model is the lack of blood flow which may significantly influence results when genes of interest are regulated by shear stress and mechanochemical signals [20]. An additional potential limitation is the source of endothelial cells which is arterial and not venous, as neovessels in vivo most commonly originate from the venous side of the microcirculation. Studies conducted to date, however, have shown robust correlation of results obtained with aortic ring cultures and in vivo models [12]. If the venous origin of the angiogenic outgrowths becomes an important consideration, the methodology used to culture aortic rings can be applied to explants of inferior vena cava which are an excellent source of venous-type vessels, particularly when stimulated with VEGF or bFGF [21].

5 Preparation of Aortic Ring Cultures

In this section we describe how to isolate the aorta, prepare aortic rings, embed the rings in gels of extracellular matrix and quantify the angiogenic response over time.

5.1 Excision of the Aorta

- Sacrifice rats or mice by CO_2 asphyxiation or intraperitoneal injection of sodium pentobarbital.
- Shave thoracic and abdominal skin and disinfect with 80% ethanol.
- Pin down the animal onto a Styrofoam board covered with a pad of impermeable paper.
- Make a Y-shaped incision with a scalpel starting laterally at the top of each shoulder, running down medially to the xyphoid process of the sternum, and continuing straight down to the lower abdomen.

- Use the scalpel to dissect the skin from the underlying muscle and small scissors to open the abdominal cavity (Fig. 1a).
- Cut the ribs directing the scissors from the side of the animal medially toward the manubrium of the sternum. This approach generates a triangular sternal plate attached to the animal at the sternal manubrium (Fig. 1b).
- Expose the thoracic cavity by displacing the sternal plate to the right side of the animal with a hemostat.
- Use forceps to gently displace intestines, stomach, spleen, and liver to the right side of the animal (Fig. 1c).
- Cut the diaphragm in the direction of the vertebral column, avoiding the diaphragmatic vessels: the thoracic aorta becomes clearly visible along the vertebral column.
- Tie a knot with a silk suture at the distal end of the thoracic aorta, just below the diaphragm. Use fine microdissection forceps to create a space for the suture between the aorta and the vertebral column (Figs 1d, e).
- While holding the suture with forceps, dissect the aorta from the posterior mediastinum with small curved scissors. Excise the aorta at the aortic arch level.
- Cut the aorta below the knot and place it into a 100 mm compartmentalized Felsen dish containing serum-free endothelial basal medium (EBM; 4 ml per quadrant) (Fig. 1f).

Critical points:

1. We recommend using the thoracic aorta for the assay because of its uniform diameter and collateral artery branching pattern. The abdominal artery (the portion of the aorta that is below the diaphragm) can be used as an additional source of rings, but its variable pattern of collaterals and tapering lumen may introduce variability in the angiogenic response. The distal most rings in particular produce fewer vessels than the thoracic aortic rings because of their smaller size.
2. Attention should be given not to damage the aorta by stretching or letting it dry. The isolation procedure should not last more than 10–15 min.
3. If veins are severed during the procedure use sterile gauze to blot the field.
4. The silk suture is used to hold the aorta and avoid damaging it during dissection. This step is optional, provided that the aorta is always handled with great care.

5.2 Preparation of Aortic Rings

- Clean the aorta of fibroadipose tissue and blood under a dissecting microscope using Noyes iridectomy scissors and curved microdissection forceps (Fig. 2a–f).
- Remove intraluminal blood by gently holding one end of the aorta with forceps and pulling the clot with a second pair of forceps at the opposite end of the tube (Fig. 2a, b).

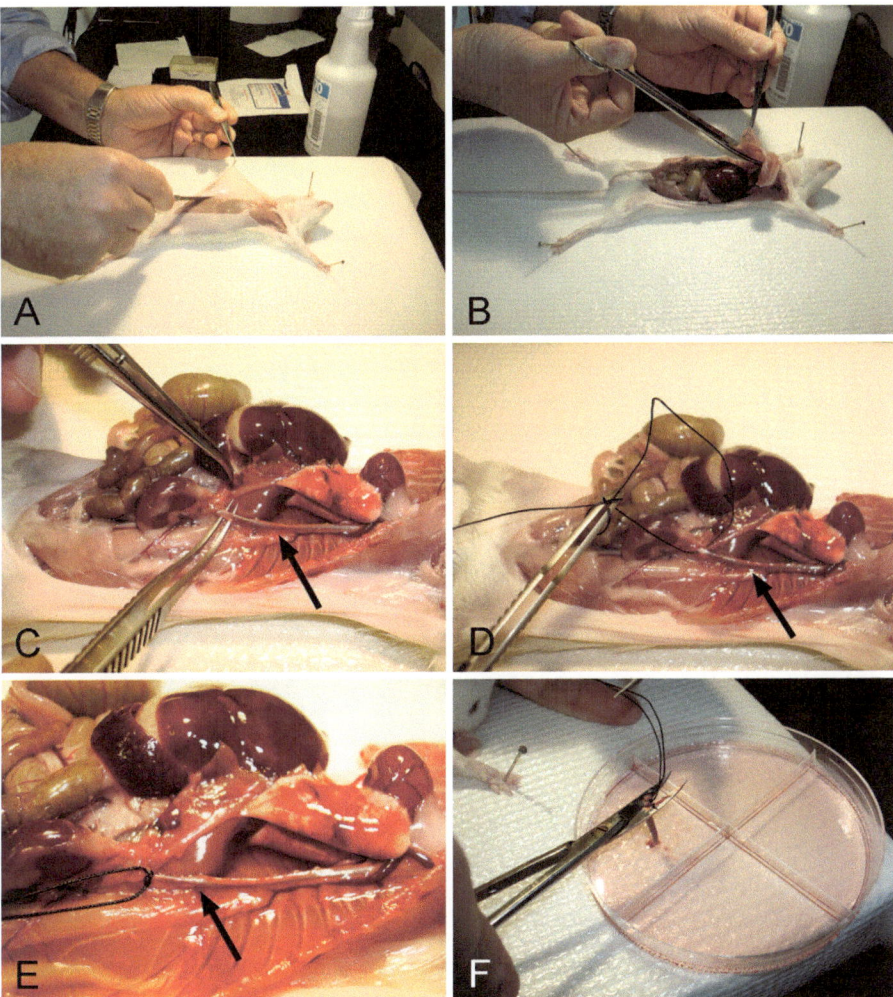

Fig. 1 Excision of rat aorta. (**a**) After the animal has been shaved, the skin is dissected from the underlying muscle with a scalpel. (**b**) The abdominal cavity is opened and ribs are cut to expose the thoracic cavity. (**c**) Following displacement of the abdominal organs and sternal plate to the right side of the animal and after cutting the diaphragm, the thoracic aorta becomes visible along the vertebral column (*arrow*); fine microdissection forceps are used to create a space between the aorta and vertebral column. (**d**) A silk suture is threaded under the aorta (*arrow*). (**e**) A double knot is tied and the suture is re-aligned along the aortic axis. (**f**) After the aorta has been cut distally to the knot, the suture is gently lifted with forceps and the aorta is dissected from the vertebral column with curved scissors and resected at the aortic arch. The suture is used to hold the explant with forceps and transfer it to a Felsen dish

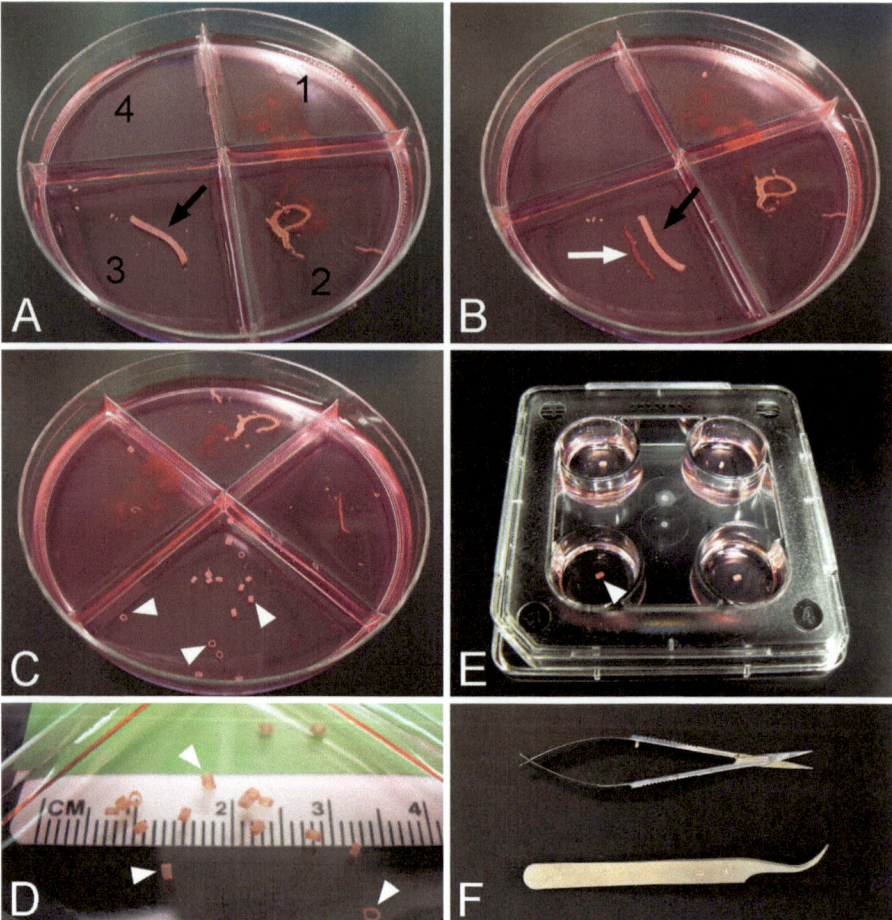

Fig. 2 Preparation of aortic ring cultures: the aorta is dissected and cleaned of blood and fibroadipose tissue using fine microdissection forceps, Noyes scissors, and sequential transfers into the compartments of a Felsen dish. (**a**) Shown here are residual blood (1), fibroadipose tissue (2), and aortic tube (3, *arrow*) following rinsing and dissection in the preceding compartments; the remaining compartment (4) is used to prepare aortic rings. (**b**) Prior to cutting the rings, the blood clot (*white arrow*) formed in the aorta (*black arrow*) is removed with forceps. (**c–d**) *Arrowheads* show aortic rings prepared by serially cross-sectioning the aorta at 1–1.5 mm intervals. (**e**) After washing, aortic rings are embedded in gels of extracellular matrix prepared in a 4-well culture dish (*arrowhead* shows representative culture) and fed with serum-free EBM. (**f**) Noyes scissors and fine microdissection forceps used for the dissection of the aorta and preparation of aortic rings

- Dissect any residual hemorrhagic fibrous tissue from the adventitial layer of the aortic wall.
- Cut stumps of collateral vessels as close to the aortic tube as possible.
- Wash away blood and dissected tissue fragments by transferring the aorta into the fresh EBM of the remaining compartments of the Felsen dish.

- Remove some of the growth medium to ensure that the aorta is kept wet but no longer floats. Cross section the aortic tube with a scalpel blade into rings of 1–1.5 mm in length (Fig. 2c, d). The thoracic aorta of a 2-month-old rat yields ~ 20–25 rings. The mouse aorta yields 10–15 rings.
- Discard the proximal and distal rings, which may have been damaged during the dissection. Add back the growth medium and let rings float in the medium.
- Clean the remaining rings in eight sequential baths of serum-free EBM using Felsen dishes (4 ml/wash). During transfers, lift and hold rings with curved microdissection forceps using the capillary action of medium trapped between the forceps prongs.
- The aortic rings are now ready to be embedded in gels of extracellular matrix.

Critical points:

1. The aortic endothelium is very delicate and can be easily damaged. Avoid stretching, cutting or crushing the aorta during the dissection procedure and the preparation of the aortic rings.
2. Dissection of the aorta and preparation of the aortic ring cultures (see below) are best carried out in a tissue culture room with HEPA-filtered air to avoid microbial contamination.

5.3   *Choice of Extracellular Matrix*

In our laboratory we culture aortic rings in gels of interstitial collagen or fibrin, which provide an excellent replica of the natural scaffold used by endothelial cells to form vessels in vivo [13]. Collagen can be produced in-house using a modification of the method originally described by Elsdale and Bard [22] (see below) or obtained from vendors. Fibrinogen is commercially available. Matrigel, a basement membrane-like matrix isolated from the Engelbreth Holm Swarm sarcoma has been used by several groups as an alternative to collagen or fibrin [23, 24]. Matrigel cultures, however, require growth factor supplements due to the limited capacity of the aorta to spontaneously sprout in this matrix. Aortic outgrowths in Matrigel are much denser than in collagen or fibrin and more difficult to quantitate due to the intricate branching morphology of the endothelial sprouts and the confounding effects of mesenchymal cells that tend to form networks in this matrix, mimicking the morphogenesis of the endothelial tubes. In addition, vessels formed in Matrigel tend to be thinner and without well defined lumens [25].

5.4   *Choice of Growth Medium*

Aortic rings are cultured in EBM, This medium, also known as MCDB131 was originally developed by Knedler and Ham for the growth of human microvascular endothelial cells. Aortic rings embedded in collagen or fibrin and cultured in EBM do not require serum or growth factor supplements [26].

5.5   Preparation of Collagen Gel Cultures

- Prepare the working collagen solution (see Sect. 12) by mixing 1 volume 10×
  Eagle's MEM (Gibco) with 1 volume 23.4 mg/ml NaHCO$_3$ and 8 volumes of
  1.3 mg/ml collagen dissolved in 0.1×MEM, pH4 (see below for collagen
  purification protocol). For optimal results, mix the 10× MEM and NaHCO3
  solutions before adding collagen. All solution must be kept on ice.
- Transfer aortic rings with microdissection forceps into 4-well Nunc dish. Place
  one ring on the bottom of each culture well. Remove medium carried over during
  transfer of aortic rings with micropipettor.
- Pipette 20–30 µl drop of working collagen solution onto each ring.
- Using a pipettor tip, reorient the rings so that the luminal axis is parallel to the
  bottom of the culture dish, and spread the collagen solution around the explant
  into a thin disc of approximately 8 mm in diameter (Fig. 2e).
- Incubate at 37°C for 10 min to induce collagen gelation.
- Add 500 µl of serum-free EBM to each culture
- Keep aortic ring cultures in a humidified CO$_2$ incubator at 37°C.
- Replace growth medium with fresh medium three times a week starting from day 3.

Critical points:

1. Given the small volume of collagen in which the aortic rings are cultured, it is
   important to remove excess medium carried over with the explants during
   transfer. Excessive dilution of collagen with culture medium may otherwise
   compromise sprouting.
2. The procedure should be carried our rapidly and without interruptions to avoid
   drying of the aortic explants during transfer.
3. The number of aortic rings handled during each step may vary from 1 to 4,
   depending on the experience and proficiency of the operator.
4. Dissection of the aorta and preparation of the rings and of the aortic ring cultures
   are best carried out in a tissue culture room with HEPA-filtered air.

5.6   Preparation of Fibrin Gel Cultures

- Prepare aortic rings as described above.
- Dispense 100 µl aliquots of 3 mg/ml fibrinogen solution in separate Eppendorf
  tubes.
- Lay rings on the bottom of 4-well-culture dish; remove with a micropipettor
  excess medium carried over with the rings.
- Add 2 µl of a 50 NIH units/ml bovine thrombin solution to one of the fibrinogen
  aliquots.
- Quickly mix and dispense with a micropipettor 20–30 µl of fibrinogen/thrombin
  solution onto each aortic ring.

- Spread fibrinogen drop with a pipette tip forming a thin wafer as described for the collagen gel method.
- Position the ring as described for the collagen gel cultures.
- Keep cultures for 10 min at 37°C in a humidified 5% CO_2 incubator.
- Add 500 μl of serum-free EBM supplemented with 300 μg/ml ε-amino-caproic acid (EACA) and return cultures to the incubator.

Critical points:

1. Fibrin forms within seconds after thrombin has been added to the fibrinogen solution. It is therefore essential to work quickly. We recommend preparing no more than four cultures at the time.
2. Aortic rings digest the fibrin gel overnight unless fibrinolysis is blocked with a plasmin inhibitor. Lysis of the gel around the explants causes premature loss of matrix support for angiogenic sprouting. Fibrinolytic activity is highest during the first days and can be blocked by adding EACA to the cultures. The dose of EACA can be tapered over time. EACA has no anti-angiogenic effects.

5.7 Mouse Aortic Ring Cultures

Aortic rings from transgenic or knockout mice can be used to test the role of specific genes in the angiogenic response. When they were first tested in this assay, mouse aortic rings failed to sprout spontaneously under serum-free culture conditions. Sprouting was obtained by adding to the culture medium homologous serum or angiogenic growth factors [17]. We later found that spontaneous sprouting was obtained by culturing the mouse rings in small volumes of growth medium using 96-well dishes [14]. More recently we have refined this method further reducing the medium to 1/10th the volume used for rat aortic ring cultures. The following protocol describes our miniaturized method for the mouse aortic ring cultures.

- Prepare mouse aortic rings as described for the rat aortic cultures.
- Place rings in wells of an Ibidi 18-well μ-slide (one ring per well).
- Remove carried over medium with micropipettor.
- Add 10 μl of collagen solution to each aortic ring (see above).
- Let collagen gel × 10 min at 37°C in a humidified CO_2 incubator.
- Add 55 μl of serum-free EBM per culture well.
- Incubate at 37°C, in a humidified CO_2 incubator.
- Monitor and analyze as described for rat aortic cultures.

Critical points:

1. The mouse aortic ring assay is more variable than the rat assay. We recommend using at least twice as many aortic rings per experimental group.
2. Fewer rings are obtained from the mouse thoracic aorta compared to the rat aorta (10–15 compared to 20–25). More mice are therefore needed for each experiment.

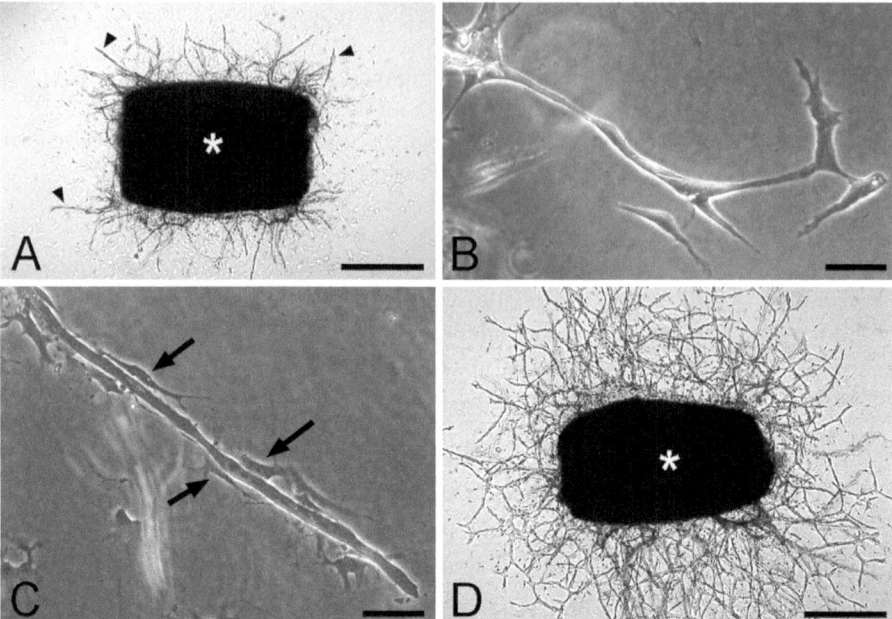

Fig. 3 Collagen gel cultures of aortic rings. (a) Control culture of rat aortic ring photographed at day 9: *arrowheads* show microvessels sprouting from the cut edges of the aorta (*asterisk*). (b) Immature endothelial sprout in day 3 culture: note the absence of pericytes. (c) Microvessel coated with pericytes (*arrows*) in day 5 culture. (d) Aortic ring (*asterisk*) culture treated with10 ng/ml VEGF: note the marked stimulation of angiogenesis compared to control (a). Magnification bars: (a) and (d): 1,000 μm; (b) and (c): 50 μm

5.8   Expected Results

Most of the vessel sprouts originate from the cut edges of the aortic rings (Fig. 3a). As they grow, branch and anastomose the newly formed vessels generate loops and networks. As the outgrowths mature, sprouts initially made of naked endothelial cords (Fig. 3b) gradually transform into patent microvessels enveloped by pericytes (Fig. 3c). The pattern and time course of angiogenic sprouting varies depending on the matrix in which the aortic rings are cultured. Angiogenesis in collagen starts at day 2–3, lasts 4–5 days and is followed by vascular regression with near complete angiolysis by the end of the second week. Vascular regression is associated with collagen lysis and formation of a periaortic halo of matrix rarefaction. Angiogenesis in fibrin is delayed (peak of growth: day 10–11) but more robust and longer lasting compared to collagen. The angiogenic response can be greatly enhanced by treating the aortic cultures with exogenous angiogenic factors (Fig. 3d).

5.9   Measurement of Angiogenesis

Angiogenesis can be measured by counting vessels [14] or by image analysis. Vessel counting is the most practical method in collagen and fibrin gel cultures, but we do not recommend it for Matrigel due to the complexity of the outgrowths in this matrix. Angiogenesis in Matrigel is best visualized by digitizing morphometry after the endothelium has been stained with specific markers. A number of image analysis methods have been developed to measure the endothelial outgrowths in aortic cultures [27–29]. Specific measurement of the angiogenic response has been obtained by excluding isolated (nonendothelial) cells with digital filters or by immunostaining the cultures with endothelial cell markers. For a detailed description of these methods the reader is referred to previous reports [27–29].

We describe here the vessel counting approach routinely used in our laboratory. The main advantage of this method is that it can be applied to living cultures and can be used to quantitatively monitor the angiogenic response over time. The curves of microvessel growth obtained with this method have allowed us to analyze the speed and extent of the angiogenic response as well as the vascular regression that follows angiogenesis [30, 31].

• Identify microvessels based on their greater thickness and cohesive pattern of growth compared to isolated cells.
• Count microvessels every 2 days starting from day 3.
• Set a visual landmark and count all the visible vessels in a clockwise or counterclockwise direction (Fig. 4).
• Use the following scoring criteria:

   – Count each branch as a separate vessel; a Y-shaped branching vessel is therefore counted as three vessels.
   – Each loop is counted as two vessels; loops most commonly originate from converging microvessel pairs.
   – When anastomosing vessels form polygonal patterns, count each side of the polygon as a separate vessel.

• Use statistical methods (t-test, analysis of variance) to analyze data and determine levels of significance between control and treated cultures.
• At the peak of the angiogenic response (days 6–8) unstimulated collagen gel cultures of rat aorta typically have 60–100 microvessels. Mouse aortic rings produce fewer vessels (20–30 per culture). Neovessels formed in fibrin tend to be more numerous, longer and more stable compared to vessels formed in collagen.

Critical points:

1. Vessels are best visualized at 4× to 10× magnification with an inverted microscope equipped with bright-field optics and 10× eyepieces.

Fig. 4 Quantitation of angiogenesis in aortic ring cultures. (a) High contrast image of unstimulated control culture illustrates vessel counting method. (b) Curve of angiogenesis and vascular regression obtained by counting vessels in collagen gel cultures of rat aorta over time

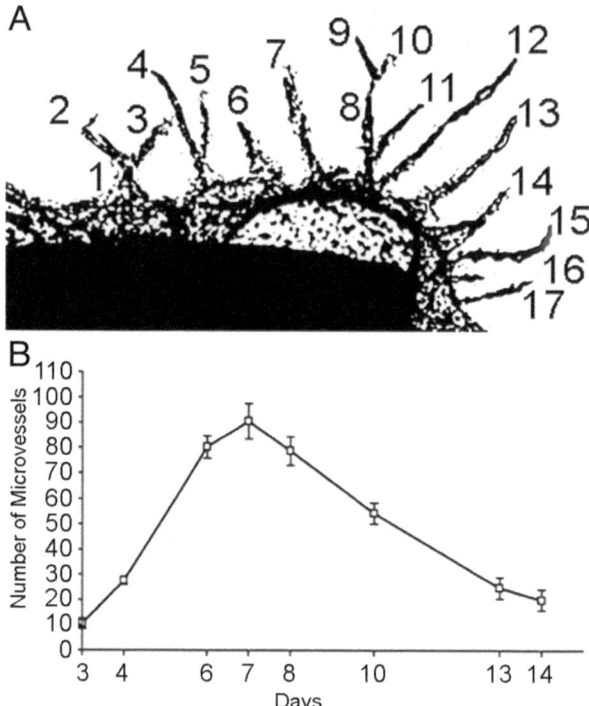

2. The condenser diaphragm should be properly adjusted to obtain optimal contrast and depth of field. If sufficient contrast cannot be achieved with bright filed optics, phase contrast is an acceptable alternative.
3. It is necessary to frequently readjust the focus in order to accurately visualize vessels growing in three dimensions.
4. Vessels should be counted over time by the same observer.
5. Proper training and experience are critical to identify genuine microvessels and to exclude nonendothelial outgrowths from the counts.
6. A control group of untreated aortic rings should be included in each experiment.
7. Each experimental group (treated or untreated) should include four aortic rings.
8. Experiments should be repeated at least three times.
9. Beware of adventitial collagen fibers or contaminating extraneous fibers which may mimic vessel sprouts.
10. The visual count method becomes challenging even to the most dedicated observer when cultures produce 250–300 or more vessels. This is an expected occurrence in cultures treated with maximal stimulatory doses of VEGF (Fig. 3d) or other potent angiogenic factors. Since outgrowths are typically symmetrical, angiogenesis in these cases can be quantitated by counting microvessels in half of the outgrowth and then doubling the score. Alternatively these cultures can be measured by image analysis.

6 Isolation of RNA

Gene expression in aortic cultures can be evaluated by RT-PCR, quantitative Real Time RT-PCR and microarray analysis. The following protocol describes how to reproducibly isolate high quality RNA for these studies. The expected yield of total RNA is 200–500 ng from each aortic culture [32].

- Transfer gels containing aortic rings into an Eppendorf tube containing 300 µl of buffer RLT + 2-Mercaptoethanol from the Qiagen RNAEasy kit. Although fewer cultures may be used, 4–8 aortic ring cultures provide optimal yield of total RNA to generate probes for microarray analysis or to serve as a template for cDNA synthesis and qRT-PCR studies.
- When all gels have been combined, snap freeze the tube by immersion in liquid N_2 for 10 s. Samples may be stored at $-80°C$ if desired.
- Invert the tube and gently tap its bottom with a spatula to knock the frozen gel solution into a biopulverizer containing liquid N_2 (BioSpec Products) for manual pulverization.
- Crush the frozen aortic cultures with a pestle and transfer the resulting powder into an Eppendorf tube.
- Allow the solution to thaw at room temperature and calculate the resulting volume of liquid.
- Follow the protocol given with the RNAEasy Micro Kit (Qiagen) for the "Isolation of total RNA from fibrous tissues".

Critical points:

1. Always use RNAse-free plasticware and reagents.
2. We recommend modifying the Qiagen final elution step by increasing the water incubation time on the column to 5 min prior to spinning out the eluate. This is critical for maximizing the RNA yield. Using this approach we have been successful in isolating sufficient amount of RNA for PCR and real time RT-PCR studies from single aortic cultures. In general we elute RNA in a volume of 10–20 µl of water.
3. Minimize freeze-thaw cycles of samples stored at -80 C.

7 Immunocytochemistry

Cell markers and other proteins of interest can be detected in whole mount preparations of aortic ring cultures using immunofluorescence or immuno-peroxidase (Fig. 5a–d) [14, 33]. Double and triple immunofluorescence protocols can be used to simultaneously visualize different cell types in the same outgrowth.

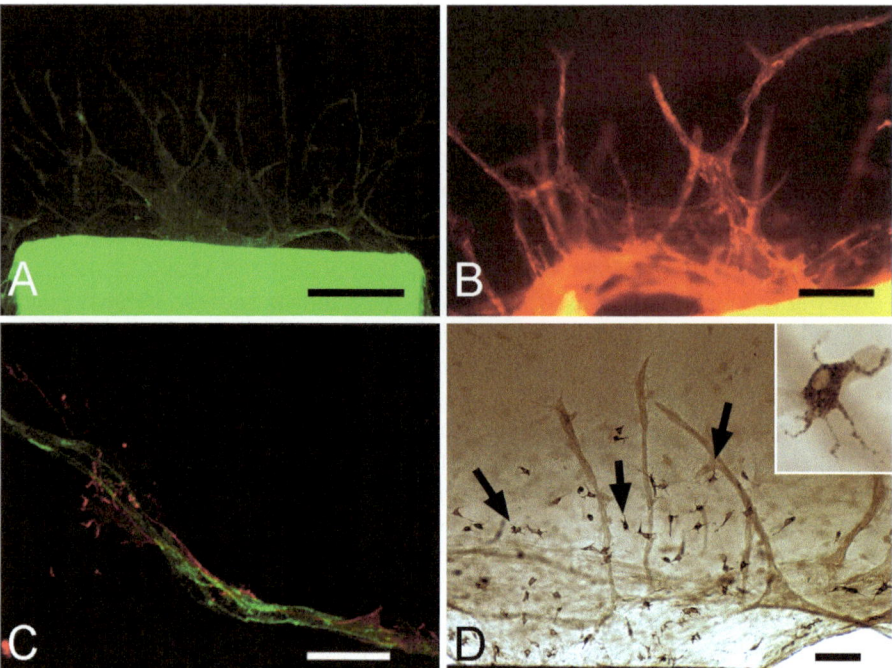

Fig. 5 Immunocytochemical stains of aortic ring cultures. (**a**) Immunofluorescent staining of endothelial outgrowths for CD31. (**b**) Branching microvessels highlighted by immunofluorescence with the endothelial cell marker *Griffonia Simplicifolia* Isolectin-B4 (I-B4). (**c**) Newly formed microvessel double stained with the endothelial marker I-B4 (*green*) and the pericyte marker anti-α smooth muscle actin antibody (*red*). (**d**) Vascular outgrowth in aortic culture stained by immunoperoxidase for the macrophage marker CD68: note the CD68+ macrophages (*arrows*) at the root of the outgrowth; Higher power image of a single CD68+ macrophage is shown in the inset. Magnification bars: (**a**) 500 μm; (**b**) and (**d**) 100 μm, (**c**) 50 μm

7.1   General Fixation Protocol

- Aspirate growth medium with transfer pipette and wash in PBS.
- Fix cultures with 10% neutral buffered formalin for 10 min.
- Remove formalin and wash three times with distilled water.
- Fixed cultures can be immunostained or stored at 4°C for future use, if needed.

7.2   Immunoperoxidase

- Remove distilled water and quench endogenous peroxidase with 0.3% hydrogen peroxide for 30 min.
- Block cultures with serum (5% in PBS) of the same species as the secondary antibody at 1 h at room temperature or at 4°C overnight.

- Rinse in PBS and react for 1 h with the primary antibody.
- Wash in PBS and incubate for 1 h with appropriate biotin-conjugated secondary antibody.
- Wash in PBS, incubate with Vectastain ABC reagent and develop with diamonobenzidine reagent (DAB), according to the manufacturer's recommendations.
- After rinsing with distilled water, gently dislodge and lift gel from bottom of the culture well with a bent spatula
- Transfer gel to a glass slide and mount with an aqueous mounting medium.
- Examine with light microscope.

7.3 Immunofluorescence

- Perform washing, blocking and antibody incubation reactions as described in the previous section.
- For double or triple staining, use cocktail of primary antibodies from different species.
- For the secondary antibody step, react cultures with cocktail of antibodies conjugated with fluorochromes of different wavelengths.
- When using directly conjugated markers such as the Griffonia Simplicifolia isolectin-B4, which recognizes rodent endothelial cells, add these reagents to the secondary antibody solution.
- Lift and mount cultures on glass slides as described above.
- Examine with fluorescent or confocal microscope.

Critical points for immunocytochemical stains:

1. Incubation in formalin should not exceed the 10 min required for optimal fixation to avoid excessive cross-linking of proteins and masking of target epitopes.
2. Special care should be taken when transferring gels from culture dishes to slides to avoid mechanical artifacts.
3. Overnight incubation with primary antibody may be needed for selected proteins of interest.
4. Cultures stained by immunofluorescence should be shielded from light to avoid quenching of fluorochromes.
5. Gel thinness is critical for optimal antibody penetration. Collagen and fibrinogen solutions should not exceed 15–20 μl in volume and should be spread as thin wafers (see preparation of aortic ring cultures).

8 ELISA of Secreted Proteins

Proteins released by the aorta its outgrowth in the growth medium can be measured by ELISA. This method is particularly recommended for secreted proteins produced at low levels such as cytokines and growth factors [14, 34].Since ELISA is performed on conditioned medium, protein levels can be measured at different time points without disturbing the aortic cultures. ELISA kits for rat and mouse proteins expressed in aortic cultures are commercially available.

9   Protein Extraction for Western Analysis

Nonsecreted proteins produced in aortic cultures can be studied by performing Western analysis on protein extracts. We have successfully used this method to evaluate the phosphorylation state of key signaling molecules during the early stages of the angiogenic response [35].

- Transfer aortic ring cultures from 4-well culture dish into 6-well culture plate using microdissection forceps.
- Wash two times in ice-cold PBS and then transfer aortic cultures to Eppendorf tubes containing 100 μl of RIPA buffer (Pierce). Pool 4–8 cultures as described for the RNA isolation procedure (see above).
- Snap-freeze the tubes in liquid N_2 for 10 s. Frozen protein extracts may be stored at −80°C if desired.
- Pulverize the frozen sample as described for the RNA isolation.
- Transfer the resulting powder into an Eppendorf tube and let it thaw on ice.
- Add ice-cold RIPA buffer into the tube to obtain the desired protein concentration (one aortic ring generates ~ 0.5–1 mg of total protein extract).
- Vortex sample for 1 min.
- Let solution stand on ice for 30–45 min and centrifuge at 10,000 g for 30 min at 4°C.
- Transfer supernatant into an Eppendorf tube.
- Perform Western analysis following standard protocols.

Critical point:

1. The RIPA buffer does not contain protease or phosphatase inhibitors. We recommend adding appropriate inhibitors to the buffer solution to avoid protein degradation.

10   Gene Transduction

Gene expression in aortic cultures can be modulated by transducing the aortic explants with viral vectors. Vectors successfully used in this model include adenovirus, retrovirus, vaccinia virus [36–38], and adeno-associated virus (Nicosia, unpublished observations). Since the collagen gel may act as a barrier to the viral vector, gene transduction should be carried out before embedding the aortic rings in collagen. Following gene transduction aortic rings are embedded in collagen or other matrices and cultured as described above. Appropriate controls for this experiment include untransduced rings, rings transduced with vector expressing a marker gene (for example green fluorescent protein), and rings transduced with empty viral vector.

11 Pitfalls

A number of variables and technical problems may affect the sprouting of the aortic rings:

- Unsatisfactory reagents. The collagen solution should not be less than 1.3 mg/ml dry weight or older than 6 months; all reagents should be tissue culture grade.
- Temperature: the angiogenic response can be impaired if the incubator's temperature drops below 35°C.
- pH: exposure to alkaline pH during dissection of the aortic explants should be minimized; neutral pH should be maintained throughout the culture period.
- Inadequate preparation and/or mixing of the collagen solution. When preparing the working collagen solution, always begin by mixing the MEM and $NaHCO_3$ prior to adding the stock collagen solution. Failure to thoroughly mix the reagents may result in uneven polymerization and defective gels.
- Inadequate preparation and/or mixing of the fibrinogen/thrombin solution. Prepolymerization of fibrin or fibrinolysis may negatively affect angiogenesis. See section on preparation of fibrin gel cultures (5.6) for tips on how to avoid these pitfalls.
- Mechanical damage to the aorta: it is important not to stretch or crush the aorta or the aortic rings at any stage of the procedure
- Drying of the endothelial lining: the aortic explants must be kept wet with growth medium throughout the dissection procedure.
- Disruption of the gel. Work quickly and position rings within 30 s of placing the drop of collagen or fibrinogen solution into the culture well. Once the gels have set, care should be taken to avoid any physical damage during transfer from bench to incubator or microscope stage and when replacing medium during the course of an experiment.
- Excessive medium carried over with the aortic ring may dilute the collagen or fibrinogen solution, causing defective gelation. The excess medium can be removed with a micropipettor before adding the collagen or fibrinogen solution.
- Excessive volume of culture medium. Optimal concentration of growth factors released by the aortic ring is required for angiogenic sprouting in serum-free EBM. Thus, miniaturization of the assay is needed to study the angiogenic response of mouse aortic rings in the absence of exogenous growth factors (see above).
- Genetic background and age of the animals can significantly affect the angiogenic response of the aortic explants. Control rings from mice of the same age and genetic makeup are therefore essential in experiments with genetically modified mice.

12   Preparation of Collagen from Rat Tails

Collagen can be purchased from a number of suppliers. Commercially available collagen may however give variable results depending on the source, the purification method and the recommended procedure to induce gelation. To ensure consistency of results we prepare collagen in-house using the following protocol.

- Excise tails and store at $-20°C$.
- Thaw 4–5 tails, disinfect with 80% ethanol and left ethanol evaporate. Keep tails in a 100 mm tissue culture dish
- Cut tail skin with a scalpel at 1–2 cm intervals (4–5 cuts).
- Hold tail with a hemostat below the first cut (distal) and pull skin and subcutaneous tissue from the underlying tendons using a second hemostat placed above the cut (proximal). This procedure removes skin and underlying soft tissues leaving tendons exposed. Cut tendons with sterile scissors and place them in a dry culture dish.
- Repeat this procedure to the last cut near the tip of the tail.
- Transfer tendons with sterile forceps to 100 mm culture dish containing 0.9% NaCl. Gently tease apart collagen fibers with fine microdissection forceps.
- Examine collagen fibers under a dissecting microscope. Collagen fibers are characteristically white, velvety and ribbon-like; discard any hemorrhagic fibers. Remove blood vessels, which are easily recognizable by their central lumen and pinkish-red color.
- Wash fibers in eight separate baths of 0.9% NaCl using 100 mm dishes.
- Sterilize fibers in 80% ethanol × 10 min in a100 mm dish; exposure to ethanol will stiffen the fibers.
- Rehydrate fibers through six baths of sterile distilled water in 100 mm dishes.
- Transfer fibers to a 250 ml Erlenmeyer flask containing 100–150 ml of 0.5 M glacial acetic acid and stir for 48 h at 4°C. Exposure to acetic acid will dissolve most of the collagen, causing the solution to become viscous.
- Pour the collagen solution through 2–3 layer-thick gauze into 50 ml centrifuge tubes. This step will remove most of the undisolved fibers.
- Centrifuge at 12,000 $\times g$ for 1 h at 4°C. Store the supernatant at 4°C.
- To calculate the concentration of collagen pipette 5 ml of acetic acid solution into an aluminum weight boat which was tarred before the procedure. Evaporate the acetic acid on a hot plate and re-weigh the aluminum boat. The difference in weight equals the dry weight of collagen in 5 ml of acetic acid solution. Divide by 5 to obtain the collagen concentration in mg/ml. Dilute with 0.5 M acetic acid to obtain a final collagen solution of 1.3 mg/ml. More concentrated solutions can be prepared if desired. Stored at 4°C.
- Prepare 1-in. diameter dialysis tubings (6,000–8,000 MWCO, 30–35 cm) for collagen dialysis. Tubings are boiled in two successive 1-l baths of sterile water (10 min each). This procedure sterilizes the tubings and removes traces of chemicals that could contaminate the collagen solution. After squeezing out the water and sealing one end of the dialysis tubing with a double knot, fill the

tubing with 25–30 ml of the 1.3 mg/ml acidic acid collagen solution. Tie off the top end of the tubing leaving an air bubble below the knot.
- Dialyze overnight at 4°C against 0.1× Minimal Essential Medium, pH 4.0 (two 25–30 ml tubings/3.5 l of dialysis solution). Repeat for an additional 24 h against fresh dialysis solution.
- Disinfect the top of the dialysis tubing at the air bubble with 80% ethanol, nick the tubing with scissors, and aspirate the collagen with a long needle. Transfer the dialyzed collagen solution into a chilled sterile bottle and keep at 4°C until use. Use collagen solution within 6 months of dialysis.

13 Materials and Reagents

13.1 Isolation of the Aorta and Preparation of Aortic Ring Cultures

- Fischer 344 Rat, 1–2 month-old, male; transgenic or knock out mice and normal (wild type) controls of the same age and genetic background.
- CO_2 Tank
- Anesthesia box
- Dissecting Styrofoam board wrapped in absorbent bench paper
- Dissecting pins
- 80% Ethanol
- Hair Clippers with size 40 A5 blade
- Instrument Sterilization Tray and Lid
- Silk sutures, size 2.0
- 1 Large curved Mayo scissors 17 cm (Fine Science Tools, Foster City, CA)
- 2x Halsted-Mosquito hemostat (Fine Science Tools)
- 1 Scalpel blade handle size 4 (Fine Science Tools)
- Surgical blades size 22
- 2–4 fine curved forceps "Dumont #7" (Roboz, Gaithersburg, MD)
- 2x small forceps (Roboz)
- 1 small curved scissors (Roboz)
- 2–4 Noyes Scissors 14 mm straight blade (Roboz)
- X-dish 4 compartment 100 mm dishes (Felsen dish, Fisher Scientific, Pittsburgh, PA)
- 4-well IVF dishes (NUNC, Rochester, NY)
- Ibidi 18-well μ-chamber slides, http://www.ibidi.de (Research Products International, Mt Prospect, IL)
- Endothelial Basal Medium (Lonza, Walkersville, MD)
- Gentamycin 50 ug/mL (Invitrogen, Carlesbad, CA)
- Rat collagen (made in-house)
- 10x MEM in water (Invitrogen)
- 23.4 mg/mL sodium bicarbonate in water (Fisher Scientific)

- ε-amino-caproic acid (Sigma)
- Humidified 5% CO_2 incubator
- Dissecting microscope

13.2 Immunocytochemistry

- 10% neutral buffered formalin (Fisher Scientific)
- Hydrogen Peroxide Solution (Sigma, St. Louis, MO)
- DAB peroxidase Substrate Kit (Vector Laboratories, Burlingame, CA)
- Vectastain ABC Standard Horse Radish Peroxidase Kit (Vector Laboratories)
- Primary antibodies, affinity purified, against rat or mouse proteins of interest, immunoglobulin isotype controls, and lectin reagents. Examples shown in Fig. 5: Mouse monoclonal antibodies against CD31 (Abcam), CD68 (AbD Serotech) and α smooth muscle actin (Sigma); Alexafluor-conjugated Griffonia Simplicifolia Isolectin-B4 (Invitrogen).
- Biotinylated or Alexafluor-conjugated secondary antibodies (Invitrogen).

13.3 RNA and Protein Extraction

- RNAEasy Micro Kit (Qiagen, Valencia, CA)
- BioPulverizer (BioSpec Products, Bartlesville, OK)
- RIPA buffer (Pierce, Rockford, IL)
- 2-Mercaptoethanol
- Liquid Nitrogen
- Microcentrifuge

13.4 Collagen Preparation

- 100 mm dishes (Fisher Scientific)
- 2x Fine curved forceps "Dumont #7" (Roboz)
- 2x Halsted-Mosquito hemostat (Fine Science Tools)
- 1 small curved scissors (Roboz)
- 1 Scalpel blade handle size 4 (Fine Science Tools)
- Surgical blades size 22
- Dialysis Tubing (Fisher Scientific)
- 3 l beaker (Fisher Scientific)
- MEM (Invitrogen)
- Glacial Acetic Acid (Fisher Scientific)
- 0.9% Sodium Chloride solution (Baxter, Deerfield, IL)
- Refrigerated centrifuge

References

1. Carmeliet P (2005) Angiogenesis in life, disease and medicine. Nature 438:932–936
2. Folkman J (2007) Angiogenesis: an organizing principle for drug discovery? Nat Rev Drug Discov 6:273–286
3. Greenblatt M, Shubi P (1968) Tumor angiogenesis: transfilter diffusion studies in the hamster by the transparent chamber technique. J Natl Cancer Inst 41:111–124
4. Folkman J (1971) Tumor angiogenesis: therapeutic implications. N Engl J Med 285:1182–1186
5. Shojaei F, Ferrara N (2007) Antiangiogenesis to treat cancer and intraocular neovascular disorders. Lab Invest 87:227–230
6. Ferrara N, Hillan KJ, Novotny W (2005) Bevacizumab (Avastin), a humanized anti-VEGF monoclonal antibody for cancer therapy. Biochem Biophys Res Commun 333:328–335
7. Carmeliet P, Jain RK (2011) Molecular mechanisms and clinical applications of angiogenesis. Nature 473:298–307
8. Auerbach R, Lewis R, Shinners B et al (2003) Angiogenesis assays: a critical overview. Clin Chem 49:32–40
9. Norrby K (2006) In vivo models of angiogenesis. J Cell Mol Med 10:588–612
10. Ribatti D, Vacca A (1999) Models for studying angiogenesis in vivo. Int J Biol Markers 14:207–213
11. Nicosia RF, Tchao R, Leighton J (1982) Histotypic angiogenesis in vitro: light microscopic, ultrastructural, and radioautographic studies. In Vitro 18:538–549
12. Nicosia RF (2009) The aortic ring model of angiogenesis: a quarter century of search and discovery. J Cell Mol Med 13:4113–4136
13. Nicosia RF, Ottinetti A (1990) Growth of microvessels in serum-free matrix culture of rat aorta. A quantitative assay of angiogenesis in vitro. Lab Invest 63:115–122
14. Aplin AC, Fogel E, Zorzi P et al (2008) The aortic ring model of angiogenesis. Methods Enzymol 443:119–136
15. Facchetti F, Monzani E, Cavallini G et al (2007) Effect of a caloric restriction regimen on the angiogenic capacity of aorta and on the expression of endothelin-1 during ageing. Exp Gerontol 42:662–667
16. Reed MJ, Karres N, Eyman D et al (2007) Culture of murine aortic explants in 3-dimensional extracellular matrix: a novel, miniaturized assay of angiogenesis in vitro. Microvasc Res 73:248–252
17. Zhu WH, Iurlaro M, MacIntyre A et al (2003) The mouse aorta model: influence of genetic background and aging on bFGF- and VEGF-induced angiogenic sprouting. Angiogenesis 6:193–199
18. Gelati M, Aplin AC, Fogel E et al (2008) The angiogenic response of the aorta to injury and inflammatory cytokines requires macrophages. J Immunol 181:5711–5719
19. Villaschi S, Nicosia RF (1994) Paracrine interactions between fibroblasts and endothelial cells in a serum-free coculture model. Modulation of angiogenesis and collagen gel contraction. Lab Invest 71:291–299
20. Garcia-Cardena G, Comander J, Anderson KR et al (2001) Biomechanical activation of vascular endothelium as a determinant of its functional phenotype. Proc Natl Acad Sci USA 98:4478–4485
21. Nicosia RF, Zhu WH, Fogel E et al (2005) A new ex vivo model to study venous angiogenesis and arterio-venous anastomosis formation. J Vasc Res 42:111–119
22. Elsdale T, Bard J (1972) Collagen substrata for studies on cell behavior. J Cell Biol 54:626–637
23. Malinda KM, Nomizu M, Chung M et al (1999) Identification of laminin alpha1 and beta1 chain peptides active for endothelial cell adhesion, tube formation, and aortic sprouting. FASEB J 13:53–62
24. Boscolo E, Folin M, Nico B et al (2007) Beta amyloid angiogenic activity in vitro and in vivo. Int J Mol Med 19:581–587

25. Nicosia RF, Ottinetti A (1990) Modulation of microvascular growth and morphogenesis by reconstituted basement membrane gel in three-dimensional cultures of rat aorta: a comparative study of angiogenesis in matrigel, collagen, fibrin, and plasma clot. In Vitro Cell Dev Biol 26:119–128

26. Knedler A, Ham RG (1987) Optimized medium for clonal growth of human microvascular endothelial cells with minimal serum. In Vitro Cell Dev Biol 23:481–491

27. Nissanov J, Tuman RW, Gruver LM et al (1995) Automatic vessel segmentation and quantification of the rat aortic ring assay of angiogenesis. Lab Invest 73:734–739

28. Blacher S, Devy L, Burbridge MF et al (2001) Improved quantification of angiogenesis in the rat aortic ring assay. Angiogenesis 4:133–142

29. Blatt RJ, Clark AN, Courtney J et al (2004) Automated quantitative analysis of angiogenesis in the rat aorta model using Image-Pro Plus 4.1. Comput Methods Programs Biomed 75:75–79

30. Zhu WH, Guo X, Villaschi S et al (2000) Regulation of vascular growth and regression by matrix metalloproteinases in the rat aorta model of angiogenesis. Lab Invest 80:545–555

31. Aplin AC, Zhu WH, Fogel E et al (2009) Vascular regression and survival are differentially regulated by MT1-MMP and TIMPs in the aortic ring model of angiogenesis. Am J Physiol Cell Physiol 297:C471–C480

32. Aplin AC, Gelati M, Fogel E et al (2006) Angiopoietin-1 and vascular endothelial growth factor induce expression of inflammatory cytokines before angiogenesis. Physiol Genomics 27:20–28

33. Zhu WH, Nicosia RF (2002) The thin prep rat aortic ring assay: a modified method for the characterization of angiogenesis in whole mounts. Angiogenesis 5:81–86

34. Zorzi P, Aplin AC, Smith KD et al (2010) Technical advance: the rat aorta contains resident mononuclear phagocytes with proliferative capacity and proangiogenic properties. J Leukoc Biol 88:1051–1059

35. Zhu WH, MacIntyre A, Nicosia RF (2002) Regulation of angiogenesis by vascular endothelial growth factor and angiopoietin-1 in the rat aorta model: distinct temporal patterns of intracellular signaling correlate with induction of angiogenic sprouting. Am J Pathol 161:823–830

36. Hajitou A, Grignet C, Devy L et al (2002) The antitumoral effect of endostatin and angiostatin is associated with a down-regulation of vascular endothelial growth factor expression in tumor cells. FASEB J 16:1802–1804

37. Alian A, Eldor A, Falk H et al (2002) Viral mediated gene transfer to sprouting blood vessels during angiogenesis. J Virol Methods 105:1–11

38. Zhu WH, Han J, Nicosia RF (2003) Requisite role of p38 MAPK in mural cell recruitment during angiogenesis in the rat aorta model. J Vasc Res 40:140–148

The Chick Embryo Aortic Arch Assay

Robert Auerbach and Veerappan Muthukkaruppan

Abstract The chick aortic arch assay represents a major modification of the rat aortic ring assay. Originally developed for the specific purpose of testing thalidomide (which had previously been shown to have limited effects in rodents but strong effects in chick embryos), the assay avoids the use of laboratory animals, is rapid, with an assay time of 1–3 days, and can be carried out in serum-free medium. Aortic arches are dissected from day 12–14 chick embryos and cut into rings similar to those of the rat aorta. When the rings are placed on Matrigel, substantial outgrowth of cells occurs within 48 h, with the formation of vessel-like structures readily apparent. If the aortic arch is everted before explanting, the time for observing endothelial cell outgrowth can be reduced to 24 h. Both growth-stimulating factors, such as VEGF-A or FGF-2, as well as inhibitory substances such as endostatin, can be added to the medium, where their effect becomes easily measured.

1 Introduction

The chick embryo aortic arch assay is based on the rat aortic ring assay originally described by Roberto Nicosia and his colleagues [1]. The abstract by Muthukkaruppan et al. in 2000 [2] has served as the principal communication of this method. Subsequently, additional details and photographs were included in several review publications [3–6]. An increasing number of research studies have now used this method, and the reader is referred to these for further documentation and photographs [7–11].

R. Auerbach
Department of Zoology and Institute on Aging, University of Wisconsin, 1117 West Johnson Street, Madison, WI 53706, USA
e-mail: rauerbac@wisc.edu

V. Muthukkaruppan
Aravind Medical Research Institute, 1, Anna Nagar, Madurai 625020, Tamil Nadu, India
e-mail: muthu@aravind.org

E. Zudaire and F. Cuttitta (eds.), *The Textbook of Angiogenesis and Lymphangiogenesis: Methods and Applications*, DOI 10.1007/978-94-007-4581-0_8, © Springer Science+Business Media Dordrecht 2012

2 Experimental Protocol

2.1 Materials

2.1.1 Chick Embryos

We incubated our eggs for 12–14 days at 37°C. [NOTE: The Hamburger/Hamilton standard staging for chick embryo development was based on an incubation temperature of 38°C [12, 13], so that our use of day 12–14 embryos would be represented by their staging of day 11–13]. We chose these days based on optimum size and handling ease, but have no reason to believe that slightly earlier or later stages would not work equally as well.

2.1.2 Medium

We used standard endothelial cell media (MCDB-131, ECM-2). Based on our experience with other chick embryo organ cultures we suspect that other tissue culture media would work equally well. We and others have successfully used serum-free media (e.g. growth factor-depleted endothelial-SFM basal growth medium from In Vitrogen) when testing various growth factors.

2.1.3 Matrigel

Our own supply of Matrigel was obtained from Dr. Hynda Kleinman at a time when Matrigel was highly variable. Subsequently, standard Matrigel (e.g. from BD BioSciences) has been used by other investigators.

2.1.4 Culture Vessels

We used standard 24-well tissue culture plates, which seemed a useful size for handling, incubation, photography and fixation.

2.2 Dissection and Preparation of Cultures

2.2.1 Sterilization of Chick Eggs

The major source of infection of cultures comes from inadvertent contamination during the opening of the egg. In our laboratory we wipe the egg surface with Betadyne. Other laboratories may use alternate methods of sterilization of the egg shell, such as immersion in 70% alcohol, to prevent this common source of contamination.

Fig. 1 13-day chick embryo-
derived aortic arches

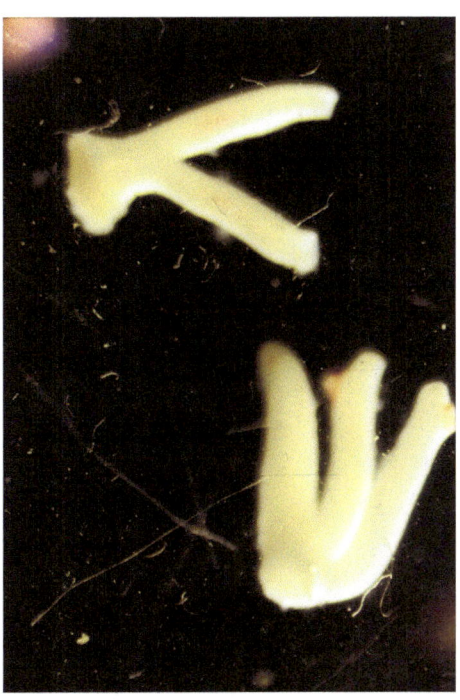

2.2.2   Isolation of Aortic Arches

Eggs are cracked by striking the shell against the sterile edge of an inverted P100 petri dish cover, then emptying the contents into the open Petri dish containing PBS. Using curved forceps and scissors, the embryo is next removed from its surroundings and transferred to a second PBS-containing P100 dish. The embryo is arranged ventral side up, and the head is removed rapidly, consistent with recommended protocols for animal experimentation. Tweezers are used to lift up the tissue above the breastbone, and the tissue above the thoracic cavity is trimmed to expose the heart and the aortic arches above it. By lifting up the heart and pulling it upwards, the aortic arches are clearly exposed and accessible. We separate the heart with its attached aortic arches from the rest of the embryo and transfer this heart/aortic arch complex to a fresh P60 dish containing PBS with penicillin/streptomycin. (This step is not essential but we prefer to err on the side of caution to prevent contamination of the cultures.) Under a dissecting microscope using $7\times$–$10\times$ magnification, (although this procedure can also be carried out without the microscope), excess tissue surrounding the arches should be removed. There should now be a pair and a threesome of arches (Fig. 1). The pair, which works best in the assay, is transferred to another P60 dish containing saline and antibiotics.

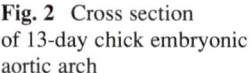

Fig. 2 Cross section
of 13-day chick embryonic
aortic arch

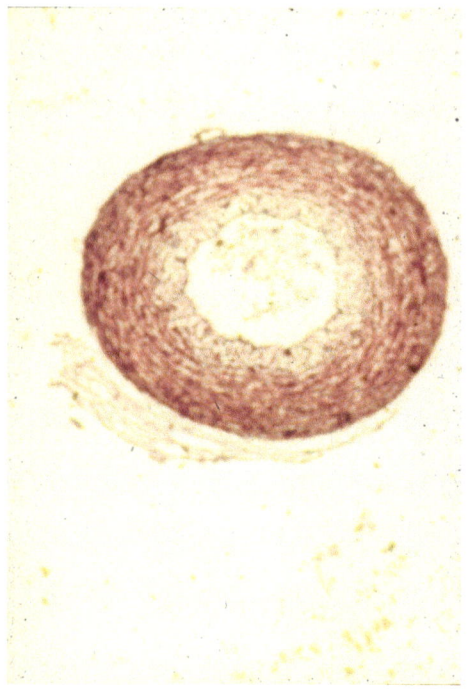

2.2.3 Preparation of Aortic Arch Fragments

The arches are trimmed cleanly and cut into 0.8 mm fragments. We use a millimeter grid on the microscope to assure uniformity. The number of pieces that can be obtained will vary depending on the age of the embryo. (We have standardized on 0.8 mm lengths but a slightly larger or smaller explant would probably work equally as well). A typical cross-section of an isolated aortic arch fragment is shown in Fig. 2. The next step is optional: Each piece can be everted so that the endothelium is on the outside instead of the inside (Figs. 3 and 4). Eversion permits more rapid evaluation of endothelial cell outgrowth which is not impeded or obscured by the proliferation and outgrowth of mesenchymal cells (cf. Figs. 4 and 5).

2.2.4 Preparation of Organ Cultures

24-well culture plates are prefilled with 1–2 µl of Matrigel. Each aortic arch ring is then placed in the center of each well. Immediately 10 µl of ice-cold Matrigel are added to surround the explant. The opening in the ring must be facing up; if on its side the ring should be repositioned before the Matrigel hardens. 500 µl of tissue culture medium should be added after the Matrigel has solidified. Cultures are then incubated at 37°C in a humidified atmosphere of 5% CO_2. Medium should be changed every 2–3 days if cultures are maintained for longer periods.

Fig. 3 Total explant area of everted aortic arch segment after 3 days in culture

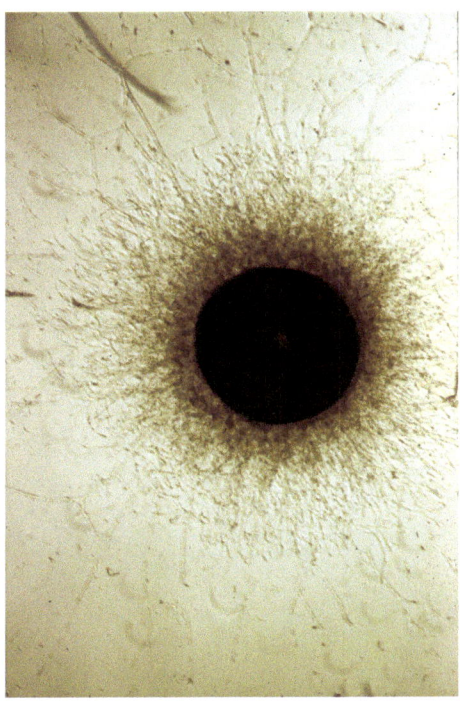

Fig. 4 Enlarged outgrowth area of everted aortic arch segment after 3 days in culture

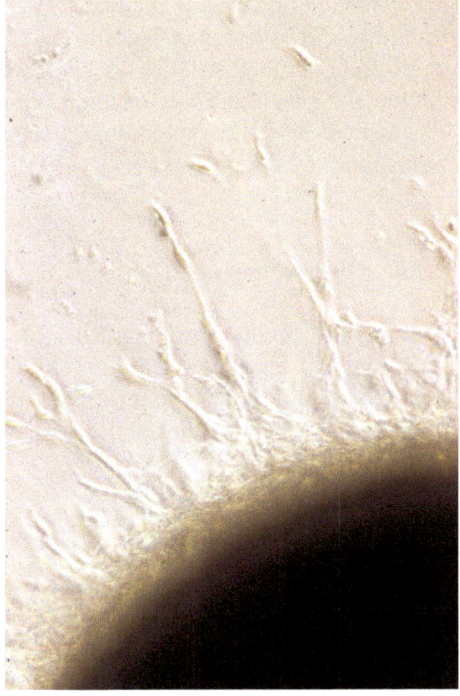

Fig. 5 Explant of non-
inverted aortic arch segment
after 3 days in culture

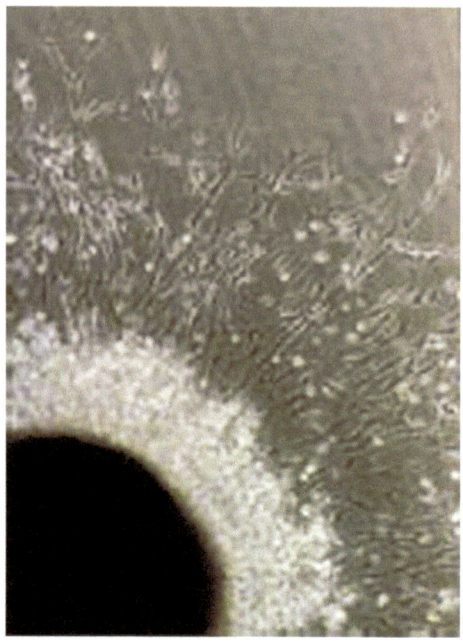

2.2.5   Experimental Variables

A key feature of the assay method, shared with other in vitro protocols, is the ability
to add test substances at specific times during the culture period. For example,
angiogenic stimulators such as bFGF or VEGF-A can be added, followed or in
combination with anti-angiogenic test substances. It is particularly useful that in
this assay it is feasible to analyze responses in growth factor-depleted and serum-
free culture media.

2.2.6   Observations

Cultures can be photographed (Figs. 3, 4, 5 and 6), followed through time-lapse
photography, stained with appropriate antibodies for endothelial cells or fibroblasts,
identified through cell-type selective lectins, fixed for histology or confocal microscopy, or harvested for biochemical analysis.

2.2.7   Other Protocols

One advantage of the Matrigel substrate is that combination cultures can be set up
involving, for example, a murine tumor fragment at some distance from the aortic
arch explant. Such a co-culture has permitted us to visualize the formation of

Fig. 6 Co-culture of aortic
arch and murine melanoma
fragments, 5 days in culture

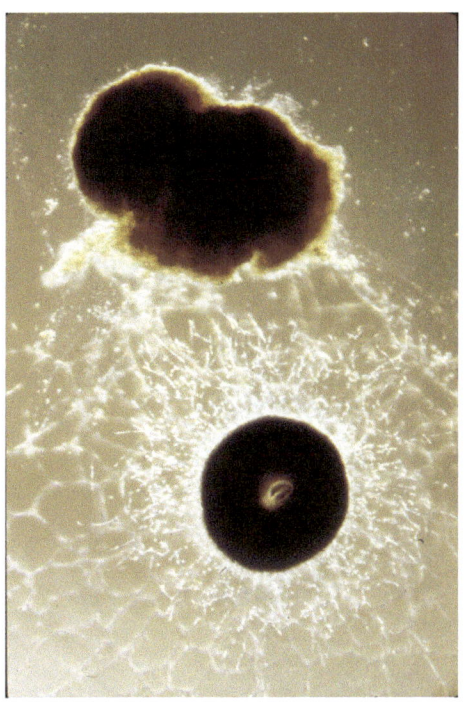

contact channels with outgrowths from the tumor making contact with vascular
sprouts originating from the chick embryo explant (Fig. 6). Sequential observations
and histology could help resolve some of the questions concerning the nature of
"pseudovessels" [14].

3   Discussion

Although historically, "tissue culture" and "organ culture" dominated in vitro
studies, the ascent of cell culture with its obvious advantages for manipulation,
observation and quantitation tended to obscure the important distinctions between
the various levels of in vitro analysis. In the field of angiogenesis, the ability to
cultivate endothelial cells revolutionized the assessment methods for evaluation of
angiogenic (and anti-angiogenic) substances. But it is the intricate relationships
among different cell and tissue types that are almost certainly critical when one is
seeking predictive value for in vivo efficacy. As a classic example, one may recall
the differences between organ culture and cell culture in studying sequential
differentiation during the immunological response to antigenic stimulation. While
isolated spleen fragments respond in vitro with production first of IgM and then
IgG, cell cultures generally fail to complete this transition [15, 16]. As we and
others have often emphasized, in vitro systems, although useful, cannot substitute

for in vivo ones, but complex organ cultures offer an intermediate approach, reflecting more accurately than cell culture what is likely to be encountered in vivo.

Paramount among the advantages of using the chick embryo aortic arch assay is the ability to carry out this angiogenesis assay in the absence of the complex limitations placed on animal studies: protocol approval, cost, need for specialized facilities, and a requirement for veterinary supervision. Since there seem little biological dissimilarities between avian and mammalian blood vessels, the argument for using mammalian organ culture models is relatively weak. On the other hand, it should be kept in mind that the embryonic vasculature involves actively proliferating cell populations, whereas adult vessels are basically quiescent or dormant. This may be an important disadvantage for some studies but highly advantageous for others, in particular for investigations into the response of a tumor-derived neovasculature to the addition of test substances.

Acknowledgments The assay method described in this manuscript was developed with the able participation of B Shinners, R Lewis, S-J Park, B Baechler and L Kubai. We thank Wanda Auerbach for editorial assistance.

References

1. Nicosia R, Ottinetti A (1990) Growth of microvessels in serum-free matrix culture of rat aorta. A quantitative assay of angiogenesis in vitro. Lab Invest 63:115–122
2. Muthukkaruppan V, Shinners B, Lewis R et al (2000) The chick embryo aortic arch assay: a new, rapid quantifiable in vitro method for testing the efficacy of angiogenic and anti-angiogenic factors in a three-dimensional, serum-free organ culture system. Proc Am Assoc Cancer Res 41:65
3. Auerbach R (2006) An overview of current angiogenesis assays: choice of assay, precautions in interpretation, future requirements. In: Staton C, Lewis C, Bicknell R (eds) Angiogenesis assays. Wiley, Chichester, pp 361–374
4. Auerbach R (2008) Models for angiogenesis. In: Figg WD, Folkman J (eds) Angiogenesis: an integrative approach from science to medicine. Springer, New York, pp 299–312
5. Auerbach R, Lewis R, Shinners Bl N et al (2003) Angiogenesis assays: a critical overview. Clin Chem 49:32–40
6. Akhtar N, Dickerson EB, Auerbach R (2002) The sponge/Matrigel angiogenesis assay. Angiogenesis 5:75–80
7. Isaacs JS, Jung YJ, Mimnaugh EG, Martinez A et al (2002) Hsp90 regulates a von Hippel Lindau-independent hypoxia-inducible factor-1 alpha-degradative pathway. J Biol Chem 277:29936–29944
8. Martinez A, Zudaire E, Portal-Nunez S et al (2004) Proadrenomedullin NH2-terminal 20 peptide is a potent angiogenic factor, and its inhibition results in reduction of tumor growth. Cancer Res 64:6489–6494
9. Wen W, Lu J, Zhang K, Chen S (2008) Grape seed extract inhibits angiogenesis via suppression of the vascular endothelial growth factor receptor signaling pathway. Cancer Prev Res (Phila) 1:554–561
10. Zhang K, Lu J, Mori T, Smith-Powell L et al (2011) Baicalin increases VEGF expression and angiogenesis by activating the ERR{alpha}/PGC-1{alpha} pathway. Cardiovasc Res 89:426–435

11. Oh S-H, Kim W-Y, Kim J-H et al (2006) Identification of insulin-like growth factor binding protein-3 as a farnesyl transferase inhibitor SCH55335-induced negative regulator of angiogenesis in head and neck squamous cell carcinoma. Cancer Res 12:653–661
12. Hamburger V, Hamilton H (1951) A series of normal stages in the development of the chick embryo. J Morphol 88:49–92
13. Wikipedia (2011) http://en.wikipedia.org/wiki/Hamburger-Hamilton_stages
14. Folberg R, Hendrix MJ, Maniotis A (2000) Vasculogenic mimicry and tumor angiogenesis. Am J Pathol 156:361–381
15. Globerson A, Auerbach R (1965) Primary immune reactions in organ cultures. Science 149:991–993
16. Mishell RI, Dutton RW (1966) Immunization of normal mouse spleen cell suspensions in vitro. Science 153:1004–1006

The P-Sp Culture System

Nobuyuki Takakura

Abstract Hematopoiesis and vascular formation are closely associated in many respects, e.g., common progenitor cells, hemangioblasts, can give rise to vascular endothelial cells and hematopoietic cells; endothelial cells function as stromal cells to support hematopoiesis, and conversely hematopoietic cells regulate angiogenesis as an accessory cell component. Therefore, for better understanding of blood vessel formation, it is valuable to recapitulate the process observed in vivo by generating both hematopoietic cells and endothelial cells concurrently in in vitro culture models. The para-aortic splanchnopleural mesoderm (P-Sp) region in the E9.5 mouse embryo contains hemangioblasts, otherwise known as hemogenic angioblasts. When the P-Sp region is dissected out and the explant co-cultured on M-CSF-deficient OP9 stromal cells, blood vessel formation mimicking vasculogenesis and angiogenesis develops, and hematopoietic stem cell proliferation and differentiation is observed in the vascular area. This P-Sp culture system can provide information on how blood vessel formation is induced in association with hematopoietic cells and stromal cells.

1 Outline of the P-Sp Culture System

Hematopoiesis is closely linked to blood vessel formation, because hematopoietic stem cells (HSCs) and endothelial cells (ECs) have common progenitors, known as hemangioblasts (hemogenic angioblasts), and the two systems interact with each other [1]. Definitive HSCs adhere strongly to ECs at several sites in the embryo, including the yolk sac [2], omphalomesenteric and vitelline artery, and dorsal aorta [1–6]. In addition, some stromal cell lines that are able to support hematopoiesis

N. Takakura (✉)
Department of Signal Transduction, Research Institute for Microbial Diseases, Osaka University, 3-1 Yamada-oka, Suita, Osaka 565-0871, Japan
e-mail: ntakaku@biken.osaka-u.ac.jp

E. Zudaire and F. Cuttitta (eds.), *The Textbook of Angiogenesis and Lymphangiogenesis: Methods and Applications*, DOI 10.1007/978-94-007-4581-0_9,
© Springer Science+Business Media Dordrecht 2012

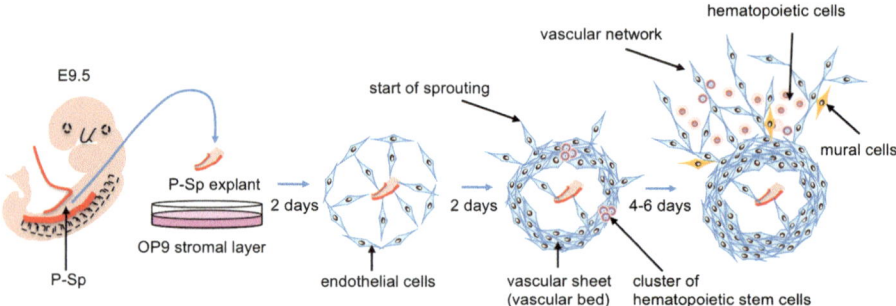

Fig. 1 Schematic of the P-Sp culture system. When the para-aortic splanchnopleural mesoderm region (P-Sp) is co-cultured with OP9 cells, outgrowth of immature endothelial cells (ECs) from the P-Sp explant, sheet-like formation of ECs mimicking vasculogenesis, and network formation of ECs mimicking angiogenesis is observed. In this system, hematopoiesis is concurrently induced. Vascular sheet formation does not associate with hematopoietic cells but network formation of ECs depends on molecular cues produced by hematopoietic cells. In addition, in the later stage of this culture, a mural cell population is observed especially next to the ECs which form the network

have been characterized as ECs [7]. Recently, it has been reported that primary ECs induced with an adenovirus gene, the early region 4-encoded open reading frame-1, which leads to constitutive activation of Akt in transfected cells, support self-renewal of HSCs in vitro [8]. These observations suggest an intimate association between hematopoiesis and vascular development. However, it is difficult to observe and visualize the crosstalk between ECs and HSCs *in vivo*. To address this issue, we established a culture system supporting blood vessel formation and hematopoiesis using the para-aortic splanchnopleural mesoderm (P-Sp) region (pre-AGM region; Aorta-Gonad-Mesonephros region) (Fig. 1), a site where definitive HSCs are committed to diverge from hemangioblasts [3, 5, 6]. When P-Sp explants from E9.5 embryos were cultured on OP9 stromal cells (Figs. 1 and 2a), ECs formed a sheet-like structure (vascular bed) and subsequently a network at the periphery of the endothelial sheet [3, 9] (Figs. 1 and 2b). The development of hematopoietic cells (HCs) was observed in this culture system. HSCs, which were initially observed at the peripheral edge of the vascular bed, migrated into the vascular network area and proliferated (Fig. 2c). As the number of HCs increased, erythroid, granulocyte-macrophage and lymphoid progenitor cells increased on days 6–10 of culture.

In this P-Sp culture system, first, ECs started to migrate from the explant into peripheral areas and formed circle-like structures around the explant on OP9 cells at day 2 of culture (Fig. 1). These ECs gradually began to proliferate and then formed cobblestone (sheet)-like structures by day 4 or 5. c-Kit$^+$CD45$^+$ HSCs then appeared on this sheet-like structure and formed clusters. Subsequently, a large number of ECs sprouted from the vascular bed and formed network-like structures at around day 8.

ECs migrating from P-Sp explants in the early stages of culture express CD31 and Flk-1 [a receptor for vascular endothelial growth factor (VEGF) known as

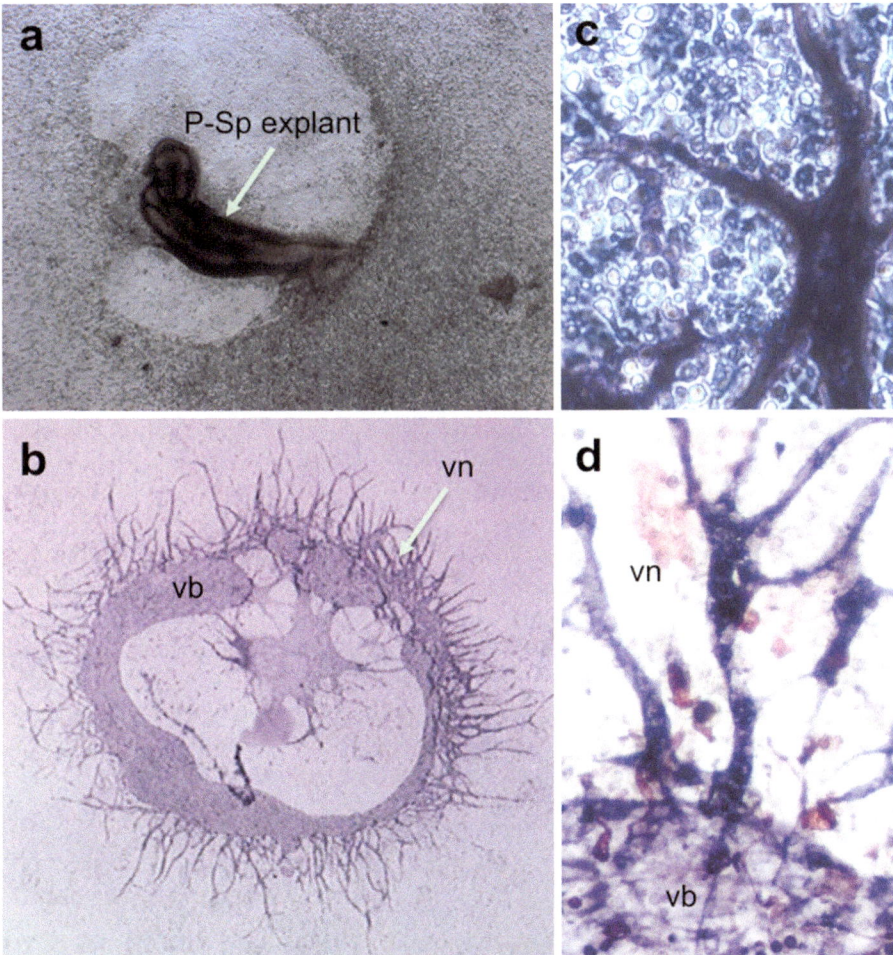

Fig. 2 Development of endothelial cells and hematopoietic cells in the P-Sp culture system.
P-Sp explants were cultured on OP9 stromal cells for 7–10 days. (**a**) Phase contrast microscopic
view of the cells developing from P-Sp explants grown on OP9 cells. (**b**) Cultured cells were fixed
and stained with anti- CD31 mAb to identify ECs. This culture system supports vasculogenesis and
angiogenesis. Vascular bed (vb), a sheet-like structure, first appeared near explants and mimics
vasculogenesis because in terms of its dependency on the VEGF-VEGFR2 system. Subsequent
vascular network (vn) formation mimics the process of fine capillary structure angiogenesis
because this process depends on the Angiopoietin-Tie2 system. (**c**) Higher magnification of
network-forming area. Many round hematopoietic cells are observed. (**d**) Next to network-forming
CD31-positive ECs (*dark blue*), α smooth muscle actin-positive mural cells (*brown*) are observed

VEGFR2], but Tie2 (a receptor for Angiopoietins) expression was very weak. ECs
forming a circle and sheet around the explant on OP9 cells did weakly express Tie2.
However, CD31⁺Flk-1⁺ sprouting ECs from vascular sheets strongly expressed
Tie2. VEGF plays roles in development and tube formation, as well as proliferation

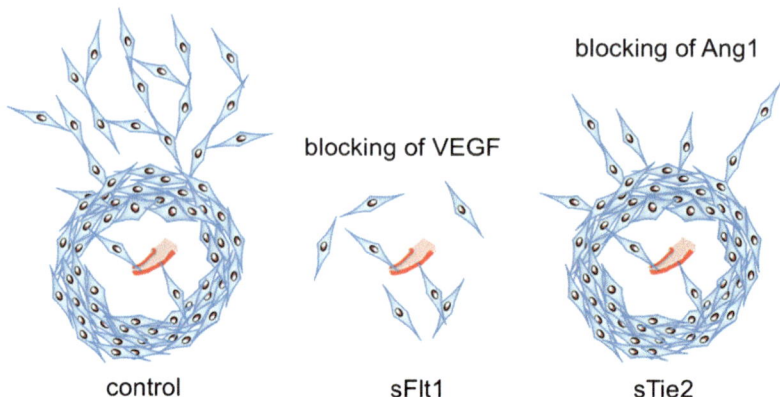

Fig. 3 Schematized P-Sp culture system in association with VEGF and Angiopoietin-1 function. Neutralization of VEGF by soluble Flt1 (sFlt1) inhibits vascular sheet formation and neutralization of Angiopoietins by soluble Tie2 inhibits network formation of endothelial cells

of ECs, and is required for vasculogenesis as well as angiogenesis. Inhibition of Flk-1 signaling by addition of soluble Flt-1 receptors to the culture for neutralization of VEGF disrupted formation of both the vascular bed and network of ECs, and addition of soluble Tie2 to neutralize angiopoietins inhibited network formation of ECs [3] (Fig. 3). Therefore, in this culture system, we define vasculogenesis as the process in which CD31$^+$Flk-1$^+$Tie2^{low}immature ECs migrate from the explant and form a vascular sheet, and angiogenesis as the process of sprouting of CD31$^+$Flk-1$^+$Tie2$^+$ ECs from such a sheet and subsequent network formation.

2   Materials and Methods

2.1   Preparation of OP9 Cells for Use as Stromal Cells

For P-Sp explant culture stromal layers, OP9 cells derived from the M-CSF-deficient *op/op* mouse [10] were selected because in long-term culture of hematopoietic cells, the predominant proliferating cells are macrophages. Macrophages migrate beneath stromal cells, resulted in peeling off of the stromal cells from the bottom of the culture dish. OP9 cells do not produce M-CSF supporting proliferation of macrophages and therefore this line is preferentially used to minimize the damage to the stromal cells. Moreover, OP9 cells can support the differentiation of immature angioblasts into mature ECs. VEGFR2-positive mesodermal cells that still do not express any other EC markers can differentiate into CD31-positive ECs adhering to each other via the VE-cadherin expressed on OP9 cells [11].

For preparation of stromal layers for P-Sp explant co-cultures in 12-well culture plates, OP9 cells were cultured in αMEM containing 20% fetal bovine serum (FBS)

(GIBCO, Gaithersburg, MD). When the cells reached confluence, they were used in cocultures. There are several ways to prepare confluent OP9 stromal feeder layers, one being that 1×10^5 OP9 cells are seeded into individual wells of 12-well culture plates and grown to confluence. In this case, a few days are required for cells to reach confluence. Alternatively, 1/12th of the amount of OP9 cells which have already reached confluence in a 10 cm culture dish are put into 12-well culture plates on the day before P-Sp culture is started. Overnight culture results in a confluent OP9 stromal layer. Which method is used makes no difference to the results of experiments on hematopoiesis and blood vessel formation using this P-Sp culture system. This is dependent on the time schedule of the researcher.

Once confluent, the OP9 stromal layer is ready and the P-Sp culture can be started. Culture medium is removed (washing is not required) and RPMI-1640 (GIBCO) with 10% FBS and 10^{-5} M 2ME (Sigma, St. Louis, MO) supplemented with SCF (50 ng/ml) (GIBCO) and Epo (2 U/ml) (GIBCO) is added. Culture plates are then incubated at 37°C in humidified 5% CO_2 and air until use. In the original description of this P-Sp culture system, IL6 and IL7 were added to the culture medium, but this is not always required. However, when B lymphocyte differentiation is desired, IL7 (20 ng/ml) is required.

2.2 Dissecting Out the P-Sp Region and Culture on OP9 Cells

1. Sacrifice mice with E9.5 embryos by permitted ways.
2. Remove uterus and cut between the deciduoma to separate the embryos and collect them in a sterile 10-cm culture dish containing 10 ml of PBS.
3. Remove the embryo from the uterus by gently peeling the uterine wall under a stereomicroscope. Yolk sac is connected to placenta; cut the portion of the yolk sac attaching it to the placenta and allantois using fine forceps. Widen the hole in the yolk sac caused by removing it from the placenta, and turn the yolk sac over. When P-Sp culture is performed using gene-manipulated mice and contamination between different embryos is to be avoided, embryos are put in a fresh 10-cm culture plate containing 10 ml of PBS and repeated at least twice to wash them. Subsequently, each individual embryo is put into a 3.5 cm culture dish containing 1 ml of PBS.
4. The omphalomesenteric artery and vein connecting yolk sac and embryo is cut close to the latter; the embryo is then cut into three parts as indicated in Fig. 4. The cutting point at the oral side is immediately at the oral portion of the omphalomesentric artery and that of the caudal side is at the end of the dorsal aorta.
5. The central part of the embryo contains the P-Sp region. Dissect out the P-Sp explant using fine forceps. It is better to bend the tip of the forceps as indicated in Fig. 4. From the bilateral side, insert the tip of the forceps into the dorsal aorta, which is visible because of the pooling of red blood cells. Grasp the explant with the forceps gently and lift it up slowly. This region is not firmly attached to other

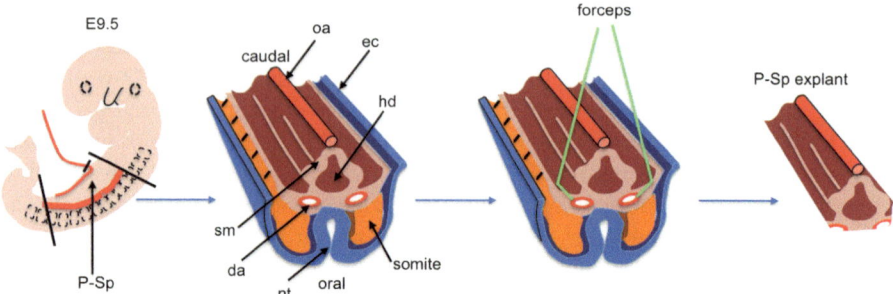

Fig. 4 Schematized procedure for P-Sp explant dissection. See text for details. *oa* omphalomesenteric artery, *ec* ectodermal layer, *hd* hindgut diverticulum, *nt* neural tube, *da* dorsal aorta, *sm* splanchnic mesoderm

    tissues at E9.5 and therefore can be easily peeled off and dissected out. Any remaining remnant can be used for genotyping.

6. Dissected P-Sp explant is picked up using a Komagome glass pasteur pipette and placed onto the prepared OP9 stromal layer. Ignore the addition of small amount of PBS in this procedure. Replace the 12-well culture plate into the incubator at 37°C in humidified 5% CO_2 in air.

2.3   *Evaluation of Vascular Formation*

For the evaluation of blood vessel formation in this culture system, immunostaining is usually performed for visualization of ECs as follows:

1. Aspirate the supernatant (culture media) from the 12-well culture dish.
2. Pour 1 ml PBS gently from the edge of the culture well and aspirate PBS to wash the plates (twice). PBS used for washing should be pre-warmed to 37°C. PBS at room temperature (RT) will also not damage stromal cells. When ice-cooled PBS (4°C) is used, there is a risk that the stromal cell monolayer detaches.
3. Pour on 1 ml of 4% paraformaldehyde (PFA) in PBS pre-warmed to RT and aspirate after 10 min.
4. Pour on 1 ml of PBS at RT and aspirate it after 10 min to wash the cells.
5. Pour on 1 ml of ice-cold 40% methanol (MeOH) in PBS and leave for 5 min at RT (we use MeOH for fixation and clearance).
6. Aspirate the 40% MeOH and add 1 ml of ice-cold 70% MeOH in PBS for 5 min at RT and then aspirate it.
7. Pour on 1 ml of ice-cold 100% MeOH in PBS for 5 min at RT and then aspirate.
8. Add 1 ml of a solution containing 100% MeOH, 10% NaN3, and 30%H_2O_2 in the proportions 50:1:0.33 for 20 min at RT to block endogenous HRP activity

and bleaching. If cells are not to be stained immediately, the culture plates can now be stored at $-30°C$ until used. Aspirate solution and pour on 1 ml of 70% MeOH in PBS for 5 min at RT and then aspirate.

9. Pour on 1 ml of 40% MeOH in PBS for 5 min at RT and aspirate.
10. Pour on 1 ml of PBS for 5 min at RT and aspirate.
11. For preventing non-specific antibody binding, add 1 ml of blocking buffer (PBSMT* + 5%NGS**/1%BSA = 4:1) for 30 min at RT. * PBSMT, 2% skimmed milk + 0.1% Triton in PBS; **NGS, normal goat serum (please change this to appropriate host animal secondary antibody). Prepare solution containing 1st antibody and put on ice. An example using anti-CD31 antibody for visualization of ECs is described in the following:
12. Aspirate blocking buffer and pour on 0.5 ml of solution containing 1 µg/ml anti-CD31 Ab (MEC13.3, rat anti-mouse monoclonal; Pharmingen, San Diego, CA.) in blocking buffer, and incubate overnight at 4°C.
13. Aspirate antibody-containing solution and wash ×3 with 1 ml PBSMT for 10 min.
14. Pour on 0.5 ml of solution containing HRP-conjugated goat anti-rat Ig antibody (Biosource, Camarillo, CA) (×300 dilution in blocking buffer) at RT for 1 h.
15. Aspirate antibody-containing solution and wash ×3 with 1 ml PBSMT for 10 min at RT.
16. Soak culture plates with PBS containing 250 mg/ml diaminobenzidine (Dojin Chem.; Kumamoto, Japan) in the presence (for blue color reaction) or absence (for brown color reaction) of 0.05% $NiCl_2$ for 10–30 min, and then add hydrogen peroxide to 0.01% for the enzymatic reaction.

2.4   Evaluation of Hematopoiesis

For evaluation of hematopoiesis in this culture system, antibody against CD45 (Pharmingen) or other antibodies for hematopoietic markers are chosen and staining performed as described above. Of course, the washing procedure in tissue culture staining results in loss of the floating hematopoietic cells. When all the hematopoietic cells developing in this culture system need to be analyzed, tissue fixation procedures must be omitted. First, collect culture media from culture plates in one tube and wash the plates with PBS. The PBS washings are also harvested and collected in the same tube. One milliliter of Dispase II (Sigma) is added to each well of the culture plate for 10 min at RT to dissociate the cells which are collected by pipetting into the same tube. Cells dissociated as above are analyzed for hematopoietic cell differentiation markers by flow cytometry. In addition, hematopoietic progenitor activity can be analyzed in colony-forming unit cultures (CFU-c) using methylcellulose semisolid culture conditions.

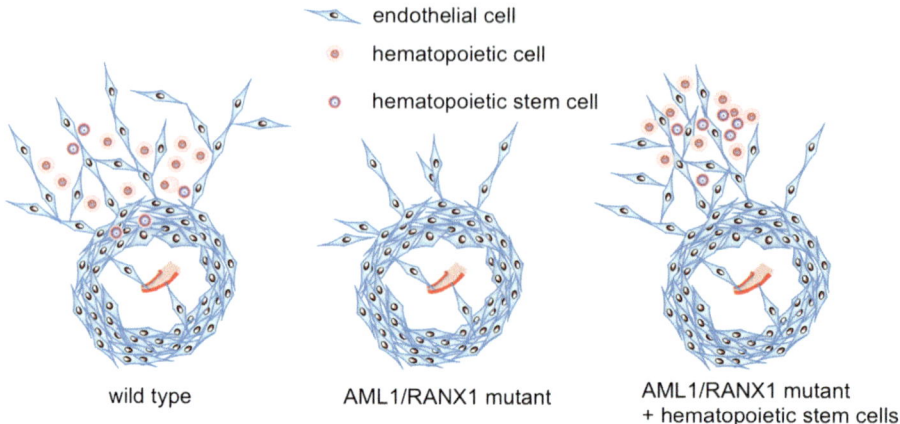

Fig. 5 Analysis of the role of hematopoietic cells in promoting angiogenesis using the P-Sp culture system. Schematized results of P-Sp cultures using AML1 mutant embryo; see text for details

3 Applications for Research

For analysis of blood vessel formation, we have used this P-Sp culture system in several studies, one of which is exemplified here [9].

AML1-deficient mice provide a tool to analyze the association between hematopoiesis and angiogenesis. *AML1* gene disruption leads to hematopoietic failure; AML mutant embryos show lethality at E12.5 [12, 13]. Mutant embryos exhibit hemorrhages in the ventricles of the central nervous system, in the vertebral canal, within the pericardial space and elsewhere. AML1-deficient mice still show the extensive wild-type (wt) vascular branching and remodeling into large and small vessels which occurs normally in the head region at E11.5. However, the number of small capillaries sprouting from anterior cardinal veins in mutant embryos is less than in wt. In mutant embryos, less branching of capillaries is observed in vessels of the pericardium and in the vitelline artery of the yolk sac. Very strikingly, severe defective angiogenesis is observed in the fetal liver. In mice, liver development occurs at E10.5 and subsequently HSCs migrate into the fetal liver, expand and differentiate into different hematopoietic lineages. Hematopoiesis changes into definitive type in the fetal liver instead of primitive one in the yolk sac. Within this time period, very fine capillary structure that is the main environment of hematopoiesis in the fetal liver needs to be established. From the phenotype of AML1-deficient embryos, we can see that HCs are the major source for induction of these very fine capillary structures in the fetal liver.

To analyze the interaction between HSCs and ECs, we observed the development of ECs in P-Sp cultures from *AML1* mutant embryos. As expected, HCs were not generated from P-Sp explants of *AML1* mutants. In contrast, explants from wt embryos developed many round HCs adhering to the presumptive vascular network area; thus, the wt explants generated vascular beds and networks (Fig. 5). On the other hand, poor vascular network formation was observed in cultures of *AML1* mutant explants (Fig. 5). To test the hypothesis that HCs promote angiogenesis, an HSC-enriched population from the bone marrow of normal mice was sorted by FACS and added to P-Sp cultures of *AML1* mutant embryos. As expected, the addition of HSCs rescued defective angiogenesis in *AML1* mutant embryos. (Fig. 5). When HSCs from Angiopoietin-1-deficient mice were used for this rescue experiment, network formation was not completely restored. Expression of Angiopoietin-1 is predominantly observed in HSCs. Thus, through this analysis, an important role of HSCs in promoting angiogenesis via production of Angiopoietin-1 has been established.

A list of analyses performed using the P-Sp culture system follows:

1. Tie2 function in HSC development associating with ECs [3].
2. Rescue of vascular defects in CREB-binding protein by VEGF [14].
3. Interaction of HCs and ECs using AML1 mutant mice as above [9].
4. Effect of VEGF-C on blood vessel formation through VEGFR2 and VEGFR3 [15].
5. EphB4 and ephrinB2 function for the development of venous ECs and arterial ECs [16].
6. The function of neuropin-1 expressed on HCs for blood vessel formation [17].
7. Directed migration of ECs by WAVE2 [18].
8. Function of presenilin-1 in blood vessel formation [19].
9. The function of c-Kit in hematopoiesis relating to blood vessel formation [20].

Acknowledgement This work was partly supported by a grant from the Ministry of Education, Science, Sports, and Culture of Japan.

References

1. Eichmann A, Corbel C, Nataf V et al (1997) Ligand-dependent development of the endothelial and hemopoietic lineages from embryonic mesodermal cells expressing vascular endothelial growth factor receptor 2. Proc Natl Acad Sci USA 94:5141–5146
2. Moore MAS, Metcalf D (1970) Ontogeny of the hematopoietic system: yolk sac of in vivo and in vitro colony forming cells in the developing mouse embryo. Br J Haematol 18:279–296
3. Takakura N, Huang XL, Naruse T et al (1998) Critical role of the TIE2 endothelial cell receptor in the development of definitive hematopoiesis. Immunity 9:677–686
4. Dieterlen-Lièvre F, Martin C (1981) Diffuse intraembryonic hematopoiesis in normal and chimeric avian development. Dev Biol 88:180–191
5. Medvinsky A, Dierzak E (1996) Definitive hematopoiesis is autonomously initiated by the AGM region. Cell 86:897–906
6. Cumano A, Dieterlen-Lièvre F, Godin I (1996) Lymphoid potential, probed before circulation in mouse, is restricted to caudal intraembryonic splanchnopleura. Cell 86:907–916
7. Lu LS, Wang SJ, Aucrbach R (1996) In vitro and in vivo differentiation into B cells, T cells, and myeloid cells of primitive yolk sac hematopoietic precursor cells expanded >100-fold by coculture with a clonal yolk sac endothelial cell line. Proc Natl Acad Sci USA 93:14782–14787

8. Kobayashi H, Butler JM, O'Donnell R et al (2010) Angiocrine factors from Akt-activated endothelial cells balance self-renewal and differentiation of haematopoietic stem cells. Nat Cell Biol 12:1046–1056

9. Takakura N, Watanabe T, Suenobu S et al (2000) A role for hematopoietic stem cells in promoting angiogenesis. Cell 102:199–209

10. Takakura N, Kodama H, Nishikawa S et al (1996) Preferential proliferation of murine colony-forming units in culture in a chemically defined condition with a macrophage colony-stimulating factor-negative stromal cell clone. J Exp Med 184:2301–2309

11. Hirashima M, Kataoka H, Nishikawa S et al (1999) Maturation of embryonic stem cells into endothelial cells in an in vitro model of vasculogenesis. Blood 93:1253–1263

12. Okuda T, van Deursen J, Hiebert SW et al (1996) AML-1, the target of multiple chromosomal translocations in human leukemia, is essential for normal fetal liver hematopoiesis. Cell 84:321–330

13. Wang Q, Stacy T, Binder M et al (1996) Disruption of the *Cbfa2* gene causes necrosis and hemorrhaging in the central nervous system and blocks definitive hematopoiesis. Proc Natl Acad Sci USA 93:3444–3449

14. Oike Y, Takakura N, Hata A et al (1999) Mice homozygous for a truncated form of CREB-binding protein exhibit defects in hematopoiesis and vasculo-angiogenesis. Blood 93:2771–2779

15. Hamada K, Oike Y, Takakura N et al (2000) VEGF-C signaling pathways through VEGFR-2 and VEGFR-3 in vasculoangiogenesis and hematopoiesis. Blood 96:3793–3800

16. Zhang XQ, Takakura N, Oike Y et al (2001) Stromal cells expressing ephrin-B2 promote the growth and sprouting of ephrin-B2(+) endothelial cells. Blood 98:1028–1037

17. Yamada Y, Oike Y, Ogawa H et al (2003) Neuropilin-1 on hematopoietic cells as a source of vascular development. Blood 101:1801–1809

18. Yamazaki D, Suetsugu S, Miki H et al (2003) WAVE2 is required for directed cell migration and cardiovascular development. Nature 424:452–456

19. Nakajima M, Yuasa S, Ueno M et al (2003) Abnormal blood vessel development in mice lacking presenilin-1. Mech Dev 120:657–667

20. Okamoto R, Ueno M, Yamada Y et al (2005) Hematopoietic cells regulate the angiogenic switch during tumorigenesis. Blood 105:2757–2763

Ex Vivo Retinal Explant Cultures to Study Angiogenic Responses

Suphansa Sawamiphak, Ioanna Bethani, Mathias Ritter, and Amparo Acker-Palmer

Abstract The mouse retina has long been regarded as an easily accessible and advantageous system to investigate important questions of developmental angiogenesis. The protocol presented here profits from the suitability of the mouse retina as experimental model and describes an *ex vivo* culture technique of mouse retina explants that allows the quantitative assessment of angiogenic responses to pharmacological manipulations. The technique involves the extraction of the retina from the intact eye, the immediate flat mounting of the tissue on a hydrophilic membrane and the acute stimulation or inhibition of angiogenic processes of the developing vessels in their physiological context *ex vivo*. The number of filopodia structures found at the growing front of the vascular plexus serves as a quantitative readout. This method offers an easily manageable and highly reproducible way to elucidate the complex molecular mechanisms that underlie tip cell function and guidance during the highly orchestrated process of angiogenic sprouting.

S. Sawamiphak
Institute of Cell Biology and Neuroscience, Buchmann Institute for Molecular Life Sciences, Goethe University Frankfurt, Max-von-Laue-Str. 15, D-60438 Frankfurt am Main, Germany

Department of Biochemistry and Biophysics, UCSF School of Medicine, San Francisco, CA 94158, USA

I. Bethani • M. Ritter
Institute of Cell Biology and Neuroscience, Buchmann Institute for Molecular Life Sciences, Goethe University Frankfurt, Max-von-Laue-Str. 15, D-60438 Frankfurt am Main, Germany

A. Acker-Palmer (⊠)
Institute of Cell Biology and Neuroscience, Buchmann Institute for Molecular Life Sciences, Goethe University Frankfurt, Max-von-Laue-Str. 15, D-60438 Frankfurt am Main, Germany

Focus Program Translational Neurosciences (FTN), Johannes Gutenberg University Mainz, Langenbeckstr. 1, D-55131 Mainz, Germany
e-mail: Acker-Palmer@bio.uni-frankfurt.de

E. Zudaire and F. Cuttitta (eds.), *The Textbook of Angiogenesis and Lymphangiogenesis: Methods and Applications*, DOI 10.1007/978-94-007-4581-0_10,
© Springer Science+Business Media Dordrecht 2012

1   Introduction

The mammalian retinal vasculature has been a valuable and widely used experimental model to address the complex mechanisms of developmental angiogenesis. The pioneer work of I.C. Michaelson in the 1940s [1] first hypothesized the existence of 'a chemical factor' that regulates angiogenesis by studying the vasculature of the fenile retina. Forty years later, this mysterious substance was purified and characterized from the bovine retina and is now known as the vascular endothelial growth factor (VEGF) [2]. In particular, the advantages of the mouse retinal model to study angiogenesis were already discovered 30 years ago [3, 4]. The postnatal development of the retinal vascular plexus within a window of 2 weeks and its well defined and highly reproducible spatiotemporal patterning make of this model an excellent system to study vascular development. Moreover, the mouse allows for genetic manipulations to assess the molecular requirements of such a coordinated and complicated process. In the mouse retina, vascularization takes place first at the superficial retinal layer (ganglion cell layer) with vascular sprouts radiating outwards from the optic nerve head to the periphery during the first 7–10 days after birth. Additional sprouting towards the deeper retinal layers occurs during the following week leading to the formation of a three-dimensional capillary network.

Even though the anatomical characteristics and the basic sequence of events occurring during developmental angiogenesis are well described, our understanding of how vessels sprout and branch remains incomplete. A break-through discovery - again with the application of the mouse retina model- was the identification of a specialized endothelial cell located at the sprouting front of the developing vascular plexus: the so called 'tip cell' [5, 6]. The primary role of these cells is to probe the surrounding microenvironment for guidance cues and determine the migratory direction of the growing vascular plexus towards avascular areas. Morphologically resembling the neuronal growth cones, tips cells are highly polarized structures, they proliferate minimally and possess numerous, highly dynamic filopodia, which in response to guidance cues, extend and retract guiding accordingly the migration of the vascular sprout. To successfully manage the orchestrated guidance of the developing vascular network, tip cells need to detect and integrate the signals from multiple molecular cues. Accordingly, tip cells express different receptor molecules at the surface and on their filopodia protrusions with vascular endothelial growth factor receptor 2 (VEGFR2) being one of the most abundant [6–9]. VEGF is a hypoxia-inducible morphogen absolutely essential for vascular patterning and in the retina it is secreted by the astrocytic network that precedes and underlies the developing vessels. VEGF is present in a concentration gradient that declines towards the periphery of the tissue and the time frame of its expression coincides with the temporal pattern of the developing retinal vessels [6, 10]. VEGF is a potent angiogenic factor and an indispensible migratory guidance cue, as indicated by the fatal phenotype of mice lacking even one VEGF isoform [11]. This angiogenic activity is also proven by the increased number of filopodial protrusions found in tip cells of the leading edge of the developing retinal vasculature in response to

VEGF stimulation [6, 12]. Despite the great importance of VEGF in vascular patterning, only the combinatorial cross-talk between VEGF and other stimulatory and inhibitory guidance cues can lead to the complex but highly-organized patterns of vascular networks observed *in vivo* (reviewed in [13, 14]). With such complex mechanisms to ensure proper regulation of neovascularization, it is likely that many of the factors involved in promoting normal vascular development are also involved in pathological neovascularization. Therefore, in-depth understanding of the molecular mechanisms that govern angiogenic processes might allow us to design effective therapeutic strategies against pathological angiogenesis and tumor development. To this end, experimental techniques that allow a great degree of exogenous manipulations but at the same time, remain close to the *in vivo* developmental settings are absolutely necessary.

As described above, the existence of tip cell filopodia extensions at the vascular front is a strong indication of active angiogenic sprouting. The current protocol is taking advantage of this phenomenon and it is describing an *ex vivo* experimental model, in which murine retinas are cultured as explants. The technique allows pharmacological manipulation to screen for potential modulators of angiogenic responses at the vascular front. These responses can be quantitatively measured by the abundance of filopodia structures on tip cells [9, 12]. In the retina model, search for molecular players involved in the angiogenic regulatory machinery had employed an intraocular injection method to apply substances into the eyes [6, 15]. Although the *in vivo* microenvironment of the retina is maintained, high pressure inside the eye allows very small volume (0.5–2 μl for the mouse eye) of additional substances. Moreover, the reflux of injected substances into the tissue causes an unreliable measurement of the amounts of the factor injected thereby affecting the reproducibility of the experiments. Although different culture techniques to explant retinas have been reported to enable the manipulation of neuronal cells which can be kept alive over a long period of time (several days) [16, 17] there is so far no reliable established *ex vivo* method that allows assessment of tip endothelial cell angiogenic responses at the vascular front. The explant culture technique described here, on the contrary, provides an easily manageable and highly flexible alternative to study retina vascularization. It allows the precise control of the concentrations of the applied substances, guaranteeing a high reproducibility among experiments, and it remains sufficiently close to the *in vivo* situation, representing a great advantage in order to get a deeper insight into the complex mechanisms that regulate vessel formation and patterning.

2 Materials

2.1 Mice

New born pups (strain: C57BL/6; age: 4–5 days old)

2.2   Reagents

- Ethanol (70% vol/vol)
- DMEM (GIBCO)
- FBS (GIBCO)
- Penicillin-streptomycin (PAA)
- Dulbecco's PBS (DPBS; GIBCO)
- VEGF-164 (ReliaTech)
- EphB4-Fc (R&D Systems)
- sFlt1 (R&D Systems)
- Dynasore (Biozol)
- PFA solution (4% wt/vol in PBS)
- BSA (Sigma)
- Fluorescein isothiocyanate (FITC)-conjugated lectin from Bandeiraea simplicifolia (Isolectin B4; Sigma)
- Triton X-100 (Roth)
- Vectashield mounting medium (Vector Laboratories)

2.3   Materials

- Millicell standing insert with polytetrafluoroethylene membrane (pore size 0.4 μm; Millipore)
- Six-well cell culture dish

2.4   Equipment

- Surgical instruments
- Sterile tissue culture hood
- Tissue culture incubator
- Dissecting stereo microscope
- Confocal microscope

2.5   Necessary Solutions

2.5.1   Explant Culture Medium Composition

50 ml of heat-conditioned FBS (final concentration 10%) are added to 445 ml of D-MEM (final concentration 89%), and 5 ml of penicillin/streptomycin

(final concentrations 100 U/ml penicillin and 100 μg/ml streptomycin, from 100× stock). The culture medium is filtered through a 0.22 μm pore sterile filter. 50 ml of medium is sufficient for each preparation. Aliquot 50 ml and store at 4°C.

2.5.2 Starving Medium

Same composition as for the culture medium but final concentration of FBS is reduced to 3%. Sterilize by filtering through a 0.22 μm membrane. Aliquot 50 ml and store at 4°C.

2.5.3 Lectin Blocking Buffer

For 500 ml of the buffer, 2.5 ml of Triton X-100 is added to 500 ml of D-PBS. Stir until the solution is homogenous. Aliquot to sufficient volume and freshly add BSA to a final concentration of 0.2% (w/v) before each use.

3 Procedure

3.1 Preparation of the Membrane Inserts

1. Fill each well of a 6-well dish with 1 ml of explant culture medium.
2. Add one Millicell standing insert into each well, taking care that the hydrophilic polytetrafluoroethylene membrane is facing the bottom of the well and is in direct contact with the medium.
3. Place the dish containing the inserts in a tissue culture incubator (37°C, 5% CO_2) to equilibrate for at least 30 min prior to the retina dissection.

3.2 Dissection of Neonatal Mouse Retinas

4. Anesthetize a mouse pup (postnatal day 4–5) by keeping on ice for several minutes and proceed with decapitation.
5. Carefully remove the eyes from both eye sockets with the use of fine forceps (Figs. 1a and 2a) and transfer them immediately into a 10-cm culture dish containing 10 ml of D-MEM medium.
6. Under a stereomicroscope, hold the eye in place with a blunt-ended forceps (grapping the optic nerve, if still attached) and insert a fine forceps or a 16-gauge needle at the cornea-scleral boundary to open a hole (Figs. 1b and 2b).

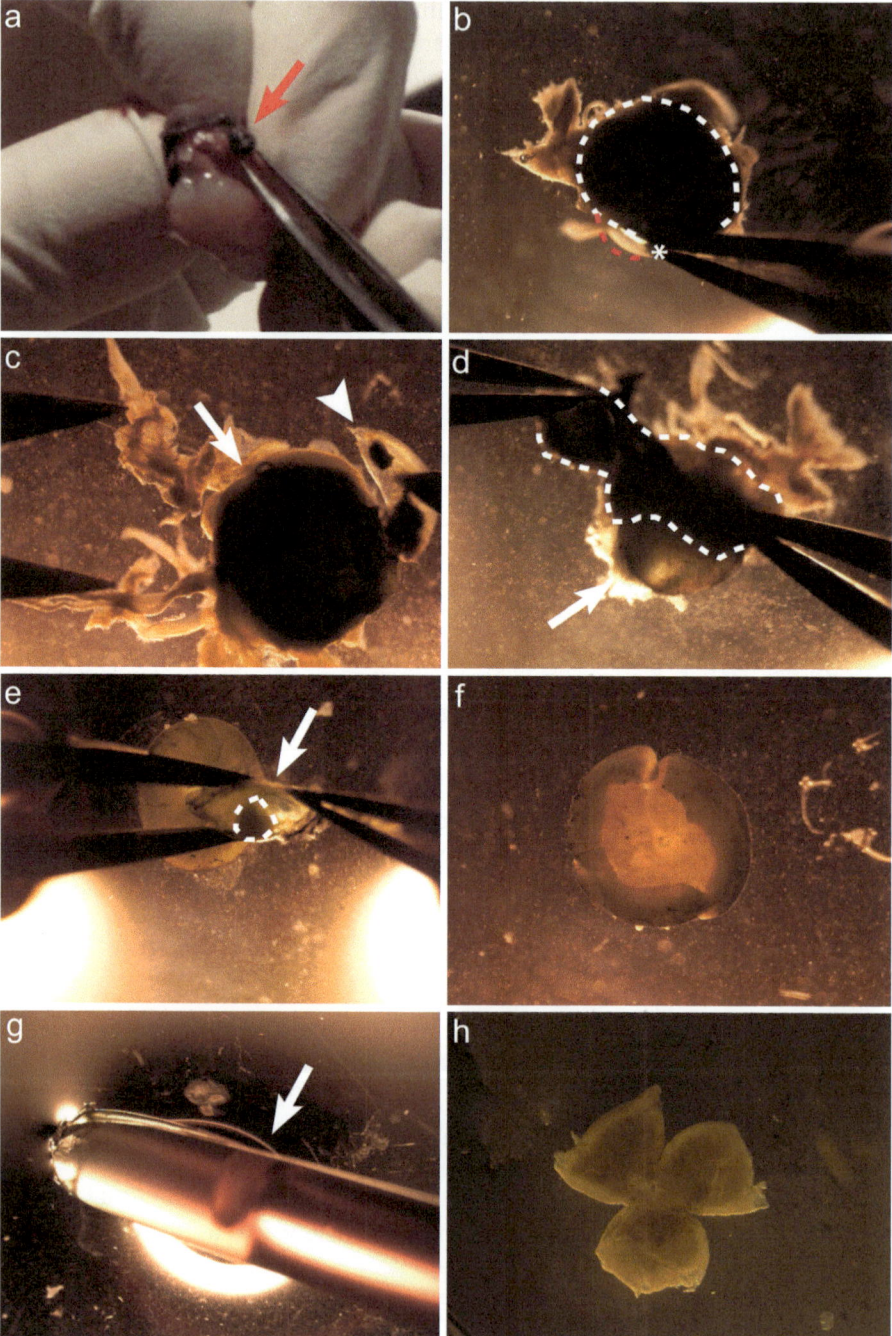

Fig. 1 Retinal dissection and explant culture. (**a**) The eye (indicated by the *red arrow*) is enucleated from the anesthetized newborn mouse. (**b**) A hole (indicated by the *asterisk*) is made at

7. Insert a fine spring scissors in the hole and make a circumferential cut around the eye, close to the limbus, to release the retina (Figs. 1c and 2c).
8. To extract the retina from the eye, insert two pairs of blunt ended forceps into the subretinal space, grab the sclera in two opposite positions and gently peal it apart by tearing away to either side (Figs. 1d and 2d).
9. Remove the iris from the retina and pierce the lens with fine forceps to pull it out, taking care that you do not touch the inner surface of the retina.
10. Vitreous bodies are often removed together with the lens during the previous step. If not, grab the vitreous with fine forceps and pull them out from the retina, by stabilizing another pair of forceps on top of the retina, to exert a counter force (Figs. 1e and 2e).
11. Divide the retina into three equal quadrants by performing fine cuts from the rim to half of its radial length (Figs. 1f and 2f).

3.3 Preparation and Stimulation of the Retina Explants

12. Transfer the retina in a small volume of DMEM medium onto the membrane of the equilibrated insert by using an 1-ml micropipette whose tip is cut to provide a wider opening (Figs. 1g and 2g). Try to orient the retina while it is in the micropipette by drawing the medium up and down till the ganglion cell layer is facing toward the opening of the micropipette tip. Keeping this orientation, drop the retina in a small amount of DMEM onto the insert.
13. Flat-mount the retina on the insert making sure that the ganglion cell layer and the vessels are facing down to the membrane (Figs. 1h and 2h). The rims of the retina can easily fold to the inner part of the tissue – to flatten them, suck the medium close to the rim and drop it back onto the center of the retina. Repeat till the explants are adhered to the membrane and remove all the remaining medium from the top of the insert. To keep the retina humidified, drop 100 μl of retina culture medium on the flat-mounted retina. One insert can accommodate one to two retinas.
14. Incubate the retina in a humidified incubator at 35°C with 5% CO_2 for 2–4 h for recovery.

◄───

Fig. 1 (continued) the cornea (*red dotted line*) – sclera (*white dotted line*) boundary using a pair of fine forceps or a thin needle. (**c**) Fine scissors are inserted in the hole and used to incise the cornea (indicated by the *arrowhead*). The rest of the eyeball is indicated by the *arrow*. (**d**) Two pairs of forceps are used to grab and tear the sclera and pigment epithelial layers (*dotted line*). The retina is then extracted. (**e**) The iris, the lens (*dotted line*) and the vitreous (*arrow*) are removed. (**f**) Fine, radial cuts are made to divide the retina into three quadrants. (**g**) The retina (*arrow*) is transferred into a cut-tip pipette from the dissection dish onto the membrane of the insert. (**h**) The retina is carefully flat-mounted onto the membrane of the Millicell insert (Reprinted in part with permission from Sawamiphak et al. [12])

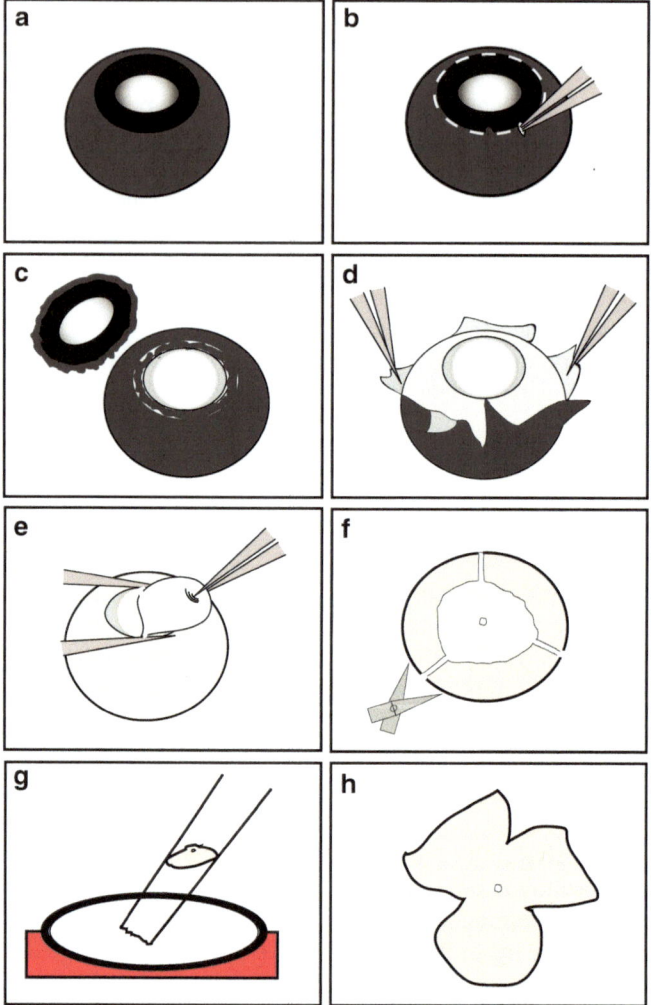

Fig. 2 Graphical representation of the dissection steps described in Fig. 1

15. Remove culture medium from the well and immediately replace it with 1 ml of prewarmed starving medium (control) or starving medium containing stimulatory or inhibitory factors at the necessary concentrations (see Table 1). Always keep the retina humidified by dropping a small amount of medium on top.
16. Incubate at 35°C in 5% CO_2 for 4 h.
17. To terminate the stimulation, remove the medium from the well and rinse the explants with 500 µl PBS.
18. Transfer each explant in a well of a 24-well culture dish containing 1 ml of pre-warmed 4% PFA. Incubate at 35°C for 4 h.

Table 1 Examples of stimulating and inhibiting factors of endothelial tip cell filopodia extensions

Agent	Concentration
VEGF164	1 µg/ml
sFlt-1	1 µg/ml
EphB4-Fc	10 µg/ml
Dynasore	320 µM

3.4   *Whole Mount Retina Staining*

19. Wash the fixed explants with PBS briefly for three times and then incubate overnight at 4°C in 500 µl blocking buffer.
20. Wash briefly with PBS containing 1% Triton X-100 and incubate the explants overnight at 4°C in 500 µl PBS containing 1% Triton X-100 and 40 µg/ml FITC-conjugated Isolectin B4.
21. Wash five times with 500 µl PBS, each wash lasting for 1 h and re-fix the explants by incubating in 500 µl of 4% PFA for 30 min at room temperature.
22. Flat-mount the explant on glass slide with the help of two fine paint brushes – the photoreceptors layer should be in contact with the glass. Add one drop of Vectashield mounting medium on the retina and place a 24 × 50 mm glass coverslip on top. Seal the coverslip with colorless nail polish to avoid leakage of the mounting medium. Samples can be stored till imaging at 4°C.

3.5   *Evaluation of Filopodial Extension at the Retinal Front*

23. Using a confocal laser-scanning microscope, image the retinal tip cell filopodia present at the sprouting front of the vascular plexus (Fig. 3a). Take as many pictures as necessary to cover the entire periphery of the retina. A 40x-magnification oil-immersion objective as well as an excitation at 488 nm, are recommended.
24. The effect of stimulatory or inhibitory factors on filopodia extensions can be addressed by quantifying the number of filopodia protrusions per 100 µm of vessel length (Fig. 3b). To achieve reliable results, approximately 20 microscopic fields from 10 to 12 retinas per conditions need to be evaluated. The averaged values from each condition are then compared.

4   Expected Results

The sprouting vascular front of the developing murine retina is well equipped with highly specialized structures, the so called 'tip cells', that are responsible for the orchestrated migration of the vascular plexus toward the avascular area. The dynamic extension and retraction of filopodia protrusions at the surface of

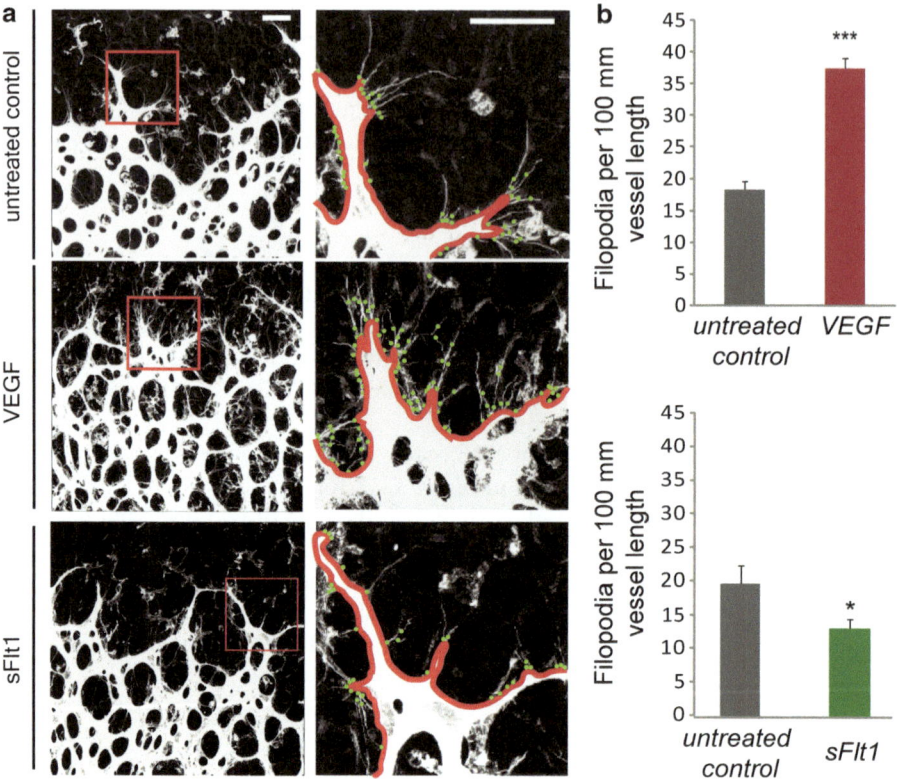

Fig. 3 Example of an explant assay to study vessel sprout outgrowth. Explanted retina vessels are stained with FITC-conjugated isolectin B4 staining and imaged at the leading front of the vascular bed. *Right panels* are enlargements of the vascular front shown in the *left panels* (*red boxes*). *Green dots* indicate filopodia extensions in 100 μm of vessel length (*red line*) (**a**) Robust sprouting activity of the explanted retinal vessels in response to VEGF treatment for 4 h. Application of soluble VEGF receptor1 extracellular domain–Fc fusion protein (s-Flt1) severely reduces angiogenic activity of explanted retina. (**b**) Quantification of filopodial extensions (*green dots*) per 100 μm of vessel length (*red line*) at the vascular front. *Scale bar* = 25 μm. *P< 0.05, ***P< 0.001 (Reprinted with permission from Sawamiphak et al. [12])

the tip cells in response to guidance cues functions as the sensory readout of the environment for the growing vascular network and accordingly determines the direction of its migration. Therefore, experimental evaluation of the number of filopodia extensions can be used to access tip cell activity at the vascular front. The described method provides an elegant but also quantitative readout of acute tip cell responses to exogenous stimuli. Explanted retinas, prepared and cultured under the conditions indicated in this protocol, have on average a number of filopodial processes at the sprouting front which is slightly lower but comparable to the number of filopodia extensions of retinal vessels developed under physiological conditions in the eye (Fig. 3). A proof of principle for our assay is the robust response of the explanted cultures to stimulation with VEGF-A. Exposure of the explants to VEGF-A led to an increase in the number of filopodial extensions (37.4 ± 1.6 filopodia/ 100 µm vessel length) compared to untreated samples (18.2 ± 1.3 filopodia/ 100 µm vessel length) (Fig. 3a, b). Accordingly, the use of an inhibitor that functions as VEGF-A trap led to severe decrease of tip cell filopodia extension activity. When retina explants were incubated with the soluble VEGFR1 extracellular domain-Fc fusion protein (soluble Flt-1), that exhibits a high affinity for VEGF-A and therefore depletes it from the medium, the amount of filopodia protrusions was compromised down to 13.0 ± 1.3 filopodia/ 100 µm vessel length (Fig. 3a, b).

The anticipated results are very reproducible between different experiments and different animals, provided that the user is following strictly the instructions of the protocol. Critical factors that can have an impact on the outcome of the experiment are proficiency and speed in handling, such as avoiding injuring the retina or delaying in adhering the retina onto the insert after dissection. We recommend a long hands-on training period before performing experiments.

5 Notes and Troubleshooting

1. Do not sacrifice all the animals that you need for your experiment at once. Proceed to the next animal only after you have prepared retina explants from both eyes of the previous animal.
2. Avoid the use of FBS-containing medium during the dissection because this will affect the adherence of the retina to the insert membrane.
3. Make sure that the vitreous bodies are totally removed as they might interfere with the accessibility of the retina to the exogenous stimuli as well as with the imaging of the developing vascular plexus.
4. While preparing the retina for flat-mounting, avoid making too deep cuts as it will make the retina easier to fold and harder to flatten.
5. If the retina is placed on the membrane in the wrong orientation, collect it back into the transfer pipette in sufficient amount of DMEM and by multiple drawing-ups and downs, re-position it on the insert with the vessels facing the membrane.
6. Ensure that the explants are adhered to the membrane – a floating retina can easily lead to vessel regression.

References

1. Michaelson IC, Campbell ACP (1940) The anatomy of the finer retinal vessels, and some observations on their significance in certain retinal diseases. Trans Ophthalmol Soc UK 60:71–112
2. D'Amore PA, Glaser BM, Brunson SK et al (1981) Angiogenic activity from bovine retina: partial purification and characterization. PNAS 78:3068–3072

3. Blanks JC, Johnson LV (1983) Selective lectin binding of the developing mouse retina. J Comp Neurol 221:31–41
4. Connolly SE, Hores TA, Smith LEH et al (1988) Characterization of vascular development in the mouse retina. Microvasc Res 36:275–290
5. Dorrell MI, Aguilar E, Friedlander M (2002) Retinal vascular development is mediated by endothelial filopodia, a preexisting astrocytic template and specific R-cadherin adhesion. Invest Ophthalmol Vis Sci 43:3500–3510
6. Gerhardt H, Golding M, Fruttiger M et al (2003) VEGF guides angiogenic sprouting utilizing endothelial tip cell filopodia. J Cell Biol 161:1163–1177
7. Gerhardt H, Ruhrberg C, Abramsson A et al (2004) Neuropilin-1 is required for endothelial tip cell guidance in the developing central nervous system. Dev Dyn 231:503–509
8. Tammela T, Zarkada G, Wallgard E et al (2008) Blocking VEGFR-3 suppresses angiogenic sprouting and vascular network formation. Nature 454:656–660
9. Sawamiphak S, Seidel S, Essmann CL et al (2010) Ephrin-B2 regulates VEGFR2 function in developmental and tumour angiogenesis. Nature 465:487–491
10. Gariano RF, Gardner TW (2005) Retinal angiogenesis in development and disease. Nature 438:960–966
11. Ferrara N, Carver-Moore K, Chen H et al (1996) Heterozygous embryonic lethality induced by targeted inactivation of the VEGF gene. Nature 380:439–442
12. Sawamiphak S, Ritter M, Acker-Palmer A (2010) Preparation of retinal explant cultures to study ex vivo tip endothelial cell responses. Nat Protoc 5:1659–1665
13. Adams RH, Alitalo K (2007) Molecular regulation of angiogenesis and lymphangiogenesis. Nat Rev Mol Cell Biol 8:464–478
14. Herbert SP, Stainier DY (2011) Molecular control of endothelial cell behaviour during blood vessel morphogenesis. Nat Rev Mol Cell Biol 12:551–564
15. Lu X, Le Noble F, Yuan L et al (2010) The netrin receptor UNC5B mediates guidance events controlling morphogenesis of the vascular system. Nature 432:179–186
16. Donovan SL, Dyer MA (2006) Preparation and square wave electroporation of retinal explant cultures. Nat Protoc 1:2710–2718
17. Kretz A, Marticke JK, Happold CJ et al (2007) A primary culture technique of adult retina for regeneration studies on adult CNS neurons. Nat Protoc 2:131–140

The Mouse Model of Oxygen-Induced Retinopathy (OIR)

Andreas Stahl, Jing Chen, Jean-Sebastian Joyal, and Lois E.H. Smith

Abstract The mouse model of oxygen-induced retinopathy (OIR) is a well established model to study retinal angiogenesis. It has been used extensively to investigate the mechanisms behind pathologic vessel formation in the eye and has helped to lay the foundations for clinical trials. For example, the current use of intravitreal anti-VEGF agents for neovascular eye disease has been significantly influenced by findings in the OIR mouse model. Originally described in 1994, many detailed reports on the OIR mouse model exist and some examples are given in this chapter. The focus of this chapter, however, lies on a novel tool to reliably quantify retinal neovascularization (NV) in the OIR mouse model. Reliable and standardized NV measurements allow better comparabiltiy between labs and will help to further establish the OIR mouse model as a standard tool in retinal angiogenesis research.

The mouse retina has long been used extensively to study both physiologic and pathologic angiogenesis. The fact that retinal vessels develop postnatally in the mouse renders the murine retina an easily accessible window into angiogenic processes *in vivo*. At birth, mouse (and other rodent) pups have an immature retinal vasculature and persistent hyaloid vessels, an embryonic vascular bed that regresses postnatally [1] through highly organized vessel pruning. Hyaloid vessel regression is paralleled by simultaneous growth of functional vessels in the retina originating from the optic nerve head and extending radially into the periphery [2]. One of the major advantages of studying postnatal retinal vascular development in mice lies in the ease of accessibility of the developing vasculature for imaging and intervention

A. Stahl (✉)
University Eye Hospital Freiburg, Freiburg, Germany
e-mail: andreas.stahl@uniklinik-freiburg.de

J. Chen • J.-S. Joyal • L.E.H. Smith
Department of Ophthalmology, Harvard Medical School,
Children's Hospital Boston, Boston, Massachusetts, USA
e-mail: lois.smith@childrens.harvard.edu

E. Zudaire and F. Cuttitta (eds.), *The Textbook of Angiogenesis and Lymphangiogenesis: Methods and Applications*, DOI 10.1007/978-94-007-4581-0_11,
© Springer Science+Business Media Dordrecht 2012

181

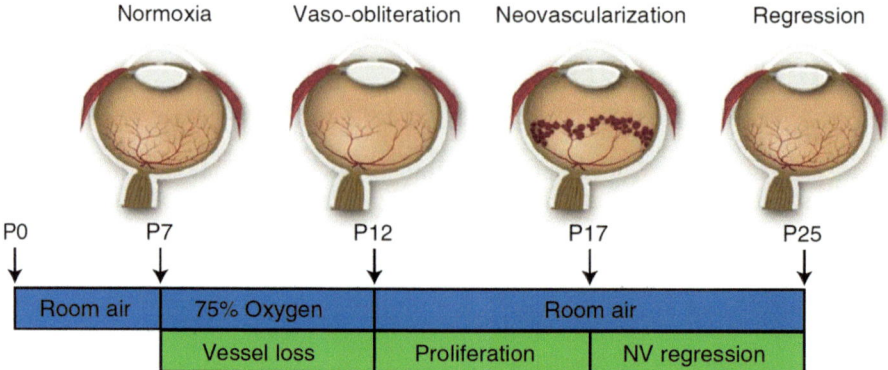

Fig. 1 Schematic representation of the mouse model of oxygen-induced retinopathy (*OIR*). Mice are exposed to hyperoxic conditions from postnatal day (P)7 to P12. During this first phase of the OIR model central vascular pruning occurs, resulting in a vaso-obliterated zone (*VO zone*) in the central retina. From P12 onwards mice are returned to room air. At this point, the 21% oxygen of normal room air represents a relative hypoxic environment to the avascular parts of the retina that had been adjusted to 75% oxygen. As a consequence, hypoxia-driven re-vascularization of the VO zone occurs. However, due to unphysiologically high levels of vaso-active growth factors, this regrowth of functional vessels is accompanied by pathologic neovessel formation, resulting in disorganized pre-retinal neovascularizations (*NV*). Peak NV is reached at P17, followed by NV regression and further repair of the retina. Vascular repair is usually complete by P25 (*Figure reproduced with permission from* [22])

without the difficulties associated with investigating embryonic development [3–6]. Thus many fundamental findings regarding vascular regulation and patterning, for example VEGF-induced vascular development [7], endothelial tip cell guidance [8] or notch [9–11] and Wnt signaling [12–15] in vascular growth have been achieved using the developing mouse retina.

The mouse retina combines not only a multitude of favorable attributes for the observation and modulation of physiologic angiogenesis, it can also serve as a reliable model for pathologic neovessel formation. Several experimental setups have been developed to investigate pathologic retinal neovascularization in the murine retina. The mouse model of oxygen-induced retinopathy (OIR) in particular has become one of the most widely used *in vivo* models used in angiogenesis research [16]. Over 20,000 publications are found on the topic of oxygen induced retinopathy in mice in Google scholar (Google Scholar search on keywords "oxygen induced retinopathy mice," July 26, 2011). Among others, the OIR mouse model has been fundamentally involved in establishing the role of VEGF in the pathogenesis of retinal neovascularization [17, 18]. Building on these early observations, the beneficial effects of anti-VEGF treatment in hypoxia-driven ocular angiogenesis has been investigated in the OIR mouse model [7, 19–21]. Results from the OIR mouse model thus helped to lay important foundations for today's clinical application of anti-VEGF therapies in human patients.

In the OIR model, neonatal mice are exposed to 75% oxygen from postnatal day (P)7 to P12 (Fig. 1). During this first phase, hyperoxia causes cessation of normal

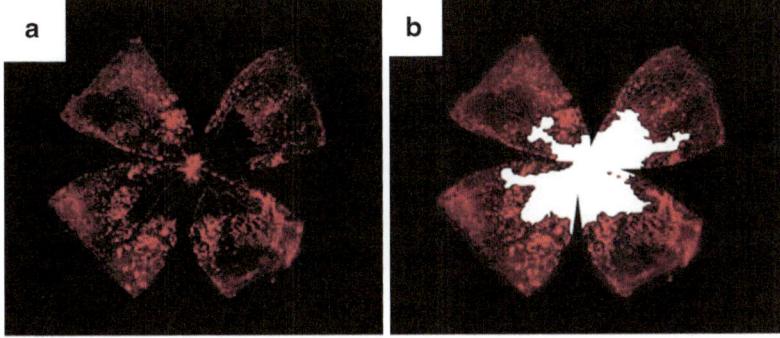

Fig. 2 The central avascular area formed during the first phase of the OIR model can be quantified by outlining the vaso-obliterated zone (*VO-zone*) on lectin-stained retinal flatmounts. The amount of VO-zone is expressed as percentage of whole retinal area (%VO) (*Figure reproduced with permission from* [23])

radial vessel outgrowth and vessel regression in the central retina. The extent of central vaso-obliteration can be determined by measuring the avascular area in retinal whole-mounts at P12 (Fig. 2). This readout is of particular interest in studies that probe vasoprotective strategies or investigate genes that are hypothesized to be involved in vascular pruning and regression [22].

The vaso-obliteration from the first phase of the OIR model leads to retinal tissue ischemia after mice return to room air at P12. The subsequent second phase is thus characterized by retinal tissue hypoxia and the expression of angiogenic factors. During this phase, two angiogenic processes can be observed in parallel: (1) the re-growth of functional retinal vessels from the existing peripheral capillaries into the central avascular zone and (2) the formation of pathologic pre-retinal neovascular tufts and clusters (NV). The second phase in the OIR model is thus similar to the vaso-proliferative phases of retinopathy of prematurity (ROP) and to proliferative diabetic retinopathy in human patients, where pre-retinal vascular proliferations can occur leading to tractional retinal detachment and blindness if left untreated.

Several approaches have been developed to measure the extent of NV in the OIR mouse model. In most cases, retinal NV is quantified at its peak at P17. Early studies have used retinal cross-sections to count pre-retinal nuclei [16, 18, 19]. This time-consuming technique has widely been replaced in the last decade by utilizing retinal whole-mounts to score retinal NV in a grading system [24, 25] or by manually measuring the area of neovascular tufts and clusters [26–28]. A detailed protocol for manual NV measurement on retinal flatmounts has recently been published [22]. In addition, a computer-aided protocol (SWIFT_NV) has been developed to allow time-efficient and nearly user-independent quantification of retinal NV [23]. This chapter will provide a detailed user-guide for the SIWFT_NV macros to measure retinal NV in the OIR mouse model.

Similar to the established manual quantification protocols, SWIFT_NV analyzes retinal NV from lectin-stained retinal whole mounts. The SWIFT_NV software consists of a set of macros that was developed to run on NIHs free ImageJ platform [23]. The macros together with a detailed manual can be obtained from the authors of

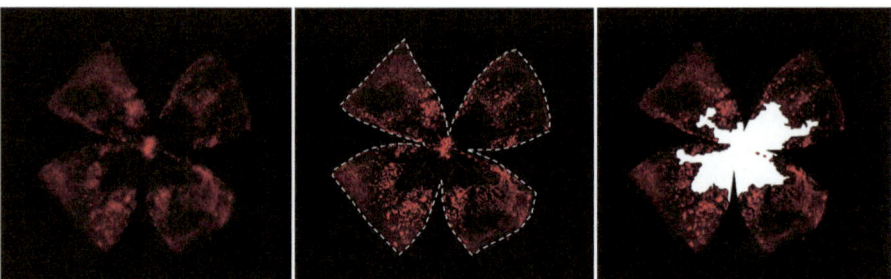

Fig. 3 The SIWFT_NV macros for computer-aided quantification of retinal NV require the original retinal image (*left*), a quantification of the total retinal area (*middle*) and an image highlighting the VO zone of the retina to be analyzed (*right*)

this chapter free of charge for all academic institutions. Two different sets of macros are provided: *SWIFT_NV all in one* and *SWIFT_NV step by step*. Both macro sets are provided in two versions, one for red retinal images (Lectin-stained) and one for green retinal images (FITC-perfused). Both macro sets require (i) the original image of the retina; (ii) a quantification of total retinal area; and (iii) an image of the same retina with the area of vaso-obliteration (VO) highlighted (Fig. 3).

The *SWIFT_NV all in one* macro set is suggested for most standard applications. *SWIFT_NV all in one* can process all images except for images having more than one VO areas that are not connected with each other or do not involve the center of the retina. These images must be measured using *SWIFT_NV step by step*. *SWIFT_NV all in one* takes approx. 2–5 min to quantify one retina. This includes displaying the results and saving a merged image of NV and original image. *SWIFT_NV step by step* allows more user input but requires more time to process one retina. *SWIFT_NV step by step* measures approx. 1% higher NV pixel numbers than *SWIFT_NV all in one*.

To run SWIFT_NV, the user copies the five SWIFT_NV macros into the plugin folder of NIHs ImageJ software and assigns an appropriate shortcut for each macro. To exclude small artifacts like vessel branch points from quantification, the SWIFT_NV macros use a cut-off value that needs to be adjusted once during the initial setup according to the image size used in the investigator's lab. In the standard version, the cut-off value is set to 100 pixels for an image size of 5217743 pixels. Adjusting the cut-off value for different image sizes can be easily done using an Excel macro supplied with the SWIFT_NV macros.

The initial setup for SWIFT_NV has to be done only once and will be saved for all further quantifications. Before quantifying NV, however, VO area and total retinal area must be quantified for each individual retina (see Fig. 3). To start NV quantification, the user opens the original retina image in ImageJ and starts the first SWIFT_NV macro. This macro automatically retrieves information on the directory and the name of the original image. It then opens the appropriate VO image. The VO image is measured for width and height and a cursor is placed in the center of the image to fill the VO area with background color. After filling the VO area, the image is cut into four quadrants. This is important as the quadrants often have

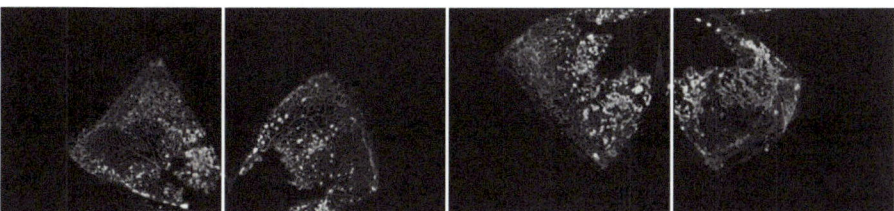

Fig. 4 The four retinal quadrants resulting from the first SWIFT_NV macro. The original retina has been cut into four quadrants and each quadrant has been individually processed to reduce background fluorescence and converted into a *black* and *white* image

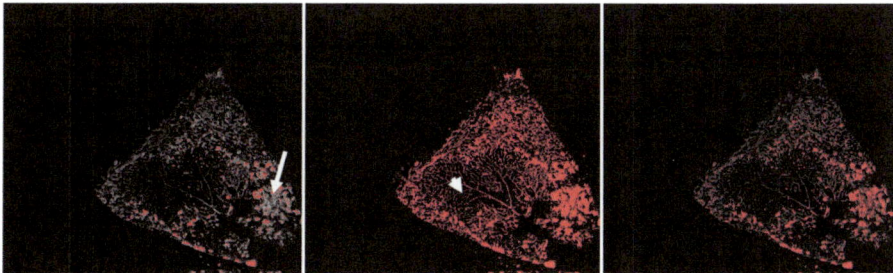

Fig. 5 Setting a fluorescence threshold for SWIFT_NV quantification. For each quadrant individually, the user chooses a threshold that is neither too low and would exclude retinal NV from the quantification (*left*), nor too high and would include normal vessels in the quantification (*middle*). The image on the *right* illustrates an appropriate threshold that highlights pre-retinal NV but not normal vasculature

different fluorescence intensities. For each quadrant, the color image is transferred into black and white and background fluorescence is removed. The four quadrants resulting from this first macro are shown in Fig. 4.

For each quadrant individually, the user can now mark any area that should be excluded from the quantification (e.g. unspecific staining of the retinal edges, remaining hyaloid vessels or artifacts within or outside the retina). The user then uses ImageJ's thresholding algorithm to mark all pixels above a certain fluorescence threshold. This step requires some judgment by the user on where to set the threshold. Usually, a good threshold is when red markings appear at the branch points of normal capillaries. These branch point markings will later automatically be removed before quantification (Fig. 5).

After setting the fluorescence threshold, the user starts the next SIWFT_NV macro. This macro automatically analyzes all pixels in the image that lie above the set threshold and are part of a group of pixels that is larger than the cut-off size (e.g. larger than 100 pixels). In addition, the macro removes small holes within tufts or clusters to obtain a solid filling of these structures. The analyzed quadrant is then saved and thresholding and quantification is repeated for the three remaining quadrants. After analyzing the last quadrant, SWIFT_NV automatically re-opens all four quadrants, quantifies the NV pixels, creates a mosaic and merges all NV quadrants onto the original image (Fig. 6). Simultaneously, a 'Results'-window

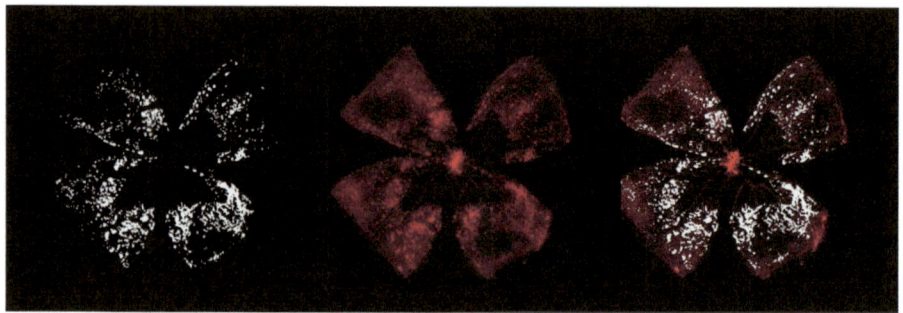

Fig. 6 In the last SWIFT_NV macros, the four NV quadrants are quantified, assembled into a mosaic and merged onto the original retinal image. The quantification results are displayed in a separate window

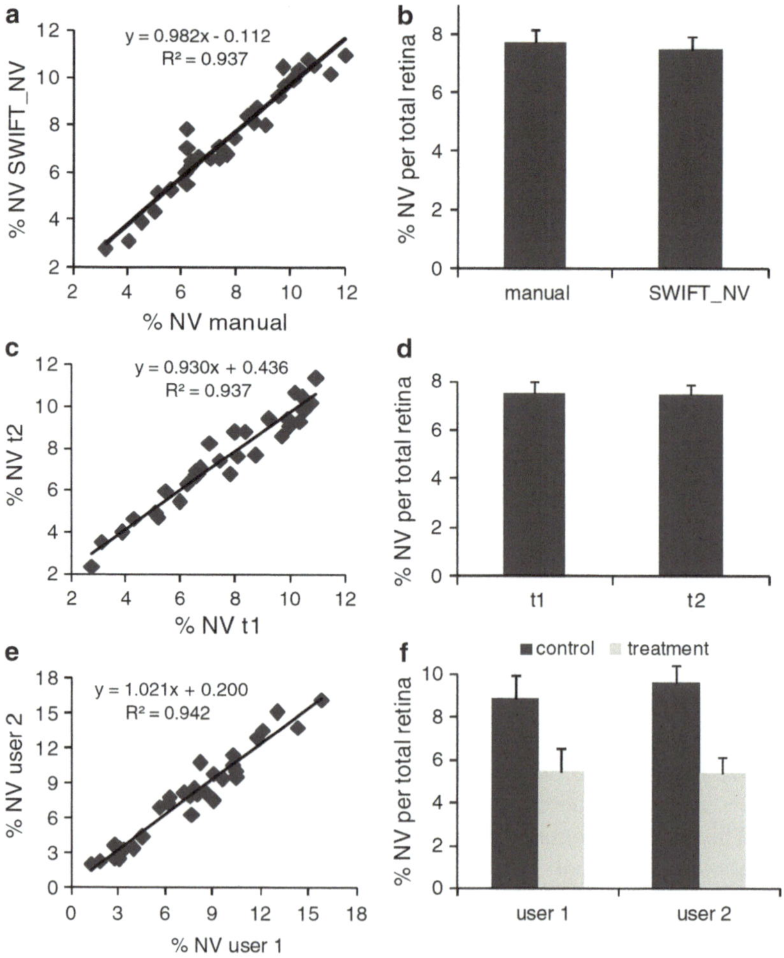

a
y = 0.982x - 0.112
R² = 0.937
% NV SWIFT_NV
% NV manual

b
% NV per total retina
manual   SWIFT_NV

c
y = 0.930x + 0.436
R² = 0.937
% NV t2
% NV t1

d
% NV per total retina
t1   t2

e
y = 1.021x + 0.200
R² = 0.942
% NV user 2
% NV user 1

f
■ control   treatment
% NV per total retina
user 1   user 2

Fig. 7 The results from SWIFT_NV quantifications correlate closely with results from manual quantifications (**a, b**). SWIFT_NV results obtained by one user are consistently reproduced after a 3 months interval (**c, d**). SWIFT_NV results obtained from two different users correlate well and retrieve similar values for both control and treatment groups (**e, f**) (*Figure reproduced with permission from* [23])

opens containing the quantification results for all four quadrants combined, thus representing total NV area of the full retinal image in pixels. This result can be copied to an Excel worksheet where retinal NV can be calculated as NV pixels/total pixels of the retina (%NV).

The accuracy of the SWIFT_NV macros has been validated by comparing SWIFT_NV results to results from manual measurements [23]. Plotting SWIFT_NV quantifications against hand measurements revealed an R^2-value of 0.9372 (Fig. 7a). Comparison of mean values and standard errors of SWIFT_NV versus manual quantification demonstrated almost exact congruence (7.76 $\pm$ 0.40 vs. 7.51 $\pm$ 0.41% NV per total retina; Fig. 7b). In order to assess the intra-individual reproducibility of SWIFT_NV quantifications, one user re-assessed his SWIFT_NV quantifications after a 3-month time interval. The repeated measurement correlated well with the initial data set ($R^2 = 0.9376$; Fig. 7c) and both mean value and standard error were almost exactly reproduced (7.51 $\pm$ 0.41 vs. 7.46 $\pm$ 0.42% NV per total retina; Fig. 7d). In order to assess the inter-individual reproducibility of SWIFT_NV measurements, two users analyzed the same images independently. The data set used for this inter-individual comparison contained retinal images from both a control and a treatment group. The two independent measurements correlated well ($R^2 = 0.9424$; Fig. 7e) and both users obtained almost identical values for the control and treatment groups (Fig. 7f).

In summary, the SWIFT_NV method represents a computer-aided quantification method for retinal NV with reduced user-dependency and favorable time-effectiveness. The robust intra- and inter-individual reproducibility of SWIFT_NV measurements is achieved by combining computer-aided image processing with limited user-interference. In addition, the SWIFT_NV method can help to standardize the quantification of retinal NV from the OIR model and therefore will add further robustness to one of the most widely used models of in vivo angiogenesis research.

References

1. Gyllensten LJ, Hellstrom BE (1954) Experimental approach to the pathogenesis of retrolental fibroplasia. I. Changes of the eye induced by exposure of newborn mice to concentrated oxygen. Acta Paediatr Suppl 43(100):131–148
2. Dorrell MI, Friedlander M (2006) Mechanisms of endothelial cell guidance and vascular patterning in the developing mouse retina. Prog Retin Eye Res 25(3):277–295
3. Aguilar E et al (2008) Chapter 6. Ocular models of angiogenesis. Methods Enzymol 444:115–158
4. Otani A et al (2002) A fragment of human TrpRS as a potent antagonist of ocular angiogenesis. Proc Natl Acad Sci USA 99(1):178–183
5. D'Amato R, Wesolowski E, Smith LE (1993) Microscopic visualization of the retina by angiography with high-molecular-weight fluorescein-labeled dextrans in the mouse. Microvasc Res 46(2):135–142
6. Ritter MR et al (2005) Three-dimensional in vivo imaging of the mouse intraocular vasculature during development and disease. Invest Ophthalmol Vis Sci 46(9):3021–3026

7. Stone J et al (1995) Development of retinal vasculature is mediated by hypoxia-induced vascular endothelial growth factor (VEGF) expression by neuroglia. J Neurosci 15(7 Pt 1):4738–4747

8. Gerhardt H et al (2003) VEGF guides angiogenic sprouting utilizing endothelial tip cell filopodia. J Cell Biol 161(6):1163–1177

9. Benedito R et al (2009) The notch ligands Dll4 and Jagged1 have opposing effects on angiogenesis. Cell 137(6):1124–1135

10. Hellstrom M et al (2007) Dll4 signalling through Notch1 regulates formation of tip cells during angiogenesis. Nature 445(7129):776–780

11. Lobov IB et al (2007) Delta-like ligand 4 (Dll4) is induced by VEGF as a negative regulator of angiogenic sprouting. Proc Natl Acad Sci USA 104(9):3219–3224

12. Xu Q et al (2004) Vascular development in the retina and inner ear: control by Norrin and Frizzled-4, a high-affinity ligand-receptor pair. Cell 116(6):883–895

13. Ye X et al (2009) Norrin, frizzled-4, and Lrp5 signaling in endothelial cells controls a genetic program for retinal vascularization. Cell 139(2):285–298

14. Stefater JA 3rd et al (2011) Regulation of angiogenesis by a non-canonical Wnt-Flt1 pathway in myeloid cells. Nature 474(7352):511–515

15. Lobov IB et al (2005) WNT7b mediates macrophage-induced programmed cell death in patterning of the vasculature. Nature 437(7057):417–421

16. Smith LE et al (1994) Oxygen-induced retinopathy in the mouse. Invest Ophthalmol Vis Sci 35 (1):101–111

17. Pierce EA et al (1995) Vascular endothelial growth factor/vascular permeability factor expression in a mouse model of retinal neovascularization. Proc Natl Acad Sci USA 92(3):905–909

18. Smith LE et al (1999) Regulation of vascular endothelial growth factor-dependent retinal neovascularization by insulin-like growth factor-1 receptor. Nat Med 5(12):1390–1395

19. Aiello LP et al (1995) Suppression of retinal neovascularization in vivo by inhibition of vascular endothelial growth factor (VEGF) using soluble VEGF-receptor chimeric proteins. Proc Natl Acad Sci USA 92(23):10457–10461

20. Rota R et al (2004) Marked inhibition of retinal neovascularization in rats following soluble-flt-1 gene transfer. J Gene Med 6(9):992–1002

21. Bainbridge JW et al (2002) Inhibition of retinal neovascularisation by gene transfer of soluble VEGF receptor sFlt-1. Gene Ther 9(5):320–326

22. Connor KM et al (2009) Quantification of oxygen-induced retinopathy in the mouse: a model of vessel loss, vessel regrowth and pathological angiogenesis. Nat Protoc 4(11):1565–1573

23. Stahl A et al (2009) Computer-aided quantification of retinal neovascularization. Angiogenesis 12(3):297–301

24. Higgins RD et al (1999) Diltiazem reduces retinal neovascularization in a mouse model of oxygen induced retinopathy. Curr Eye Res 18(1):20–27

25. Lange C et al (2007) Intravitreal injection of the heparin analog 5-amino-2-naphthalene-sulfonate reduces retinal neovascularization in mice. Exp Eye Res 85(3):323–327

26. Chen J et al (2008) Erythropoietin deficiency decreases vascular stability in mice. J Clin Invest 118(2):526–533

27. Connor KM et al (2007) Increased dietary intake of omega-3-polyunsaturated fatty acids reduces pathological retinal angiogenesis. Nat Med 13(7):868–873

28. Ritter MR et al (2006) Myeloid progenitors differentiate into microglia and promote vascular repair in a model of ischemic retinopathy. J Clin Invest 116(12):3266–3276

Quantitative Analysis of Angiogenesis in the Allantois Explant Model

Laure Gambardella, Enrique Zudaire, and Sonja Vermeren

Abstract The murine allantois represents a powerful system for the analysis of developmental angiogenesis *ex vivo*, producing a complex vascular network in less than a day's time. This can be visualised by appropriate antibody staining. Angiogenesis in the explants occurs in the absence of external, confounding factors. Since the allantois is taken at an early developmental stage, it is a useful system for studying embryonic lethal mutations with no need for breeding conditional knockouts. In addition, allantois explants are useful for testing inhibitors without any need for animal procedures. Manual quantification of angiogenesis in the complex vascular networks is time and labour intensive. We developed AngioTool, a piece of software which allows the quick, hands-off and reproducible quantification of microscopic images of vascular networks. AngioTool is available free of charge and was developed for analysis of angiogenesis in the allantois explant, although it is also suitable for other systems. Parameters measured include the overall size of the vascular network, average and total vessel length, percentage of area covered by vessels and number of endpoints. In addition, AngioTool calculates the so-called "branching index" (branch points/unit area of the explant), measuring the sprouting activity of an explant whilst correcting for its overall size. The entire experimental protocol including analysis takes ≈5 days to complete.

L. Gambardella • S. Vermeren (✉)
Inositide Laboratory, The Babraham Institute, Cambridge CB22 3AT, UK
e-mail: sonja.vermeren@babraham.ac.uk

E. Zudaire
Angiogenesis Core Facility, National Cancer Institute (NCI), National Institutes of Health (NIH), Gaitersburg, MD 20892-465, USA

E. Zudaire and F. Cuttitta (eds.), *The Textbook of Angiogenesis and Lymphangiogenesis: Methods and Applications*, DOI 10.1007/978-94-007-4581-0_12, © Springer Science+Business Media Dordrecht 2012

1   Introduction

Angiogenesis defines the biological process that generates new blood vessels from a pre-existing vascular plexus. It requires tight regulation of cell proliferation, migration and differentiation. Angiogenesis is necessary to the growth of the organism during development but in the adult, it is also implicated in an array of pathological situations (e.g. cancer, diabetic retinopathy, inflammation) [9]. Many experimental systems, *in vitro* or *in vivo,* have been developed which allow studying of the different aspects of angiogenesis. *In vitro* assays use cultured cells and study endothelial cell properties like motility or tube formation in a matrix. *In vivo* assays are particularly used with genetically modified animal models and provide details about the physiology of the endothelial system. The murine post-natal retina is perhaps the most comprehensive, commonly used *in vivo* experimental system today [5, 13]. It allows analysis of angiogenesis in the presence or absence of intravitreally or systematically administered substances (activators, inhibitors; in suitable mouse models agents inducing the deletion of a gene of interest). Other systems include the embryonic hindbrain, a system that is not amenable to drug treatments, as well as a number of wounding assays and tumour models.

Mouse knock-out strategies have in recent years permitted the identification of many crucial players in angiogenesis [1]. The inactivation of such molecules often leads to embryonic lethality around mid-gestation. This limits the information that can be gained from the analysis of germ-line knock out mice to early embryogenesis and means, that conditional models (such as inducible, and/or cell type-specific knock-out models) need to be established to allow studying the full function of a protein of interest. This requires a floxed mouse and lengthy breeding steps.

Embryonic explants are taken before the onset of embryonic wasting due to cardiovascular defects has set in. They can be carried out without any need for conditional mouse models, and allow an immediate analysis of developmental angiogenesis even in embryonically lethal mutants. Growing in defined conditions in a tissue culture incubator, embryonic explants also evade potentially deleterious external influences such as placental defects. Two embryonic regions are commonly used as explants: the para aortic splanchnopleura (P-Sp) and the allantois. The P-Sp explant requires a layer of OP9 feeder cells and develops over 2 weeks. In the first week it generates a vascular bed; from this, a vascular network sprouts during the second week [12]. P-Sp explants permit to distinguish between defects in vasculogenesis (the establishment of the primary vascular network during embryogenesis) and angiogenesis. The second system is the allantois explant [8]. Allantois explants grow on fibronectin-coated tissue culture dishes and directly produce a vascular network over 18 h by angiogenesis. Allantois explants allow the immediate analysis of cellular events such as proliferation, apoptosis, migration and sprouting.

The murine allantois appears during early gastrulation. By embryonic day 11(E11) it has transformed into the umbilical artery and vein that constitute the vascular link between the placenta and the foetus. Vascular cells inside the allantois arise at the early headfold unturned stage (4–5 somite pairs) [2]. By E7.5–E8 (5–7 somite pairs) a PECAM-positive vascular primordial network can be detected in the allantois [3].

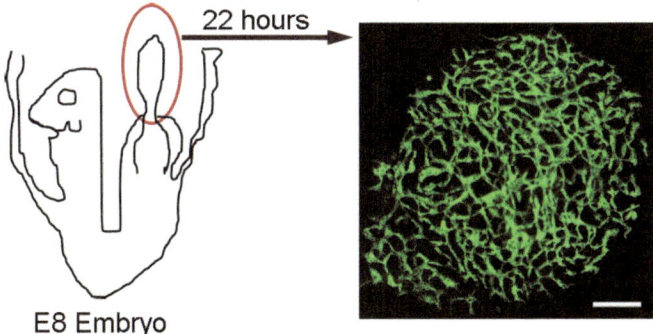

Fig. 1 Schematic representation of the allantois explant technique. The allantois is harvested from a E8 embryo and set up in culture for 22 h in a fibronectin-coated plate. The *right panel* shows the aspect of the explants after an anti-VE-cadherin and anti-Endomucin immunofluorescence detection. The scale bar = 0.5 mm

If harvested between this stage and 8 somite pairs (E7.5–E8.5) the allantois is capable of extending its primordial vasculature into an enlarged network by angiogenesis *ex vivo* (Fig. 1). This network is amenable to staining with antibodies specific for endothelial cells, and can then be visualized microscopically. Analysing the complexity of such networks manually is time-intensive and fastidious.

We therefore designed AngioTool, a light-weight user-friendly software for the quantitative assessment of angiogenesis in allantois explants. The software uses a fast multi-scale hessian based algorithm [4, 7, 11] for vessel segmentation and a skeletonization approach [6] for individual segment identification. AngioTool is written in Java and takes advantage of the new concurrency features in Java Standard Edition 7. It uses and extends the open source libraries ImageJ (image processing and analysis; http://rsbweb.nih.gov/ij/), Apache POI (Excel compatibility; http://poi.apache.org/), JIDE Common Layer (Swing components; http://www.jidesoft.com/) and JGraph (Graph analysis; http://www.jgraph.com/). AngioTool can be downloaded free of charge as at http://angiotool.nci.nih.gov.

Parameters measured by AngioTool include the overall size of the vascular network, average length and diameter of vessels and number of endpoints. In addition AngioTool calculates the so-called "branching index" (branch points/unit area of the explant), measuring the sprouting activity of an explant whilst correcting for its overall size. Whilst the system was developed for analysis of angiogenesis in the allantois explant, it has also been useful for analysis of retinal angiogenesis.

1.1 Advantages and Disadvantages of the Use of Allantois Explants

Allantois explants have several advantages compared to alternative assays:

- Allantois explants can be carried out with embryonically lethal mutants, where death typically occurs from around E10 when foetal-maternal exchange begins to be important. Since the allantois is taken at E8 (E7.5–E8.5), before embryonic wasting occurs, this method allows assessing sprouting angiogenesis even from severe mutants without the need for lengthy breeding of inducible conditional knock-out models.

- Being *ex vivo*, allantois explants assay sprouting angiogenesis in the absence of confounding factors such as heart or placental defects. Since the explants are set up before foetal-maternal exchange starts, circulatory defects cannot interfere with the observed phenotype.
- The assay is easy to learn, fast and inexpensive.
- Allantois explants are amenable to testing of pro- or anti-angiogenic substances, which can be added to the culture medium. Testing of such substances in the allantois system represents improved animal welfare, as no animal procedures need to be carried out [10].
- Allantois explants can be monitored in real-time using time-lapse imaging, provided the endothelial cells are rendered clearly identifiable (for example by use of an endothelial-specific GFP transgene or by intravenous injection of FITC-dextran of the mother prior to setting up of the explants). Allantois explants are informative about many aspects of endothelial cell behaviours. For example, cell proliferation can easily be monitored by phosphohistone H3 staining.

However, allantois explants also have some disadvantages compared to alternative *in vivo* angiogenesis assays:

- Allantois explants represent angiogenesis in the absence of a gradient of VEGF. This does happen under physiological conditions (e.g. in the yolk sac) but represents an exception rather than the norm.
- Blood flow is known to shape angiogenesis, but there is no blood circulation in the vessels generated by allantois explants.

1.2 Experimental Design

Generation of the embryos. It is crucial to generate sufficient embryos for a successful experiment. To sacrifice the pregnant female at the right time, one needs to know the time of mating between the male and the female. This can be achieved by checking for vaginal plugs, which are made up of coagulated secretions from the vesicular glands of the male and persists in the female's vagina for 10–12 h after copulation. A good practice is to set up the pairs late in the afternoon, such that mating will occur around midnight and the vaginal plug will still be apparent in the morning. To see the plug, lift the female by the base of her tail and examine her vaginal opening for a white mass. The presence of the plug is only proof of mating. If fertilization has followed the mating, the female will be with embryos at the right stage for the experiment 7–8 days later.

Harvesting allantois at the right stage. As described above, the timeframe in which the allantois is suitable for *ex-vivo* angiogenesis is quite narrow. Even with a correct plug date, it is important to be able to recognize that the embryos are at the right developmental stage. The allantois should look like that in Fig. 2g. If the allantois looks smaller than in the picture, with a cone shape and some red blood cells circulating inside, the embryos are too old.

Suitable number of explants. A suitable number of explants have to be set up in culture to be able to run a reasonable comparison between several treatments or genotypes. We suggest a minimum of six explants for each condition to be analysed, taken from at least two independent experiments. It is best to use large litters as it is important to compare individual embryos from the same litter: there may be subtle differences between litters. It is also advisable to generate more explants for a statistically sound analysis. Keep in mind that some explants could get lost during the process: they may not attach to the tissue-culture dish, and fail to make the endothelial network or adhere in a corner of the well rendering the microscopical analysis impossible. Since plugged females are not necessarily pregnant, it is advisable to set up slightly more females than would be required if everything went to plan.

Controls. To analyse the effect of a drug, it is important to know in which substance it has been reconstituted (e.g. DMSO). If a defined volume of a drug solution has been added to the treated explants, a similar volume of the vehicle should be added to the culture medium of the control explants.

Image acquisition and processing. During the image acquisition process, it is important to ensure that the intensity level of the labelled tissues is adequate. Both over- and underexposed images will be difficult to analyze whilst low levels of back- and/or foreground noise will help making the analysis easier and more accurate. Since AngioTool is an image analysis, rather than an image processing tool, images will need to be processed with an image processing software prior to being analyzed with AngioTool. AngioTool accepts 8 and 16 as well as 3×8 bit images.

2 Materials

2.1 Reagents

- Pregnant females 7.5–8.5 days post-coitum. Animals must be maintained under supervised conditions according to the relevant national and local authorities' rules and regulations
- Ethanol (70% (vol/vol))
- 10X and 1X PBS (use 1X unless indicated 10X)
- Bovine-fibronectin (e.g. Sigma-Aldrich, cat. no. F4759)
- Tissue culture grade distilled water (e.g. GIBCO, cat. no. 15230)
- Heat inactivated fetal bovine serum (FBS, e.g. from GIBCO)
- DMEM (e.g. GIBCO, cat. no. 22320)
- 100X Penicillin-streptomycin (e.g. PAA, cat. no. P11-010)
- DMSO (Dimethyl sulfoxide; Sigma-Aldrich, cat. no. 154938)

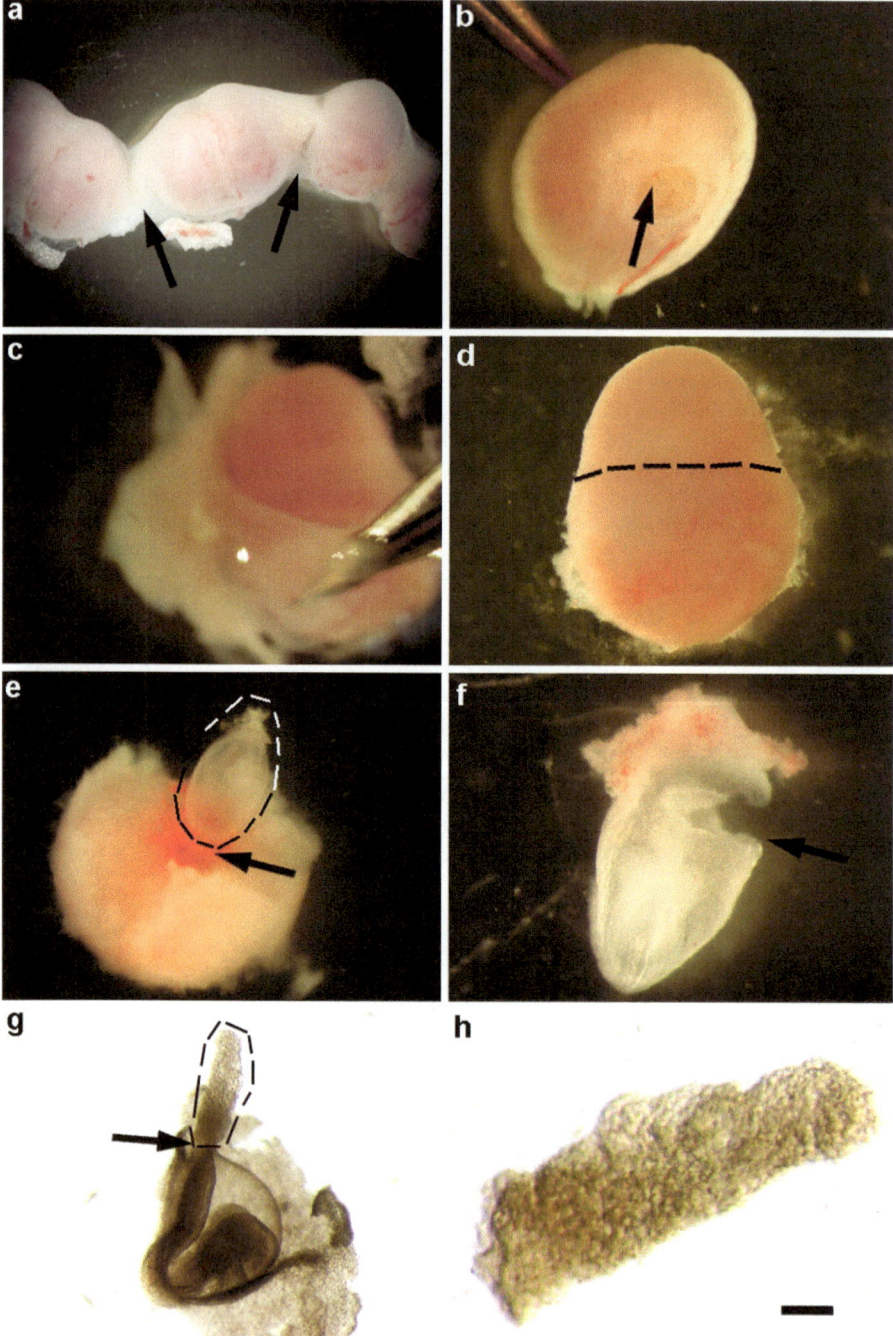

Fig. 2 Allantois dissection. (**a**) The uterus horns under the dissecting microscope. The *arrows* indicate where to cut to separate the embryos. (**b**) Embryo still surrounded by the uterus. The *arrow*

- Paraformaldehyde (PFA; e.g. Sigma-Aldrich, P6148): PFA is highly toxic if inhaled or swallowed and irritating to eyes, respiratory system and skin. Wear appropriate protective equipment and manipulate in a chemical fume hood
- Sodium hydroxide pellets (e.g. Merck, cat. no. 106498): This can cause severe burns
- Hydrochloric acid (HCl, 37% wt/vol; e.g. BDH): This causes severe burns and needs to be handled in a fume hood
- Triton X-100 (Sigma-Aldrich, cat. no. T9284)
- Skimmed milk powder (e.g. Marvel)
- Monoclonal rat anti-mouse endomucin (V.7C7; Santa Cruz, cat. no. sc65495)
- Monoclonal rat anti-mouse VE-cadherin (BD Bioscience, cat. no. 550548)
- Alexa Fluor 488 goat anti-rat IgG (H + L) antibody (Invitrogen, cat. no. A11006)
- Glutaraldehyde (8% (wt/vol)) aqueous solution (Fluka, cat. no. 49624) This compound is harmful by inhalation and if swallowed. It is irritating to respiratory system and skin and can cause serious damage to eyes
- Tris base (e.g. Melford)
- Glycerol (e.g. Sigma-Aldrich, ≥99%, ACS grade)
- DABCO (1.4-Diazabicyclo[2.2.2]octane; Sigma-Aldrich, cat. no. D2522) This compound is harmful if swallowed and irritating to eyes, respiratory system and skin

2.2 Equipment

- Chemical fume hood
- Whatman 3M paper
- Hot plate magnetic stirrer
- pH test strips 6–7.7 (Sigma-Aldrich, cat. no. P3536)
- Sterile tissue culture hood
- Water-bath
- Dissection instruments: standard surgical scissors, one pair of Dumont tweezers #5 and one pair of Dumont tweezers 'biology tips' #5
- Cell culture dishes (6 or 10 cm)
- Plastic pasteur pipettes (1 ml)
- Dissecting microscope equipped with light source in the base (e.g. Nikon SMZ800)
- 8-wells tissue-culture treated μ-slides (Ibidi, cat. no. 80826)
- Tissue culture incubator (humidified; 6% CO_2)
- 500 ml sterile filter units, pore size 0.22 μm (e.g. Millipore, cat. no. SCGPT05RE)

Fig. 2 (continued) indicates the hole where to introduce the forceps to remove the uterine muscles. (**c**) Removal of the uterine muscles. (**d**) The decidua. The *dotted line* indicates where to open it. (**e**) Opened decidua. The *dotted circle* delineates the embryo in its yolk sac. The *arrow* shows where to detach. (**f**) The embryo in its yolk sac. The *arrow* shows how to open the yolk sac. (**g**) The embryo once the yolk sac is removed. The *dotted circle* surrounds the allantois. (**h**) High magnification of the allantois. The scale bar represents 500 μm in (**a**), 250 μm in (**b-e**), 120 μm in (**f** and **g**) and 30 μm in (**h**)

- Inverted microscope equipped with wide-field epi-fluorescence (e.g. Cell^R from Olympus) coupled to a digital camera
- Computer with a Windows Platform operative system (32-bits), 3 GB (or more) of RAM with Java^{TM} Platform Standard Edition 7 and AllantoisAnalysis software installed. Multiprocessor systems will perform significantly better
- Photoshop CS3 software (Adobe)

2.3 Reagent Setup

Bovine fibronectin stock. Dissolve the fibronectin in distilled water at 1 mg/ml and stored at $-20°C$ in small aliquots (20 μl) to avoid repeated freeze-thawing. Once reconstituted in water do not keep longer than 6 months.

Heat-inactivated FBS. Incubate FBS at 56°C for 30 min. Aliquot in 50 ml tubes under sterile conditions and freeze at $-20°C$.

Explant culture medium. Add 50 ml of heat-inactivated FBS and 5 ml of penicillin-streptomycin 100X to 445 ml of DMEM using sterile technique. Aliquot the medium and store at 4°C for a maximum of 2 months.

4% PFA. Under a chemical hood weigh 20 g of PFA powder and transfer in a 500-ml beaker. Add 250 ml distilled water and heat under agitation on a hot plate. Add 5–10 μl of NaOH 10 N to the obtained solution and it should become transparent. Filter on Whatman paper above a 500 ml-graduated cylinder. Add 50 ml of 10X PBS. Make up with distilled water to 500 ml. Check that the pH is around 7.3. If not then adjust with HCl or NaOH. Store aliquotted at $-20°C$.

10% Triton X-100. Pour 50 g of Triton X-100 in a beaker and make up with distilled water to 500 ml. Mix using a magnetic stirrer and filter sterilise. Keep at room temperature.

Rinsing buffer. To make 50 ml, add 1 ml 10% Triton X-100 to 49 ml PBS.

Blocking buffer. Put 150 mg of skimmed milk powder in a suitable tube and pour rinsing buffer until you obtain a 5 ml solution.

Primary antibodies mix. Add the anti-VE-cadherin (dilution 1/50) and the anti-endomucin (dilution 1/100) antibodies together in blocking buffer.

Secondary antibody mix. Add the Alexa Fluor 488 goat anti-rat (1/1000) in blocking buffer.

Post-fixative. To make 2.8 ml of post-fixative, add 1.4 ml of 4% PFA and 35 μl of glutaraldehyde to 1.365 ml PBS.

1 M Tris–HCl pH 8.0 Add 12.1 g Tris base in 100 ml H_2O. Adjust pH to 8.0 with concentrated HCl.

DABCO. Weigh 12.5 g DABCO and add 450 ml glycerol, 25 ml 1 M Tris–HCl pH 8.0 and 25 ml distilled water. Dissolve by warming on a hotplate to 70°C. Mix by swirling and gentle inverting. Store at +4°C in a lightproof bottle.

2.4 Equipment Setup

Dissecting instruments. Ensure that dissecting instruments are cleaned and sterilized with 70% ethanol.

Installation of the software. AngioTool requires Java Standard Edition 7 which can be downloaded from the official website at http://dlc.sun.com.edgesuite.net/jdk7/ binaries/index.html. Download and install the (self-extracting) Windows Platform, Multi-Language JDK version (jdk-7-ea-bin-b130-windows-i586-18_feb_2011.exe in the website above). To ensure that your computer is running the appropriate java version go to http://www.java.com/en/download/installed.jsp. The test should confirm "Your Java version is 1.7.0-ea". To install the AllantoisAnalysis software, download AllantoisAnalysisSetup.exe from http://angiotool.nci.nih.gov to your computer. Double click and follow the installation Wizard.

3 Procedure

3.1 Preparation of the Allantois Explants TIMING 24 h

1. Euthanize a female mouse 7.5–8.5 day post-coitum employing a humane method. Animals must be manipulated according to the relevant national and local authorities' rules and regulations.
2. Place the animal on its back facing you. Open the abdominal cavity using surgical scissors. Locate the two uterine horns at the lower extremity of the cavity. Gently pull one of the horns by pinching it with a regular pair of Dumont #5 tweezers between two implantation sites and cut the upstream extremity (located under the gut) where it is attached to the ovary. Repeat with the second horn. Finally extract the entire uterus by cutting at the place where the two horns are joining. Place the uterus in sterile cold PBS.
3. Transfer the uterus into a dish containing ice-cold sterile PBS. Count the number of concepti keeping the dish on ice.
4. Coat the ibidi μ-slides with a 5 μg/ml fibronectin solution. Plan one well per conceptus and use 200 μl of fibronectin solution per well (the fibronectin solution is made by diluting the stock solution (200 X) into sterile distilled water). Leave the slides at room temperature for 30 min.
5. Prepare the media to culture the explants. You need 250 μl per well. The treatment medium contains stimulatory or inhibitory substances diluted at the required concentration in the culture medium. The control medium contains a volume of the vehicle equivalent to the volume of active agent added in the medium to make the treatment condition (see Table 1 for examples).
6. When the coated ibidi μ-slides are ready, pour the suitable medium in each well and take the μ-slides to the dissecting microscope area.

Table 1 Troubleshooting advice

Step	Problem	Possible reason	Solution
7	The allantois is not flushed in the μ-slide well	The allantois is stuck to the tip	Look carefully at the tip under the microscope to locate the allantois. Aspirate all the medium contained in the μ-slide well and flushed again. Try to do several times or until the allantois detaches and falls in the well
26	Photoshop is not able to assemble the jigsaw	Not known	Select 'Interactive Layout' instead of 'Auto' in the Photomerge window and do the jigsaw yourself manually
34	AngioTool does not generate an initial outline	The software may not be able to discern the border of the explant, because (i) the explant extends to the border of the image or (ii) the background is noisy	Prepare your images in photoshop. Using the lasso tool define the outline of the explant, and artificially darken or widen the background as required to allow AngioTool to discriminate between explant and background pixels

7. Look at the uterus under the dissection microscope and separate the embryos by cutting between the individual implantation sites (Fig. 2a). Put the dish back on ice. From now on use the Dumont tweezers 'biology tips' #5. Pick one of the embryos under the microscope using a fresh dish containing ice-cold PBS. The conceptus is still inside the uterus, which now has a hole on each side due to the previous cutting (Fig. 2b). Insert the two pairs of tweezers in one of these holes and remove the uterine muscles surrounding the conceptus (Fig. 2c). At this developmental stage, the embryo is surrounded by the decidua (Fig. 2d). The decidua is ovoid with one larger and darker extremity (the side of the developing placenta) (Fig. 2d). Open it gently between these two regions. The embryo is now visible inside its yolk sac that looks like a transparent bag (Fig. 2e). Gently separate the yolk sac from the developing placenta and delicately open it as shown (Fig. 2f). The allantois is now identifiable (indicated by a dotted circle in Fig. 2g). Separate it from the embryo. To collect the allantois (Fig. 2h), cut one millimetre from the extremity of a blue tip and very gently aspirate the allantois with a P1000 pipette. Transfer it into one of the wells of a prepared μ-slide. Check under the dissecting microscope that the allantois has indeed been transferred into the μ-slide. Position it in the middle of the well by orientating it gently with the forceps. Be careful not to destroy the fibronectin coating when doing this.
8. Repeat this with all remaining embryos, then transfer the μ-slides in a tissue culture incubator set to 37°C, 6% CO_2 for a period of 18–22 h.

3.2   Staining of the Allantois Explants (2 Days)

9. Aspirate the culture medium and rinse explants once with PBS.
10. Fix in 4% PFA at 4°C for 4 h.
11. Rinse twice with PBS

At this point, explants can be stored in PBS at 4°C for a few weeks as long as they do not dehydrate. To avoid dehydration, place the μ-slides in a humid chamber (closed container with damp tissues) and regularly check if PBS needs to be added in the wells.

12. Aspirate PBS and block nonspecific antibody binding with Blocking buffer for 2 h at 4°C.
13. Aspirate the Blocking buffer and add 120 μl primary antibodies mix to each well. Incubate over night (ON) at 4°C
14. Rinse several hours at 4°C in Rinsing buffer changing the solution at least 6 times.
15. Aspirate the Rinsing buffer and add secondary antibody mix to each well. Incubate ON at 4°C in the dark.
16. Rinse 2 h at 4°C in the dark with Rinsing buffer changing the solution at least 6 times.
17. Post-fix the explants in Post-fixative for 30 min at RT in the dark (300 μl per well)
18. Rinse twice with PBS.
19. Add 200 μl of DABCO to each well and store at 4°C in the dark until imaging.

At this point, samples can be stored at 4°C in the dark for up to 2 weeks as long as they do not dehydrate. To limit dehydration place the μ-slides in a humid chamber and regularly check if DABCO needs to be added in the wells.

3.3   Imaging

20. To get sufficiently detailed images to run the analysis software, image the entire explants at 10x magnification by taking several pictures.
    It is important to make sure there is some overlap between any adjacent images to allow a smooth assembly of the entire explant. Adjust the camera's exposure for each individual picture to make sure each branch of the explants will be clearly visible but not overexposed. Be particularly attentive at the edge of the explants where the labelling tends to be less bright.
21. Save the pictures in TIFF format.
    Once the pictures have been taken you could wait as long as you wish before analysing them. However it is advisable to make sure that the photographs taken cover the entire explants without gaps before the explants deteriorate.
22. Open Adobe Photoshop CS3

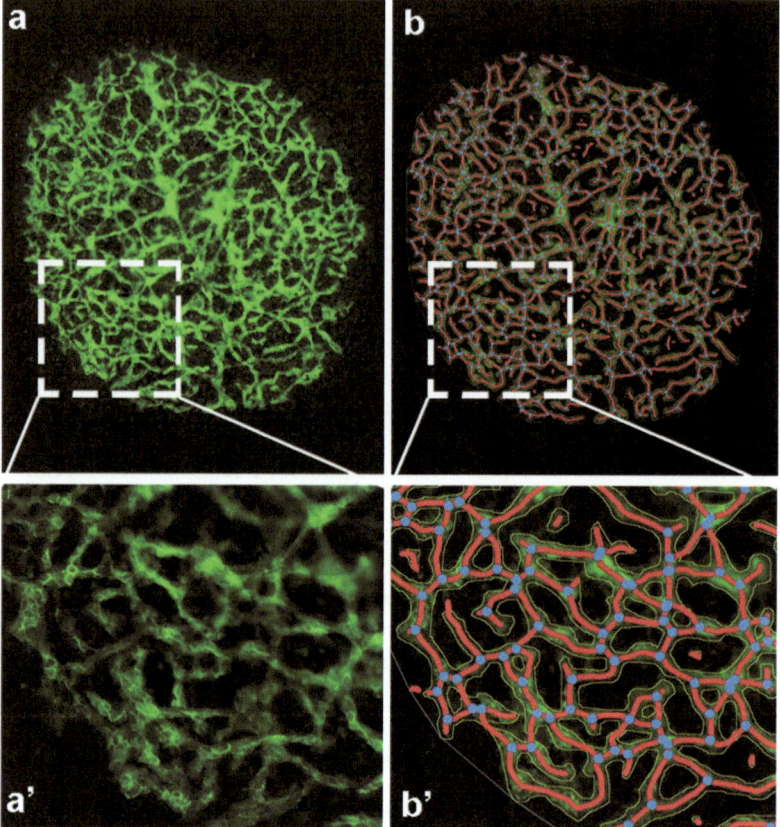

Fig. 3 A stained allantois explant before and after skeletonising. (a) Image of a representative explant. **(b)** Skeleton of the explant shown in **(a)**. **(a', b')** Enlarged sections indicated in **(a)** and **(b)**. The present allantois explant formed part of an analysis performed in the context of work published elsewhere [14]

23. Go to File > Automate > Photomerge…
24. In Photomerge window, select Auto for Layout
25. Browse to select the different images taken for a particular explant.
26. Then click "load". Photoshop aligns the different parts of the explants to reconstitute it as a jigsaw.
27. Once the jigsaw is assembled, check that all the pieces of the explants are present and that the exposure is suitable for analysis. If not, go back to the microscope and take new images.
28. Display Layers (Go to window > Layers)
29. In the Layer window, right click and select 'Merge visible'.
30. Save the "panorama" in Tiff format. (See Fig 3 for examples of assembled jigsaws.)

3.4 Computational Analysis of the Explants

31. Click on the AngioTool icon on the computer desktop or directly in the executable file (AngioTool.exe) located in the folder where the program was installed.
32. Click on 'Settings'. The following options are to be chosen before opening an image to be analysed:

    - The *Resize Image* control allows the user to downsize the image. Resizing of images has to be done prior to opening the image. To utilize this control, click the "Resize Image" checkbox. Now use the spinner to adjust the resizing factor. Once opened, the image will be resized by the factor indicated in the spinner. For instance, if the resizing factor is 2, the software will create an image with half the width of the original (the height is resized accordingly). While this reduces the resolution of the image, results are often not affected and the analysis is done significantly faster. Once an image is opened the resize feature will lock; it can be unlocked using the associated button.
    - The *Calibration* control allows setting a correlation between number of pixels and one unit length of the image. This control must be used before running the analysis. In the first box, type the desired length in pixels and in the second box the corresponding distance in millimeters. Once a calibration is set the program will save measurements both in pixels and the defined units.
    - The *Overlay Settings* allow the user to customize the appearance (stroke width or diameter and colour) of the different components of the overlay: outline, skeleton, branching points and boundary. This control can be used before or after the analysis has been run. You can choose to include or exclude any feature of the overlay by clicking on or off the checkbox for that feature. Use the spinner to the right of the feature's name to adjust the width. To change the colour, click on the colour box. The *Save Image* button allows the user to save any changes in the overlay's appearance. The user can also change the file format of the image to be saved.

33. To open an image to be analysed, click on the *Open Image* button. In the open dialog, navigate to the folder that contains the image/s to analyze. Only image files in formats recognized by the software (tiff, jpg, etc.) will be shown. Choose a file to preview the image and then click *Open* to open it in AngioTool. Only one image can be analyzed at a time and images are expected to contain brightly labelled tissue over a dark background.
34. AngioTool will open the image in a separate window and will estimate an initial set of analysis parameters. The initial set of parameters can be tuned to optimize the analysis of the allantois. Two main controls have been included in the *Analysis tab* to help with optimising the detection of the allantois, parameters can be adjusted as described below. An outline of the selection

will be displayed as an overlay on the test image. This outline will dynamically update its shape in response to the adjustments done to the controls. This enables the user to visually tune the selection of the allantois. The **Hide Overlay** button toggles the overlay on and off, assisting the user to evaluate the accuracy of the selection at every moment during the analysis.

35. The **Vessel Diameter** slider allows the user to detect vessels based on their diameter. The units in the controls are always shown in pixels. The user selection is shown as red marks on the control, which can be individually toggled on and off in order to adjust the selection. Clicking on higher values of thickness will select thicker vessels and vice versa. Since allantois explants can contain vessels with a wide range of diameters, multiple user selections are allowed. If necessary, the upper limit of this range can be dynamically modified by adjusting the spinner next to the control.

36. The **Vessel Intensity** slider allows for detection of vessels based on their intensity. Lower and higher intensity values in the control represent respectively dimmer and brighter pixels in the image. Weakly stained vessels can be detected by lowering the left knob and unspecific bright spots in the image can be eliminated by lowering the upper knob.

37. Two additional controls are provided for removal of background and foreground particles based on size. In order to adjust these controls, first click on the checkbox for the control. The **Remove small particles** slider allows removal of outlines with sizes below the value chosen in the slider. Adjust the slider to the desired size, remembering to consult the actual changes in the outline overlay of the test image. The range of sizes shown in the slider can be dynamically modified using the spinner next to the slider. Finally, the **Fill Holes** slider allows filling areas within the tissue which are not included in the selection (for instance due to low intensity). Move the slider to the desired size and if needed, modify the range of the slider with the associated spinner.

38. Once the adjustments are complete and the outline overlay best matches the test image, click the **Run Analysis** button. The analysis starts with the computation of the skeleton of the selected area. The skeleton is a one-dimensional representation of the vessels in the allantois and allows the computation of branching points, vessels lengths, etc. The progress bar visually updates the user on the progress of the analysis. Once the analysis is complete, the resulting image overlay will show the allantois outline, the skeleton and the branching points. Branching points are defined as points in the image where there is a confluence of more than two vessel segments.

39. After running an analysis, a summary of the data generated is saved to an Excel file. The directory and name of the results Excel file can be set by using the **Save To** button under the **Analysis** tab and are shown in the associated text field. If the user does not chose a directory and a name, an Excel file will be generated and saved with a unique name (ResultsXXXX.xls) in the folder where the test image is located. By default, the resulting image containing the overlay will be saved to the same location of the original image and this setting can be customized using the **Save result image** checkbox.

The user-defined parameters are saved together with the name of the image, the time of the analysis and the computed metrics. The data generated for an image can be appended to an existing Excel file by selecting the Excel file, which contains the old data. If several images are being analyzed during a session, all the data will be saved to the same Excel file unless specified otherwise.

40. If the results of an image analysis are not satisfactory the controls can be tuned again and the analysis rerun. A new set of data will be appended to the existing Excel file in use (or in a new file if the filename and directory has been changed). A new image can be analyzed by opening it using the ***Open Image*** button.

    Analysis of large images requires allocation of a substantial amount of memory. The ***memory monitor*** shows the used and total available memory. The monitor shows a red background when the system is running out of memory resources or green otherwise. The application may run out of memory if the image being analyzed is very large. Resizing images by a factor of 2 will help the system save substantial memory resources, speed up the analysis, often without effect on the results.

41. To exit the application click on the ***Exit*** button.

4   Anticipated Results

An example of a stained allantois explant before and after skeletonising is shown in Fig. 3. Examples of inhibitor-treated explants and of other applications for AngioTool (analysing angiogenesis in the embryonic hindbrain and murine retinal systems) and statistical analysis of data obtained have been published elsewhere [14].

Acknowledgements This work was funded by UK Medical Research Council grant G0700740. SV is a Biotechnology and Biological Sciences Research Council David Phillips Fellow (BB/C520712).

References

1. Adams RH, Alitalo K (2007) Molecular regulation of angiogenesis and lymphangiogenesis. Nat Rev Mol Cell Biol 8:464–478
2. Downs KM, Gifford S, Blahnik M, Gardner RL (1998) Vascularization in the murine allantois occurs by vasculogenesis without accompanying erythropoiesis. Development 125:4507–4520
3. Drake CJ, Fleming PA (2000) Vasculogenesis in the day 6.5 to 9.5 mouse embryo. Blood 95:1671–1679
4. Frangi AF, Niessen WJ, Vincken KL, Viergever MA (1998) Multiscale vessel enhancement filtering. In: Wells WM, Colchester A, Delp S (eds) Medical image computing and computer-assisted intervention, Lecture notes in computer science. Springer, Berlin, pp 130–137
5. Fruttiger M (2007) Development of the retinal vasculature. Angiogenesis 10:77–88
6. Lee TC, Kashyap RL, Chu CN (1994) Building skeleton models via 3-D medial surface axis thinning algorithms. CVGIP-Graphical Model Image Process 56:462–478

7. Manniesing R, Viergever MA, Niessen WJ (2006) Vessel enhancing diffusion: a scale space representation of vessel structures. Med Image Anal 10:815–825

8. Perryn ED, Czirok A, Little CD (2008) Vascular sprout formation entails tissue deformations and VE-cadherin-dependent cell-autonomous motility. Dev Biol 313:545–555

9. Potente M, Gerhardt H, Carmeliet P (2011) Basic and therapeutic aspects of angiogenesis. Cell 146:873–887

10. Russell WMS, Burch RL (1959) The principles of humane experimental technique. Methuen & Co.Ltd, London

11. Sato Y, Nakajima S, Shiraga N, Atsumi H, Yoshida S, Koller T, Gerig G, Kikinis R (1998) Three-dimensional multi-scale line filter for segmentation and visualization of curvilinear structures in medical images. Med Image Anal 2:143–168

12. Takakura N, Huang XL, Naruse T, Hamaguchi I, Dumont DJ, Yancopoulos GD, Suda T (1998) Critical role of the TIE2 endothelial cell receptor in the development of definitive hematopoiesis. Immunity 9:677–686

13. Uemura A, Kusuhara S, Katsuta H, Nishikawa S (2006) Angiogenesis in the mouse retina: a model system for experimental manipulation. Exp Cell Res 312:676–683

14. Zudaire E, Gambardella L, Kurcz C, Vermeren S (2011) A computational tool for quantitative analysis of vascular networks. PLoS One 6:e27385

The Murine Hindbrain as a Model to Study the Molecular and Cellular Mechanisms of Angiogenesis in Intact Tissues

Charlotte Maden and Christiana Ruhrberg

Abstract This chapter describes methodology to exploit the murine hindbrain as a powerful and versatile model system with three main advantages for studying developmental angiogenesis. Firstly, the hindbrain model permits the accurate assessment of the intricate behaviour of endothelial cells within a natural multicellular microenvironment. Secondly, due to the early embryonic vascularisation of the hindbrain, it is particularly useful to study genetic mutations that result in embryonic lethality, reducing the need to breed conditional knockout mice, as for postnatal models. Thirdly, the unique architecture of the vascular network in this tissue allows high resolution imaging of angiogenic molecules and vascular structures that are readily quantifiable.

Keywords Angiogenesis • Hindbrain • Mouse • Embryo

Abbreviations

HRP	Horseradish peroxidase
IB4	Isolectin B4
PECAM	Platelet endothelial cell adhesion molecule
NGS	Normal goat serum
NRS	Normal rabbit serum
PBS	Phosphate buffered saline
PBT	PBS + 0.1% TritonX100

C. Maden • C. Ruhrberg (✉)
UCL Institute of Ophthalmology, University College London, 11-43 Bath Street,
London EC1V 9EL, UK
e-mail: c.ruhrberg@ucl.ac.uk

E. Zudaire and F. Cuttitta (eds.), *The Textbook of Angiogenesis and Lymphangiogenesis: Methods and Applications*, DOI 10.1007/978-94-007-4581-0_13,
© Springer Science+Business Media Dordrecht 2012

PFA   Paraformaldehyde
SVP   Subventricular vascular plexus
VEGF   Vascular endothelial growth factor

1 Introduction

Hindbrain vascularisation is normally initiated around embryonic day (E) 9.5 in
the mouse, when a few vascular sprouts emerge from the perineural vascular
plexus (Fig. 1a, step 1) and invade the hindbrain parenchyma (Fig. 1a, step 2)[1].
At E10.25, these radially growing vessel sprouts change direction, turning at near
right angles to extend parallel to the ventricular hindbrain surface (Fig. 1a, step 3)[1].
As these sprouts meet and anastomose, the subventricular vascular plexus (SVP)
begins to form (Fig. 1a, step 4)[1, 2]. VEGF isoform signalling regulates these first
steps of vessel recruitment into the brain (Fig. 1b)[2, 3], with neuronal progenitors in
the ventricular zone being the source of VEGF (Fig. 1b)[4, 5].

At E11.5 in the mouse, the process of SVP formation is most readily observed,
as numerous filopodia-studded tip cells extend and begin to fuse with neighbouring tip
cells to establish a complex SVP of sufficient capillary density to supply the growing
brain with blood (Fig. 1a, step 4)[1]. This fusion process is optimised by yolk sac-
derived tissue macrophages that invade the hindbrain independently of blood vessels,
but have a high affinity for endothelial cells and interact with opposing tip cells
(Fig. 1b, c)[1]. By E12.5, an extensive vascular network has been established in the

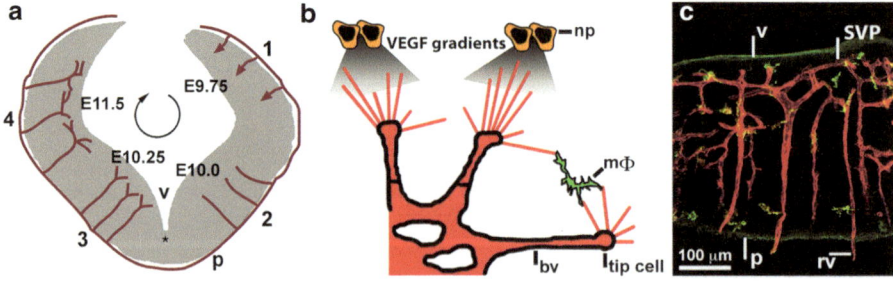

Fig. 1 Time course and mechanism of blood vessel growth in the mouse embryo hindbrain.
(a) Vessels begin to sprout from the perineural vascular plexus and invade the neural tissue at
E9.75 (step 1) and continue to grow radially at E10.0 (step 2). Vessels begin to branch near the
ventricular surface from E10.25 onwards (step 3); between E10.25 and E12.5, a the SVP forms
(step 4). (b) Tip cells in the sprouting blood vessel (*bv*) extend filopodia, which detect VEGF
gradients produced by neuronal progenitors (*np*) to guide the growing vessels. Anastomosis of
neighbouring vessels creates a network, a process that is optimised by yolk sac-derived tissue
macrophages (mΦ) interacting with tip cells. (c) Vibratome section through an E11.5 hindbrain
immunofluorescently labelled with IB4 in *red* to highlight blood vessels and F4/80 in *green* to
identify macrophages. Scale bar: 100 μm. Abbreviations: *v* ventricular brain surface, *p* pial brain
surface, *SVP* subventricular vascular plexus, *rv* radial vessel. * indicates the midline

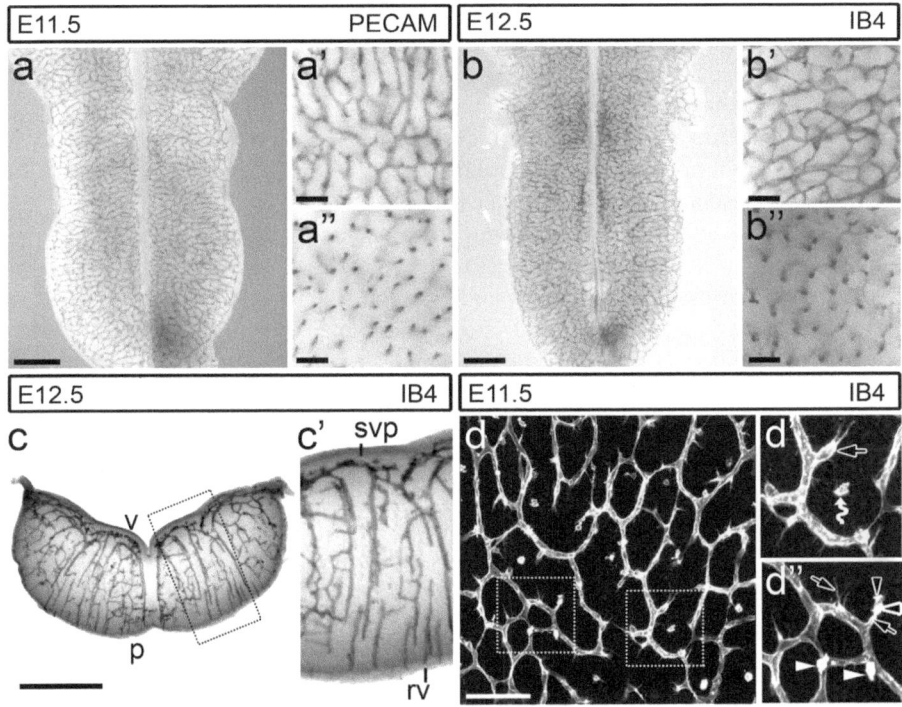

Fig. 2 Mouse embryo hindbrains immunolabelled to visualise the developing blood vessels.
(a) Blood vessels in a flatmounted E11.5 hindbrain, immunolabelled for PECAM, visualised by HRP histochemistry and mounted ventricular face up; higher magnification of SVP vessels on the ventricular side (**a'**) and radial vessels on the pial side (**a"**); size of field: 500 μm × 500 μm (0.25 mm²). (b) Blood vessels in a flatmounted E12.5 hindbrain, immunolabelled for IB4, visualised by HRP histochemistry and mounted ventricular face up; higher magnification of SVP vessels on the ventricular side (**b'**) and radial vessels on the pial side (**b"**); size of field: 500 μm × 500 μm (0.25 mm²) Note that IB4 also detects tissue macrophages, but few are left on the hindbrain surface at this stage. (c) A 100 μm vibratome section through an E12.5 hindbrain, with blood vessels immunolabelled for IB4 and visualised by HRP histochemistry; radial vessels (*rv*) extend from the pial side (*p*) of the hindbrain and form the SVP on the ventricular side (*v*); the boxed area in c is shown at a higher magnification in (**c'**). (d) SVP blood vessels and tissue macrophages in a flatmounted of an E11.5 hindbrain, immunofluorescently labelled for IB4 and mounted ventricular face up; boxed areas are shown at higher magnifications in (**d'** and **d"**), indicating examples of filopodia-studded tip cells (*clear arrows*), tissue macrophages in the hindbrain parenchyma (*wavy arrow*), tissue macrophages in association with tip cells (*clear arrowheads*) and tissue macrophages at vessel junctions (*solid arrowheads*). Scale bars: **a, b, c** – 500 μm; **a', a", b', b", d** – 100 μm

hindbrain, consisting of radial vessels originating from the perineural plexus and the SVP that is placed orthogonally to the radial vessels (Fig. 2a–c)[2].

At E11.5 and E12.5, both the pial and subventricular vessels and their associated cells can be readily visualised by wholemount immunolabelling of flatmounted hindbrains, once they have been dissected away from perineural tissue. Specifically, the radial vessels entering the brain can be visualised on the pial aspect of the flatmounted hindbrain (Fig. 2a", b"), whilst the SVP vessels are apparent in the ventricular view (Fig. 2a, a', b, b') [1]. Vibratome sectioning of the dissected hindbrains illustrates the relationship of pial and SVP vessel segments (Fig. 2c) [1]. Once the SVP has formed, vascular sprouting proceeds to deeper brain layers, which consequently accumulate tissue macrophages that increase vessel fusion [1]. In the deeper tissue parts of hindbrains from E12.5 onwards, tissue penetration of antibodies begins to decrease, and vibratome sectioning followed by immunolabelling is most helpful to demonstrate vascular structures (Fig. 2c,c).

A variety of established rat anti-mouse monoclonal antibodies and lectins are available to study the endothelial cells of blood vessels and their associated cell types. Amongst these, PECAM [6] and isolectin B4 (IB4) [7] are commonly used endothelial markers, whilst F4/80 is a specific macrophage marker [8]. Various polyclonal antibodies directed to endothelial, pericytes and macrophage proteins are also available for double immunolabelling, however their usefulness varies with batch and they therefore require testing for suitability on an individual basis. We provide here a detailed protocol for hindbrain dissection and immunolabelling for PECAM (Fig. 2a–c) and IB4 (Fig. 2d), and additionally include generic protocols for double labelling with rabbit and goat antisera.

2   Materials

2.1   Dissection and Fixation

Plastic cell culture dishes (5 cm diameter, e.g. from Nunc)
24-well cell culture plates (e.g. from Nunc)
Watchmaker forceps, no. 5 (e.g. from Dumont)
Phosphate buffered saline (PBS)
PBT (PBS containing 0.1% Triton X100)
Plastic Pasteur pipettes (to transfer embryos and hindbrains between dishes; cut off
    the tip to create an appropriately sized opening)
Formaldehyde solution: Dissolve 4% w/v paraformaldehyde (PFA) in PBS at 65°C,
    aliquot for storage at −20°C; defrost at room temperature and cool in ice prior to
    application.
Absolute methanol and 25%, 50% or 75% methanol in PBT

2.2   Immunolabelling

2.2.1   Horseradish Peroxidase (HRP) Immunolabelling of PECAM or IB4

2.0 ml round-bottomed reagent tubes (safe-lock; e.g. Eppendorf)
Hydrogen peroxide
Diaminobenzidine and urea hydrogen peroxide tablets (SigmaFast, cat. no.
    D4293, Sigma)

HRP-labelling of PECAM:
Rat anti-mouse PECAM monoclonal antibody (CD31; cat. no. 553370, BD Pharmingen)
Heat inactivated normal rabbit serum (NRS; Gibco)
HRP-tagged rabbit anti-rat secondary antibody (cat. no. P0450, Dako)
Note: For PECAM staining, it is best to stain freshly fixed tissues and avoid fixation longer than 2 h or methanol storage.

HRP-labelling of IB4:
Biotinylated isolectin B4 (IB4; cat. no. L2140, Sigma)
Heat inactivated NRS or normal goat serum (NGS; Sigma)
HRP-tagged streptavidin (cat. no. P0397, Dako)
Note: IB4 is a lectin and binds α-D-galactosyl residues on glycoproteins, which are generally more stable than protein epitopes; IB4 staining is also compatible with immunolabelling after in situ hybridisation.

2.2.2 Fluorescent Immunolabelling of PECAM or IB4

Fluorescent labelling of PECAM:
AlexaFluor488-conjugated goat anti-rat secondary antibody (cat. no. A11006, Invitrogen)

Note: It is possible to choose an alternative species-appropriate fluorophore-conjugated antibody.

Fluorescent labelling of IB4:
AlexaFluor488-conjugated streptavidin (cat. no. S32354, Invitrogen)
Note: It is possible to choose an alternative streptavidin-conjugated fluorophore.

2.2.3 Fluorescent-Labelling of Primary Antibodies Raised in Rabbits

AlexaFluor488-conjugated goat anti-rabbit secondary antibody (cat. no. A11008, Invitrogen)
Note: It is possible to choose an alternative species-appropriate fluorophore-conjugated antibody.

2.2.4 Fluorescent-Labelling of Primary Antibodies Raised in Goats

Ready-to-use serum-free protein block solution (cat. no. X0909, Dako)
Cy3-conjugated donkey anti-goat Fab fragment (cat. no. 305-116-003, Jackson ImmunoResearch)

Note: We recommend to not substitute this secondary antibody, except where other fluorophore-conjugated Fab fragments are available that were raised in a species other than goat.

2.3   Mounting, Sectioning and Imaging

Vibratome (e.g. Vibratome 1000Plus Sectioning System)
Disposable plastic moulds (e.g. from Electron Microscopy Sciences)
Agarose (e.g. from BDH Electran)
Glycerol
Glass slides and glass coverslips
SlowFade kit (cat. no. S2828, Invitrogen)
Stereomicroscope with suitable software
Epifluorescence stereomicroscope with suitable software
Confocal laser scanning microscope with suitable operating software

3   Methods

3.1   Embryo Isolation

1. Place pregnant uterus (E10.5 onwards) into a clean plastic dish containing PBS.
2. Under a dissecting stereomicroscope, remove each embryo sac from the uterus by rupturing the muscular layer that extends between all embryo sacs.
3. Rupture each embryo sac until the embryo emerges, sever the umbilical cord, collect the embryos with a plastic Pasteur pipette into a dish with clean PBS and place the dish on ice.
4. Transfer one embryo at a time with a plastic Pasteur pipette to a fresh dish containing ice-cold PBS for hindbrain dissection.

3.2   Hindbrain Dissection

1. By squeezing with forceps, sever the head above the shoulders (Fig. 3a, line 1) and remove the front of the head (Fig. 3a, line 2).
2. Rupture the thin and translucent roof-plate overlying the hindbrain by pulling gently with sharp forceps (Fig. 3b). Continue to rip the membrane up towards the midbrain, then down towards the spinal cord (Fig. 3b, small arrows) to open up the hindbrain (Fig. 3b, curved arrows). The hindbrain tissue can now be seen clearly under the stereo dissecting microscope. Note: Do not insert the forceps too deep, as this will damage the hindbrain tissue.

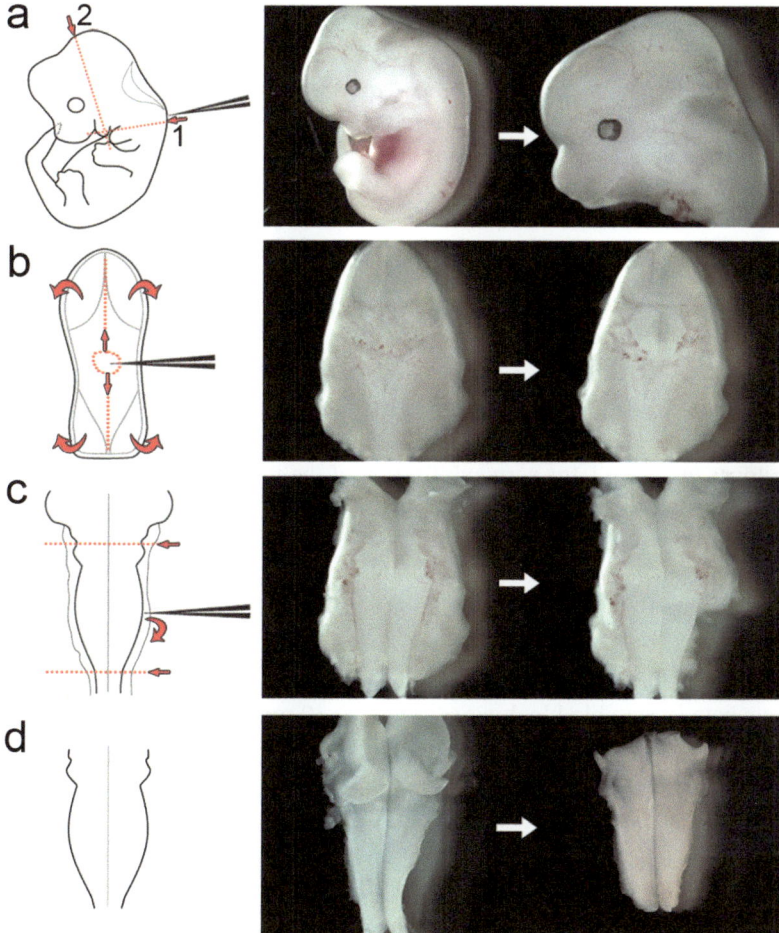

Fig. 3 Dissection of the hindbrain from a mouse embryo. Schematics are shown on the *left hand side* of the figure, dissected tissues on the *right*. (**a**) The embryo is removed from the uterus with forceps; body and face are removed by tearing along lines 1 and 2 with forceps. (**b**) A hole is made in the roof-plate with forceps, and the roof-plate is then torn along the dotted lines with forceps to expose the hindbrain and allow it to unfurl (*curved arrows*). (**c**) The pial membranes and surrounding non-neural tissues are removed from beneath the hindbrain with forceps (*curved arrow*). The midbrain and spinal cord are removed by squeezing the brain tissue with forceps along the dotted lines. (**d**) The isolated hindbrain

3. Tease off the pial membrane from underneath the hindbrain (Fig. 3c); it usually comes off easily after opening the roof-plate, and it is simply a matter of tugging a bit here and there, like peeling an orange – some regions give way easier than others.
4. Remove the midbrain and spinal cord to allow the hindbrain to unfurl (Fig. 3c, lines).
5. Place the isolated hindbrains (Fig. 3d) with a plastic Pasteur pipette flat onto the bottom of a well in a 24-well plate.

3.3 Fixation

1. For most antibodies, freshly prepared or freshly thawed formaldehyde is a suitable fixative. Add 4% cold formaldehyde (see Sect. 2.1) to the hindbrain.
2. Fix for 2 h on ice with gentle agitation, rinse three times in PBS and proceed to staining protocol. Alternatively, transfer through a rising methanol gradient to store at −20°C for up to 3 months (gradient: 25%, 50%, 75%, absolute methanol).

3.4 Immunolabelling

1. Wash the fixed hindbrains in PBS for non-permeabilised tissue staining, for example for visualisation of extracellular epitopes rather than cytoplasmic epitopes, or in PBT for permeabilisation of the tissue to detect both cell surface and cytoplasmic epitopes. For tissue stored in methanol, transfer through a decreasing methanol gradient (75%, 50%, 25%, PBS).
2. Hindbrains can either be immunofluorescently (Fig. 2c) or immunohisto-chemically (Fig. 2a, b) labelled in 2.0 ml round-bottomed safe-lock tubes, as follows.

(a) Immunofluorescence:

3a. Incubate in suitable blocking solution for 30 min with gentle rolling:

- For PECAM labelling, use 10% normal rabbit serum in PBS or PBT.
- For IB4 labelling, use 10% normal goat or rabbit serum in PBS or PBT.

4a. Incubate overnight at 4°C in blocking solution containing a primary antibody:

- Rat anti-mouse PECAM, diluted 1/200.
- Biotinylated IB4, diluted 1/200.

5a. Thoroughly wash the hindbrains five times in PBS or PBT for an hour.
6a. Incubate overnight at 4°C in blocking solution containing a secondary antibody diluted 1/200:

- For PECAM detection, use an AlexaFluor488-tagged goat anti-rat antibody.
- For biotinylated IB4 detection, use AlexaFluor488-tagged streptavidin.

7a. Thoroughly wash the hindbrains five times for 1 h each in PBS or PBT.
8a. Post-fix in 4% formaldehyde for 30 min.

(b) Immunohistochemistry:

3b. Bleach the tissue to remove endogenous peroxidase activity by incubating in 1% hydrogen peroxide in PBS for 30 min with gentle rolling.
4b. Incubate in blocking solution for 30 min with gentle rolling:

- For PECAM labelling, use 10% normal rabbit serum in PBS or PBT.
- For IB4 labelling, use 10% normal goat or rabbit serum in PBS or PBT.

5b. Incubate overnight at 4°C in blocking solution containing a primary antibody:

- Rat anti-mouse PECAM, diluted 1/200.
- Biotinylated IB4, diluted 1/200.

6b. Thoroughly wash the hindbrains five times for an hour in PBS or PBT.
7b. Incubate overnight at 4°C in blocking solution containing a secondary antibody diluted 1/200:

- For PECAM detection, use horseradish peroxidase (HRP)-tagged rabbit anti-rat antibody.
- For biotinylated acronym IB4 detection, use HRP-tagged streptavidin.

8b. Thoroughly wash the hindbrains five times for an hour in PBS or PBT.
9b. Incubate the hindbrains in a solution of diaminobenzidine for 20 min, then in a solution of diaminobenzidine and urea hydrogen peroxide for 5 min, or until colour develops.
10b. To stop the reaction, wash the hindbrains briefly in distilled water and then PBS and post-fix in 4% formaldehyde for 30 min.

3.5 Vibratome Sectioning

1. Melt 3% agarose in distilled water in a glass bottle in the microwave.
2. Place the hindbrain in a square plastic mould and remove the liquid surrounding the hindbrain with paper tissue or a pipette.
3. When the molten agarose has cooled sufficiently to comfortably touch the glass, pour it into the mould and quickly position the hindbrain before the agarose begins to set.
4. After the agarose has set, remove the agarose block containing the hindbrain from the mould (which can be re-used) and cut 100 µm transverse sections with a vibratome.

3.6 Imaging

1. For immunofluorescently labelled hindbrains, stick three layers of black electrical tape onto a glass slide and cut rectangular pockets to create a spacer. Flatmount each hindbrain into one of the pockets, using SlowFade and a glass coverslip. The hindbrains can then be imaged with a fluorescence stereomicroscope or by confocal scanner laser microscopy.
2. For immunohistochemically-labelled hindbrains, melt 3% agarose in distilled water in a glass bottle in a microwave, leave to cool and pour an approximately 1 mm thin layer into a plastic dish (5 cm diameter) to set; place each hindbrain individually on the agarose, remove excess liquid and cover with glass coverslip.

Table 1 Examples of cell type-specific primary antibodies for vessel-associated cell types

Cell type	Antibody	Dilution/block
Pericytes	Rabbit anti-mouse NG2 (cat. no. AB5320, Millipore)	1:200 in 10% NGS
Macrophages	Rat anti-mouse F4/80 (cat. no. MCA497G, Serotec)	1:500 in 10% NGS
Capillary/venous EC	Rat anti-mouse endomucin (clone V.7C7; cat. no. sc65495, Santa Cruz)	1:50 in 10% NGS

The hindbrains are imaged with a standard stereomicroscope equipped with a suitable camera. Note: It may be necessary to wet the coverslip before taking the image, to remove dust.

3. To image HRP-labelled vibratome sections, mount them on a glass slide in 90% glycerol in PBS with a glass coverslip. The sections are imaged with a standard stereomicroscope equipped with a suitable camera.

3.7 Alternative Immunolabelling Protocols

1. Double labelling of vessels and other cell types can be achieved using a combination of appropriate primary and secondary antibodies. Some examples of antibodies that may be of interest to label macrophages, pericytes or capillary and venous endothelium are described in Table 1. It is important to choose blocking serum compatible with the primary and secondary antibody. For primary antibodies raised in goats, protein-free serum block (see materials) should be used; it is not necessary to include this block in the antibody dilutions.

Acknowledgements We thank Dr Alessandro Fantin for Fig. 1 and critical reading of the manuscript. We are grateful to L. Denti, K. Davidson and the staff of the Biological Resources Unit at the UCL Institute of Ophthalmology for help with mouse husbandry. We gratefully acknowledge the Imaging Facility of the UCL Institute of Ophthalmology for maintenance of the confocal microscopes. C.M. was funded by an MRC doctoral training account (ref. G0601093).

References

1. Fantin A, Vieira JM, Gestri G et al (2010) Tissue macrophages act as cellular chaperones for vascular anastomosis downstream of VEGF-mediated endothelial tip cell induction. Blood 116(5):829–840
2. Ruhrberg C, Gerhardt H, Golding M et al (2002) Spatially restricted patterning cues provided by heparin-binding VEGF-A control blood vessel branching morphogenesis. Genes Dev 16(20):2684–2698
3. James JM, Gewolb C, Bautch VL (2009) Neurovascular development uses VEGF-A signaling to regulate blood vessel ingression into the neural tube. Development 136(5):833–841

4. Raab S, Beck H, Gaumann A et al (2004) Impaired brain angiogenesis and neuronal apoptosis induced by conditional homozygous inactivation of vascular endothelial growth factor. Thromb Haemost 91(3):595–605

5. Haigh JJ, Morelli PI, Gerhardt H et al (2003) Cortical and retinal defects caused by dosage-dependent reductions in VEGF-A paracrine signaling. Dev Biol 262(2):225–241

6. Albelda SM, Muller WA, Buck CA, Newman PJ (1991) Molecular and cellular properties of PECAM-1 (endoCAM/CD31): a novel vascular cell-cell adhesion molecule. J Cell Biol 114(5):1059–1068

7. Laitinen L (1987) Griffonia simplicifolia lectins bind specifically to endothelial cells and some epithelial cells in mouse tissues. Histochem J 19(4):225–234

8. Austyn JM, Gordon S (1981) F4/80, a monoclonal antibody directed specifically against the mouse macrophage. Eur J Immunol 11(10):805–815

Visualization and Quantification of *De Novo* Angiogenesis in *Ex Ovo* Chicken Embryos

A. Zijlstra and John D. Lewis

Abstract The formation of new blood vessels through *de novo* angiogenesis is a fundamental process in developmental biology and pathogenesis, during which changes in the vasculature can be highly dynamic and involve proliferation, migration, sprouting, and remodeling. In this chapter, methodologies are described which allow for the precise temporal and spatial quantification of angiogenesis in an *ex ovo* chicken embryo model. A protocol is detailed that allows for the enumeration of blood vessels in newly vascularized collagen onplants placed on the chorioallantoic membrane. Onplants can be infused with growth factors, cytokines, pro- or anti-angiogenic factors or even living cells to determine the precise impact on *de novo* angiogenesis. A complementary intravital imaging approach is also described that allows for the visualization and evaluation of angiogenesis in *ex ovo* chicken embryos as it occurs in real time. These highly scalable assays evaluate new vessel growth in an intact tissue where angiogenesis naturally occurs, and importantly, allow one to evaluate factors that inhibit vascularization without impacting pre-existing vessels.

1 Introduction

Our understanding of neovascularization has been transformed by advances in our ability to visualize and quantitate the process of angiogenesis. One model organism in particular, the chorioallantoic membrane (CAM) of the avian embryo, has

A. Zijlstra
Department of Pathology, Vanderbilt University, 1161 21st Ave. S., C-2104A MCN, Nashville, TN 37232-2561, USA

J.D. Lewis, Ph.D. (✉)
Department of Oncology, University of Alberta, 5-142C Katz Group Building, 114th St and 87th Ave, Edmonton, AB T6G 2E1, Canada
e-mail: jdlewis@ualberta.ca

E. Zudaire and F. Cuttitta (eds.), *The Textbook of Angiogenesis and Lymphangiogenesis: Methods and Applications*, DOI 10.1007/978-94-007-4581-0_14,
© Springer Science+Business Media Dordrecht 2012

contributed significantly to this understanding by virtue of its unique advantages over other *in vivo* models. The CAM is a highly vascularized membrane containing mature vessels that is immediately proximal to the eggshell membrane. The accessibility of the CAM, its ability to accept orthotopic grafts and the self-sustained nature of the avian egg makes the CAM an ideal model for the study of angiogenesis. Furthermore, the lack of an intact immune system makes the embryo capable of accepting grafts containing normally immunogenic tumor cells and/or extracellular matrix proteins from other species.

While the CAM has been employed for the study of angiogenesis for the better part of a century [1, 2], several recent advances have enhanced the utility of this model both for detailed biological investigation and high throughput analyses. The use of shell-less embryo culture techniques (Fig. 1) have improved reproducibility and scalability of assays [1, 3, 4]. Sequencing of the chicken genome [5, 6], along with recent improvements in chick transgenics [7], have expanded its potential as a model to understand vertebrate gene function. Moreover, recent advances in imaging technologies have made it possible to visualize vascular perfusion, vascularization of the CAM and the distinct steps of angiogenesis [3, 8, 9]. New contrast and imaging agents that selectively label developing vessels allow for the selective visualization of vascular structures at the microscopic level [3, 8, 10]. This chapter describes angiogenesis assays in the shell-less chicken embryo model that are currently used in our laboratories.

2   The Chicken Embryo CAM Model

While the CAM assay is probably the most widely used *in vivo* assay for studying angiogenesis [11], lack of standardization has resulted in a wide variation of techniques being utilized by different laboratories. *In ovo* methods are straightforward to conduct, but can be subject to increased variability due to inflammation from eggshell dust. Furthermore, it is difficult to discern between new and pre-existing vasculature, and the assays are sensitive to changes in oxygen tension. Improvements in the *ex ovo* or shell-less culture of chicken embryos have allowed for the development of highly reproducible, cost-effective and quantitative *de novo* angiogenesis assays that overcome many of these challenges. We detail methods in Sect. 4 that allow the reliable *ex ovo* culture of chicken embryos to developmental day 20. When embryos are cultured in this way, the CAM develops on the top surface, providing a large "canvas" of mature vasculature that is structurally accessible for experimental manipulation and imaging. The onplant angiogenesis assay (Sect. 5) allows for quantitative analysis of vessel growth into a collagen matrix and is particularly useful for determining the influence of exogenously added compounds on vessel development. In this assay, the collagen matrix placed on top of the CAM remains superficial and accessible during its vascularization by the CAM vessels. New blood vessels can be visualized and scored with a basic stereomicroscope, and if desired, onplants can be easily removed and analyzed in more detail

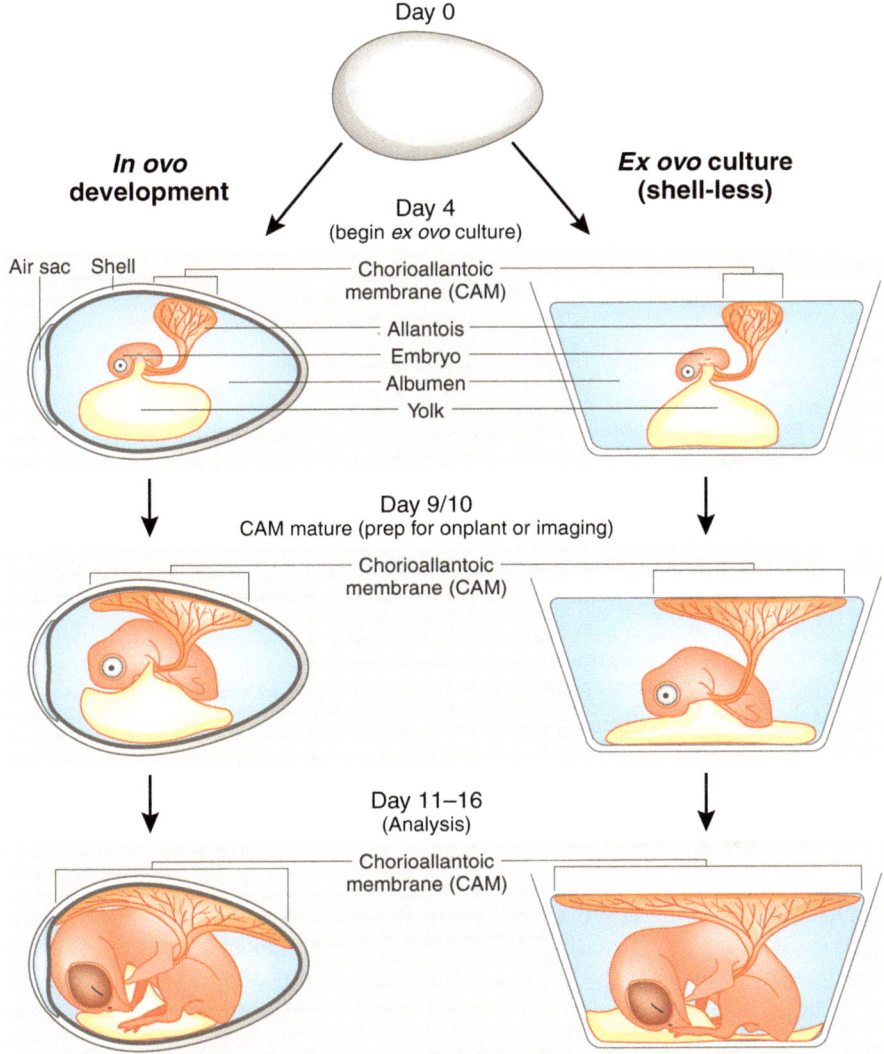

Fig. 1 Comparison of *in ovo* and *ex ovo* chick embryo development. The basic structure of the chick embryo is depicted at key developmental and experimental stages *in ovo* and in *ex ovo* culture. The yolk is primarily composed of protein and fat to provide food to the developing embryo. The albumen functions to cushion the embryo from the shell and will be absorbed as the embryo grows. The allantois is the respiratory structure for the embryo and fuses with the developing chorion to form the highly vascularized chorioallantoic membrane (*CAM*). In the *ex ovo* culture system, the CAM is sufficiently developed at day 9/10 to place onplants for angiogenesis assays or to implant xenograft tumor cells for intravital imaging. Experiments are then typically analyzed between day 11 and day 18, when the CAM has fully developed and is comprised of mature vasculature. A key advantage of *ex ovo* culture is the accessibility of the CAM for manipulation, treatment and visualization

using immunohistochemistry or similar techniques. In a modification of this assay, human tumor cells can be mixed into the collagen onplant to provide the pro- or anti-angiogenic signal. Thus, the relative "angiogenic potential" of different cell lines can be evaluated. A second variation of this assay allows for the direct visualization of vascular dynamics during development or disease using intravital imaging (Sect. 6). In this assay, human cancer cells implanted directly into the CAM establish tumors that become vascularized by host endothelium. The use of specialized dyes and nanoparticles facilitates the visualization and quantitation of angiogenic neovasculature in real time.

3 Assay Design Considerations

3.1 Timing

Because angiogenesis is a dynamic process it is important to establish and adhere to a defined timeline. When purified VEGF or HT1080 fibrosarcoma cells are used as the angiogenic stimulus, maximum angiogenesis will be seen after 72 h. Consequently, the inhibition of this response should be tested within the same time frame. Other angiogenic stimuli or inhibitors may have a different time-frame and it is important to explore the efficacy of any new compound across a full time-course of 2–5 days to establish the optimum time frame.

3.2 Dosing and Treatment

Both the angiogenic stimulus and the treatment must be carefully defined and, when possible, experimentally defined. Angiogenic stimuli as well as treatments are readily incorporated in the onplants. However, in many instances the treatment must be administered repeatedly or simply cannot be incorporated into the onplant. In such an instance, treatments can be applied topically to the onplant at defined time points during the trial. Alternatively, they can be administered intravenously or injected into the allantoic fluid underneath the CAM from where they will be absorbed by the chick and distributed systemically.

Treatment with an experimental agent can be applied in a number of ways and the method one chooses is guided by timing: Does the animal have to survive treatment for extended periods of time, is the compound soluble in aqueous solutions, is frequent treatment necessary or is it a one-time treatment?

Treatments methods suitable for short-term survival:

- I.V. injection by means of a standard or insulin needle. This will cause significant bleeding but the animal may survive for several hours. Sometimes a low number (10–30%) of animals will survive several days.

- Intra-amnion injection. Injection into the amniotic sac is easy and rapid. The animal can survive for several hours to days. However the amniotic sac will collapse over time preventing the protection offered by this structure.
- Intra-peritoneum (or other fleshy parts of the body). As surprising as it is, intrabody injections are possible and relatively easily sustained by the animal. Unfortunately this requires puncturing the amniotic sac resulting in its collapse and the possible death of the animal.

Treatments useful for long-term survival:

- I.V. injection by means of a glass capillary needle. While initially technically challenging, this method works quite well and allows for long term survival (until developmental day 20).
- Topical treatment on the CAM is also effective and can be repeated frequently. However, the compound must be soluble in aqueous solutions.
- Intra-yolk injections are straightforward and easily taken by the animal. Unfortunately, almost nothing is known about processing and uptake of injected material in the yolk. This is frequently very difficult to assess. The yolk is also very large and viscous resulting in slow diffusion.

3.3   Choice of Controls

In many cases, angiogenesis assays are designed to determine the contribution of a specific experimental parameter. To quantitatively define the impact of an angiogenic stimulator or inhibitor, it is essential to define the dynamic range of the assay. Consequently, each experiment must include the appropriate negative controls and positive controls. Typically, a negative control group should be included that consists of onplants containing PBS and/or the diluent used for the experimental compounds. Typically, we recommend using positive control groups that include well-defined angiogenic stimulators such as VEGF and FGF-2, or angiogenic tumor cells such as HT1080. At the same time, the response of the system can be further defined by using an established angiogenesis inhibitor such as Avastin.

3.4   Additional Analyses

While direct scoring of angiogenic neovasculature is the most straightforward analysis used for these assays, several other techniques can be employed to gather additional information.

1. Histological evaluation of tissue sections [12].
2. Vascular patency can be evaluated through the visualization of injected vascular dyes or circulating particles.
3. Vascular function can be evaluated through vascular leak assays [13].

Furthermore, in the case of cancer, the vascular network is developed to support cancer growth and metastasis. This ability can be assessed secondarily by allowing the assay to progress beyond the traditional 3-day time frame and allowing tumor cells within the onplant to grow, develop a tumor, and subsequently metastasize [14].

The onplant angiogenesis assay provides a consistent, reproducible and high-volume analysis of angiogenesis. It does not provide, however, the dynamic perspective of vascular formation, blood flow, tissue perfusion, and cell motility that is known to exist within an angiogenic microenvironment. When these parameters are of interest, an intravital imaging angiogenesis assay can be performed as described in Sect. 6.

4 Shell-Less Cultivation of Chicken Embryos

We detail here a robust method for the cultivation of shell-less (*ex ovo*) chicken embryos that is relatively cost-effective. *Ex ovo* culture effectively exposes the abundant vasculature of the CAM and provides a large surface area on which to perform angiogenesis assays. Both the onplant (Sect. 5) and intravital imaging (Sect. 6) angiogenesis assays described herein utilize this model. The following protocol addresses the generation, timing and maintenance of the *ex ovo* chick embryo platform.

4.1 Positioning the Eggs

The positioning of the intact eggs is more important for *in ovo* work than it is for *ex ovo* work. The embryo attaches to the eggshell membrane after developmental day 4, making it impossible to remove the embryo from the shell. Because the embryo is removed from the egg at day 4, the eggs can be incubated in any position during the first 4 days. However, if the eggs are used *in ovo*, they should always be incubated on their side.

To simplify the instructions we describe positioning of the egg in space as follows: the blunt end which contains the air-sac is considered the upper pole while the tapered end is the lower pole. The long axis of the egg runs from pole to pole.

4.2 Experimental and Developmental Timelines

Experimental outcomes are influenced significantly by the developmental stage of the chick embryo at each step of the experiment. Thus, distinct experimental and developmental timelines must be considered. The experimental timeline reflects the intervals between experimental steps during which the experiment is performed

Table 1

Timeline	Eggs delivered	Crack eggs	Start: onplants/ cells	Score: angiogenesis	Metastasis: harvest
Experimental	0	3/4	9	12	10–17
Developmental	1	4/5s	10	13	11–18
Weekly	Sunday	Wednesday/ Thursday	Tuesday	Friday	Week

while the underlined developmental timeline indicates the developmental stage of the embryo (Fig. 1). The principle reason for this distinction is that the developmental timeline can be significantly affected by the egg temperature during storage and transport. Although the eggs are laid on day 0, they arrive in the lab 24–48 h after they are harvested at the hatchery. While it is assumed that some developmental time passes while the eggs are collected and transported, the day that the eggs are put into the laboratory incubator is considered developmental day 1. It is important to note that in warmer weather, the eggs will have developed more rapidly during shipment and therefore they must be cracked after only 3–3.5 days of incubation. We organize our experimental timelines according to Table 1.

4.3   Protocol for Ex Ovo *Chick Culture*

- Order just laid eggs and upon arrival place eggs in fridge or wine cooler (12°C) until use. Eggs at this stage are considered developmental day 1. When ready, transfer eggs to a rotating egg incubator. Place the eggs on their side during the incubation, such that the long axis of the egg is parallel with the ground, and place a mark on the side of the shell that faces up. The length of incubation is dictated by the developmental stage of the eggs. Typically, eggs are incubated for 4 days before cracking. In warmer weather, however, the eggs will have developed during shipment and 3–3.5 days of incubation will be sufficient.
- Sterilize the weigh-boats (which will hold the embryo) in 70% EtOH by dipping them in a bath of EtOH and leaving them to dry in a laminar flow hood (2 h-O/N). The tops and bottoms of the Petri dishes which will act as lids and should be ordered sterile. CRITICAL STEP: Proper sterilization of eggs, consumables and equipment ensures that *ex ovo* embryos do not become contaminated.
- Place cover (top or bottom of Petri dish) over weigh-boats and label (date) the cover. It is best to place these covered weigh-boats on a tray of some kind so as to minimize handling of individual boats.
- Set up the Dremel tool with a cut-off wheel in a clamp on a stand such that the tool is parallel to the bench and the height is comfortable. It is best if the tool is set up in a hood or behind a shield that can protect the handler from dust and flying yolk.
- Remove eggs from the incubator without rotating them and place them on their side in carton egg racks with their long axis parallel to the surface on which they stand (keep the mark facing upwards). The racks we use are cut from the original

5×6 egg racks that the eggs come in. At this point the eggs should be lying in their racks the two poles pointing to the left and right and the mark put on the shell facing upwards. This is important because the embryo is free-floating at this point. CRITICAL STEP: Ensure that the egg is kept lengthwise in the same position for at least 10 s so that the embryo rotates to the top.

- Work with only 6–12 eggs at a time: keep eggs in a stationary incubator until ready to crack.
- Douse the cut-off wheel with 70% EtOH and place a small weigh-boat with 70% EtOH next to the Dremel in the stand.
- Wipe your hands or gloves with EtOH before handling the eggs. (Handing the eggs with gloves can be awkward and it is possible to carefully handle them with clean hands)
- Turn the Dremel tool to its lowest setting.
- Pick up an egg without rotating it (keeping the long axis parallel to the ground and the mark up), dip its lower half in the boat with 70% EtOH then dab it onto a clean tissue to remove excess EtOH.
- While maintaining the egg with the long axis parallel to the ground, position the egg over the cut-off wheel (which is perpendicular to the ground) and scour/cut the egg shell on the lower half of the egg (Fig. 2a). The cut should be equatorial, that is, it should be right through the center separating the egg into two halves. You can cut through the egg-shell, don't mind fluid leakage, but don't cut into the embryo or the yolk. Cut 90–180° around the axis of the egg. How much you'll have to cut depends on the egg and your technique.
- Remove the lid from the weigh-boat and gently place the cut surface of the egg into the weigh-boat. Hold the egg with thumb and forefingers place on either hemisphere and gently apply pressure downwards until you can feel the egg crack along the cut. When the shell feels ready to give way completely, quickly pull apart the two halves of the shell pulling your fingers slightly upwards. CRITICAL STEP: a visible heartbeat indicates a healthy embryo. If it is not present, the egg should be discarded. (Fig. 2b).
- Place a lid on the weigh-boat (Fig. 2d).
- When a set of embryos has been transferred and placed onto a tray, move the tray into a 37.5°C incubator with very high humidity (>70%). In dry climates it would be best to move the shell-less embryos directly into a secondary humidity container before placing them in an incubator to prevent any dehydration. CRITICAL STEP: *Ex ovo* embryos must be kept in a consistently humid environment with > 70% humidity to ensure viability and assay consistency.
- Incubate the eggs for an additional 4 days (until developmental day 9; 8 days incubation) (Fig. 2c).

4.4   Tips for Successful Ex Ovo *Culture*

Transferring embryos is as much art as it is science. Everyone develops a slightly different technique. Find the one that best suits you and practice before starting an

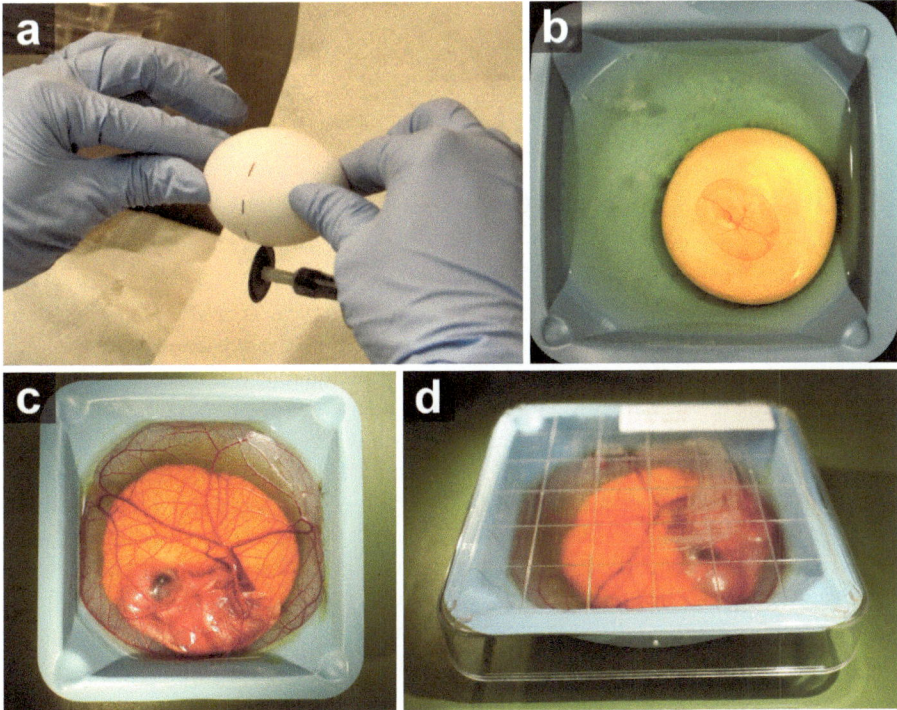

Fig. 2 Preparation of shell-less chicken embryos. (**a**) The shell is removed at day 4 using a Dremel tool. Several superficial equatorial cuts are made into the shell, and then the cut surface of the egg is gently placed in a sterilized weigh boat and the two halves of the shell are removed after pressing down until the egg begins to crack. (**b**) Freshly cracked day 4 embryo. At this stage, the embryo should be on the top surface of the yolk, and should be about the size of a silver dollar. The heartbeat should easily be observed. Embryos without a heartbeat should be discarded. (**c**) Day 11 avian embryo with highly vascularized chorioallantoic membrane completely covering the top surface. (**d**) Day 11 embryo shown covered with Petri dish lid. Lid should be kept on at all times when the embryo is not being manipulated directly. Embryos with lids should be kept in a secondary plastic container at all times (not shown)

experiment. With a successful transfer, the yolk will remain intact and the embryo will be untouched and floating on top of the yolk. If the embryo sticks to the egg-shell then the egg has been incubated too long. If the yolk or the embryo is too fragile then the egg may not have been incubated long enough. A heartbeat indicates a healthy embryo. If it is not present, the embryo should be discarded. Incubation conditions can dramatically affect the health and viability of the developing embryo, and consequently the results of angiogenesis assays. We find that for successful *ex ovo* culture, high and consistent levels of humidity (>70%) must be maintained throughout the experiment. Because the humidity in a typical incubator drops precipitously when the door is opened, we employ secondary plastic containers (Rubbermaid containers with lid) with 1″ diameter holes drilled into each side that each house 18 embryos each (6 stacks of 3). High local humidity is maintained by keeping 1/4–1/2″ of water in the bottom of each container. Inverted Petri dish lids placed upside-down at the bottom of each container keep the embryos separated from the standing water.

5 Onplant CAM Angiogenesis Assay

The traditional CAM assay was developed to identify compounds that possess angiogenic activity [15]. During such assays the compounds would be added directly to the CAM to induce a local increase in vascularization. With the subsequent development of anti-angiogenesis therapies the same assay was frequently employed, only now a local decrease in vascularization was expected. Consequently, an assay was developed where a filter disk was applied to the CAM [16]. It was soon apparent that the endogenous vasculature in the CAM could confound the analysis of pro/anti-angiogenic therapies. We detail here a modified CAM angiogenesis assay in which a collagen containing onplant is applied to the CAM of an *ex ovo* chicken embryo to provide an avascular environment adjacent to the richly vascularized capillary bed. The inclusion of a defined angiogenic stimulus (VEGF and FGF) reproducibly induces angiogenesis which can be suppressed by molecular therapies [17]. This model has been the prevailing method to provide quantitative analysis of vascularization *in vivo*.

5.1 Angiogenesis Assay

- Adjust the Vitrogen collagen (henceforth referred to interchangeably as Vitrogen or collagen) to neutral pH. This is a critical step as this material will form the solid substrate for each onplant. Vitrogen is pepsinized and thus does not gel as well as intact collagen. Check the pH after neutralization and again when the collagen is mixed with growth factors. CRITICAL STEP: Establishing a neutral pH is absolutely required to obtain consistent results. Bring the pH to neutral after each step.
- Neutralize 8 vol. Vitrogen with 1 vol. 10x PBS, 1 vol. 0.1 M NaOH and 0.2 vol. 1 M Hepes. Adjust pH to 7.4–7.6 using indicator paper and extra NaOH or HCl if necessary. Pipette up and down thoroughly to mix but don't introduce bubbles. CRITICAL STEP: Keep all solutions on ice unless otherwise indicated.
- Example: 4,000 μL Vitrogen, 500 μL 10X PBS, 500 μL 0.1 M NaOH, and 100 μL Hepes.
- Prepare Petri dishes by laying down Parafilm inside the dish and rinsing with 70% EtOH. Allow to dry in the sterile hood.
- Using sterile forceps arrange 4 × 4 (bottom) meshes on Parafilm (Fig. 3a) and then place a 2 × 2 mm (top) mesh at a slight angle in the center of each bottom mesh (Fig. 3b).

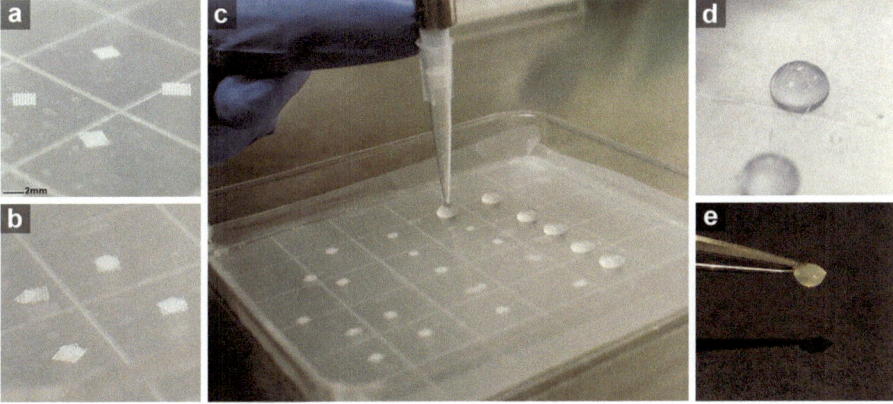

Fig. 3 Preparation of collagen mesh onplants. (**a**) 4 mm^2 bottom meshes are first arranged on Parafilm in a Petri dish lid, and then (**b**) 2 mm^2 top meshes are overlaid on the bottom meshes at a slight angle. (**c**) 30 μL of collagen solution is then pipetted onto each mesh by gently holding the pipette tip down onto the top mesh while expelling the solution. (**d**) The collagen solution is then left to solidify in a humidified incubator (37°C) for 1.5 h. (**e**) While the onplants remain quite soft after solidification, they can be picked up by grabbing onto the bottom mesh with needle nose forceps. If the gel lifts up then it is solid enough

- Prepare labeled tubes for experimental and control solutions of PBS/BSA with or without growth factor, etc. to be combined with double the volume of neutralized Vitrogen (that is, two parts of neutralized Vitrogen to one part growth factor or control diluent).
- First add one volume of 1x PBS calculated to reach final volume for each tube.
- Add PBS/BSA at a 1:50 ratio to all tubes.
- Add growth factor (e.g., bFGF at final concentration 16.7 μg/mL) and experimental or control compounds to appropriate tubes.
- Mix solutions thoroughly by pipetting and re-check pH of each solution and adjust if necessary. CRITICAL STEP: Again, maintaining neutral pH is essential in this assay.
- Turn off the blower in the hood. Pipette 30 μL of solution onto each mesh by gently holding the pipette tip down onto the top mesh while expelling the solution (Fig. 3c). Then draw back without shifting the top mesh.
- Place Petri dish in a humidified incubator (37°C) for 1.5 h or until solidified (but not dried) (Fig. 3d). Note that these gels/onplants are always soft even when solidified. To check, pick one up by grabbing onto the bottom mesh with needle nose forceps (Fig. 3e). CRITICAL STEP: When is the onplant sufficiently polymerized? If the gel lifts up then it is solid enough, if it stays behind it has not gelled sufficiently.
- Once solidified, use sterile forceps to transfer the gels to the CAM of developmental day 9 eggs (Fig. 4). Each egg should receive four onplants in total; two control onplants and two experimental onplants.
- Return the eggs to the incubator. If you did not put them already in a secondary humidified container, do so now.

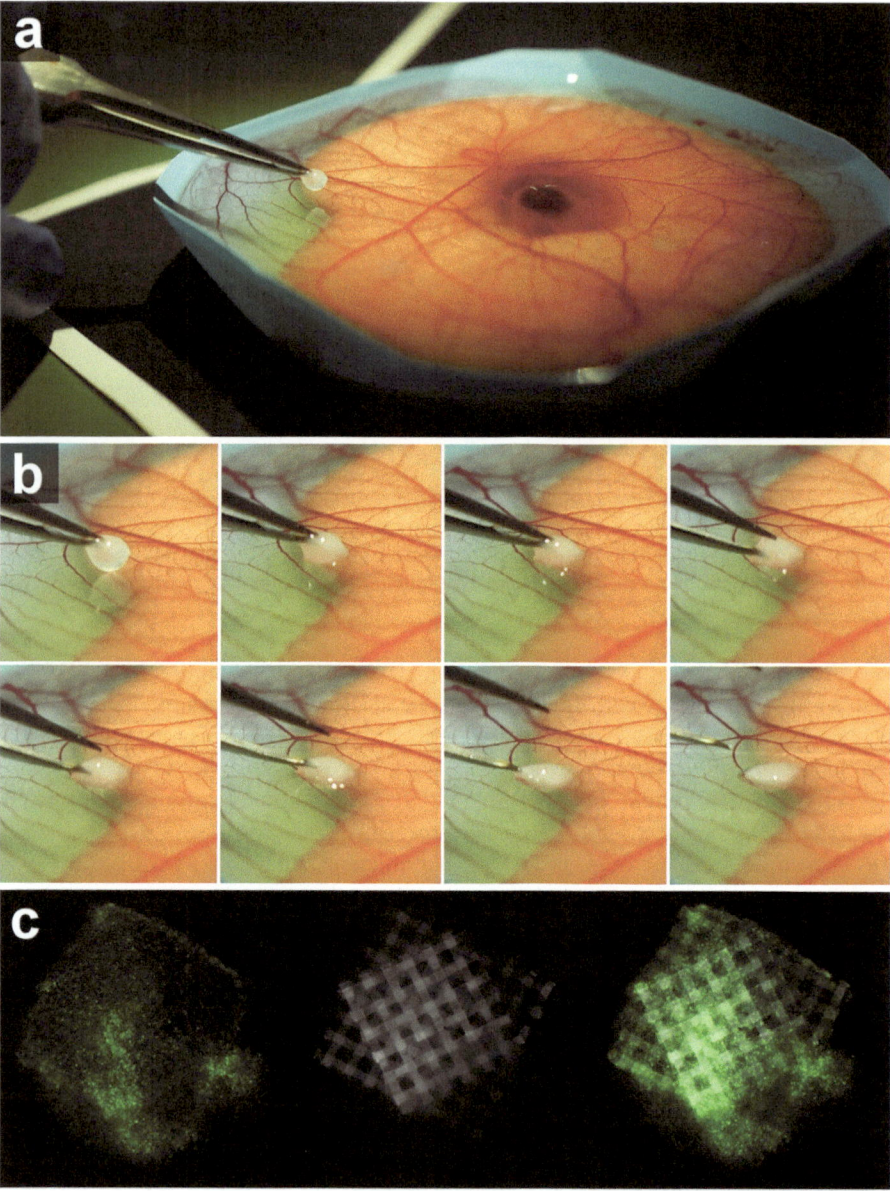

Fig. 4 Placement of collagen onplants onto the *ex ovo* CAM. (**a**) Solidified onplants are transferred to the CAM of developmental day 9 eggs using sterile forceps. (**b**) Gently set one corner of the onplant on the CAM and then lower the opposite corner. It is important that the collagen remain associated with the mesh. Each egg will generally receive two control onplants and two experimental onplants. (**c**) As a positive angiogenic stimulus, tumor cells can be incorporated into the collagen onplant. The distribution of GFP-expressing tumor cells can be visualized with a fluorescence stereomicroscope

- Allow the embryos to sit undisturbed in the humidified incubator for 3 days (66 h is generally optimal). Remove and score as outlined in the following section using a binocular dissecting microscope.

5.2   Scoring the Assay

Use a high quality binocular dissecting scope with a dual gooseneck illuminator such as "Flexi-lights" or a similar light source that is easily manipulated. Light should be directed from above down onto the onplant for best visibility. Focus on the surface of the upper mesh and scan the mesh grid visually. Identify new blood vessels by their color (red) and the presence of red blood cells (Fig. 5). Use the mesh grid as a scoring array, counting the number of grid squares that contain new blood vessels. Divide this number into the total number of grid squares on the onplant, so that each onplant is scored as the # of squares that have blood vessels/total # squares. We find that this scoring method is the easiest to master and provides extremely consistent results. Scoring visually in real time can be quite tedious, so workflow can be improved by capturing high resolution color images of each onplant and analyzing later. CRITICAL STEP: It is important to score only those vessels that have sprouted through the onplant and are visible at the top surface. Therefore, when scoring the onplants, focus up and down to verify that the vessels that you are looking at are truly entering the onplant and are not just part of the CAM located directly below the onplant. You can best position yourself by focusing first on the bottom mesh (against the CAM) and then the top mesh.

5.3   Additional Tips

Planning a well structured experiment can prevent frustration and give the results relevant context. Choosing the appropriate positive and negative controls is extremely important! This is a sensitive *in vivo* assay where small errors in handling can cause false positive or false negative results. Thus, handle both the eggs and the samples with extreme care and confirm that the controls elicit the appropriate response. Small differences in temperature and humidity can significantly affect the rate of angiogenesis. It may be useful to monitor the progress of the positive controls during the incubation period to determine the optimal incubation time. If onplants are left on the CAM for too long, the negative control onplants may also become vascularized thereby reducing the effective dynamic range of the assay. Positioning of the onplant on the CAM can also affect the scoring. Avoid placing the onplants directly over large vessels. The best position is in between two larger vessels. Try to place the onplants at equal distance from each other, and at least 1 cm apart.

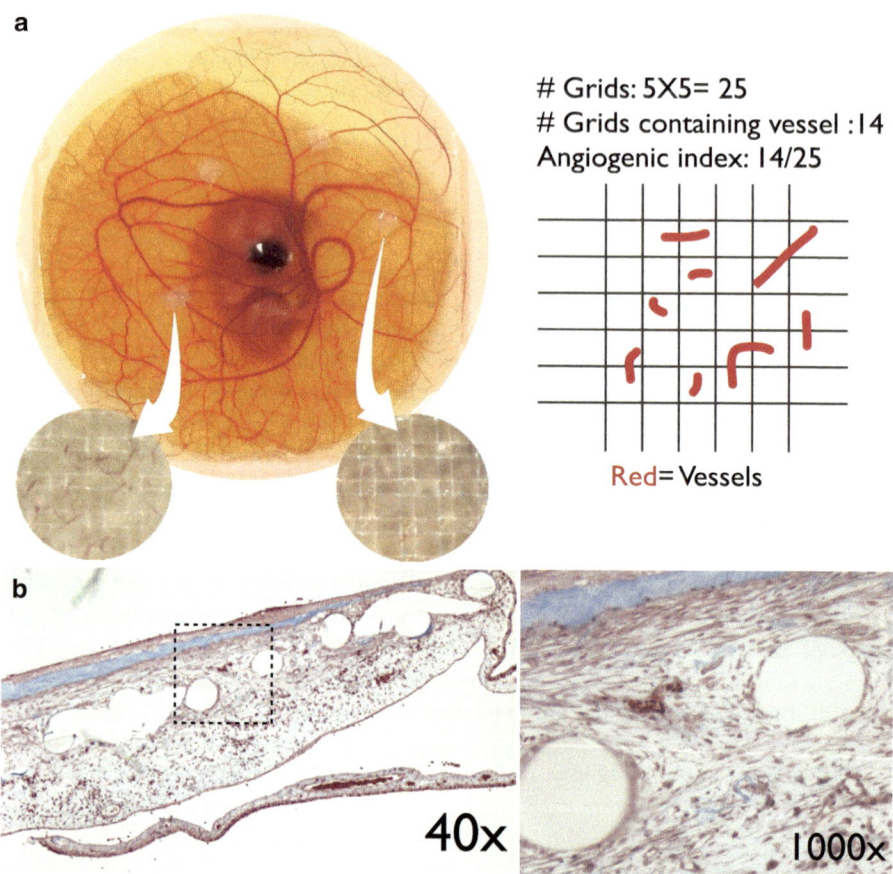

Grids: 5X5= 25
Grids containing vessel :14
Angiogenic index: 14/25

Red= Vessels

40x

1000x

Fig. 5 Scoring the onplant angiogenesis assay. (a) Onplants are visualized on the living embryo using a dissecting stereomicroscope and a good gooseneck illluminator. New blood vessels are identified by their *red color*. The mesh grid is used as a scoring array, where the number of grid squares that contain new blood vessels are divided into the total number of grid squares on the onplant to produce the "angiogenic index". (b) Onplants can be removed from the CAM and further analyzed by fixation, sectioning and hematoxylin and eosin (*H&E*) staining. New blood vessels and a number of infiltrating cells are clearly seen in these *low* and *high* resolution images. The clear *circles* are the nylon mesh

5.4   Expected Results

Typically, the background angiogenesis that occurs in control (PBS-containing) onplants is approximately 15%. Angiogenic stimuli can induce angiogenesis to ~85% in 72 h. It is important to note that 0% angiogenesis in the case of the negative control or 100% angiogenesis in the case of the positive control is not a desirable effect because this will effectively reduce the dynamic range of the assay. Anti-angiogenesis therapies such as Avastin at an appropriate concentration typically inhibit angiogenesis by 75–85%. Some therapies will reduce angiogenesis below the background level of the negative control. This indicates that the therapy is inhibiting both the introduced angiogenic stimulus (VEGF, FGF or tumor cells) and the inherent pro-angiogenic environment of the collagen onplant. In some cases, an inhibitor can have an impact on the normal vasculature below the onplant, from where the new capillaries are sprouting.

Because the assay has a broad dynamic range, it is desirable in many cases to establish a dose–response for angiogenic stimuli or angiogenic inhibitors. The scalability of the assay is well-suited to the analysis of multiple factors simultaneously, such as combination therapeutic interactions and/or synergies of independent inhibitors.

6 Intravital Imaging of Angiogenesis

In order to better understand the underlying biology of neovascularization, it may be useful to directly visualize blood vessels to evaluate their morphology, function and dynamics. In this case, high-resolution intravital imaging can be utilized to visualize blood vessels as they develop or regress in the CAM or to visualize neovascularization in human tumor xenografts in the *ex ovo* chicken embryo model. The chicken embryo model is particularly well-suited for intravital imaging because it supports the growth of human tumors, is relatively inexpensive and does not require anesthetization or surgery. Tumor cells form fully vascularized xenografts within 7 days when implanted directly into the CAM. The resulting tumor vasculature can be visualized using non-invasive real-time imaging for periods of more than 72 h with little impact on either the host or tumor systems. Vascular imaging and labeling agents administered distal to the vessel bed of interest can be visualized as they bind endothelium, flow through the bloodstream, or extravasate from leaky tumor vasculature. This technique provides a method to both visualize and quantify the dynamics of angiogenesis in the CAM vasculature and in human xenograft tumors implanted in the CAM.

6.1 Considerations for Intravital Microscopy

While snapshots of the CAM can be taken in a living embryo without immobilizing the tissue, immobilization is critical for intravital or timelapse microscopy. This is accomplished in our laboratories through the use of an embryo imaging unit of our own design placed under an upright epifluorescence or spinning disk confocal microscope. The embryo imaging unit maintains the embryo at high humidity (>70%) and should ideally be placed in a heated microscope environmental chamber (at 37°C) with no CO_2. Embryos mounted in this unit continue to develop normally and remain optically stable for periods of 72 h or more [3, 4, 18–22].

The embryo imaging unit can be obtained from Quorum Technologies (Guelph, ON, Canada), or by contacting the authors directly. The choice of upright microscope for this type of imaging is important, as it must have sufficient working distance to fit the embryo imaging unit. For widefield epifluorescence microscopy, we use a Zeiss Examiner Z1 upright microscope fitted with a Hamamatsu 9100–02 EM-CCD camera. For confocal microscopy, we use a Zeiss Examiner Z1 upright microscope fitted with a Yokogawa spinning disk, Hamamatsu ImagEM 9100–12 EM-CCD camera, and diode lasers (405, 491, 561, 647 nm).

6.2 Inoculation of Tumors into the CAM

- A cancer cell line of interest should be cultured in the absence of antibiotic or selective medium to 80% confluency. CRITICAL POINT: Cells grown to confluency greater than 80% (beyond exponential growth phase) will usually fail to form vascularized tumors.
- Cells must first be trypsinized. Wash twice with 1X PBS pH 7.4. Aspirate remaining PBS then add 0.5% Trypsin-EDTA (e.g. 2 mL to T75 flask, 3 mL to T175 flask, 3 mL to 150 mm culture dish) and incubate at 37°C for 2–5 min until the cells detach.
- Add 7–8 mL of growth medium containing serum to neutralize the Trypsin, and then transfer the cell suspension to a 15 mL Falcon tube. Centrifuge at room temperature at $200 \times g$ for 5 min.
- Aspirate supernatant and resuspend with 10 mL of PBS. Pour out supernatant and resuspend cells with 1,000 μL of PBS and transfer to a 1.5 mL Eppendorf tube.
- Take 10 μL of suspension and dilute into 490 μL of PBS. The cells should now be counted using a hemacytometer.
- For implantation into the CAM, it is generally best to start with a working concentration of 1×10^7 cells/mL. CRITICAL POINT: Once the working suspension is made, it is necessary to proceed immediately to the implantation procedure.
- Prepare a micro-injection syringe by fitting an 18 gauge needle onto 1 mL syringe, and then slide a 5–6″. piece of Tygon tubing over the end of the needle (Fig. 6a). The needle should be inserted snugly into the bore of the tubing, leaving about 4–5″ of tubing extending from the tip of the needle.
- Dip the end of the Tygon tubing into the cell suspension and draw it into the syringe. Invert and tap to remove air bubbles and depress the plunger until the cell suspension fills the tubing.
- Microinjection needles drawn from borosilicate glass capillary tubes are prepared ahead of time using a pipette puller and stored in a sterile dish. CRITICAL POINT: Make needles that are long and tapered, but thick enough to resist bending when applied to the CAM. First, use forceps to break the tip off of the microinjection needle (Fig. 6b). Twisting the forceps slightly will help to create a sharp needle tip. The needle is then inserted into the open end of the Tygon tubing.

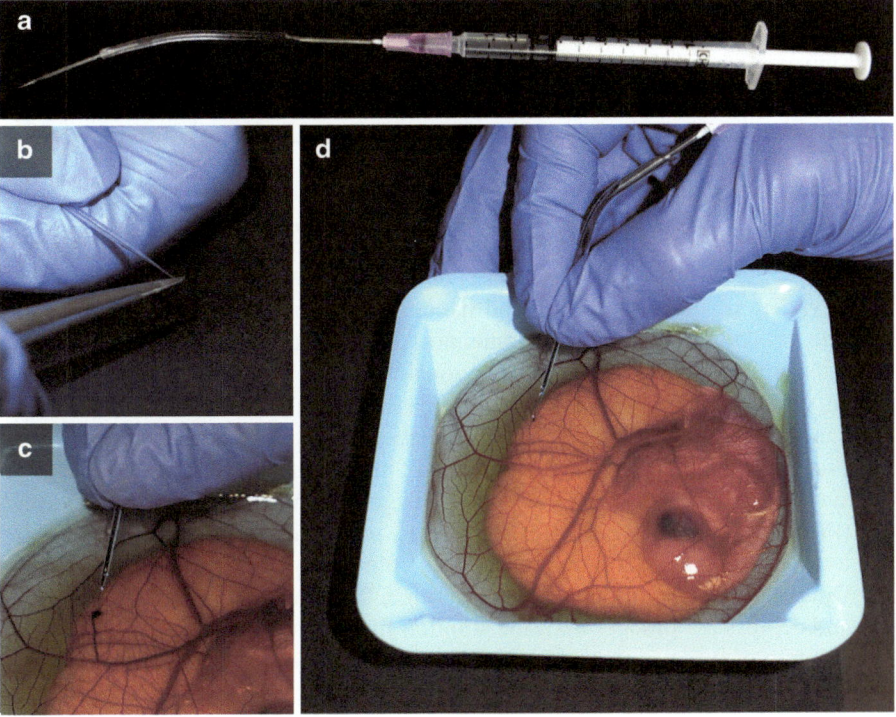

Fig. 6 Microinjection into the CAM. (**a**) Assembled micro-injection syringe consisting of (*left to right*) a pulled borosilicate needle, Tygon tubing, an 18 G needle and a 1 cm³ syringe. (**b**) Prior to insertion in the Tygon tubing, the tip of the pulled borosilicate needle is broken off using forceps. A twisting motion makes the tip sharper. (**c, d**) Microinjection needle tip entering CAM vein and cannulating the vessel lumen. In this example, Evan's *blue* dye is being injected and can be observed filling the distal vasculature

- Depress the plunger lightly to fill the glass needle with cells. If the cells clump within the syringe or clog the needle, it may be necessary to re-break the tip. If this does not work, remove the glass needle using a razor blade to cut the tubing, re-mix the cell suspension, and insert a new needle.

- For optimal results, utilize embryos that are between day 8 and day 10 of development. Keep embryos in their secondary container until ready for injection. Immediately return them once the injection is complete.

- Using a dissection stereomicroscope with a gooseneck illuminator, select an area of CAM that is adjacent to a larger vessel. The CAM is slightly thicker in these areas, and this will improve injection success. Carefully push the end of the needle against the surface of the CAM until it just penetrates the epidermis without passing through the CAM completely. Slowly depress the plunger to create a bolus of 10,000–100,000 cancer cells within the CAM adjacent to the vessel. Be careful not to inject the tumor cells into the vessel! CRITICAL POINT: Avoid leaving cells on the CAM surface as this significantly compromises imaging. If this occurs, dab with a clean Kimwipe or cotton-tipped applicator.

- Once implanted, tumors will typically take 5–10 days to develop and become sufficiently vascularized. Cell lines that perform well in this model include HT1080, HEp3, and HT-29 among others. For improved imaging, utilize cell lines that constitutively express fluorescent proteins such as GFP or tdTomato.
- Return embryos to humidified incubator and allow tumor to grow (up to 10 days).

6.3   Intravenous Injection of Fluorescent Dyes or Nanoparticles

- A variety of fluorescent dyes and nanoparticles can be used to visualize blood vessels. To label the global endothelium, fluorescent *lens culinaris* lectin (Vector Labs, Burlingame, CA) works exceptionally well [4, 18]. Fluorescent dextrans can be used to fill the vessel lumen and visualize blood flow [13]. Specialized nanoparticles can be used to selectively label neovasculature [3, 8, 22].
- First, prepare a micro-injection syringe by fitting an 18 gauge needle onto 1 mL syringe, and then slide a 5–6″ piece of Tygon tubing over the end of the needle (Fig. 6a). The needle should be inserted snugly into the bore of the tubing, leaving about 4–5″ of tubing extending from the tip of the needle.
- Draw the injection solution into the syringe. Invert and tap to remove air bubbles and depress the plunger until the solution fills the tubing.
- A microinjection needle drawn from sodium borosilicate glass capillary tubes is now inserted into the open end of the Tygon tubing. Microinjection needles are prepared ahead of time using a pipette puller and stored in a sterile dish. CRITICAL POINT: Make needles that are long and tapered, but thick enough to resist bending when applied to the CAM.
- Depress the plunger lightly to fill the glass needle with solution.
- Microinject CPMV into a CAM vein distal from the desired site to be visualized (Fig. 6c, d). Position the needle parallel to the blood flow and inject in the same direction as the flow. A successful cannulation of CAM vein is evident by clearing of blood in the path of the injection flow. Injection volumes can range from 25 to 100 μL.
- Intravital imaging will generally start immediately after injection of the imaging agent.

6.4   Intravital Imaging

- Sterilize the embryo imaging unit in a dilute bleach solution and dry completely.
- Preheat the microscope enclosure and embryo imaging unit to 37°C.
- Remove the lid of the imaging unit and add 1/4″ of water to the bottom and the outer jacket (Fig. 7a).
- Apply a thin layer of vacuum grease around the circumference of the coverslip seating port on the underside of the lid of the embryo imaging unit and then fit an 18 mm glass coverslip onto it (Fig. 7b). Ensure that the coverslip is well-seated and is free from dust or grease.

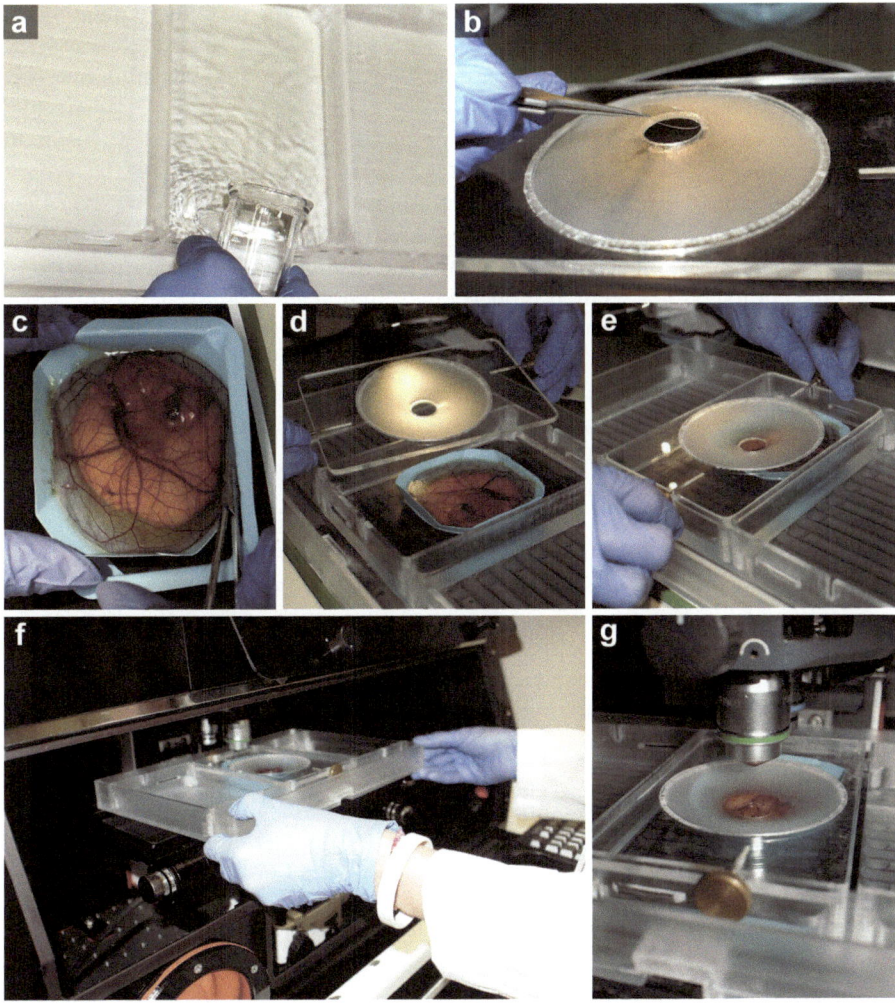

Fig. 7 Assembly of the chick embryo imaging unit. To assemble the microscope-mounted embryo imaging unit for intravital imaging, (**a**) a 1/4″ of water is poured into the bottom of the chamber. (**b**) After a thin film of vacuum grease is applied to the periphery of the coverslip mount, the coverslip is lowered into place with forceps to form a tight seal. (**c**) The edge of the weigh boat is then trimmed with sterile scissors and (**d**) the embryo is placed in the imaging unit so that the area of interest is in the centre of the chamber. (**e**) The lid is then slowly lowered over the embryo until the coverslip just makes contact with the CAM. (**f**) The embryo imaging unit is then placed on stage of the upright confocal microscope with a heated enclosure set to 37°C, (**g**) and the objective is lowered over the coverslip for intravital imaging

- Trim the edge of the weigh boat with sterile scissors (Fig. 7c).
- Please the embryo in the imaging unit such that the coverslip can be lowered directly onto the area of interest (Fig. 7d). Carefully lower the lid until the coverslip makes contact with the CAM without applying force to the tissue surface, ensuring that no air bubbles form between the CAM and the coverslip

(Fig. 7e). Carefully tighten the adjustment screws to fix the embryo imaging unit lid in place. CRITICAL POINT: Please the imaging unit on a flat, level surface and ensure the lid is level using a mini-bubble level. This will minimize focal plane artifacts during imaging.

- Place the embryo in the embryo imaging unit onto microscope stage inside the environmental chamber equilibrated to 37°C (Fig. 7f). Secure the embryo imaging unit to the stage with laboratory tape to minimize movement artifacts or accidental movements.
- Move the desired objective lens in place and begin imaging (Fig. 7g). Generally, if the embryo is not secured properly then significant field of view drift will be apparent within 15–20 min. If this occurs, loosen the lid and reposition the coverslip on the CAM. This can be repeated several times until a completely stable imaging field is obtained.
- An imaging protocol can now be utilized that best addresses the particular experiment. As the embryo imaging unit will keep the field of view fixed, one is free to collect three-dimensional Z stacks, timelapse imagery and/or multiple fields of view. Acquisition software with the capability to collect three-dimensional and timelapse data (we use Perkin Elmer's Volocity) are strongly recommended. To visualize the dynamics of blood flow, a rapid succession of fluorescence images can be acquired in a single plane. For detailed structural analyses at specific time-points, a high-resolution three-dimensional image stack can be acquired. Detailed structural changes in vasculature can be mapped over time by acquiring three-dimensional stacks at regular time-points.

6.5   Expected Results

Intravital imaging of tumor neovasculature has the potential to reveal new insight into tumor angiogenesis, which is important not only for supporting tumor growth but also to provide metastatic cancer cells with access to the circulation. Whether standard widefield epifluorescence microscopy or sophisticated confocal microscopy techniques are used, the visualization of tumor neovasculature is possible using this technique (Fig. 8). When a bolus of GFP-expressing HT1080 fibrosarcoma tumor cells is injected within the CAM at 10 days of development, the surrounding vasculature is clearly delineated after the intravenous administration of rhodamine *lens culinaris* lectin (Fig. 8a). After a tumor cell bolus has been incubated in the CAM for 24 h, the vascular label reveals extensive angiogenesis at the site of the tumor bolus (Fig. 8b) when widefield epifluorescence microscopy is used. A similar technique can be employed to assess vascular function and permeability when fluorescent dextrans are injected [13]. Confocal microscopy of a growing HT1080 tumor allows the endothelium and vessel lumen to be readily distinguished, including their close juxtaposition to the tumor cells (Fig. 8c).

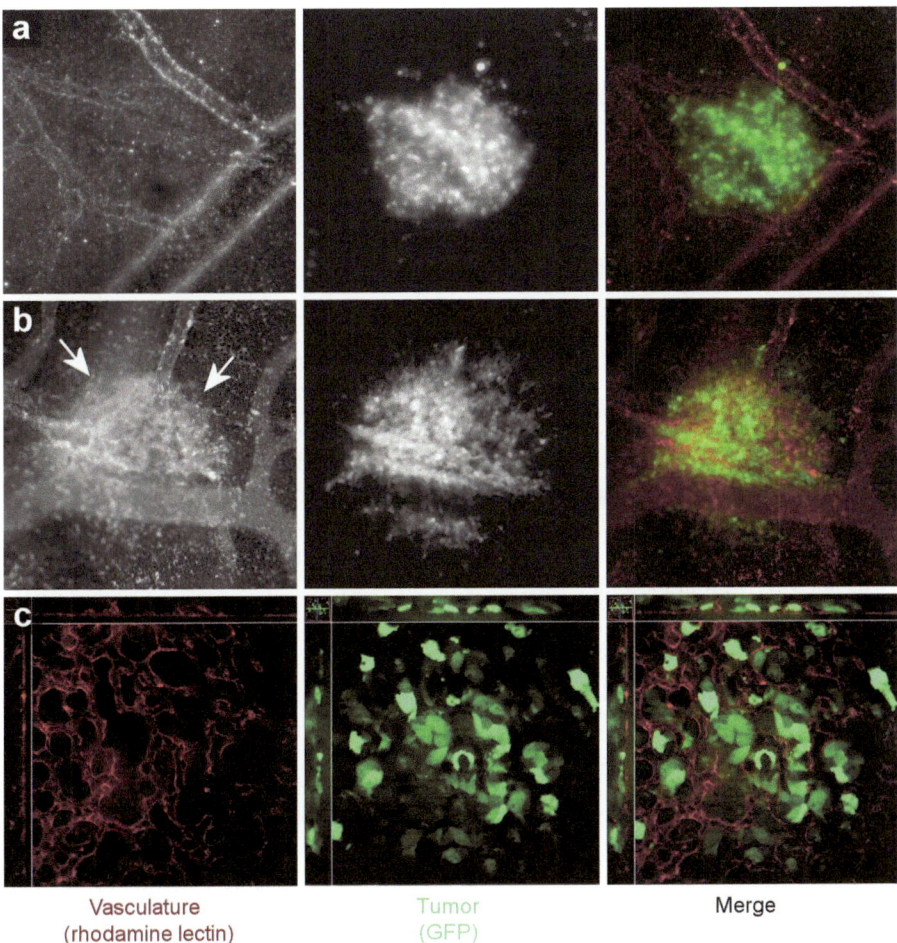

Vasculature (rhodamine lectin)	Tumor (GFP)	Merge

Fig. 8 Intravital imaging of tumor neovasculature in the CAM using widefield and confocal microscopy. HT1080 fibrosarcoma tumor cells express cytoplasmic GFP (*green*) and rhodamine lectin (*red*) is injected intravenously to label the vasculature. (**a**) A bolus of GFP-expressing tumor cells are visualized using widefield epifluorescence imaging shortly after injection into the CAM. (**b**) A bolus of tumor cells are visualized 24 h after injection into the CAM. Note the formation of angiogenic sprouts (*white arrows*) and the invasion of tumor cells into the surrounding stroma. (**c**) A three dimensional stack is acquired within a GFP-expressing tumor using an upright spinning disk confocal microscope. The endothelium and vessel lumen are readily distinguished, and are closely juxtaposed to the tumor cells

7 Materials and Reagents

7.1 Shell-Less Cultivation of Chicken Embryos

- All experiments were performed in accordance with the regulations and guidelines of the Institutional Animal Care and Use Committee at the University of Alberta and Vanderbilt University.
- SPAFAS (Specific Pathogen Tested) just laid fertile chicken eggs, 45–55 g, standard grade from a 33–55 week old flock.
- Rotating and stationary humidified egg incubator (37°C, 70% humidity)
- 80 mm^2 weigh-boats (medium size, VWR cat.#12577-027)
- 100 × 100 × 15 mm square Petri dishes (sterile, Fisher cat.#08-757-11A)
- Rectangular containers w. lid (VWR 70690–226) which can hold 6 weigh boats + lids to make a humidity chamber. Make sure the lid leaves room for air exchange.
- 37°C humidified incubator maintaining > 70% humidity (without CO_2)
- Dremel tool with cut-off wheel
- Unvented hood for egg drilling (optional; VWR cat.#30181-775)
- 70% EtOH and/or iodine

7.2 Onplant CAM Angiogenesis Assay

- Type I collagen (Vitrogen, Cohesion cat.# 100691)
- 0.1 M NaOH
- 0.1 M HCl
- 1 M Hepes
- 10X PBS
- 1X PBS
- 1X PBS, 5% protease-free BSA (Sigma A3059)
- Tetko Nitex nylon mesh (SEFAK America cat.#03-180/44), cut into 2 × 2 and 4 × 4 mm^2 and autoclaved (5 min on dry cycle)
- Parafilm and 100 × 100 × 15 mm square Petri dishes (sterile, Fisher cat.#08-757-11A)
- 70% EtOH
- Growth factor or alternate angiogenic stimulus (e.g. bFGF)
- Sterile forceps (needle nose)
- Laminar flow hood
- Dissecting stereomicroscope
- Flexi-lights or some light source that is readily focused and positioned

7.3 Intravital Imaging Angiogenesis Assay

- Dulbecco's PBS (pH 7.2–7.4) (Invitrogen, Cat. no. 14190250)
- DMEM/F12 media (Invitrogen, Cat. no. 11330–057)

- Heat inactivated fetal bovine serum (Invitrogen, Cat. no. 12483020)
- Penicillin Streptomycin liquid 100 mL (Invitrogen, Cat. no. 15140122)
- 2.5x Trypsin-EDTA (Invitrogen, Cat. no. 25300054)
- 1 M HEPES buffer (Bioshop, Cat. no. HEP003.100)
- 70% ethanol
- Vacuum grease (VWR, Cat. no. 59344–055)
- Dissecting microscope
- Epi-fluorescence widefield microscope (Quorum, Zeiss Axio Examiner)
- Spinning disk confocal fluorescence microscope (Quorum, Yokogawa CSU 10)
- Temperature enclosure unit for microscope (Precision Plastics Inc.)
- Sportsman Hatcher (Berry Hill Limited, Cat. no. 1550HA)
- Sportsman incubator (Berry Hill Limited, Cat. no. 1502EA)
- Rubbermaid container with lid (Guillevin Int., Cat. no. RH3-228-00-BLU) with 1″ diameter holes drilled into sides for air exchange
- Vertical pipette puller (David Kopf Instruments, Model 720)
- Sodium borosilicate glass capillary tubes O.D.: 1.0 mm, I.D.: 0.58 mm, 10 cm length (Sutter Instrument, Cat. no. BF100-58-10)
- 18 mm circular glass coverslips No. 1 (VWR, Cat no. 16004–300)
- Fine point forceps (VWR, Cat. no. 25607–856)
- Pyrex Petri dish 100 × 20 (VWR, Cat. no. 89000–306)
- Tygon R-3603 laboratory tubing 50 ft for injections (1/32″ inner diameter, 3/32″ outer diameter, 1/32″ wall thickness, VWR Cat. no. 63009–983)
- 1 mL syringes for injections (BD, box of 100 Cat. no. 309602)
- 20 mL syringes (BD, box of 100 Cat. no. 521906)
- 18 and 27.5 gauge hypodermic needles for injections (BD, Cat. nos. 305195 and 305196)
- Intellitemp Heat Mat 19.7″ × 11.8″ (30 W, Big Apple Herpetological, Inc.)
- Dremel tool with #36 cut off wheels (Dremel, Cat. no. 409))
- Polystyrene weigh boats (VWR, Cat. no. 12577–01)
- Square Petri dishes (Simport, VWR Cat. no. 25378–115)
- Glass Pasteur pipettes (VWR, Cat. no. 14673–010)
- Embryo imaging unit (Quorum Technologies, Inc.)

Acknowledgements The authors would like to acknowledge Dr. Desmond Pink, Dr. Hon Leong and Amber Ablack for their wonderful contributions to the figure images. We greatly appreciate editorial input from Dr. Kristin Kain. This work was supported by grant 700537 from the Canadian Cancer Society Research Institute to JDL and NIH/NCI grants CA120711-01A1 and CA120711-01A1 to AZ.

References

1. Auerbach R, Kubai L, Knighton D, Folkman J (1974) A simple procedure for the long-term cultivation of chicken embryos. Dev Biol 41(2):391–394
2. Ribatti D (2004) The first evidence of the tumor-induced angiogenesis in vivo by using the chorioallantoic membrane assay dated 1913. Leukemia 18(8):1350–1351

3. Leong HS, Steinmetz NF, Ablack A, Destito G, Zijlstra A, Stuhlmann H et al (2010) Intravital imaging of embryonic and tumor neovasculature using viral nanoparticles. Nat Protoc 5(8):1406–1417
4. Zijlstra A, Lewis J, Degryse B, Stuhlmann H, Quigley JP (2008) The inhibition of tumor cell intravasation and subsequent metastasis via regulation of in vivo tumor cell motility by the tetraspanin CD151. Cancer Cell 13(3):221–234
5. Wallis JW, Aerts J, Groenen MA, Crooijmans RP, Layman D, Graves TA et al (2004) A physical map of the chicken genome. Nature 432(7018):761–764
6. Consortium (2004) Sequence and comparative analysis of the chicken genome provide unique perspectives on vertebrate evolution. Nature 432(7018):695–716
7. van de Lavoir MC, Diamond JH, Leighton PA, Mather-Love C, Heyer BS, Bradshaw R et al (2006) Germline transmission of genetically modified primordial germ cells. Nature 441(7094):766–769
8. Lewis JD, Destito G, Zijlstra A, Gonzalez MJ, Quigley JP, Manchester M et al (2006) Viral nanoparticles as tools for intravital vascular imaging. Nat Med 12(3):354–360
9. MacDonald IC, Schmidt EE, Morris VL, Chambers AF, Groom AC (1992) Intravital videomicroscopy of the chorioallantoic microcirculation: a model system for studying metastasis. Microvasc Res 44(2):185–199
10. Jilani SM, Murphy TJ, Thai SN, Eichmann A, Alva JA, Iruela-Arispe ML (2003) Selective binding of lectins to embryonic chicken vasculature. J Histochem Cytochem 51(5):597–604
11. Ribatti D, Vacca A, Roncali L, Dammacco F (1996) The chick embryo chorioallantoic membrane as a model for in vivo research on angiogenesis. Int J Dev Biol 40(6):1189–1197
12. Zijlstra A, Seandel M, Kupriyanova TA, Partridge JJ, Madsen MA, Hahn-Dantona EA et al (2006) Proangiogenic role of neutrophil-like inflammatory heterophils during neovascularization induced by growth factors and human tumor cells. Blood 107(1):317–327
13. Pink D, Fung L, Zijlstra A, Lewis JD (2012) Real-time visualization and quantitation of vascular permeability in vivo: implications for drug delivery. PLoS One 7:e33760
14. Zijlstra A, Mellor R, Panzarella G, Aimes RT, Hooper JD, Marchenko ND et al (2002) A quantitative analysis of rate-limiting steps in the metastatic cascade using human-specific real-time polymerase chain reaction. Cancer Res 62(23):7083–7092
15. Ausprunk DH, Knighton DR, Folkman J (1975) Vascularization of normal and neoplastic tissues grafted to the chick chorioallantois. Role of host and preexisting graft blood vessels. Am J Pathol 79(3):597–618
16. Brooks PC, Montgomery AM, Cheresh DA (1999) Use of the 10-day-old chick embryo model for studying angiogenesis. Methods Mol Biol 129:257–269
17. Zijlstra A, Aimes RT, Zhu D, Regazzoni K, Kupriyanova T, Seandel M et al (2004) Collagenolysis-dependent angiogenesis mediated by matrix metalloproteinase-13 (collagenase-3). J Biol Chem 279(26):27633–27645
18. Arpaia E, Blaser H, Quintela-Fandino M, Duncan G, Leong HS, Ablack A et al (2011) The interaction between caveolin-1 and Rho-GTPases promotes metastasis by controlling the expression of alpha5-integrin and the activation of Src, Ras and Erk. Oncogene 31:884–896
19. Cho CF, Ablack A, Leong HS, Zijlstra A, Lewis J (2011) Evaluation of nanoparticle uptake in tumors in real time using intravital imaging. J Vis Exp (52)
20. Leong HS, Lizardo MM, Ablack A, McPherson VA, Wandless TJ, Chambers AF et al (2012) Imaging the impact of chemically inducible proteins on cellular dynamics in vivo. PLoS One 7(1):e30177
21. Steinmetz NF, Ablack AL, Hickey JL, Ablack J, Manocha B, Mymryk JS et al (2011) Intravital imaging of human prostate cancer using viral nanoparticles targeted to gastrin-releasing Peptide receptors. Small 7(12):1664–1672
22. Steinmetz NF, Cho CF, Ablack A, Lewis JD, Manchester M (2011) Cowpea mosaic virus nanoparticles target surface vimentin on cancer cells. Nanomedicine (Lond) 6(2):351–364

Developing an *Ex Vivo* Model of Ischemia Using Early Chick-Embryo: A Model to Study Ischemia Related Angiogenesis

Syamantak Majumder, Sree Rama Chaitanya Sridhara, and Suvro Chatterjee

Abstract Ischemia created in the animals were long been used as a model to study ischemia mediated effects *in vivo*. Pathological angiogenesis is the key hallmark of various ischemic diseases where blood vessel formation was compromised due to low blood flow. New blood vessels form in order to compensate the low blood perfusion in the ischemic area. This neovascularization and remodeling of the existent vessels protect from the consequences of ischemia associated diseases like myocardial infarction and stroke. A better understanding of the mechanisms of functional vessel formation is a pre-requisite to improve the treatment of ischemic pathologies. Therefore, the research area warrants an easily accessible model in which vessel formation can be both manipulated and studied. However, a limited number of efforts have been put forward yet to develop an ischemia models where ischemia mediated remodeling of vessels can be studied in real time. In present study, we used 4 day grown chick embryo to ligate right vitelline artery and create partial ischemia in the vascular bed of the embryo. The model has been developed based on the principle that blocking blood flow in the vascular bed will stop the nutrient and oxygen supply to the adjacent vessels and thus creating an ischemia like condition. Additionally, ischemia related changes in angiogenesis can be followed and tracked in real time in the vascular bed of the chick embryo. The present *ex vivo* model can be utilized in studying ischemia related angiogenesis in specific and hypoxia and/or low oxygen mediated angiogenesis in general.

S. Majumder • S.R.C. Sridhara • S. Chatterjee (✉)
Vascular Biology Lab, Life Sciences Division, AU KBC Research Centre,
MIT Campus, Anna University, Chennai 600 044, Tamil Nadu, India
e-mail: soovro@yahoo.ca

E. Zudaire and F. Cuttitta (eds.), *The Textbook of Angiogenesis and Lymphangiogenesis: Methods and Applications*, DOI 10.1007/978-94-007-4581-0_15,
© Springer Science+Business Media Dordrecht 2012

1 Introduction

In many aspects embryonic chick heart resembles the developing human heart and so has been exploited as a model since many decades [1, 2]. An intervention model for the chick embryo was designed to obtain insight into the long-term hemodynamic effects of altered venous return patterns on cardiac morphogenesis and malformations. Specific cardiac malformations were induced by permanently obstructing the right lateral vitelline vein with a microclip (venous clip model; [3]), thereby altering the intra-cardiac blood flow patterns. A spectrum of outflow tract anomalies can be induced by this intervention. Hogers et al. [3] postulated that alterations in hemodynamic parameters could lead to changes in shear stress and thus alter the expression of shear-stress-responsive genes further leading to developmental perturbations finally resulting in cardiac malformations. Alterations in intra-cardiac blood flow pattern during clipping (visualized by injected India ink) suggested that hemodynamics is influenced by clipping [4]. Recently we also showed that low flow in chick embryo vascular bed (controlled by uplifting the right vitelline artery using a surgical suture and height manipulation unit) leads to low nitric oxide bio-availability further impairing angiogenesis in the vascular bed [5]. Other studies have also shown that alterations in hemodynamics can precede the onset of structural defects [6].

Ischemia is a condition in which blockade in blood flow leads to restricted oxygen and nutrient supply to the affected area of the body wherein cardiac ischemia is compromised blood flow and oxygen supply to heart muscle. However, pathological angiogenesis due to ischemic stress is one of the key phenomenon that occurs as a defense mechanism of the body to perfuse blood in the ischemic area. Neo-angiogenesis strictly regulates the survival of the cells and tissues that are present in the ischemic proximity. Therefore, an ischemic model that could help us to observe the remodeling of vessels in real time will be crucial to develop and study. The principle aim of developing an ischemic animal model is to study the basic processes or potential therapeutic interventions in ischemia associated diseases, and the extension of patho-physiological knowledge, which will lead to improve medical treatment of human ischemic stroke [7]. Chick embryo vascular bed has long been used to study the effect of different angiogenic modulators on neo-angiogenesis. We manipulated this model to develop ischemia in the embryonic vascular bed and therefore ischemia associated vascular remodeling can be followed in real time. Moreover, an intact chick embryo with yolk can be maintained under controlled conditions that make this model more flexible for manipulation and real-time observation.

The present study established a novel *ex vivo* partial ischemia model by blocking the right vitelline artery of chick embryo. The model was validated by measuring several physiological (deteriorated cell and tissue functions) and biochemical parameters (elevated level of HIF-1α) in the vascular bed of chick embryo. Additionally we were able to visualize real time vascular remodeling in the ischemic area.

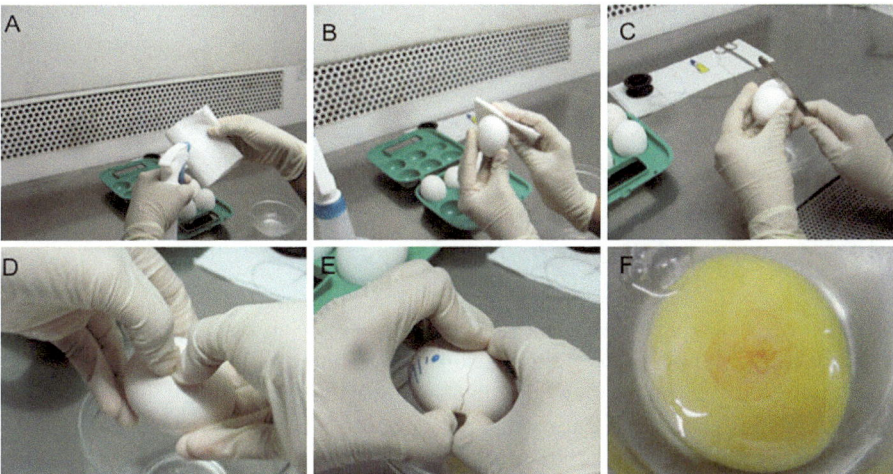

Fig. 1 Preparing shell less chick embryo for experimentation: after 96 h of incubation, the egg shells were cleaned with 70% alcohol solution (**a**, **b**), and carefully broken by gently hitting the in the *middle* of the egg shell with a sharp knife (**c**). Next, the entire contents of the egg were transferred to a sterile glass bowl with extreme care taken to keep the embryo intact (**d–f**)

2 Methodology and Discussion

2.1 Selection of Chick Embryos

Fertilized white Leghorn chick (*Gallus domesticus L.*) eggs were obtained from Poultry Research Station, Nandanam, Chennai, India and incubated at 37°C, with the blunt end up and at a relative humidity of 70–80%. The embryos were exposed by creating a window in the shell and removing the overlying membranes. Embryos at day 4 or stage HH17 [8], that showed no bleeding or deformities, were selected.

2.2 Preparation of Chick Embryos: Shell-Less Ex Vivo Cultures

Fertilized White Leghorn chicken eggs were incubated at 37°C and 70–80% relative humidity. At day 4, the egg shells were cleaned with 70% alcohol solution, and carefully broken by gently hitting the shell in the middle of the egg with a sharp knife. Next, the entire contents of the egg were transferred to a sterile glass bowl with extreme care taken to keep the embryo intact (Fig. 1). However, damaged eggs with ruptured vascular bed or with any visible abnormalities were not used for this study. Figure 1 of the present chapter demonstrated the whole process of preparation of shell less chick embryos to carry out the experiments.

The glass bowls with the egg contents are covered with round shaped sterile glass lids and placed back in incubator (at 37°C and 70–80% relative humidity). To avoid any stress to the embryos, the embryos were used for the manipulation after 1 h of incubation.

2.3 Ligation of Right Vitelline Artery

In the early stages of chick embryo development, two distinct circulatory systems are developed, one for the embryo itself maintaining the blood circulation inside the embryo body and another vitelline system extending into the egg sac, which helps the embryo to get a constant supply of nutrients and oxygen for the embryo survival [7]. Blocking the circulation of any of the vitelline artery can interfere with nutrient and oxygen diffusion and can create a partial ischemia. In the present study, ischemia was created in the vascular bed of chick embryo by using a modified protocol as mentioned by Vos et al. [9] where they studied the hemodynamics due to blockade in right vitelline vein [9].

The egg shell was cleaned with 70% ethanol and broke open in "U" shaped sterile glass bowl in sterile conditions. All manipulations were performed macroscopically. Above the intended ligation site, the vitelline membrane was removed, and a small incision was made with a sterile tungsten needle in the yolk sac membrane, adjacent to the vitelline artery. A surgical suture of 0.1 mm thickness was tied with the tungsten needle or copper wire with a bio-friendly adhesive in the end of the needle/wire. The tungsten needle or the copper wire should be of a minimum thickness (<0.5 mm) to avoid serious injury to the vascular bed. As demonstrated in Fig. 2, the tungsten needle or the copper wire was then passed through the vitelline membrane in one side of the artery to the other side by approaching through yolk sac (Fig. 2a, b). During the surgery, minimum disturbance was made in the yolk sac to avoid the leak of yolk material. Forceps were used to hold the needle while passing trough the yolk sac (Fig. 2c–f). Care has been taken to avoid the contact of gloves covered fingers to the yolk sac. The surgical suture was then tied gently avoiding any injury to the vessels (Fig. 2g, h). Figure 3 shows the close up images of the knot in the vascular bed of the chick embryo. The cessation of blood flow in artery proximal to the microclip was confirmed by observing the blood flow in the adjacent vessels under stereo microscope surveillance. Only embryos that showed no bleeding or deformities were selected. The material was subdivided into artery ligated embryos and sham operated control embryos. From the 233 ligated embryos, 62 did not survive, leaving a survival rate of approximately 73.4% (171 of 233). During the experiments, all eggs were placed in an incubator (at 37°C and 70–80% relative humidity) while the bowl containing the embryos were covered with a sterile glass lid. During the time of incubations the embryos were least disturbed to avoid any more stress to the embryos.

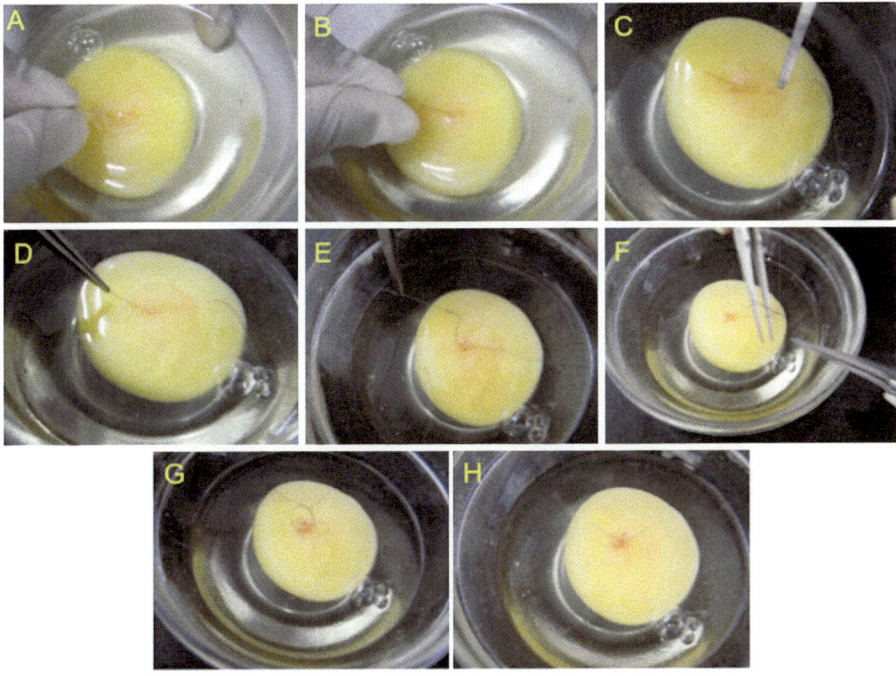

Fig. 2 Ligation of the right vitelline artery: the copper wire of 0.5 mm thickness was passed through the vitelline membrane in one side of the artery to the other side by approaching through yolk sac (**a, b**). Next, forceps were used to hold the needle while passing trough the yolk sac (**c**). Slowly the needle was passed through the yolk (**d**) followed by extending the surgical suture to have enough length to put the knot (**e, f**). The surgical suture was then tied gently avoiding any injury to the vessels (**g, h**)

3 *Ex Vivo* Chick Embryo Ischemia Model

3.1 *Observation of Blood Flow*

Upon opening the chick embryo in sterile glass bowl, the whole vascular bed of chick embryo can be seen as shown in Fig. 3a. Before and after ligation of the vitelline artery, the blood flow in the vascular bed of chick embryo was observed under stereomicroscope (using 4X objective). Before ligation, we were able to observe the flow of RBC in the vessels while upon ligation the RBC flow stopped. Due to transparent nature of the vessels in chick embryo vascular bed, it is quite easy to follow the blood flow in the vascular bed and confirming the manipulation in blood flow. A sensitive micro flow meter can also be adapted to the model to track the flow critically.

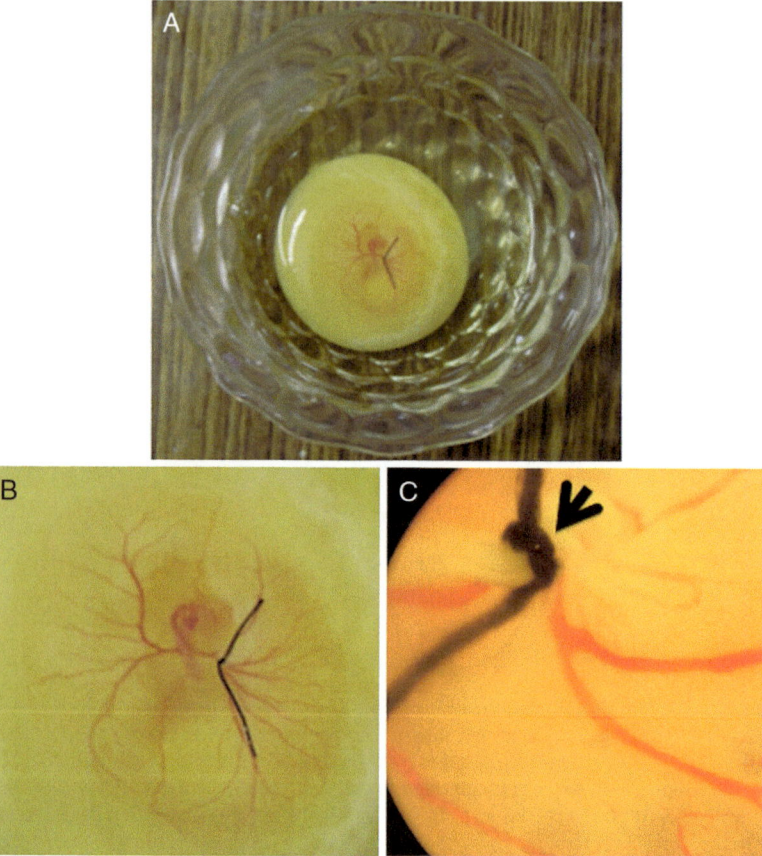

Fig. 3 Ligation of right vitelline artery in chick embryo vascular bed. At 4th day of embryo growth, the entire egg content was broke open in sterile glass bowl. The right vitelline artery of the egg was ligated using a surgical suture and the images of the embryo were taken with Nikon Cool Pix Camera (**a**, **b**) while close up images of the knot in the artery were taken using a Stereomicroscope adapted with a Nikon Cool Pix camera (**c**). The *arrow head* showed the presence of knot in the vessel

3.2   Measurement of Hypoxia Inducible Factor-1α (HIF-1α) in the Vascular Bed of Chick Embryo

Overexpression of HIF-1α (a key determinant of oxygen-dependent gene regulation in angiogenesis) may be beneficial in cell therapy of hypoxia-induced pathophysiological processes, such as ischemic heart disease [10]. HIF-1α get stabilized under hypoxic or ischemic conditions and therefore can be measured and used as a marker of hypoxic or ischemic stress [11, 12]. We measured HIF-1α in the ischemic vessels of chick embryo to confirm ischemia in the vascular bed of chick embryo. HIF-1α level elevated with the time of ischemic stress reaching maximum after 1 h of

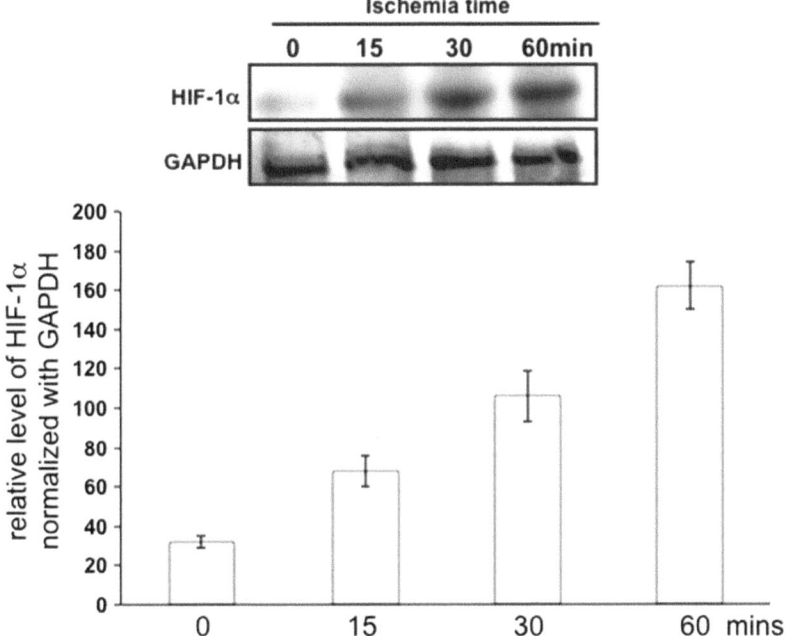

Fig. 4 Elevated level of HIF-1α in ischemic vessels. HIF-1α was measured in the ischemic vessels and quantified using densitometry analysis. GAPDH was used as an internal loading control

ligation (Fig. 4) thus demonstrating the presence of ischemic stress in the vascular bed of chick embryo.

3.3 Studying the Angiogenesis Pattern After Implementing Ischemia

Cell death during ischemia is mostly known to be caused by necrosis and apoptosis of the cells. Recent evidence suggest that apoptosis is a major contributor to I/R-induced cell death [13]. Similarly, ischemia mediated necrosis and apoptosis of cells in the vascular bed of chick embryo can facilitate vascular deformities and have been considered as a phenotypic response due to ischemia. Therefore, we studied the vascular deformities in the vascular bed of chick embryo. Upon ligation of the right vitelline artery we observed necrosis in the ischemia affected vascular bed. Due to necrosis, deformed vessel structures and destruction in physiological phenotype of the vessels can be observed in the vascular bed. Necrotic lesions in the vascular bed of chick embryo can be visualized due to the destruction in vessel integrity (Fig. 5, refer to arrow heads in the illustration).

Before ischemia   After 1h Ischemia

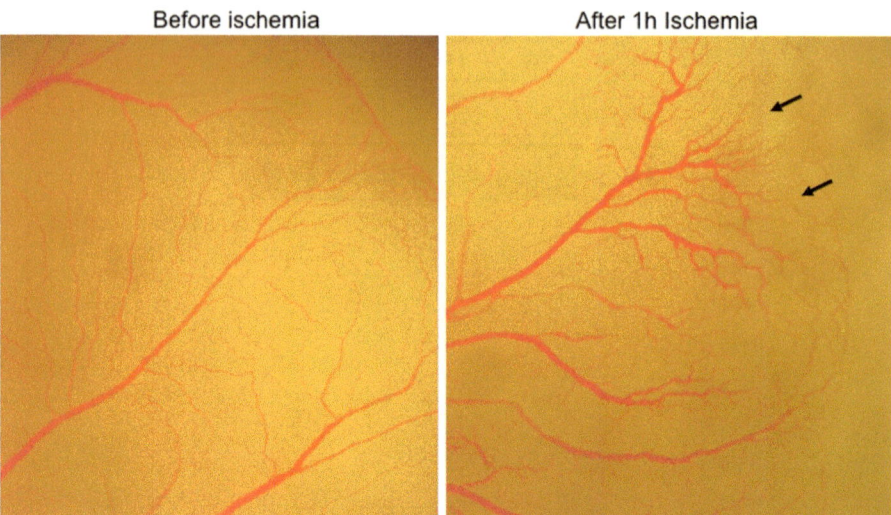

Fig. 5 Images of the normal and 60 min post ischemic vessels were taken using a stereomicroscope. *Arrow heads* showed the presence of deformed vessels in the ischemic area

Secondly, we also tried to follow the angiogenesis pattern of the non-ischemic vessels that lie in close proximity to ischemic area [7]. We speculated that ischemia mediated release of HIF-1α from ischemic vessels could potentially induce angiogenesis in the adjacent vessels to support recover from the ischemic stress. Surprisingly, a significant level of neo-vasculature and/or vascular remodeling was observed in areas close to primary ischemic site. Further, we observed that remodeling of the vessels is significantly faster than that observed in control area. The observation propounds that ischemia in surrounding vessels remodeled the non-ischemic vessels to grow faster to recover from the ischemic stress. Images were analyzed using the Angioquant software [14] as demonstrated (Fig. 6a). Imaging of the adjacent vessel demonstrated a significant change in the vessel density and length as can be observed from the images (Fig. 6b).

4   Advantages and Disadvantages of the Model

4.1   Studying Angiogenesis in Hypoxia and/or Low Oxygen Conditions

Egg yolk angiogenesis model has long been used as a model to study angiogenesis and embryonic development [15–17]. As like zebrafish model and unlike animal models, angiogenesis can be followed in real time in chick embryo vascular bed. Previous studies showed that blood flow can be manipulated in the chick embryo

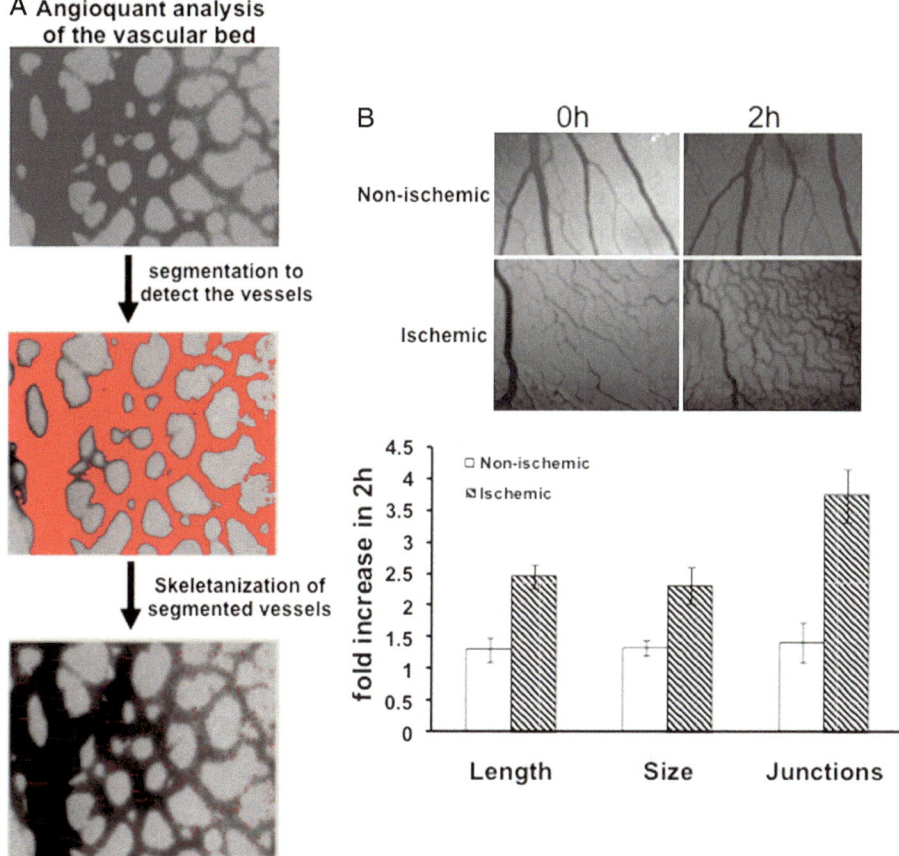

Fig. 6 Images of the vessels that lies adjacent to the ischemic area and of the normal non-ischemic vessels were acquired. Images were analyzed using Angioquant software as demonstrated by the panel (**a**). Drastic remodeling of the vessels that exist in the adjacent of ischemic area was observed and analyzed (**b**)

vascular bed using a venous clip model or by lifting the vessels using tungsten wire and thus manipulating the blood flow in the vessels. Therefore, we manipulated with the blood flow using surgical suture which can permanently block the blood flow in the vascular bed of chick embryo. We performed the experiments when the whole egg content is maintained in "U" shaped sterile glass bowl thereby visualizing the whole vascular bed (that can be followed macro and microscopically) to study the pattern of angiogenesis in the whole vascular bed of the chick embryo under ischemic stress. This specific reason makes the present model most flexible and useful than other well established ischemic models to study the angiogenesis remodeling in real time under hypoxic and/or low oxygen and ischemic stress. The model can be most useful in studying the effect of new ischemic drugs which target angiogenesis and its associated phenomenon to recover from the ischemic stress. Additionally, the model can also be used in parallel with the already existing models to study the anti-ischemic activity of different potential drugs. In a recent study [7] we have used ranolazine, a relatively new anti-ischemia drug, to study the potential of partial ischemia model as a platform for testing the efficacy of anti-ischemia drugs. Present model will be most useful in exploring the effectiveness of several drug candidates which are in very initial stage of screening for their anti-ischemic activity thereby reducing the number of animal experiments. We also deem that this model will help in comprehending angiogenesis signaling in hypoxic and/or low oxygen conditions on a general. Per se, in cancer conditions the core of the outgrowth is hypoxic, dragging extensive angiogenic perfusion towards core. Although this model is not a complete hypoxia model, we extrapolate that this model may be used to study 'hypoxia mediated angiogenesis' and drugs affecting hypoxia mediated signaling in angiogenesis.

4.2   Other General Considerations

The present model offers some unique features for routine ischemia related drug testing such as (1) less ethical issues (2) easy to develop model (3) real time observation of physiological parameters such as blood flow and angiogenesis and (4) culturing of cells and tissues of one's choice on the ischemic bed makes the model more feasible and flexible. However, the present model has some inherent limitations like; (1) the model cannot be tissue or organ specific as it is possible in animal models. (2) The model cannot produce intermittent ischemia or ischemic reperfusion injury, which is prevalent in human. (3) Unlike adult tissues, the avian embryo during early stages of development obtains oxygen through direct diffusion into the tissues (until day 5–6). In subsequent stages the oxygen diffuses into blood circulating through the extra-embryonic chorioallantoic membrane. However, due to blockade in artery, blood flow to that specific area is blocked which will partially hinder with the diffusion properties of free oxygen to the vessels. Therefore, a partial blockade in both oxygen and nutrient diffusion was obtained in the present model and thus termed as partial ischemia model.

5   Conclusion

The present ex vivo chick embryo model helps in understanding the angiogenic response in ischemia conditions in specific and hypoxia and/or low oxygen conditions in general.

Acknowledgement This work was supported by grants from KB Chandrasekhar-Research Foundation and from Indian Council Of Medical Research (No. 51/2/2009-Ana-BMS to SC).

References

1. Clark EB, Hu N (1982) Developmental hemodynamic changes in the chick embryo from stage 18 to 27. Circ Res 51:810–815
2. Nakazawa M, Miyagawa S, Nishibatake M et al (1988) Hemodynamic characteristics in neural crest cell-excised chick embryo. Heart Vessels 4:136–140
3. Hogers B, DeRuiter MC, Gittenberger-de Groot AC et al (1997) Unilateral vitelline vein ligation alters intracardiac blood flow patterns and morphogenesis in the chick embryo. Circ Res 80:473–481
4. Hogers B, DeRuiter MC, Gittenberger-de Groot AC et al (1999) Extraembryonic venous obstructions lead to cardiovascular malformations and can be embryolethal. Cardiovasc Res 41:87–99
5. Kolluru GK, Sinha S, Majumder S et al (2010) Shear stress promotes nitric oxide production in endothelial cells by sub-cellular delocalization of eNOS: a basis for shear stress mediated angiogenesis. Nitric Oxide 22:304–315
6. Stewart DE, Kirby ML, Sulik KK (1986) Hemodynamic changes in chick embryos precede heart defects after cardiac neural crest ablation. Circ Res 59:545–550
7. Majumder S, Ilayaraja M, Seerapu HR et al (2010) Chick embryo partial ischemia model: a new approach to study ischemia ex vivo. PLoS One 5:e10524
8. Hamburger V, Hamilton HL (1951) A series of normal stages in the development of the chick embryo. J Morphol 88:49–92
9. Vos SS, Ursem NTC, Hop WM et al (2003) Acutely altered hemodynamics following venous obstruction in the early chick embryo. J Exp Biol 206:1051–1057
10. Jiang M, Wang B, Wang C et al (2008) Angiogenesis by transplantation of HIF-1 alpha modified EPCs into ischemic limbs. J Cell Biochem 103:321–334
11. Lee SH (2000) Early expression of angiogenesis factors in acute myocardial ischemia and infarction. N Engl J Med 342:626–633
12. Bergeron M, Yu AY, Solway KE et al (1999) Induction of hypoxia-inducible factor-1 (HIF-1) and its target genes following focal ischaemia in rat brain. Eur J Neurosci 11:4159–4170
13. Wang WZ, Fang XH, Stephenson LL et al (2008) Ischemia/reperfusion-induced necrosis and apoptosis in the cells isolated from rat skeletal muscle. J Orthop Res 26:351–356
14. Niemistö A, Dunmire V, Yli-Harja O et al (2005) Robust quantification of in vitro angiogenesis through image analysis. IEEE Trans Med Imaging 24:549–553
15. Tamilarasan KP, Kolluru GK, Rajaram M et al (2006) Thalidomide attenuates nitric oxide mediated angiogenesis by blocking migration of endothelial cells. BMC Cell Biol 7:17
16. Majumder S, Tamilarasan KP, Kolluru GK et al (2007) Activated pericyte attenuates endothelial functions: nitric oxide-cGMP rescues activated pericyte-associated endothelial dysfunctions. Biochem Cell Biol 85:709–720
17. Deryugina EI, Quigley JP (2008) Chick embryo chorioallantoic membrane models to quantify angiogenesis induced by inflammatory and tumor cells or purified effector molecules. Methods Enzymol 444:21–41

The Zebrafish/Tumor Xenograft Angiogenesis Assay

Marco Presta, Giulia De Sena, and Chiara Tobia

Abstract Zebrafish (*Danio rerio*) represents a powerful model system in cancer research. Recent observations have shown the possibility to exploit zebrafish to investigate tumor angiogenesis, a pivotal step in cancer progression and target for anti-tumor therapies. Novel genetic tools and high resolution *in vivo* imaging techniques are also becoming available in zebrafish. It is anticipated that zebrafish will represent an important tool for chemical discovery and gene targeting in tumor angiogenesis. Here we describe a method to study tumor angiogenesis in zebrafish (*Danio rerio*) based on the injection of proangiogenic mammalian tumor cells into the perivitelline space of zebrafish embryos at 48 h post-fertilization. Within 24–48 h, proangiogenic tumor grafts induce a neovascular response originating from the developing sub-intestinal vessels. Angiogenesis inhibitors added to the fish water or to the injected cell suspension prevent tumor-induced neovascularization. Also, gene inactivation by antisense morpholino oligonucleotide injection in zebrafish embryos may allow the rapid identification of genes involved in tumor angiogenesis. The assay represents a novel tool for investigating tumor angiogenesis and for antiangiogenic drug discovery.

1 Introduction

1.1 Zebrafish as a Platform for Angiogenesis Studies

The teleost zebrafish (*Danio rerio*) has exceptional utility as a human disease model system and represents a promising alternative model in cancer research [1]. Zebrafish

M. Presta (✉) • G. De Sena • C. Tobia
General Pathology and Immunology, Department of Biomedical Sciences and Biotechnology, University of Brescia Medical School, Viale Europa 11, 25123 Brescia, Italy
e-mail: presta@med.unibs.it

E. Zudaire and F. Cuttitta (eds.), *The Textbook of Angiogenesis and Lymphangiogenesis: Methods and Applications*, DOI 10.1007/978-94-007-4581-0_16,
© Springer Science+Business Media Dordrecht 2012

embryo allows disease-driven drug target identification and *in vivo* validation, thus representing an interesting bioassay tool for small molecule testing and dissection of biological pathways alternative to other vertebrate models [2]. Indeed, when compared to other vertebrate model systems, zebrafish offers many advantages, including ease of experimentation, drug administration, and amenability to *in vivo* manipulation. Also, zebrafish is suitable for forward genetic screens and transient or permanent gene inactivation via antisense morpholino oligonucleotide (MO) injection or "targeting-induced local lesions in genes" (TILLING), respectively [3]. Moreover, the possibility to introduce targeted heritable gene mutations into the zebrafish germ line using engineered zinc-finger nucleases has been reported [4]. Importantly, zebrafish is suitable for high-throughput screening of chemical compounds using robotic platforms [4, 5].

Zebrafish possesses a complex circulatory system similar to that of mammals [6]. The basic vascular plan of the developing zebrafish embryo shows strong similarity to that of other vertebrates [7]. At the 13 somite-stage, endothelial cell precursors migrating from the lateral mesoderm originate the zebrafish vasculature and a single blood circulatory loop is present at 24 hours post-fertilization (hpf). Blood vessel development continues during the subsequent days by angiogenic processes. In particular, angiogenesis occurs in the formation of the intersegmental vessels (ISVs) of the trunk that will sprout from the dorsal aorta at 20 hpf (Fig. 1a). Also, the subintestinal vein vessels (SIVs) originate close to the duct of Cuvier area at 48 hpf and will form a vascular plexus across most of the dorsal-lateral aspect of the yolk ball during the next 24 h [7] (Fig. 1b).

Various animal models have been developed in rodents and in the chick embryo to investigate the angiogenesis process and for the screening of pro- and anti-angiogenic compounds, each with its own unique characteristics and disadvantages [8]. Previous studies had shown that developmental angiogenesis in the zebrafish embryo, leading to the formation of the ISVs of the trunk [9] and of the SIV plexus [10], represents a target for the screening of anti-angiogenic compounds [11, 12]. In these assays, low molecular weight compounds dissolved in fish water are investigated for their impact on the growth of new blood vessels driven by the complex network of endogenous, developmentally regulated signals. Recently, a novel zebrafish yolk membrane (ZFYM) assay has been proposed based on the injection of an angiogenic growth factor [e.g. recombinant fibroblast growth factor-2 (FGF2)] in the perivitelline space of zebrafish embryos in the proximity of developing SIVs. FGF2 induces a rapid and dose-dependent angiogenic response from the SIV basket, characterized by the growth of newly formed, alkaline phosphatase-positive blood vessels [13]. The ZFYM assay differs from the previous zebrafish-based angiogenesis assays since the angiogenic stimulus is represented by a well-defined, topically delivered exogenous agent that leads to the growth of ectopic blood vessels. This allows the screening of low and high molecular weight antagonists targeting a specific angiogenic growth factor and/or its receptor(s) [13].

However, the study of vascular development and on the effects of positive or negative modulators of the embryonic angiogenic process may have important limitations when translated to cancer research. Indeed, tumor-induced vessels show profound morpho-functional alterations when compared to the normal

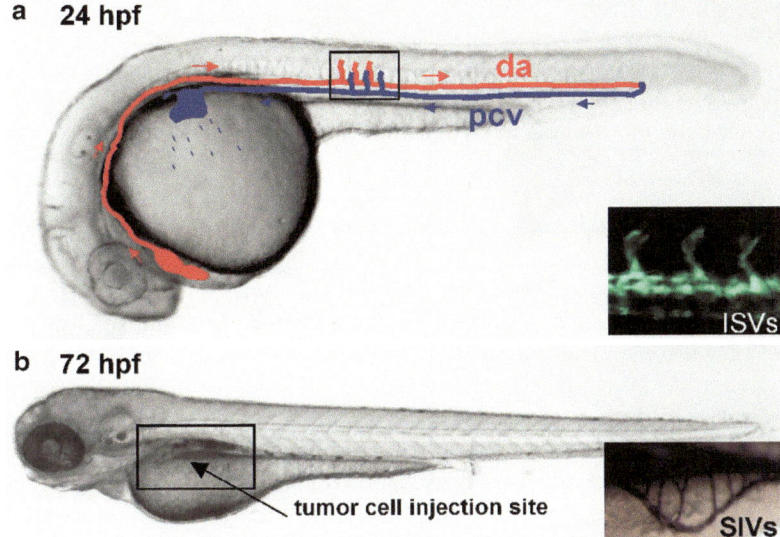

Fig. 1 Blood vessel development in zebrafish embryo. A single blood circulatory loop is present at 24 hpf (**a**). Angiogenesis occurs in the formation of the intersegmental vessels (ISVs) of the trunk sprouting from the dorsal aorta (da) and posterior cardinal vein (pcv). The subintestinal vein vessels (SIVs) originate close to the duct of Cuvier area at 48 hpf and will form a vascular plexus across most of the dorsal-lateral aspect of the yolk ball during the next 24 h (**b**). *Insets*: in (**a**), ISV sprouts from a *VEGFR2:G-RCFP* transgenic zebrafish embryo; in (**b**), alkaline phosphatase staining of a SIV basket at 72 hpf. The site of tumor cell grafting is indicated by the *arrow*. Lateral view, head on the *left*

vasculature [14]. This is reflected by significant differences in gene expression profiling between normal and tumor-derived endothelium [15, 16]. Thus, the identification of therapeutic targets and the assessment of the efficacy of anti-angiogenic compounds require the development of appropriate animal models in which tumor vasculature can be investigated. To this respect, tumor models have been established in zebrafish (see below) that may be suitable for studying the tumor angiogenesis process and its modulators [17, 18]. The availability of inbred, transgenic, gene knock-out/knock-in animals, of a wide array of antibodies, as well as of bioinformatic genomic, transcriptomic and proteomic information represent important tools for tumor angiogenesis studies. Several of these tools have been becoming available also for zebrafish.

1.2   Tumor Angiogenesis Models in Zebrafish

The use of tumor cell syngrafts or xenografts in animal models may allow the continuous delivery of angiogenic factors produced by a limited number of tumor cells, thus mimicking the initial stages of tumor angiogenesis and metastasis.

Various animal models have been developed in rodents and in the chick embryo to investigate the angiogenesis process and for the screening of pro- and anti-angiogenic compounds, each with its own unique characteristics and disadvantages [8]. Recent studies have shown the feasibility of injecting mammalian tumor cells in zebrafish adults, juveniles and embryos [19].

1.2.1 Tumor Angiogenesis Models in Zebrafish Adults

Zebrafish spontaneously develops almost any type of tumor. Also, several approaches have been developed to induce cancer in zebrafish. They include treatment with chemical carcinogens, forward genetic screening, target-selected inactivation of tumor suppressor genes, and expression of mammalian oncogenes. An overview of these approaches and of their main advantages and disadvantages has been published recently [17]. Also, transplantable tumor cell lines have been generated in clonal zebrafish and maintained for several passages in syngeneic and isogeneic adults [20]. Interestingly, microarray analysis has shown that gene expression signatures are conserved in fish tumors when compared to their human counterpart [1]. Relevant to tumor angiogenesis studies in adults, a transparent *casper* zebrafish line that lacks all types of pigments has been generated, allowing the rapid identification of transplanted tumor cells [21]. Also, crossing of the *casper* mutant with transgenic lines that label vasculature or internal organs with fluorescent tags may represent an useful approach to study tumor-host interactions in zebrafish by epifluorescence stereomicroscopy, confocal microscopy, and dual-photon confocal microscopy.

Noninvasive imaging in non-transparent zebrafish adults has been attempted. Ultrasound biomicroscopy has been used to follow the growth of liver tumors, their vascularity, and response to treatment [22]. Other imaging techniques, including microcomputerized axial tomography, micromagnetic resonance imaging, and optical projection tomography are beginning to be applied in zebrafish and will help to investigate tumor growth and vascularization in adult zebrafish [23].

1.2.2 Tumor Angiogenesis Models in Zebrafish Juvaniles

Human cancer cells have been successfully transplanted in the peritoneal cavity of 30 day-old zebrafish [24]. This has allowed the study of the dynamics of microtumor formation and neovascularization using high resolution imaging techniques, leading to a detailed description of the interaction among fluorescent tumor cells and the green fluorescent protein (GFP)-labeled vasculature of the host by three-dimensional reconstruction of confocal microscopy images. The results of these studies have shown that tumor cells secreting human vascular endothelial growth factor (VEGF) promote fish vessel remodeling and angiogenesis and that the human metastatic gene *RhoC* drives the initial steps of the metastatic process.

Due to the fact that juvenile zebrafish has a functional immune system, dexamethasone administration is required to prevent the rejection of the tumor engraftment. Also, at variance with zebrafish embryos (see below), the MO gene targeting approach is unfeasible in zebrafish juveniles. On the other hand, the impact of the tumor graft on the mature vasculature of juvenile fishes may recapitulate more closely the events that occur during tumor angiogenesis in adult animals and cancer patients. Indeed, developing vessels of zebrafish embryos may respond differently to tumor grafts compared to the fully developed vasculature of juvenile animals [18].

1.2.3   Tumor Angiogenesis Models in Zebrafish Embryos

The optical transparency and ability to survive for 3–4 days without functioning circulation make the zebrafish embryo especially amenable for vascular biology studies. Also, because of the immaturity of the immune system in zebrafish embryos, no xenograft rejection occurs at this stage [6]. Moreover, transient gene inactivation via MO injection represents a powerful tool for the identification of target genes in zebrafish embryo [3].

Recent studies have shown the feasibility of injecting human melanoma cells in zebrafish embryos to follow their fate and to study their impact on zebrafish development [25]. In these studies, tumor cells were injected at the blastula stage to explore potential bidirectional interactions between cancer cells and embryonic stem cells. The results indicate that developing zebrafish can be used as a biosensor for tumor-derived signals. However, grafting of tumor cells at this stage, well before vascular development, results in their reprogramming toward a non-tumorigenic phenotype, thus hampering any attempt to investigate tumor-driven vascularization. At variance, injection of melanoma cells into the hindbrain ventricle or yolk sac of 48 hpf embryos results in the formation of tumor masses within 4 days [26]. Immunostaining analysis of the grafts reveals the presence of blood vessels within the brain and abdominal lesions, even though the high vascularity of the invaded regions may not allow easy discrimination between developmental and tumor-induced angiogenesis [26].

Hypoxia represents an important driving force for tumor angiogenesis, mainly mediated by VEGF upregulation via activation of the hypoxia inducible factor (HIF) signaling pathway [27]. When cancer cells are injected in zebrafish embryos maintained in hypoxic water (7.5% air saturation), invasion into neighbouring tissues, dissemination, and metastasis of labelled tumor cells was greatly enhanced when compared to cells injected under normoxic conditions [28]. Consistent with increased tumor cell dissemination, hypoxia significantly stimulated neovascularization and tortuosity of the tumor vasculature via tumor cell-derived VEGF upregulation. Accordingly, VEGF receptor blockade by sunitinib administration in the fish water or by MO injection inhibited hypoxia-mediated pathological angiogenesis, early dissemination of malignant cells, invasiveness and metastasis.

The possibility to study the metastatic behavior of primary human tumors in zebrafish embryos has been confirmed also when human tumor tissue samples or primary tumor cells are injected into the yolk of 2 day old embryos or are organotopically implanted into the liver of zebrafish larvae [29]. This may represent a tool for investigating the role of neovascularization in the metastatic process.

Here, we describe an experimental procedure that allows the study of tumor angiogenesis triggered by mammalian tumor cells grafted in the zebrafish embryo [30, 31]. This method is suitable for assessing the effect of antiangiogenic chemicals and for the identification of non-redundant gene products involved in tumor angiogenesis.

2 The Zebrafish/Tumor Xenograft Angiogenesis Assay

The zebrafish/tumor xenograft angiogenesis assay is based on the grafting of mammalian tumor cells in the proximity of the developing SIV plexus at 48 hpf. Pro-angiogenic factors released locally by the tumor graft affect the normal developmental pattern of the SIVs by stimulating the migration and growth of sprouting vessels towards the implant. One-two days after tumor cell grafting, whole mount phosphatase alkaline staining allows the macroscopic evaluation of the angiogenic response. The use of transgenic zebrafish embryos, in which endothelial cells express GFP under the control of endothelial-specific promoters [[32] and references therein], represents an improvement of the method, allowing the observation and time-lapse recording of newly formed blood vessels in live embryos by epifluorescence microscopy as well as by *in vivo* confocal microscopy [30, 31]. Also, quantum dots may be used as labeling agents of the zebrafish embryo vasculature for long-lasting intravital time-lapse studies [33].

The identification of genes essential for blood vessel formation is of pivotal importance for the understanding of the angiogenesis process and for the discovery of novel therapeutic targets. In zebrafish embryos, MO injection induce a translational block in gene function [34]. Gene inactivation by this approach is easy and fast (3–4 days) when compared to the generation of knock-out mice (several months). Also, the simultaneous injection of different MOs may allow the inactivation of more than one gene at the same time. This represents a paramount advantage compared to any mammalian assay available and it can be exploited for the identification of novel gene(s) involved in tumor neovascularization. For instance, MO-induced inactivation of *VE-cadherin* [31] or *calcitonin receptor-like receptor* [35] zebrafish gene orthologs results in a significant inhibition of the angiogenic process triggered by the tumor graft in zebrafish embryos.

2.1 Materials

2.1.1 Reagents

- Wild-type AB zebrafish strain and/or transgenic zebrafish lines;
- Fish water (1.2 mM $NaHCO_3$, 1.1 mM $CaSO_4$, 0.1 g/l Instant Ocean);

- Agarose-modified microinjection plates;
- Matrigel (Cultrex® Basement Membrane Extract, R&D Systems);
- Tricaine® (Sigma-Aldrich);
- 1-Phenyl-2-thiourea (PTU, Sigma-Aldrich);
- Tumor cell suspension (from 4×10^6 to 1×10^7 cells resuspended on ice in 20–50 μl of 12.0 mg/ml of Matrigel solution). If required, angiogenesis inhibitors can be dissolved in Matrigel before cell resuspension and injection;
- Standard solutions for paraffin embedding, including; ethanol series; toluene; embedding paraffin (Sigma-Aldrich);
- Standard solutions for gelatin embedding, including sucrose series and embedding gelatin (Sigma-Aldrich);
- 4′,6-Diamidino-2-phenylindole (DAPI, Sigma-Aldrich). Use a 0.25 μg/ml aqueous solution.

2.1.2   Equipment

- Air incubator set at 28°C (International PBI S.p.A, Milan, Italy);
- Curved-tip or straight-tip forceps;
- Borosilicate capillaries for microinjection (1 mm O.D. × 0.58 mm I.D., GC100F-10, Harvard Apparatus);
- Magnetic glass microelectrode horizontal puller (PN-30, Narishige, Japan);
- FemtoJet 5247 and Inject Man NI2 microinjection apparatus (Eppendorf);
- Stereomicroscope (MZ75, Leica);
- Epifluorescence stereomicroscope (LeicaMZ16F) equipped with digital camera (DFC480, Leica);
- Epifluorescence Axiovert 200 M microscope (Zeiss);
- Software for computerized image analysis (Image-Pro Plus; MediaCybernetics).

2.2   Step-by-Step Tumor Cell Injection Protocol

- Incubate fertilized wild-type AB zebrafish or transgenic eggs at 28°C in fish water for 24 h.
- At 24 hpf, collect the embryos, remove the chorion by forceps, soak the dechorionated embryos in fish water added with 0.2 mM PTU, and incubate them for further 24 h at 28°C.
- Prepare agarose-modified Petri dishes as described in Fig. 2, suitable for holding the embryos during microinjection.
- Pull borosilicate capillaries to prepare microinjection needles (indicative puller settings: heater = 95, magnet sub = 57, magnet main = 60).
- Anesthetize 48 hpf zebrafish embryos in fish water added with 0.2 mg/ml Tricaine®, collect each embryo with a Pasteur pipette and place it on the agarose-modified Petri dish taking advantage of the slits in the gel.

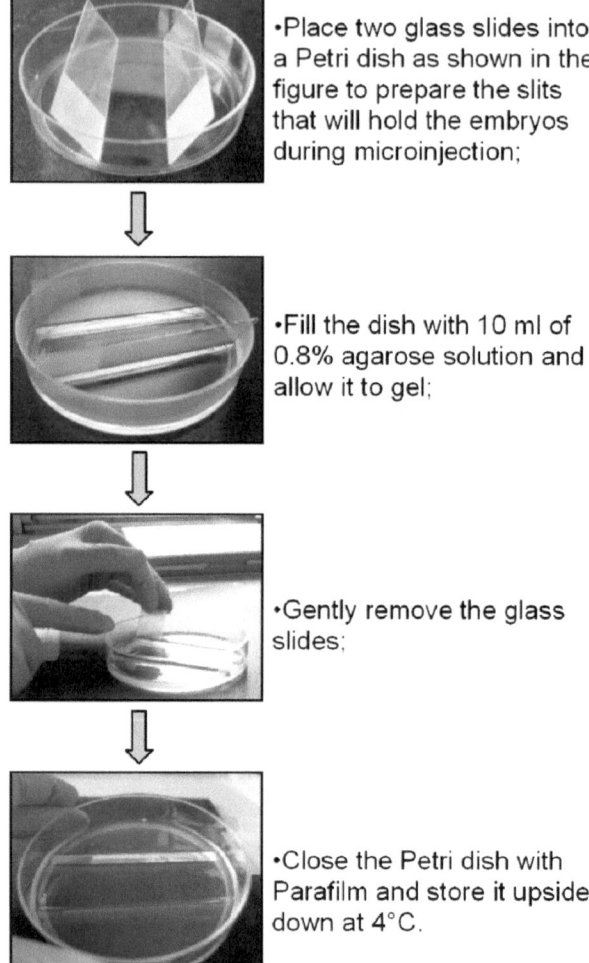

Fig. 2 Preparation of the agarose-modified microinjection plates. Glass slides are used to form the slits in the agarose that will hold the embryos during microinjection

•Place two glass slides into a Petri dish as shown in the figure to prepare the slits that will hold the embryos during microinjection;

•Fill the dish with 10 ml of 0.8% agarose solution and allow it to gel;

•Gently remove the glass slides;

•Close the Petri dish with Parafilm and store it upside down at 4°C.

- Remove excess water from the dish using a syringe and orient the embryos with the yolk on a flank (Fig. 3). **Technical tip**: a proper orientation of the embryos is of pivotal importance for a subsequent correct cell injection, embryos oriented with the yolk on the top being difficult to inject into the perivitellin space. To maintain the correct orientation of the embryo and its position in the agarose dish at the time of injection, remove most of the water until a light shaking of the Petri dish will not cause any movement of the embryo.
- Load pre-cooled borosilicate needles with 5 µl of tumor cell suspension in Matrigel (a solubilized basement membrane extract) by using a 20 µl-Eppendorf pipette. **Technical tips**: maintain a low temperature throughout the whole injection procedure by using ice bath and pre-cooled equipment to avoid gelling of the tumor cell suspension in the Matrigel solution. Also, avoid air bubble formation in the Matrigel suspension that may obstruct the needle.

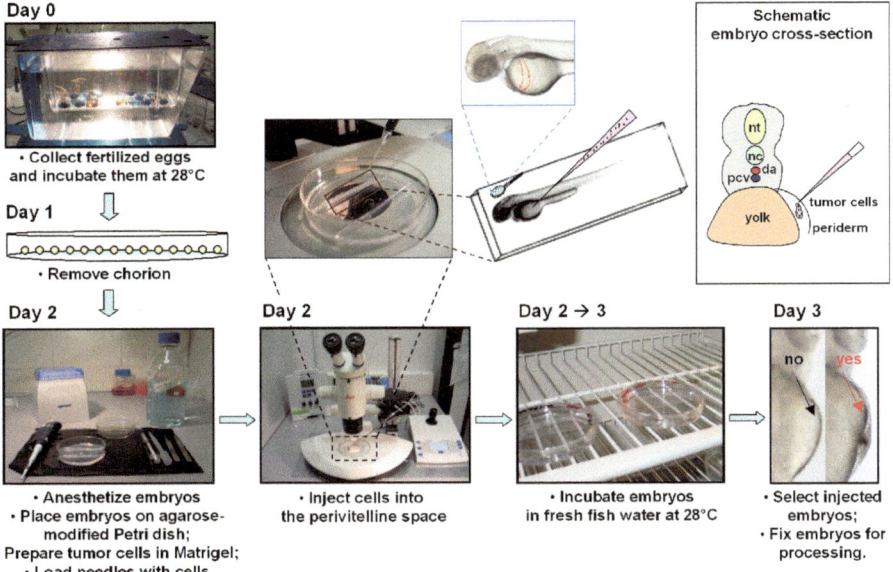

Fig. 3 Workflow diagram of the zebrafish/tumor xenograft angiogenesis assay. See text for details. *nt* neuronal tube, *nc* notochord, *da* dorsal aorta, *pcv* posterior cardinal vein

- Put the Petri dish with the orientated embryos under the stereomicroscope at 50× magnification and point the cell-loaded needle toward the region of the embryo body between the yolk and the duct of Cuvier area, close to the sinuous venous (Figs. 1b and 3). By using the microinjector, insert the tip of the needle into the yolk and pull it back slightly in order to create an artificial space between the periderm and the yolk syncytial layer where the cells will be grafted. Then, inject a drop of the cell suspension (approximately 4–10 nl) by setting the proper pressure (500–1,000 hPa) and time (0.5–1.0 s) parameters of the apparatus. Injection of the same volumes of the Matrigel solution or of a non tumorigenic cell suspension may be used as a negative control. **Technical tips**: (i) warming of the tip of the needle under the light of the stereomicroscope may cause the gelling of the Matrigel solution, thus hampering the injection of the cell suspension; (ii) the injection of the cell suspension in the correct region of the embryo body is of pivotal importance to obtain a positive angiogenic response. Given the close distance between the grafting site and the duct of Cuvier, take care of avoiding injection of tumor cells into the blood stream.
- Incubate the injected embryos for 24–48 h at 28°C in fresh fish water added or not with angiogenesis inhibitors. **Technical tip**: small organic molecules, including synthetic angiogenesis inhibitors, may require organic solvents. Dimethylsulfoxide may be used at concentrations up to 0.2% without adverse effects on embryo development and survival.
- At the end of the incubation, collect and select the injected embryos. **Technical tip**: take care that the injected cell pellet causes a protrusion of the periderm

when the embryo is observed from a dorsal view (Fig. 3). Discard all the embryos in which cells were injected into the yolk sac.

- Fix the embryos with 4% paraformaldehyde (PFA) in PBS (pH 7.0) for 2 h at room temperature, rinse them in PBS and store them at 4°C in PBS until further use. **Technical tip**: PFA fixation will cause autofluorescence of the yolk with consequent masking of the SIV-associated GFP fluorescence signal in the EGFP transgenic zebrafish embryos. Thus, assessment of the macroscopic angiogenic response in these embryos must be performed and recorded under an epifluorescence stereomicroscope before fixation.
- Perform whole-mount staining of the fixed embryos for endogenous alkaline phosphate activity according to standard procedures [36]. **Technical tip**: better staining is obtained with 72 hpf-old embryos.
- Embryos are mounted in agarose-coated petri dishes on a white background and photographed at 11.5× magnification under a stereomicroscope equipped with optical fibers and digital camera. **Technical tip**: alkaline phosphate-stained embryos can be fixed with 4% PFA and stored at −20°C in 100% methanol for several months.
- For histological analysis, dehydrate fixed embryos in an ethanol series, clear them in toluene, and immerse them in embedding paraffin for 2 h. For cryosections, treat fixed embryos with a scale of sucrose for cryopreservation before gelatin embedding accordingly to standard procedures [31].
- Cut 8-μm serial transversal sections of paraffin embedded embryos in the region of the tumor graft and observe them under a light microscope. Stain gelatin embedded sections with DAPI for 5 min at room temperature and observe them under an epifluorescence microscope.
- Injection of tumor cell suspension in 50 zebrafish embryos requires 1–2 h. Histological processing and staining of the samples requires 24 h.

2.3   Evaluation of the Angiogenic Response

Macroscopic evaluation of the vasoproliferative response triggered by mammalian tumor xenografts in zebrafish embryos can be performed by whole-mount analysis of the modifications of SIV development as evidenced by alkaline phosphatase staining in wild-type AB strain embryos (Fig. 4) and/or by fluorescence microscopy in GFP-labeled transgenic embryos (see below). These modifications are characterized by the convergence of SIVs toward the graft with different angiogenic morphological features depending, at least in part, upon the site of injection of the xenograft (Fig. 5). In details: (i) when the xenograft is located approximately 50–100 μm apart from the developing SIV plexus, new blood vessels will sprout and migrate toward the graft. These vessels will eventually reach, surround, and penetrate the graft within 24–48 h after injection. (ii) When the graft is closer to the plexus it becomes rapidly invaded by SIVs with a consequent local increase in blood vessel branching and density when compared to mock and control embryos.

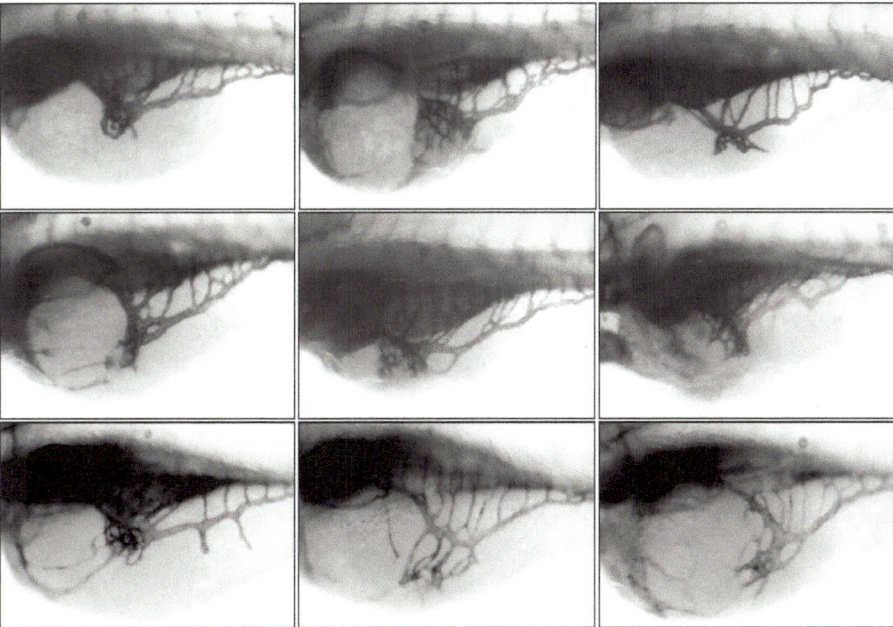

Fig. 4 Macroscopic appearance of the angiogenic responses triggered by mammalian tumor xenografts in zebrafish embryos. Wild-type AB strain embryos were grafted with tumor cells at 48 hpf. After 24 h, embryos were stained for alkaline phosphatase activity and photographed. Note the different morphological features of the angiogenic response. Lateral view, head on the *left*

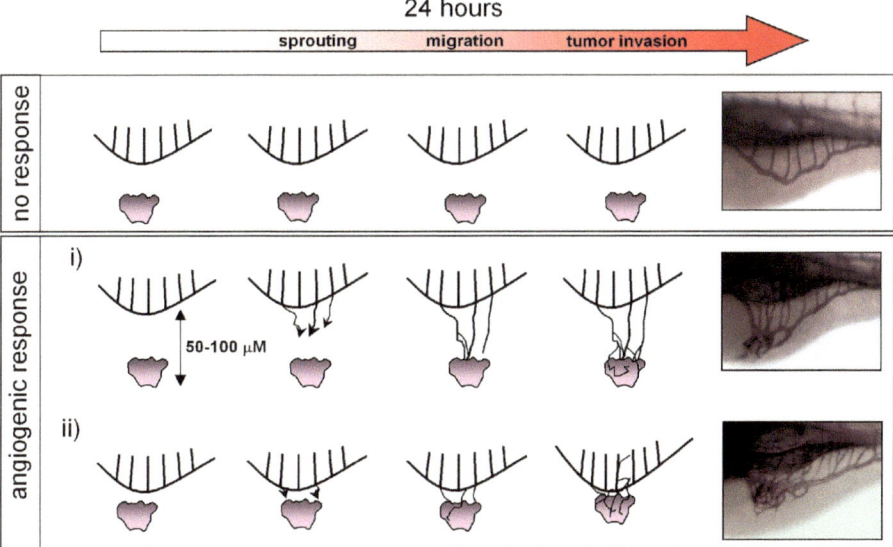

Fig. 5 Schematic representation of different angiogenic responses. Note the different angiogenic response when the tumor graft is located 50–100 μm apart or proximal to the SIV plexus. In both cases, a positive angiogenesis score is assigned to the embryo. Lateral view, head on the *left*

In both cases, a positive angiogenic score is assigned to the embryo. Routinely, each experimental point consists of 20 embryos and each experiment is repeated three times. Data are expressed as the ratio between positive and total grafted embryos. Also, computerized image analysis can be performed on lateral view images of alkaline phosphatase-stained embryos using Image-Pro Plus software [13].

Histological analysis of tumor grafts shows the presence of numerous blood vessels infiltrating the implant with a consequent significant increase in vascular density. Microscopic quantitative evaluation of the angiogenic response, expressed as microvessel density, can be obtained by applying a computerized image analysis on transversal embryo sections. This can be performed on digitized images acquired at $600\times$ magnification under a light microscope for alkaline phosphatase-stained embryos or under an epifluorescence microscope for GFP-transgenic embryos. However, we recommend the usage of the transgenic embryos due to the high specificity and higher sensitivity of the GFP fluorescence signal in zebrafish endothelial cells that can be merged with the DAPI nuclear staining of the whole graft section. For each section, vascular density is calculated as the ratio between the blood vessel area and the total area of the graft. Routinely, five sections are analyzed for each xenograft. Then, mean values ± 1 standard deviation are determined for each zebrafish specimen. In a typical experiment, tumor grafts originating by the injection of highly angiogenic, FGF2-overexpressing murine tumor cells show a vascular area equal to approx. 20% of total tumor area [31].

The macroscopic evaluation of the angiogenic response requires about 60 min for 50 embryos whereas the microscopic evaluation of microvessel density requires about 1 h for each specimen. In a typical experiment, between 60% and 100% of injected embryos showed a positive response characterized by the migration of new blood vessels from the SIV plexus toward the implant or by an increased in blood vessel density within the plexus. The percentage of positive response depends upon the grafted tumor cell line. Also, depending on their angiogenic potency, tumor cell grafts may provide a stronger response when observed 24–48 h after injection. No positive responses are observed in embryos injected with Matrigel alone whereas the injection of poorly angiogenic cells results in a percentage of positive embryos lower than 5%.

2.4   Benefits and Caveats of the Assay

When compared to other *in vivo* tumor angiogenesis assays, the zebrafish/tumor xenograft model presents both advantages and disadvantages that should be considered:

- The zebrafish/tumor xenograft model allows the delivery of a very limited number of cells, thus mimicking the initial stages of tumor angiogenesis and metastasis.
- Because of the immaturity of the immune system in zebrafish embryos at 48–72 hpf, no graft rejection occurs at this stage.

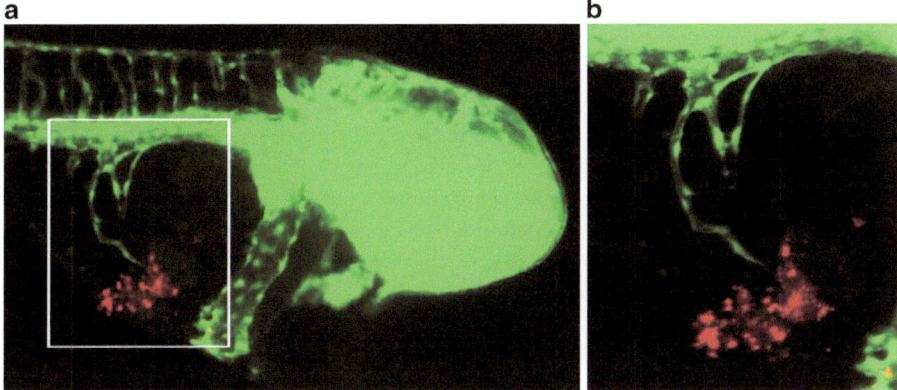

Fig. 6 Epifluorescence microscopy of tumor cell grafts in transgenic zebrafish embryos. *tg*
(fli1:EGFP)^{y1} transgenic embryos (endothelium in *green*) were injected with fluorescent dye-
loaded tumor cells (in *red*). After 24 h, embryos were photographed under an epifluorescence
stereomicroscope (**a**). Note a long, newly-formed blood vessel converging towards the graft (at
higher magnification in **b**). Lateral view, head on the *right*

- Labeled tumor cells (e.g. GFP-transduced or fluorescent dye-loaded cells) can be
  easily visualized within the embryo. When labeled cells are injected in transgenic embryos with fluorescent vasculature, *in vivo* analysis of the spatial/
  temporal relationship among tumor cells and newly formed blood vessels may
  represent an important feature of this model (Fig. 6).
- Several techniques can be applied within the constraints of paraffin or gelatin
  embedding, including histochemistry and immunohistochemistry. Electron
  microscopy can also be used in combination with light microscopy. Moreover,
  reverse transcriptase-polymerase chain reaction analysis with species-specific
  primers allows the study of gene expression by grafted tumor cells and by the
  host under different experimental conditions [31].
- When compared to the rabbit cornea and chick embryo chorioallantoic membrane assays, the zebrafish/tumor xenograft model has shown a similar capacity
  to discriminate between highly-angiogenic and poorly-angiogenic tumor cell
  lines [31]. However, only the parallel screening of a variety of cell lines will
  establish whether the zebrafish-based assay may provide results fully over-
  imposable with those obtained with the classic rodent and chick embryo angiogenesis assays.
- Because of the permeability of its embryos to small molecules, zebrafish allows
  disease-driven drug target identification and *in vivo* validation, thus representing an interesting bioassay tool for small molecule testing and dissection of
  biological pathways alternative to other vertebrate models [2]. Accordingly, addition of an angiogenesis inhibitor to the cell suspension or in the fish water [31] will
  result in the impairment of the angiogenic response triggered by the tumor graft.
  This can be observed both at macroscopic level, as a decrease in the percentage of
  positive embryos, and at microscopic level, as a decrease in vascular density of the

graft. Thus, the zebrafish/tumor xenograft model may represent a short-term assay suitable for the identification of novel tumor angiogenesis inhibitors. In this context, it is interesting to note the rapid response of this model to angiogenesis inhibitors (24–48 h) when compared to other tumor graft/angiogenesis assays, including the chick embryo chorioallantoic membrane assay (3–4 days), the s.c. murine Matrigel plug assay (5–7 days), the murine (1 week) and rabbit (2–3 weeks) cornea assays, and the s.c. mouse syngraft and xenograft assays (several weeks) [8]. Also, a large number of zebrafish embryos can be injected and maintained in 96 well-plates, thus allowing systemic in vivo treatment of the animals with minimal amounts of compound. Therefore, dose-response experiments can be easily performed and numerous compounds can be tested in an effective manner. This is usually unfeasible in most of the laboratories when using mammalian models (mice, rats, rabbits). Last but not least, a zebrafish facility is much cheaper and its logistic is much simpler than a mammalian facility.

• Because of the expected variability frequently observed in any *in vivo* assay, at least 20 embryos should be injected per experimental point.

• Timing of analysis of the angiogenic response is of importance, best results being obtained 24–48 h after tumor cell grafting, depending upon the cell line used. At these times, macroscopic evaluation of the angiogenic response can be performed, even though histological evaluation of vessel density represents an unbiased measurement of vascularity.

• A possible drawback of the zebrafish/tumor xenograft model may be represented by the possibility that the metabolic fate of the drug (either in terms of its activation or inactivation) may differ in zebrafish embryo in respect to mammalian species.

• It should be pointed out that zebrafish embryos are maintained at 28°C. This may not represent an optimal temperature for mammalian cell growth and metabolism, even though we have observed mitotic figures with no sign of apoptosis in grafted tumors throughout the whole experimental period [31]. Also, we have found that temperature can be increased to 33°C with no developmental defects in zebrafish embryos.

• At variance with low molecular weight compounds, high molecular weight antiangiogenic molecules (e.g. neutralizing antibodies and protein inhibitors) can not be delivered in the fish water. However, they can be dissolved in the tumor cell suspension before injection.

• The availability of inbred, transgenic, gene knock-out/knock-in animals, of a wide array of antibodies, as well as of bioinformatic genomic, transcriptomic and proteomic information represent important tools for tumor angiogenesis studies performed in murine models. Several of these tools have been becoming available also for zebrafish.

• The morpho-functional differences between normal and tumor blood vessels are well-known [14]. The zebrafish/tumor xenograft model allows to study the impact of angiogenesis inhibitors or of MO-induced gene inactivation on tumor-driven neovascularization as well as on physiological angiogenesis in the same embryo (e.g. angiogenesis that occurs in developing intersegmental vessels of the trunk) [31, 35].

Acknowledgements This work was supported by Ministero dell'Istruzione, Università e Ricerca (Centro IDET, PRIN projects), Associazione Italiana per la Ricerca sul Cancro (grant n° 10396), Fondazione Berlucchi, and Fondazione Cariplo (grant 2008–2264 and NOBEL Project) to M. Presta.

References

1. Lam SH, Wu YL, Vega VB et al (2006) Conservation of gene expression signatures between zebrafish and human liver tumors and tumor progression. Nat Biotechnol 24:73–75
2. Pichler FB, Laurenson S, Williams LC et al (2003) Chemical discovery and global gene expression analysis in zebrafish. Nat Biotechnol 21:879–883
3. Thisse C, Zon LI (2002) Organogenesis–heart and blood formation from the zebrafish point of view. Science 295:457–462
4. Meng X, Noyes MB, Zhu LJ et al (2008) Targeted gene inactivation in zebrafish using engineered zinc-finger nucleases. Nat Biotechnol 26:695–701
5. Funfak A, Brosing A, Brand M, Kohler JM (2007) Micro fluid segment technique for screening and development studies on Danio rerio embryos. Lab Chip 7:1132–1138
6. Weinstein B (2002) Vascular cell biology in vivo: a new piscine paradigm? Trends Cell Biol 12:439–445
7. Isogai S, Horiguchi M, Weinstein BM (2001) The vascular anatomy of the developing zebrafish: an atlas of embryonic and early larval development. Dev Biol 230:278–301
8. Hasan J, Shnyder SD, Bibby M et al (2004) Quantitative angiogenesis assays in vivo–a review. Angiogenesis 7:1–16
9. Cross LM, Cook MA, Lin S et al (2003) Rapid analysis of angiogenesis drugs in a live fluorescent zebrafish assay. Arterioscler Thromb Vasc Biol 23:911–912
10. Serbedzija GN, Flynn E, Willett CE (2000) Zebrafish angiogenesis: a new model for drug screening. Angiogenesis 3:353–359
11. Belleri M, Ribatti D, Nicoli S et al (2005) Antiangiogenic and vascular-targeting activity of the microtubule-destabilizing trans-resveratrol derivative 3,5,4'-trimethoxystilbene. Mol Pharmacol 67:1451–1459
12. Yeh JC, Cindrova-Davies T, Belleri M et al (2011) The natural compound n-butylidenephthalide derived from the volatile oil of Radix Angelica sinensis inhibits angiogenesis in vitro and in vivo. Angiogenesis 14:187–197
13. Nicoli S, De Sena G, Presta M (2009) Fibroblast Growth Factor 2-induced angiogenesis in zebrafish: the zebrafish yolk membrane (ZFYM) angiogenesis assay. J Cell Mol Med 14:2109–2121
14. Carmeliet P, Jain RK (2000) Angiogenesis in cancer and other diseases. Nature 407:249–257
15. Ghilardi C, Chiorino G, Dossi R et al (2008) Identification of novel vascular markers through gene expression profiling of tumor-derived endothelium. BMC Genomics 9:201
16. St Croix B, Rago C, Velculescu V et al (2000) Genes expressed in human tumor endothelium. Science 289:1197–1202
17. Feitsma H, Cuppen E (2008) Zebrafish as a cancer model. Mol Cancer Res 6:685–694
18. Stoletov K, Klemke R (2008) Catch of the day: zebrafish as a human cancer model. Oncogene 27:4509–4520
19. Taylor AM, Zon LI (2009) Zebrafish tumor assays: the state of transplantation. Zebrafish 6:339–346
20. Mizgireuv IV, Revskoy SY (2006) Transplantable tumor lines generated in clonal zebrafish. Cancer Res 66:3120–3125
21. White RM, Sessa A, Burke C et al (2008) Transparent adult zebrafish as a tool for in vivo transplantation analysis. Cell Stem Cell 2:183–189

22. Goessling W, North TE, Zon LI (2007) Ultrasound biomicroscopy permits in vivo characterization of zebrafish liver tumors. Nat Methods 4:551–553
23. Spitsbergen J (2007) Imaging neoplasia in zebrafish. Nat Methods 4:548–549
24. Stoletov K, Montel V, Lester RD et al (2007) High-resolution imaging of the dynamic tumor cell vascular interface in transparent zebrafish. Proc Natl Acad Sci USA 104:17406–17411
25. Topczewska JM, Postovit LM, Margaryan NV et al (2006) Embryonic and tumorigenic pathways converge via Nodal signaling: role in melanoma aggressiveness. Nat Med 12:925–932
26. Haldi M, Ton C, Seng WL, McGrath P (2006) Human melanoma cells transplanted into zebrafish proliferate, migrate, produce melanin, form masses and stimulate angiogenesis in zebrafish. Angiogenesis 9:139–151
27. Pugh CW, Ratcliffe PJ (2003) Regulation of angiogenesis by hypoxia: role of the HIF system. Nat Med 9:677–684
28. Lee SL, Rouhi P, Dahl Jensen L et al (2009) Hypoxia-induced pathological angiogenesis mediates tumor cell dissemination, invasion, and metastasis in a zebrafish tumor model. Proc Natl Acad Sci USA 106:19485–19490
29. Marques IJ, Weiss FU, Vlecken DH et al (2009) Metastatic behaviour of primary human tumours in a zebrafish xenotransplantation model. BMC Cancer 9:128
30. Nicoli S, Presta M (2007) The zebrafish/tumor xenograft angiogenesis assay. Nat Protoc 2:2918–2923
31. Nicoli S, Ribatti D, Cotelli F, Presta M (2007) Mammalian tumor xenografts induce neovascularization in zebrafish embryos. Cancer Res 67:2927–2931
32. Baldessari D, Mione M (2008) How to create the vascular tree? (Latest) help from the zebrafish. Pharmacol Ther 118:206–230
33. Rieger S, Kulkarni RP, Darcy D et al (2005) Quantum dots are powerful multipurpose vital labeling agents in zebrafish embryos. Dev Dyn 234:670–681
34. Nasevicius A, Ekker SC (2000) Effective targeted gene 'knockdown' in zebrafish. Nat Genet 26:216–220
35. Nicoli S, Tobia C, Gualandi L et al (2008) Calcitonin receptor-like receptor guides arterial differentiation in zebrafish. Blood 111:4965–4972
36. Serbedzija GN, Flynn E, Willett CE (1999) Zebrafish angiogenesis: a new model for drug screening. Angiogenesis 3:353–359

Quantitative Study of *In Vivo* Angiogenesis and Vasculogenesis Using Matrigel-Based Assays

Kaustabh Ghosh, Mehrdad Khajavi, and Avner Adini

Abstract Neovascularization, the formation of new blood vessels, is critical for various physiological and pathological events such as wound repair and tumor growth as well as for functional tissue engineering as it ensures proper nutrient and oxygen supply to newly forming tissues. Neovascularization can occur via angiogenesis, where new capillaries sprout from pre-existing vessels, and/or vasculogenesis, where endothelial cells spontaneously self-assemble into vascular structures. Because neovascularization is the rate-limiting step in new tissue formation, there is great interest in developing robust quantitative approaches that facilitate greater understanding of this complex process. Matrigel-based assays are particularly useful for studying neovascularization as it provides a more natural environment for endothelial cell recruitment and capillary formation. In addition, Matrigel-based assays are reproducible and easy to perform when compared to other available *in vivo* angiogenesis assays. We have recently developed a novel matrigel assay that permits quantification of the endogenous angiogenic response *in vivo* by determining the number of recruited endothelial cells using fluorescence-activated cell sorting (FACS). We have also developed a technique to quantitatively evaluate vasculogenesis where endothelial cells and mesenchymal stem cells are suspended together in Matrigel prior to subcutaneous injection and vascular network formation is quantitatively analyzed from whole-mount confocal images of lectin-perfused Matrigel implants. Here, we outline the utility of these Matrigel assays for quantitative analyses of both angiogenic response and vasculogenesis and provide a detailed description of the methodology involved.

K. Ghosh
Department of Bioengineering, University of California, Riverside, CA 92521, USA

M. Khajavi • A. Adini (✉)
Vascular Biology Program, Department of Surgery, Children's Hospital,
Harvard Medical School, Boston, MA, USA
e-mail: avner.adini@childrens.harvard.edu

E. Zudaire and F. Cuttitta (eds.), *The Textbook of Angiogenesis and*   269
Lymphangiogenesis: Methods and Applications, DOI 10.1007/978-94-007-4581-0_17,
© Springer Science+Business Media Dordrecht 2012

1 Introduction

Angiogenesis and vasculogenesis are the two major processes responsible for the development of new blood vessels (i.e., neovascularization) [1–7]. Angiogenesis is a process where new capillaries sprout from pre-existing mature vessels. In response to tissue ischemia or injury, the angiogenic process begins with the "activation" of endothelial cells within a parent vessel, followed by the disruption of basement membrane and extracellular matrix, and the subsequent migration and outgrowth of endothelium into interstitial space. Subsequent endothelial cell proliferation, pericyte recruitment, and production of a new basement membrane matrix complete the angiogenic process [8]. Angiogenesis is a morphogenetic event that plays a major role in the development of a vascular supply in adult tissue remodeling and disease. Accordingly, angiogenesis is critical in reproduction (placenta, uterus, formation of the corpus luteum), wound healing, bone repair, as well as in pathological conditions as rheumatoid arthritis, ischemic heart disease, ischemic peripheral vascular disease, diabetic retinopathy, and tumor growth and metastasis. In the last four decades, significant advancements have been made in our understanding of angiogenesis, shedding light on how this normal physiological process can contribute to pathophysiological events such as wound repair and cancer [9–12].

In contrast, vasculogenesis refers to the *in situ* formation of blood vessels from spontaneous self-assembly of endothelial cells or their precursors (endothelial progenitor cells or angioblasts). During embryogenesis, this process begins with the formation of angioblast clusters or blood islands. Growth and fusion of multiple blood islands in the embryo ultimately give rise to the capillary network structure. After the onset of blood circulation, this network differentiates into an arteriovenous vascular system. Evidence of vasculogenesis in adults was first reported in 1997 by Isner et al. [13], who implicated the critical role of circulating endothelial progenitor cells (EPCs) in neovascularization. Since then, EPCs have been widely believed to also play a crucial role in vascular homeostasis and repair [14]. The important therapeutic implications of angiogenesis necessitate the development of robust quantitative assays that permit identification of optimal conditions that regulate this complex process. Animal models are the preferred choice for testing angiogenic conditions as they closely recapitulate the complex tissue environment seen in humans. Recently, the Matrigel plug assay has become a popular strategy for screening biochemical factors that exhibit *in vivo* angiogenic activity [15–19]. It is a relatively quick assay, easy to set up, and highly reproducible. Further, it can be used to test one or two conditions, or it can quickly be scaled up to screen hundreds of compounds [20, 21]. Typically, when known angiogenic factors are mixed with matrigel and injected subcutaneously into mice, host endothelial cells migrate into the gel plug and form capillary structures [22]. The conventional method to determine neovascularization response is to histologically analyze thin implant sections that are immunostained with endothelial-specific antibodies. However, this technique fails to provide an accurate quantification of the various cells that contribute to blood vessel formation (e.g. endothelial cells, pericytes etc.). This lack of proper quantification significantly limits our ability to accurately evaluate novel therapeutic agents in angiogenesis therapy.

In this chapter we present a modification of the traditional matrigel assay for the accurate quantification of angiogenic response *in vivo* by directly counting the number of endothelial cells that recruited into the plug [23]. The harvested plugs are subjected to a mild protease treatment, yielding intact cells. The liberated cells are then stained using endothelial cell-specific antibodies and quantified by flow cytometry. The novel combination of FACS analysis with the traditional matrigel assay significantly improves the ability to quantify the recruitment of host endothelial cells, and perhaps other host cells, that participate in *in vivo* angiogenic response. Further, the extracted single cells from Matrigel implants can be used for a variety of genetic, molecular and functional studies that together will provide a more detailed understanding of the angiogenic potential of the implant.

We also report a new method to quantitatively evaluate the *in vivo* vasculogenic potential of implanted endothelial cells. In this technique, which is particularly suitable for determining the efficacy of vascular tissue engineering approaches, endothelial cells are suspended in Matrigel together with human bone marrow-derived mesenchymal stem cells (MSC) and implanted subcutaneously into nude mice [24, 25]. Vascular networks formed within these cell-laden Matrigel implants are then labeled using high molecular weight fluorescent lectin that is infused intravenously prior to implant harvest. Whole-mount confocal imaging of these lectin-stained Matrigel plugs, followed by detailed image analysis for the measurement of vessel number, length, area, total fluorescence (pixel counts), etc. together provide a very robust quantitative evaluation of the vasculogenic potential of implanted cells as well as the carrier scaffold (e.g. synthetic polymer biomaterials). Thus, we believe that the matrigel plug assay can be developed as a standardized, comprehensive technique to obtain both quantitative and qualitative in-depth evaluation of neovascularization strategies as it offers several key advantages such as the ease of preparation, recovery, and imaging, thereby providing a useful addition to the current repertoire of angiogenesis assays.

2 Host Endothelial Cell (EC) Recruitment into Cytokine-Supplemented Matrigel

2.1 Animals

Male, 8-week, C57BL/6J or nude mice are typically used.

2.2 Reagents

Matrigel- Phenol Red-free, 1 vial of 10 ml stock (BD. Bioscience, cat. No 356237)
– Fibroblast Growth Factor (bFGF), (Peprotec, cat. No.100-18B)

- Avertin (tribromoethanol; Fisher, cat. No. AC42143-0100)
- 1 M HEPES buffer
- Liberase (DIPASE HIGH) 10 MG (2X5MG) (Roch diagnostics, cat. No. 0540105400)
- FACS tubes containing filter lid; 5 ml polystyrene tube with cell-stainer cap (BD bioscience, cat. No. 352235)
- FACS buffer (1% BSA, 0.01% sodium azide in PBS)
- Antibodies:CD31,CD45

2.3   Equipment

- Pre-chilled 15 ml Falcon tubes and 1 ml pipet tips
- Pre-chilled tuberculin syringes with removable needles
- 25 G needles for injection
- Centrifuge

2.4   Reagent Setup

2.4.1   Avartin

2.5% (vol/vol) avertin A 100% stock of avertin is prepared by mixing 10 g of tribromoethanol with 10 ml of tertiary amyl alcohol (Fisher A7301). For use, dilute to 2.5% (vol/vol) in PBS or isotonic saline. Both stock and 2.5% avertin solutions should be stored wrapped in foil at 4°C. The dose will vary slightly with different preparations. A dose of 0.017 ml/g body weight is a good starting point, although we often find that we must use a slightly higher dose.

2.4.2   Liberase

- Dilute the Liberase in 25 mM HEPES buffer to a final concentration of 0.25 mg/ml.

2.4.3   Matrigel

All the process of matrigel preparation for injection will be done on ice. Thaw the frozen matrigel vial one day in advance at 4°C.

2.5   Procedure

2.5.1   Matrigel: (Preparation for Ten Mice)

– Mix the Matrigel with bFGF (50 ng/ml) to injection
– Calculate 800 ul matrigel needed for one injection (Actuall amount used is 500 ul but extra amount needed due to viscosity and mixing losses)
– Each animal will receive a total of two matrigel implants, one 500 ul implant on each flank
– Load the Matrigel mix into the pre-cooled 1 ml syringes. To do so, collect the mix in a 1 ml pipette tip, fit the tip into the mouth of the 1 ml syringe, and slowly pull the syringe piston back to withdraw the Matrigel mix from the pipette tip and into the syringe
– Bring the Matrigel-loaded syringes to the animal facility on ice
– Anesthetize mice using isofluorane and transfer it onto a clean disposable surface such that it lies on its ventral side
– Next, gently lift and hold the mouse skin on one flank, insert the needle into the skin and then gently spread out the skin against the back of the mouse. Then, slowly inject 200 μl of the Matrigel mix subcutaneously
    Note: During the subcutaneous injection of this Matrigel mix, a noticeable bulge will appear under the skin, which will persist until the day of harvest.
– Repeat the previous step for the other flank on the same mouse
    Critical step: Since the anesthetizing effect of isofluorane is short-lived, subcutaneous injections must be completed within 3–5 min post-isofluorane exposure.
– After subcutaneous injection of the Matrigel mix, transfer the mice back into their cages
    Note: Do not leave more than two treated mice per cage as that will ensure minimal physical disturbance to the subcutaneous implant.
– Inject matrigel on left and right center back (Fig. 1)
    Critical: Typical maximal response is 7 day post-injection. More than 10 days can bring to negative results.

2.5.2   Matrigel Harvest Procedure

– Sacrifice animals with CO_2 and dissect off skin to expose matrigel plugs.
– Take gross pictures to record peripheral angiogenesis.
– Use forceps to transfer the matrigel plugs directly to 15 ml tube containing 1 ml liberase working solution.
– Rotate the implant/liberase mix for 1 h at 37°C. After incubation you will still observe some Matrigel clumps in the sample. These clumps are easily separated through filtration.
– Filter the liberated cells using the 5 ml tube with cell-strainer cap.
– Wash the liberated cells three times with FACS buffer.

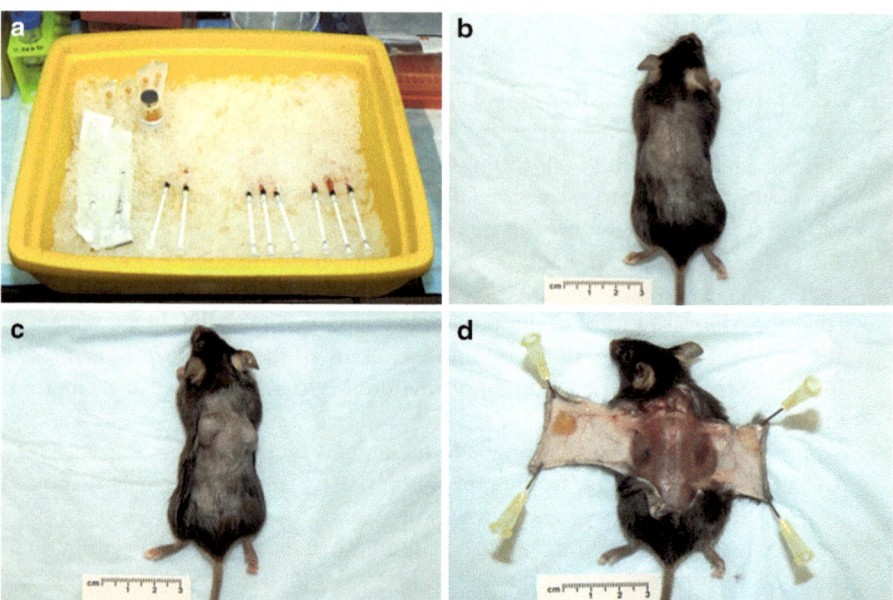

Fig. 1 Matrigel preparation and procedures (**a**) Syringes and needles are pre-cooled before loading with Matrigel mix. Matrigel-loaded syringes are kept on ice at all time prior to injection. (**b**) The hind flank is gently shaved prior to injection. (**c**) Matrigel is slowly injected (200 μl) subcutaneously on left and right center back. Upon body temperature, Matrigel will solidify and a noticeable bulge will appear under the skin. (**d**) After euthanizing the animal, dissect off the skin to expose Matrigel plugs. (The injection was performed by Amy Birsner at the department of Vascular Biology Program, Children's Hospital Boston, MA. Photography by Kristen Johnson)

– Incubate cells with the antibodies (typical dilution of 1:50 in FACS buffer) for 30 min on ice.
– Wash the labeled cells three times with FACS buffer.
   Note: After washing, it is possible to fix the cells with 1% paraformaldehyde and store them at 4°C till FACS analysis.
– Analyze by FACS.

2.6   Timing

2.6.1   Injections

1–2 h (for 24 Matrigel plugs)

2.6.2   Suspending Cells in Matrigel

~ 2 h (for 20 Matrigel implants)

2.6.3   Injection of Matrigel to the Mice

~5 min per mouse (for two implants post isofluorane exposure)

2.6.4   Harvesting Matrigel Implants

~20–30 min per mouse (for two implants). This time can be shorter depending on the user skill level.

2.6.5   Liberate and Evaluate the Implant's Cells

~ 2–3 h per 20 Matrigel's implants

2.7   Trouble Shooting

2.7.1   Air Bubbles in the Matrigel

(a) As Matrigel is a high viscosity liquid, you should be very careful when mixing it with your cells and injecting it in the mice. Pipetting it too vigorously will lead to air bubbles in the mixture. These air bubbles are very difficult to get rid of once they form in the cell-Matrigel suspension within the syringe. Air bubbles in the syringe causes incorrect injection volume as well as Matrigel disintegration when implanted *in vivo*.
(b) Based on our experience, Matrigel thawed overnight at 4°C facilitates easier handling.

2.7.2   Clogging of the Needle

(a) If the Matrigel/cell mixture reaches room temperature while in the syringe, it will clog the needle. Keep the syringe and needle on ice before use.

2.7.3   Very Low Levels of Cytokines

(a) Cytokines are relatively unstable and generally have a short half-life. To prevent degradation of cytokines in your Matrigel/cell mixtures, keep them on ice as long as possible during preparation. This reduces necrotic cell death and the subsequent release of proteases that may degrade the cytokines.
(b) Aliquot the cytokines before freezing to avoid a reduction in their activity due to repeated freezing and thawing.

3 In Vivo Vasculogenesis by Human Microvascular Endothelial Cells (EC) Seeded in Matrigel Implants

3.1 Animals

Male or Female, 6- to 8-week old, Nude or SCID mouse.

3.2 Materials

– Matrigel- Phenol Red-free, 1 vial of 10 ml stock (BD. Bioscience, cat. no. 356237)
   Note: Phenol red-free Matrigel is preferable as it allows gross examination of the extent of implant neovascularization (by virtue of its redness).
– Cultured human microvascular endothelial cells (HMVEC), preferably between passages 2 and 5
   (In principle, microvascular endothelial cells isolated from any organ can be used. However, the protocol described here is based on the authors' experience with cells isolated from human neonatal foreskin.) (Lonza cat. no. CC-2505)
– Cultured human bone marrow-derived mesenchymal stem cells (MSC) (Lonza cat. no. AA-2501-16 07/11)
– Rhodamine *ulex europaeus* agglutinin I (Rho-UEA-1) to label human endothelial cells red (2 mg/ml, Vector Labs, Cat. No. RL1062)
– Pre-cooled 1 ml syringes
– 22 G ½ needles
– Isofluorane (Fisher, cat. No. AC42143-0100).
– Sterile ethanol pads

3.3 Major Equipment

Centrifuge
Laser confocal microscope
Cryostat

3.4 Material Setup

3.4.1 Matrigel

Thaw the frozen bottle (or aliquots) of phenol red-free Matrigel at 4°C 12–18 h prior to experiment and, for the entire duration of the experiment, place the thawed Matrigel on ice.

3.4.2 Syringe

Place the 1 ml syringes on ice before starting cell trypsinization.

3.4.3 Cells (Preparation for Four Implants, Two Implants per Mouse)

Total number of cells required per 200 μl Matrigel implant is 2×10^6 cells (i.e. 10×10^6 cells/ml). For four implants, total volume of Matrigel required = 800 μl. Because Matrigel is viscous, prepare 200 μl excess of the cell-Matrigel suspension. Therefore, the final volume of Matrigel required is 1 ml and the total number of cells required is 10×10^6. HMVECs and MSCs can be mixed at a ratio of 2:1 i.e. 6.5×10^6 HMVECs mixed with 3.5×10^6 MSCs. Therefore, to prepare four implants for subcutaneous injection, at least four 10 cm dishes of 80% confluent HMVECs and three 10 cm dishes of 80% confluent MSCs should be available.

3.4.4 Rhodamine *Ulex europaeus* Agglutinin I (Rho-UEA-1)

At the time of harvest 7 days post-subcutaneous Matrigel injection, dilute the 2 mg/ml Rho-UEA-1 stock solution to 400 μg/ml using Ca^{2+}, Mg^{2+}-supplemented PBS (1:5 dilution).

3.5 Procedure

3.5.1 Suspending Cells in Matrigel (Preparation for Four Implants, Two Implants per Mouse)

– For each cell culture dish, aspirate out the medium, rinse the cells twice with calcium-, magnesium-free PBS, and add trypsin-EDTA solution for 3–5 min at 37°C to detach cells.
– Following cell detachment, neutralize trypsin-EDTA by adding the respective cell culture medium. Spin down the detached cells in a centrifuge at 1,000 rpm for 5 min to obtain cell pellet.
– Re-suspend cell pellets in the respective culture medium and count total cell number.
– In a new 15 ml conical tube, add appropriate volumes of the individual cell suspensions to obtain 6.5×10^6 HMVECs and 3.5×10^6 MSCs.
– Spin down this mixed suspension of HMVECs and MSCs and aspirate out the supernatant to obtain a mixed cell pellet containing 10×10^6 cells.
– Resuspend this cell pellet in 1 ml Matrigel solution.
  Critical Step: Be very careful while resuspending the mixed cell pellet in Matrigel- try not to generate bubbles! Failure to do so will result in a significant loss of the cell-Matrigel suspension.

3.5.2 Injecting Cell-Matrigel Suspension Subcutaneously in Mice

– First, load the cell-Matrigel mix into the pre-cooled 1 ml syringes. To do so, collect the mix in a 1 ml pipette tip, fit the tip into the mouth of the 1 ml syringe, and slowly pull the syringe piston back to withdraw the cell-Matrigel mix from the pipette tip and into the syringe.
   Critical step: Never load the cell-Matrigel mix into the syringe using a needle as it will cause excessive shear-induced damage to the cells.
– Bring the Matrigel-loaded syringes to the animal facility on ice.
– Anesthetize mice using isofluorane and transfer it onto a clean disposable surface such that it lies on its ventral side.
– Next, gently lift and hold the mouse skin on one flank, insert the needle into the skin and then gently spread out the skin against the back of the mouse. Then, slowly inject 200 µl of the cell-Matrigel mix subcutaneously.
   Note: During the subcutaneous injection of this cell-Matrigel mix, a noticeable bulge will appear under the skin, which will persist until the day of harvest.
– Repeat the previous step for the other flank on the same mouse.
   Critical step: Since the anesthetizing effect of isofluorane is short-lived, subcutaneous injections must be completed within 3–5 min post-isofluorane exposure.
– After subcutaneous injection of the cell-Matrigel mix, transfer the mice back into their cages.
   Note: Do not leave more than two treated mice per cage as that will ensure minimal physical disturbance to the subcutaneous implant.

3.5.3 Harvesting Matrigel Implant

– 30 min prior to sacrifice, inject 125 µl of the Rho-UEA-1 working solution intravenously to label the implanted human endothelial cells red.
   Note: The UEA-1 working solution of 400 µg/ml is obtained from 1:5 dilution of the 2 mg/ml stock solution using Ca^{2+}, Mg^{2+}-supplemented PBS. Assuming the total circulating blood volume of an adult mouse to be 2 ml, the final Rho-UEA-1 concentration will be approximately 25 µg/ml.
   Critical step: The UEA-1 working solution (400 µg/ml) must be prepared in Ca^{2+}, Mg^{2+}-supplemented PBS as it promotes strong Rho-UEA-1 binding to human endothelial cells.
– After allowing Rho-UEA-1 perfusion for 30 min, sacrifice mice using CO_2, transfer them to a disposable surface with their dorsal side facing up, stretch out their limbs and immobilize them using pins/tape.
– Using tweezers, gently grab and lift the skin near the cervical end and dissect the skin along its spinal cord, finally extending the dissection onto the sides to expose the Matrigel plugs.
– Using a high-resolution digital camera, take pictures of the Matrigel implants to grossly evaluate *in situ* neovascularization (by virtue of the implant redness), as shown in Fig. 2a.

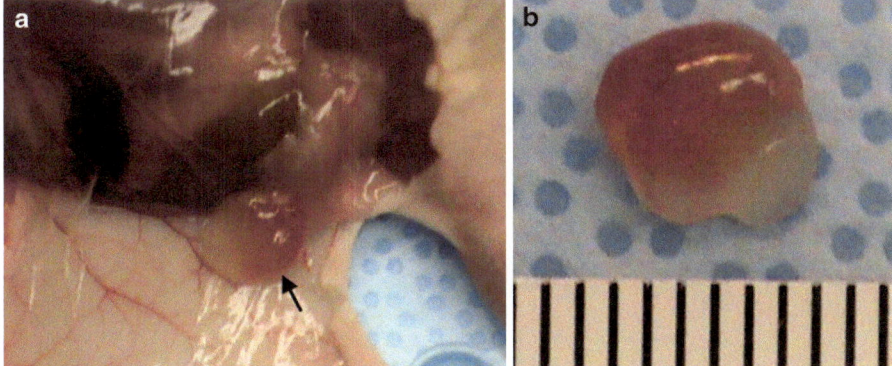

Fig. 2 Photomicrographs of harvested cell/Matrigel plugs at 9 days post-implantation. (**a**) Matrigel plugs containing EPC/MSC mixture undergo vasculogenesis when implanted subcutaneously in nude mice to form functional blood vessels that carry blood, which imparts *red color* to the otherwise colorless Matrigel (*arrow*). (**b**) Magnified view of the harvested cell/Matrigel implant showing its size and degree of functional vascularization

– Harvest the Matrigel implants using tweezers and scissors and take pictures of the harvested implant alongside a ruler (Fig. 2b).
– Fix the implants using 4% paraformaldehyde (PFA) overnight at 4°C.

3.5.4   Evaluation of Neovascularization in the Harvested Matrigel Implant

– Post-fixation, rinse the samples with PBS. The samples are now ready for whole-mount imaging for direct evaluation of neovascularization by the implanted human endothelial cells.
– For whole-mount confocal fluorescent imaging, transfer the PFA-fixed Matrigel implant on to a glass coverslip and place the sample on the microscope stage.
– Use a 20× objective to visualize the Matrigel implant up to a depth of 150–200 μm.
   Note: To obtain high resolution fluorescent images of the neovasculature, it is recommended that multiple 5 μm-thick z-sections are acquired in each field of view.
   Critical step: The harvested Matrigel implant may likely contain some residual subcutaneous fat. It is best to completely remove all fat residue from the implant for optimal image quality. However, if that is not possible, orient the implant such that the transparent Matrigel end faces the objective.
– An automated image analysis software (e.g. Volocity, Perkin Elmer; MetaMorph®, Molecular Devices) can then be used to process the sequential fluorescent z-sections to obtain a final three reconstructed dimensional (3D) image (Fig. 3).
– This 3D image can then be converted into a consolidated 2D image that contains information from the entire stack of z sections.

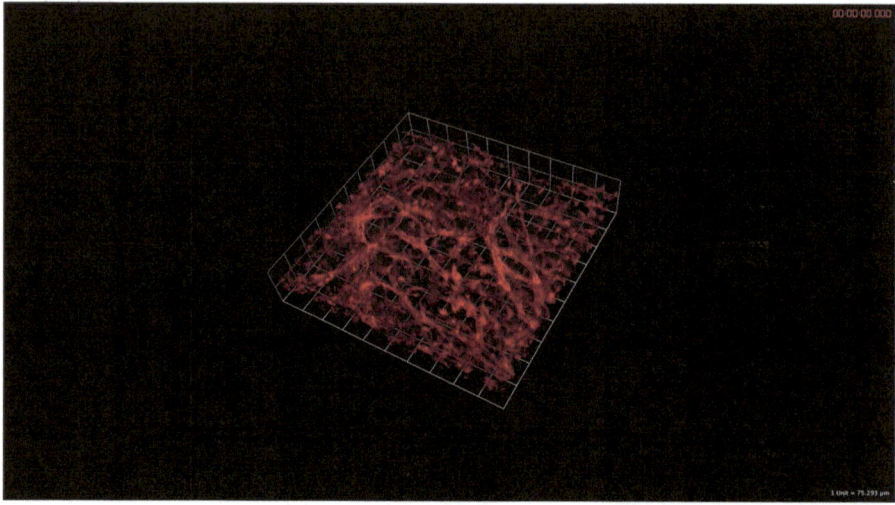

Fig. 3 Whole mount imaging of lectin-perfused cell/Matrigel implants. Intravenous injection of rhodamine-conjugated human-specific Rho-UEA-1 labels the functional human EPC-derived blood vessels *red*, which renders them suitable for whole mount imaging using a fluorescence confocal microscope. An automated image analysis software (e.g. Volocity, Perkin Elmer) can be used to process the sequential fluorescent z-sections to obtain a final reconstructed three dimensional (3D) image (scale bar: 1 unit = 75 um)

- The NIH-based Image J software can then be used to quantify neovascularization by measuring the total length of vascular structures, number of branch points, etc.
- Following whole mount imaging, embed the implant in 30% sucrose overnight at 4°C.
- Next day, embed the implants in OCT and store the samples at −80°C prior to cryosectioning.
- The OCT-embedded frozen samples can be sectioned on a cryostat to obtain ~10 μm-thick sections, which can then be air-dried and stained for either the conventional H&E or for other human endothelial cell-specific markers.

3.6   Timing

3.6.1   Suspending Cells in Matrigel

~ 2 h (for four Matrigel implants)

3.6.2   Subcutaneous Injection into Mice

~5 min per mouse (for two implants post isofluorane exposure)

3.6.3 Injection of Human-Specific Lectin

~1–2 min per mouse

3.6.4 Harvesting Matrigel Implants

~20–30 min per mouse (for two implants). This time can be shorter depending on the user skill level.

3.6.5 Imaging of Harvested Matrigel Implants

~ 2–3 h per Matrigel implant

3.6.6 OCT Embedding and Cryosectioning

~30–45 min Matrigel implant for OCT embedding and ~1 h for cryosectioning

3.7 Trouble Shooting

3.7.1 Clogging of the Syringe Needle

If the cell-Matrigel mixture reaches room temperature while inside the syringe, the Matrigel will begin to solidify, which in turn will clog the needle. Keep the syringe and needle on ice before use.

3.7.2 Poor Neovascularization in the Cell-Matrigel Implant *In Vivo*

(a) Matrigel that has been left at 4°C for more than 2–3 days should be avoided for this assay as it may have lost some potency and may lead to poor implant neovascularization. For best results, Matrigel should be freshly thawed from −20°C.

(b) A major factor causing poor neovascularization of the cell-Matrigel implant is the lower vasculogenic potential of higher passage HMVECs. For best results, HMVECs must be used between passages 2 and 5. Further, it may be prudent to always check HMVEC quality prior to *in vivo* implantation by plating them on Matrigel and monitoring their 2D capillary network formation *in vitro*. With increasing passage, the number of HMVECs needed to form a robust capillary network increases, thus indicating that a higher HMVEC density may also be needed for *in vivo* neovascularization.

(c) Likewise, for best results, use MSCs between passages 2 and 7 as higher passage cells may lose their potential to differentiate into pericytes, which is an important determinant of robust neovascularization.

3.7.3 Poor Visualization of the Vascular Network in Harvested Matrigel Implant

(a) This may be either due to poor Rho-UEA-1 binding to human endothelial cells or the presence of ruptured or leaky vessels that cause non-uniform distribution of Rho-UEA-1 throughout the implant. In the former case, make sure the Rho-UEA-1 working solution was prepared in Ca^{2+}, Mg^{2+}-supplemented PBS. Increasing the final Rho-UEA-1 concentration to 35–50 µg/ml (from the previous 25 µg/ml) may also provide a solution to this problem. However, if the neovessels are malformed, neither one of these steps will improve the imaging outcome so use another implant for confocal visualization.
(b) Ensure that confocal imaging is being done using a 20× objective. A higher magnification will prevent from going deep into the implant (as it's working distance is shorter) and will result in an incomplete vascular network image. In contrast, using a 10x objective will result in a weak fluorescent signal (and a concomitant strong noise) detected by the detector.

4 User Experience

Angiogenesis can be studied by using many different methods that are available to the investigator. Choosing the most appropriate assay can be based on a number of factors, such as reagent availability, equipment, time it will take to perform the assay, cost, and skills needed. One of the widely accepted assays is the corneal implant assay. However, this assay is difficult to perform since it requires special equipment and a highly-skilled person (usually a trained ophthalmologist) to implant the pellets in the eyes of animals. In contrast to the technically-challenging mouse corneal angiogenesis assay, the matrigel plug assay is very easy to perform. Matrigel containing the cells or factors of interest is injected subcutaneously, following which it solidifies to form a plug. This plug can be recovered 7–21 days post-implantation in the animal and examined histologically for qualitative evaluation of neovascularization. Further, our recently developed protocol, which we have described above in details, offers a distinctly simplified approach for quantitative analysis of new vessel formation by combining the strengths of FACS and confocal imaging. While FACS directly quantifies the number of recruited vascular cells into Matrigel plugs, an image analysis software can be used to quantitatively analyze the confocal z-stacks from these Matrigel samples for parameters such as vessel number, length, area and branching. We believe that the Matrigel plug assay, owing to its relative ease of preparation, recovery and imaging, will be widely used as a standardized, quantitative technique to study *in vivo* blood vessel formation, thus providing a useful addition to the current repertoire of *in vivo* neovascularization assays.

References

1. Isner JM, Asahara T (1999) Angiogenesis and vasculogenesis as therapeutic strategies for postnatal neovascularization. J Clin Invest 103:1231–1236
2. Murohara T (2003) Angiogenesis and vasculogenesis for therapeutic neovascularization. Nagoya J Med Sci 66:1–7
3. Folkman J (2006) Angiogenesis. Annu Rev Med 57:1–18
4. Folkman J (1999) Angiogenesis research: from laboratory to clinic. Forum (Genova) 9:59–62
5. Folkman J, Shing Y (1992) Angiogenesis. J Biol Chem 267:10931–10934
6. Folkman J (1992) Angiogenesis–retrospect and outlook. EXS 61:4–13
7. Folkman J (1985) Angiogenesis and its inhibitors. Important Adv Oncol 1985:42–62
8. Ribatti D, Nico B, Crivellato E (2011) The role of pericytes in angiogenesis. Int J Dev Biol 55:261–268
9. Folkman J (1996) Tumor angiogenesis and tissue factor. Nat Med 2:167–168
10. Folkman J (1974) Tumor angiogenesis. Adv Cancer Res 19:331–358
11. Folkman J (1971) Tumor angiogenesis: therapeutic implications. N Engl J Med 285:1182–1186
12. Folkman J (1998) Therapeutic angiogenesis in ischemic limbs. Circulation 97:1108–1110
13. Asahara T, Murohara T, Sullivan A, Silver M, van der Zee R, Li T, Witzenbichler B, Schatteman G, Isner JM (1997) Isolation of putative progenitor endothelial cells for angiogenesis. Science (New York, NY) 275:964–967
14. Khakoo AY, Finkel T (2005) Endothelial progenitor cells. Annu Rev Med 56:79–101
15. Francescone RA 3rd, Faibish M, Shao R (2011) A matrigel-based tube formation assay to assess the vasculogenic activity of tumor cells. J Vis Exp. Vol. 55
16. Malinda KM (2009) In vivo matrigel migration and angiogenesis assay. Methods Mol Biol 467:287–294
17. Akhtar N, Dickerson EB, Auerbach R (2002) The sponge/Matrigel angiogenesis assay. Angiogenesis 5:75–80
18. Kragh M, Hjarnaa PJ, Bramm E, Kristjansen PE, Rygaard J, Binderup L (2003) In vivo chamber angiogenesis assay: an optimized Matrigel plug assay for fast assessment of anti-angiogenic activity. Int J Oncol 22:305–311
19. Janiak M, Hashmi HR, Janowska-Wieczorek A (1994) Use of the Matrigel-based assay to measure the invasiveness of leukemic cells. Exp Hematol 22:559–565
20. Norrby K (2006) In vivo models of angiogenesis. J Cell Mol Med 10:588–612
21. Auerbach R, Lewis R, Shinners B, Kubai L, Akhtar N (2003) Angiogenesis assays: a critical overview. Clin Chem 49:32–40
22. Kleinman HK, Martin GR (2005) Matrigel: basement membrane matrix with biological activity. Semin Cancer Biol 15:378–386
23. Adini A, Fainaru O, Udagawa T, Connor KM, Folkman J, D'Amato RJ (2009) Matrigel cytometry: a novel method for quantifying angiogenesis in vivo. J Immunol Methods 342:78–81
24. Melero-Martin JM, Khan ZA, Picard A, Wu X, Paruchuri S, Bischoff J (2007) In vivo vasculogenic potential of human blood-derived endothelial progenitor cells. Blood 109:4761–4768
25. Melero-Martin JM, Bischoff J (2008) Chapter 13. An in vivo experimental model for postnatal vasculogenesis. Methods Enzymol 445:303–329

Corneal Pocket Assay

Marina Ziche and Lucia Morbidelli

Abstract This chapter is a methodological contribution on the corneal pocket angiogenesis assay. The avascular cornea presents advantages for the study of angiogenesis in vivo. At the same time, the cornea tissue reaction to different noxa (physical, chemical, biological) evolves in pathological neovascularization which can be controlled by antiangiogenic drugs. It consists in the creation of a corneal pocket, where test substances, cells or tissues are introduced, eliciting the growth of new vessels from the peripheral limbic vasculature. Although the corneal pocket assay is a powerful in vivo angiogenesis assay, technical challenges as well as advantages and drawbacks of using different species models remain a hurdle in its practical application. A detailed methodological description of all the assays which can be run on different species and their advantages/disadvantages is reported.

1 Introduction

Angiogenesis is practically a synonym of vascular sprouting. The sprouting of new vessels follows a well-defined program: degradation of basement membrane, endothelial cell migration and proliferation, formation of solid sprouts connecting a neighboring vessel, differentiation of the sprout into a lumen lined by endothelial cells and integration in the vascular network [1].

The term neovascularization is used to describe the formation of new vascular structures in previously avascular areas. Within the eye, neovessels are directly responsible for most of the destructive events that are characteristic for certain diseases, such as proliferative retinopathies and age-induced macular degeneration [2–6].

M. Ziche, M.D. (✉) • L. Morbidelli, Ph.D.
Laboratory of Angiogenesis, Section of Pharmacology, Toxicology and Chemotherapy,
Department of Biotechnology, University of Siena, Istituto Toscano Tumori,
Via Aldo Moro 2, 53100 Siena, Italy
e-mail: ziche@unisi.it

E. Zudaire and F. Cuttitta (eds.), *The Textbook of Angiogenesis and Lymphangiogenesis: Methods and Applications*, DOI 10.1007/978-94-007-4581-0_18,
© Springer Science+Business Media Dordrecht 2012

The cornea is the transparent portion of the anterior eye with the following functions: barrier to protect the eye against external environment, and optic for light refraction. Corneal clarity is necessary for vision. Its anatomy guarantees these functions: an external epithelial layer, a intermediate connective tissue rich in collagen fibrils, strictly organized, and interconnected by glycosaminoglycans, stromal cells (cheratocytes and fibroblasts), and an inner endothelial layer facing the aqueous humor.

Cornea insult, such as trauma, chemical injury, infections, immune or metabolic disorders, can lead to corneal vascularization and loss of corneal clarity, present in 4% of ophthalmic patients [7, 8]. Occasionally, corneal angiogenesis is helpful to wound healing or to control infection, but this reparative angiogenesis is transient [9]. Clinical management of corneal neovascularization is challenging, and preexisting neovascularization in acceptor eye significantly increases the risk of xenograft rejection [10].

The absence of vessels (both blood and lymphatic vessels) within the cornea (also termed corneal angiogenic privilage) is explained by several considerations: the compact structure of the stroma, which the vessels cannot penetrate (proven by the presence of vessels within the edematous cornea), and/or the abundance of mucopolysaccharides that, together with proteins and water, form a barrier against the penetration of vessels. Various known angiostatic molecules have been found in the cornea as angiostatin, endostatin, interleukin-1 receptor antagonist, pigment epithelium-derived factor and thrombospondin [7, 11, 12], although they may not be essential for cornea avascularity. Recently, the preservation of the avascular phenotype of the cornea has been definitely associated to high abundance of soluble vascular endothelial growth factor receptor (sVEGR1), able to neutralize the VEGF-A present in the cornea which is physically close to vascularized tissues as the sclera [13]. Thus, vascularization occurring during different conditions is the result of the perturbed balance among redundant inhibitory mechanisms.

Several factors influence neocapillary growth. Diffusible mediators, able to initiate directional capillary growth, are found in normal as well as pathological/diseased tissues. Indeed upregulation of these factors (VEGF-A, fibroblast growth factor-2 – FGF-2, erythropoietin, insulin-like growth factor-1, etc.) contribute to the corneal vascularization [7, 11, 12]. Furthermore, competition for oxygen and nutrients among cells, together with the accumulation of large amounts of metabolites, may stimulate the release of growth factors capable of inducing directional vascular growth, with the purpose of correcting these imbalances. An important factor seems hypoxia, resulting in the up-regulation of VEGF [14]. Edema causes a decrease in the stromal density, favoring vessel sprouting. With the development of neovascularization, the oxygen supply increases, suppresses the proangiogenic stimuli, favouring the regression of the new vessels. The consequences of corneal vascularization are important. In the early stages, it may be seen as a defense mechanism, because it increases the supply of oxygen and metabolites, activating the metabolic exchanges required for tissue repair. Later, vascularization carries immune elements such as phagocytes, immunocompetent cells, and antibodies. The intense blood flow in this tissue triggers a cascade of events, which supports and increases corneal angiogenesis.

Several studies have analyzed the evolution of corneal hemangiogenesis in association with lymphangiogenesis in different conditions. Lymphangiogenesis in particular is stimulated by VEGF-C and D through the activation of VEGFR-3 [11, 15–17].

Based on the above findings, a variety of agents such as corticosteroids, cyclosporine, indomethacine, metrotrexate, rapamycin, low molecular weight heparin and thalidomide have been evaluated in animal modeles of angiogenesis to treat corneal vascularization. However, these molecules have had limited success in the clinic [18–24]. Bevacizumab, the prototype of anti-VEGF antibodies, has recently been successfully tested in patients with corneal neovascularization [25, 26].

2 The Corneal Angiogenesis Assay

As the cornea recapitulates all the molecular and cellular events required for new vessel formation in the adult, the corneal angiogenesis assay is still considered one of the most reliable in vivo assays for the study of new blood vessel formation [27]. Any vessels seen within the cornea after stimulation by angiogenesis-inducing tissues or factors ought to be considered as new vessels, because the cornea is normally avascular. The original method was developed on rabbit eyes [28, 29], but has been adapted to rats and mice [30, 31]. It involves creation of a corneal pocket, where test stimuli or tissues or cells are introduced, eliciting the growth of new vessels from the peripheral limbic vasculature. The gold standard for angiogenesis is VEGF-A, a factor capable of promoting corneal angiogenesis in the different species. Other angiogenic stimuli are FGF-2, PGE2, interleuchins, which we have contributed to characterize [32, 33]. However, the literature reports the action of several inducers of angiogenesis (VEGF, FGF-2, substance P, bradykinin, prostaglandin E2, interleukin-8) through the upregulation of endogenous effectors such as FGF-2 or VEGF [34–39]. Neovascularization of the cornea is also observed following thermal or chemical injury. In these experimental settings VEGF has been shown to be up-regulated in inflamed and vascularized cornea, as well as in human pathological conditions [40–42].

3 Protocol Details

The cornea assay consists in the placement of an angiogenesis stimulus (tumor tissue, cell suspension, growth factor) into a micropocket produced in the cornea thickness in order to evoke vascular outgrowth from the peripherally located limbal vasculature. Vessel sprouting can also be induced by chemical or thermal burns. Neovascular development and progression can be modified by the presence of locally released or applied inhibitory factors or by systemically given antiangiogenic drugs.

Materials and instruments required for the assay are reported in Tables 1 and 2 and Fig. 1.

Table 1 Reagents and drugs required for the rabbit cornea assay

Pellet preparation	Recombinant growth factors or drugs in water or phosphate buffered saline (PBS) or ethanol or methanol in highly concentrated solutions (0.1–1 mg/ml). Avoid DMSO since not volatile and high salt buffers
	Ethylen-vinyl-acetate copolymer (Elvax-40) (DuPont de Nemours, Wilmington, DE, www.dupont.com)
	Organic solvents: absolute alcohol, methanol, methylen-chloride
Drugs for surgery	Ketamine
	Xylazine
	Sodium pentothal in saline solution (0.1 g/ml)
	Local anesthetic (0.4% benoxinate)
	Ether
	Cloral-hydrate
Corneal tissue manipulation for histology	Organic solvents: absolute alcohol, isopentane, acetone
	Fixative: 4% paraformaldehyde in PBS, pH 7.4
	Liquid nitrogen
	OCT tissue-teck medium or similar
	Haematoxylin and eosin
	Phosphate buffered saline (PBS)
	Bovine serum albumin (BSA)
	Primary antibodies: i.e. for markers of neovascularization (anti-CD31 Ab, Dako), inflammation (anti-RAM11 Ab, Dako) and adhesion molecule (anti α5β1 integrin Ab, Chemicon)
	Reagents for immunohistochemistry: goat anti mouse IgG (Sigma), Mouse Peroxidase anti-Peroxidase (PAP, Sigma), hydrogen peroxide in PBS, 3 3' diaminobenzidine tetrahydrochloride (DAB, Sigma)
	Aquatex medium (Merck)

Table 2 Equipment and instruments required for corneal assay

Cell cultures	Vertical laminar flow hood and autoclave
	Sterile instruments to prepare the pellets; stain less steel spatulae (1 for each substance to be tested), Dumont no. 5 twezzers, Vannas scissors, teflon plate (10 × 10 cm)
	6 cm glass Petri dishes
Animal facility	Sterile surgical room
	Sterile instruments for surgical implantation: Disposable scalpels for ocular microsurgery (n° 10/11, Aesculap), 1.5 mm pliable silver spatula with smooth edge blade, Dumont no. 5 twezzers, iris microforceps
	Latex dental dam for endodontic procedures (DentalTrey, www.dentaltrey.com)
	Insulin syringes
	Contention box for the different species
	Slit lamp stereomicroscope equipped with a digital camera
Histology and histochemistry	Cryostat or microtome
	Slides and glassware for histology
	Chemical hood
	Microscope equipped with a digital camera

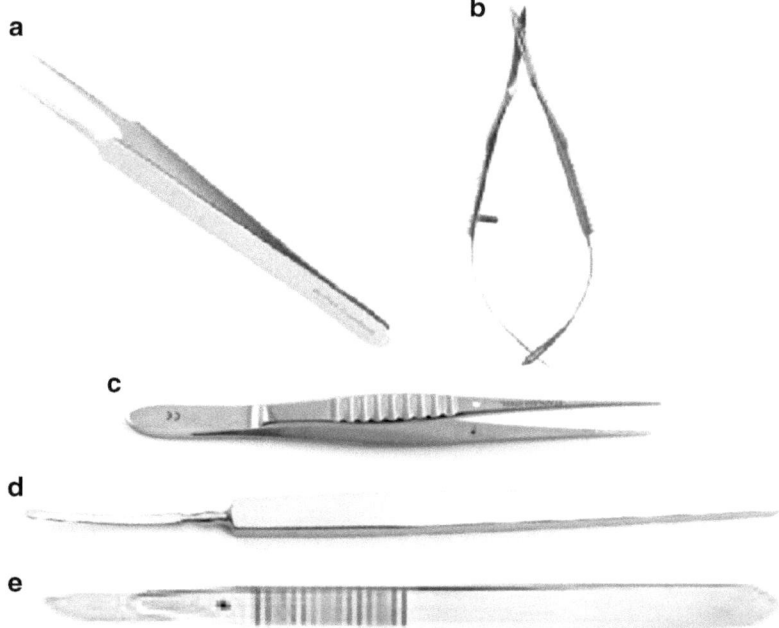

Fig. 1 Instruments for pellet preparation and implant: (**a**) Dumont tweezers, (**b**) Vannas scissors, (**c**) microforceps to keep open the edge of the corneal micropocket during pellet implantation, (**d**) pliable iris spatula, (**e**) disposable scalpel no. 10

3.1   Animals and Anesthesia

The corneal assay is mainly performed in our lab in New Zealand white male rabbits (1.5–2.5 kg), since the absence of a pre-existing vascular pattern. In this strain, the newly formed vessels (red) are clearly visible on the background of the underlying iris (pink) by stereomicroscope examination. Corneal angiogenesis for experimental purposes is also performed in other animal species like C57BL mice and Sprague-Dawley rats.

All treatments must be approved by the local laboratory animal ethics board and the national agencies, according to the current laws (i. e. European Directive 2010/63/EU) since the surgical procedure requires general anesthesia. Sample preparation, surgery and monitoring of angiogenesis require the skill of qualified operators.

3.2   Sample Preparation

Since pathogens are the most frequent cause of corneal angiogenesis, sterility of materials, samples and instruments is crucial to avoid unspecific responses. Drugs and instruments are listed in Tables 1 and 2 and shown in Fig. 1.

3.2.1 Angiogenic Factor Preparation

In order to be implanted in the cornea, angiogenic factors (i.e. VEGF, FGF-2, cytokines or other molecules) have to be brought in a semisolid state, enabling surgical implantation and gradual release of the factor. Pellets (implants) bearing molecules to be tested are prepared under sterile conditions according to the following steps.

(a) Recombinant growth factors or drugs solubilized at highly concentrated solutions (0.1–1 mg/ml) in water or phosphate buffered saline (PBS) (not in highly concentrated salt buffers) or ethanol or methanol (not DMSO since it is not volatile).
(b) Ethylen-vinyl-acetate copolymer (Elvax-40) (DuPont de Nemours, Wilmington, DE, www.dupont.com) preparation and testing (to be done some weeks before everyday use of casting solution):

  – wash Elvax-40 original pellets (20 g) in absolute alcohol (200 ml) 100 fold at 37°C. Store the washed pellets in glass dishes.
  – Weight 1 g of Elvax-40 and dissolve in 10 ml of methylen-chloride in a glass vial to prepare 10% casting stock solution. Incubate Elvax-40 in methylen-chloride at 37°C for 30–60 min to speed up solubilization.
  – Test the Elvax-40 preparation for its biocompatibility [43]. The casting solution is eligible for use when no implant performed with this preparation show inflammatory reaction at microscopic and histological exam of the rabbit cornea.

  Some authors use Hydron (Interferon Science, New Brunswick, NJ) as a casting solution (12% w/v), prepared dissolving the polymer in absolute alcohol at 37°C [43].

(c) Preparation of slow release pellets: Recombinant growth factors are prepared as slow-release pellets by incorporating the substance under test in Elvax-40. For testing, a pre-determined volume of Elvax-40 casting solution is mixed by the use of stainless steel spatula with a given amount of the compound to be tested previously dried on a flat teflon surface. The polymer and the compound are homogeneously mixed under a laminar flow hood by the use of spatula. After drying, the film sequestering the compound is cut into $1 \times 1 \times 0.5$ mm pieces under a stereomicroscope by the use of Vannas scissors and Dumont no. 5 tweezers. The pellets (in glass Petri dishes) are left under vacuum at 4°C overnight to remove residual solvent.

Pellets of Elvax-40 implants should always be used as negative controls, while, depending on the experimental design, VEGF or FGF-2 containing pellets are implanted as positive controls.

3.2.2 Cell and Tissue Sample Manipulation

Sample tissue from humans and experimental animals have been successfully implanted into the rabbit cornea to produce angiogenesis [44–47]. The intrinsic angiogenic potential due to different stages of tumor progression or to the expression of genes or gene products have been documented by our group as well as by others [33, 48–51].

Prepare a cell suspension by trypsinization of confluent cell monolayers to a final dilution of 2–5 × 10^5 cells in 5 μl. When implanting cells, angiogenic response can be graded based also on the number of cell implanted into the corneal stroma.

When tissues are tested, fragments are removed within 2 h from patients or animals and kept at 4°C in complete medium. Samples of 2–3 mg are obtained by cutting the fresh tissue fragments under sterile conditions by the use of microdissection instruments under a stereomicroscope.

3.3   Anesthesia and Surgical Implantation

The surgical manipulation of the cornea should be performed in a sterile environment under general anesthesia. Sterility is necessary to avoid unspecific reaction due to lipoxin. A surgical table with an appropriate illumination is used.

3.3.1   Rabbits

Immobilized rabbits, are anesthetized by intramuscular injection of ketamine hydrochloride (50 mg/kg) and xylazine hydrochloride (10 mg/kg), or, alternatively, by intravenous slow injection of sodium pentothal (30 mg/kg). The deepness of anesthesia is checked as reflex to pressure. Each eye is enucleated by the use of a dental dam and a local anesthetic (i. e. 0.4% benoxinate) is instilled intraocularly just before surgery. The pellet implantation procedure starts with a linear intrastromal incision, parallel to the corneoscleral limbus (linear keratotomy), using a surgical blade (disposable scalpel n°10). The preparation of the corneal pocket for the pellet implant is made with a 1.5 mm pliable silver spatula with smooth edge blade in the lower half of the cornea (see Fig. 2).

– Pellet implant: The implant is introduced through the keratotomy line, parallel to the corneal epithelium and under it, in the external third of the stroma, up to 2 mm from the limbus. One single pellet is selected from the Petri dish using Dumont n° 5 tweezers and then introduced in the corneal pocket. Microforceps are used to keep open the edge of the cut. Locate the implant at 2 mm from the limbus to avoid false positives due to mechanical stress and to favor the gradient diffusion of test substances in the tissue, toward the endothelial cells at the limbal plexus.

When two factors are tested simultaneously, make two independent and parallel micropockets.

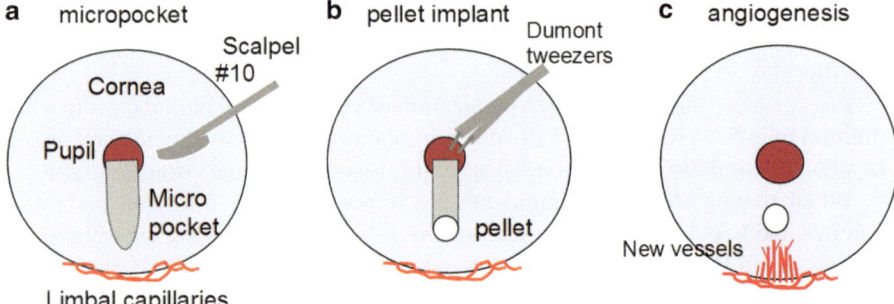

Fig. 2 Schematic representation of technical execution of surgical implant. After eye enucleation by the use of a dental dam, a cut is produced by the surgical scalpel and the micropocket is created in the cornea deepness by the use of pliable spatula (**a**). The implant is inserted by Dumont tweezers into the micropocket kept open by microforceps (**b**). The neovascular response starts from the limbal vessels, reaching the implanted pellet in approximately 1 week (**c**)

– Cell or tissue implant: A small amount (20–50 µl) of the aqueous humour can be drained from the anterior chamber with an insulin syringe when reduced corneal tension is required (i. e. for cell or tissue samples).

By using a micropipette introduce 5 µl containing 2–5 × 10^5 cells in medium supplemented with 10% serum in the corneal micropocket. When the overexpression of growth factors/inhibitors by stable transfection of specific cDNA is studied, one eye is implanted with transfected cells and the other with the wild type or vector transduced cell line. Suitable cell lines for these experiments are mammary carcinoma cells (MCF-7), lymphoma Burkitt's cells (DG75), chinese hamster ovary cells (CHO) [33, 49, 51]. It might be necessary to evaluate the angiogenic potential of drug-treated cells. In these experiments cell monolayers are pharmacologically treated before the implant (18–24 h). One eye is implanted with treated cells and the controlateral with control cells.

Tissue fragments are inserted in the corneal pocket with the aid of Dumont n° 5 tweezer. The angiogenic activity of tumor samples is compared with healthy tissue [46].

3.3.2 Mouse

The mouse cornea micropocket assay was firstly described by Muthukkaruppan and Auerbach [30].

(a) Anaesthetise animals by intramuscular injection of ketamine/xylazine (80 mg/kg and 8 mg/kg, respectively).
(b) Use Hydron (Interferon Science, New Brunswick, NJ) as a casting solution (12% w/v), prepared by dissolving the polymer in absolute alcohol at 37°C [43]. When peptides are tested, sucralfate (sucrose aluminium sulphate, Bukh Meditec,

Copenhagen, Denmark) is added to stabilize the molecule and to slow its release from Hydron [52, 53].

(c) The corneal micropocket is produced in each eye with a cyclodialysis spatula.

3.3.3 Rat

The rat is the least suitable animal for the cornea assay for angiogenesis, since in the rat corneal vascularity is frequently found due to infectious agents or injury.

Purified growth factors are combined 1:1 with Hydron as described by Polverini and Leibovich [31].

Pellets are implanted 1–1.5 mm from the limbus of the cornea of anaesthetised rats (sodium pentobarbital, 30 mg/kg, i.p.).

3.4 Protocols for Corneal Injury

Burn injury is induced within the cornea deepness to induce inflammatory cell infiltrate and neovascularization, reproducing the wound healing process. In this model, lymphangiogenesis has been clearly associated with inflammation index [54]. A critical point is the deepness of the injury which influences the time of healing and is difficult to be reproduced between and within experiments.

From the ethical point of view a particular attention should be paid to the definition of detailed experimental protocols. Local anesthetics and antibiotics should be applied during each manipulation and in the case of excessive sufferance and infection the animal should be sacrified.

3.4.1 Chemical Burns

After anesthesia, central corneal wounding in rabbit is performed by applying a 6 mm diameter filter paper (Whatman #3) soaked 1 N NaOH onto the central cornea for 10 s and the cornea should be washed with 5–10 ml of sterile saline solution [55].

The silver nitrate cauterization technique described by Mahoney and Waterbury in the rat [56] is also used to induce corneal neovascularization. Briefly, under the operating microscope, an applicator stick coated with 75% silver nitrate and 25% potassium nitrate with a diameter of 1.8 mm is pressed on the central cornea for 10 s. Excess chemical reagent is removed by rinsing the eye with 5–10 ml of sterile saline solution.

3.4.2 Thermal Burns

The test performed on rabbits requires general and local anesthesia as above, and the application for 15 s of a concave brass dowel (8 mm diameter) previously cooled in liquid nitrogen on the central cornea. Saline irrigation is used to thaw the aqueous iceball which is forming at the interface dowel tip/cornea. The most relevant parameter (cornea swelling) has been previously described [57].

4 Follow Up of Angiogenesis and Inflammation

The evaluation of angiogenesis evolution and appearance of inflammation in the different species is monitored by the use of different techniques. In rabbits, the evolution of angiogenesis can be easily monitored during time in the same living animal.

4.1 Light Microscopy

After the pellets are implanted, rabbits are followed-up without general anesthesia for the all duration of the study. Each animal eye is examined with the slit-light stereomicroscope every 2–3 days. The clinical evolution of the implants and of the ocular lesions is recorded and the presence of corneal reactions, such as redness, corneal edema, the intensity of the corneal cellular infiltrate, the total area of neovascularization are scored. Representative images of the cornea response to different experimental conditions (implant of pellets, cells, tissue, and alkali burning) are shown in Figs. 3 and 4.

In mice and rats light microscopy is possible but only after general anesthesia, with all the risks connected.

4.1.1 Quantification of Angiogenesis

An angiogenic response is scored positive when budding of vessels from the limbal plexus occurs within 3–4 days and capillaries progress to reach the implanted pellet in 7–10 days. Implants that fail to produce a neovascular growth within 10 days are considered negative, while implants exhibiting a pronounced inflammatory reaction after 4 days are discarded.

During each observation the number of implants showing neovascular growth (positive implants) is scored over the total number of surgical implants performed.

The angiogenic activity is evaluated on the basis of the number and growth rate of newly formed capillaries. The angiogenic score is calculated by the formula [vessel density × distance from limbus] [33, 58]. A density value of 1 corresponds to 0–25 vessels per cornea, 2 from 25 to 50, 3 from 50 to 75, 4 from 75 to 100 and 5 for more

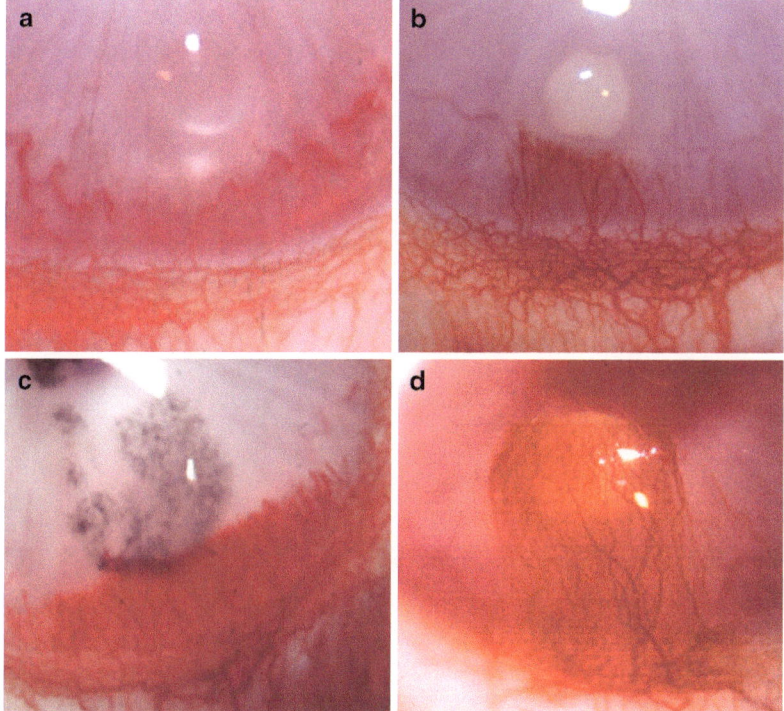

Fig. 3 Examples of corneal response to implant of (**a**) empty Elvax-40 pellets, (**b**) VEGF releasing pellet, (**c**) B16 melanoma cells, and (**d**) head and neck tumor specimen. In (**b**) the dose of VEGF is 200 ng/pellet. Note the presence of newly formed functioning vessels within the corneal stroma in (**b–d**). See references [33] and [46] for details. Pictures are taken through the stereomicroscope after at least 10–14 days. 18× original magnification

than 100 vessels. The density is calculated with the aid of an ocular grid inserted in the stereomicroscope, while the distance from the limbus is graded by the use of a caliper.

In mice the vascular response measured as the maximal vessel length and number of clock hours of neovascularization is scored at fixed time (usually on postoperative 5 and 7 days, under anesthesia) using a slit-lamp stereomicroscopy and photographed. To quantify the section of the cornea in which new vessels are sprouting from the pre-existing limbal vessels, the circumference of the cornea is virtually divided into the equivalent of 12 clock hours. The number of clock hours of neovascularization for each eye is measured during each observation.

4.1.2 The Computerized System of Image Recording, Processing, and Analysis

The anterior ocular pole images are computer analyzed at fixed times on animals under anesthesia. An advanced video camera (Sony 450X) connected to a color

Fig. 4 Example of corneal
response to alkali burns
(**a**) and the corresponding
histological analysis (**b**). Note
the injured tissue (the *white
spot*), the florid angiogenic
response and corneal opacity
in panel (**a**), and the presence
of newly formed vessels lined
by endothelial cells (V)
within the corneal stroma,
accompanied by
inflammatory cells (*arrows*)
in panel (**b**). Ep, corneal
epithelium. 18× and 20×
original magnification,
respectively in (**a** and **b**)

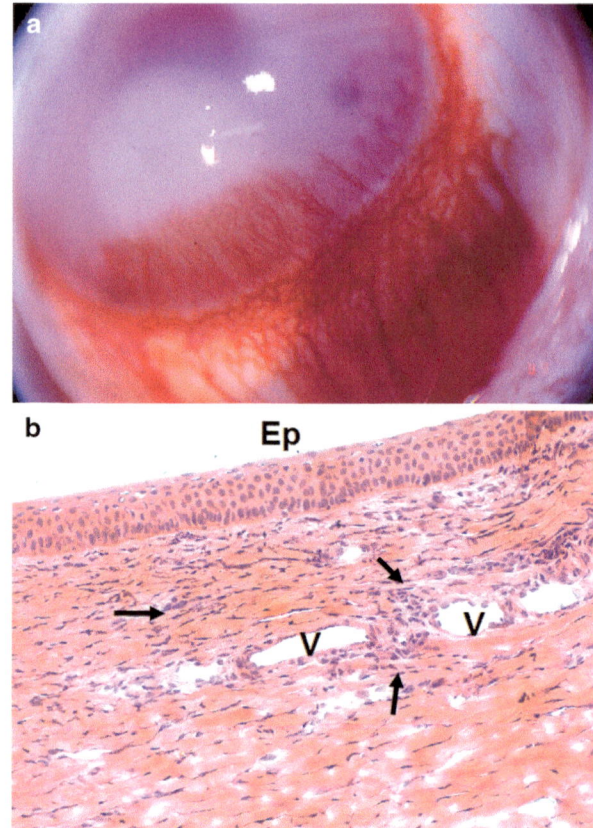

video monitor and a computer with video-bluster and special capture software are
used to record corneal responses. In order to extract the vascular tree from every
image, the following graphic processing are required:

- adjustment of contrast and brightness, in order to highlight the vascular tree;
- image conversion in a gray scale format (8 bytes for a pixel);
- image extraction of the vascular tree (skeletonisation).

Commercially available softwares (i.e. Corel Photo Paint and Corel Draw;
Adobe Photoshop and National Institute of Health Image J1.38X) can be used for
these purposes. Some manual skeletonisation is used, because the software can
produce errors. The trajectory of the vascular tree is traced over the processed BMP
(Bit Map) image, retaining only the image corresponding to the corneal vessels.
A special software can also been created, in order to process the skeletonisated
image to generate data (length, width, number of pixels) [59, 60].

4.2   India Ink or Carbon Perfusion

This is a postmortem procedure, which requires a large number of animals to monitor the time course of the corneal response. At predetermined time periods, mice are anesthetized with ketamine/xylazine and perfused with 1 ml waterproof black India Ink for drawing. The eyes are removed and the corneas (with limbal regions) are isolated, radially cut and flat mounted. The samples are reviewed under a light microscope for new vessels and photographed [61].

Similarly, neovascularization in rat cornea is assessed at fixed days (usually 3, 5 and 7 days and over depending on the type of protocol). Anesthetized animals are perfused with colloidal carbon solution to label vessels, eyes are enucleated and fixed in 10% neutral buffered formalin overnight. The following day, corneas are excised, flattened and photographed.

A positive neovascularization response is recorded when sustained directional growth of capillary sprouts and hairpin loops toward the implant is observed. Negative responses are recorded when either no growth is observed or when only an occasional sprout or hairpin loop showing no evidence of sustained growth is detected.

4.3   Fluorescein Angiography

After the angiogenesis process is completed, animals are anesthetized and intraperitoneally injected with 0.04 ml of 25% sodium fluorescein. Staining is recorded by a digital fundus camera or, after euthanasia of the animals, in flat corneas using a microscope [61].

4.4   Histopathological Examination

At predefined time periods, the animals (rabbits, rat, mouse) are humanely sacrificed by the intraperitoneal injection of a toxic dose of chloral-hydrate associated with ether inhalation, or by intravenous bolus injection of pentobarbital (30 mg/kg). Shortly after death, in order to prevent as much as possible the onset of postmortem events, the ocular globes are enucleated. Corneas are removed with a portion of underneath sclera by the use of a surgical scalpel and dissection sharp scissors, and processed for histomorphological fixation, in order to stop vital tissue phenomena and capture changes present within the tissue at sampling. The final step is the inclusion in paraffin blocks, solidified in ice. These blocks are cut, with a microtome, into serial sections of 5 μm each. These sections are then placed on albumin–glycerin lubrified slides. Hematoxylin–Eosin and Van Gieson stains are routinarily performed. The sections are placed, using Canada balm, between slides, resulting in a large number of samples. The samples are examined with a microscope, and color microphotographs are taken [34, 62]. An example of histological analysis is reported in Fig. 4.

In order to perform immunohistochemical analysis, freshly isolated cornea samples are snap frozen. The corneas are removed, oriented and marked (with a cotton thread), immediately frozen in isopentane cooled in liquid nitrogen for 10 s, in optimal temperature cutting compound (OCT) on a wood support. The frozen tissue blocks are maintained at $-80°C$. Tissue sections (7 μm) are cut with a cryostat. Sections are placed on microscope Superfrost Plus slides (Fisher) and maintained frozen at $-80°C$ until staining. For immunohistochemical staining, after fixation in absolute acetone at $-20°C$ for 5 min, sections are washed in phosphate buffered saline (PBS) and then treated with 1.5% hydrogen peroxide in PBS for 8 min in order to perform quenching of endogenous peroxidases. Aspecific binding sites are then blocked in 3% bovine serum albumin (BSA) in PBS for 45 min. Sections are incubated overnight with the primary antibodies diluted in 0.5% BSA in PBS. Primary antibodies validated on rabbit samples are anti-CD31 Ab (Dako, 200 ug/ml) (marker of neovascularization), anti-RAM11 Ab (Dako, 1.2 ug/ml) (marker of inflammation) (anti α5β1 integrin Ab, Chemicon, 1:50) (adhesion molecule expressed in epithelial and endothelial cells). For co-localization serial and adjacent sections can be labelled with different antibodies. Sections are extensively washed in 0.5% BSA in PBS and then incubated in goat anti mouse IgG (Sigma,1:40) for 1 h. After washing in 0.5% BSA in PBS, sections are incubated in mouse peroxidase anti-peroxidase (PAP, Sigma 1:35) for 45 min. Immunoreaction is developed in 3 3' diaminobenzidine tetrahydro-chloride (DAB, Sigma) for 8 min. Sections are then extensively rinsed in distilled water counterstained in haematoxylin and mounted in Aquatex medium (Merck). Samples are observed at the microscope (at 10–40× magnification) and digital images are taken [33].

5 Advantages and Disadvantages in Different Species

5.1 Species and Strains

The rabbit size (approx. 2 kg) lets an easy manipulation of the animal; the eye is accessible, may be easily enucleated and surgically manipulated.

Rabbit cornea has been found avascular in all strains examined so far, although the New Zealand strain offers the advantage of having an easily detectable neovas-cularization. In some strains of rats the presence of preexisting vessels within the cornea and the development of keratitis are serious disadvantages. Further-more, rabbits are more docile and amenable to handling and experimentation than mice and rats. In case of inflammatory reactions, these are easily detectable in rabbits by daily stereomicroscopic examination.

5.2   Measurements

In mice and rats 1 or 2 fixed time point observations are usually recorded. The evolution of the angiogenic response over time in the same animal is not recommended, because each time the cornea is observed the animal has to be anesthetized (with the connected risks of the overexposure to the anesthetic) to allow accurate observation and recording. Experiments require a large number of animals for each experimental point [6–10] and vessel growth can be visualized and measured after perfusion with colloidal carbon solution in individual animals.

Multiple observations are easily performed in rabbits, thus reducing the number of animals required for statistical evaluation. The use of slit lamp stereomicroscope and of awake animals allows the observation of newly formed vessels for prolonged time monitoring, up to 1–2 months in the same living animal.

5.3   Different Experimental Procedures

In the rabbit eye, due to its wide surface, stimuli in different forms can be placed. In particular the activity of specific growth factors can be studied in the form of slow-release pellets [32, 34, 49, 62], or of tumor and non-tumor cell lines stably transfected for the over-expression of angiogenic factors [33]. Cells with double transfection can also be studied [49, 50].

The modulation of the angiogenic responses by different stimuli can be assessed in the rabbit cornea assay (i) by implanting single pellets releasing both the angiogenic stimulus and the inhibitor [63, 64], (ii) by implanting in the same cornea two pellets placed in parallel micropockets and releasing different molecules or tissues or cells or combinations of all [38, 65], and (iii) through the removal or addition of multiple pellets [65].

The protocols (ii) and (iii) are particularly used to evaluate regression of angiogenesis. First an angiogenic stimulus is implanted (VEGF or FGF-2 releasing pellets). After 5–7 days when neovacularization has occurred, the inhibitor is systemically given or is locally added in an adjacent empty pocket performed before. Angiogenesis regression is monitored thereafter in the following days.

The implant of tumor samples can be performed both in corneal micropockets and in the anterior chamber of the eye to monitor angiogenesis produced by hormone-dependent tissues or tumors (i. e. human breast or ovary carcinoma in female rabbits) and it allows the detection of both the iris and the corneal neovascular growth [48, 66].

5.4   Treatment with Drugs

The effect of local drug treatment on corneal neovascularization may be studied in the form of ocular drops or ointment [67] or microinjection in the corneal thickness [68]. The effect of systemic drug treatment on corneal angiogenesis may be also evaluated [33, 46, 58, 69]. Given the size of the animals, systemic drug treatment in rabbits requires an higher amount of drugs than smaller animals.

6 Discussion

Several experimental models have reproduced corneal vascularization; they include various means of damaging the cornea (chemical agents, physical means, exogenous infections, immunological reactions, intracorneal inoculation of different agents, toxic states, or nutritional deficits). Such models vary in their reproducibility. Two different ways of inducing neoangiogenesis have been extensively used in our lab: implantation of slow release polymer pellets containing angiogenic molecules, such as VEGF or FGF-2, or tumor/tissue samples and engineered or drug treated cells.

The angiogenesis model is considered reliable, because the angiogenic stimulus is directly induced by purified factors and so the activity of angiogenesis inhibitors may be easily monitored. A large number of serial observations and measurements of the neovascular reactions in the cornea are performed by the use of slit-lamp stereomicroscope, without anesthesia.

6.1 Critical Issues

Some experimental procedures are however critical and should be carefully considered.

6.1.1 Skill of Operators

Different steps are critical for the execution of the assay and require a solid experience of the operators: work under sterility, manipulation of the different animal species, microsurgery of the eye, quantification of angiogenesis, histomorphology.

Moreover, in order to provide objective measurements and data, at least two independent operators should be enrolled in order to perform the critical steps (surgical implant and evaluation of the response) in a blind manner.

6.1.2 Distance from the Limbus

We noticed that the onset of corneal neovascularization depends on the proximity of the implanted pellet or burn lesion to the corneoscleral limbus. The more proximal is the implant or the lesion, the highest is the presence of inflammatory infiltrate and the occurrence of neovascularization is fast. This event can be exploited within the experimental protocol, but it should be considered cautiously since it can cause unspecific response. In contrast, when the implant or the injury is too far from the corneoscleral limbus (>3 mm), the implant is invariably negative. The explanation is related to the gradient of molecules reaching the endothelial cells of the limbal capillaries, being too high in the first case and too low in the second one.

6.1.3 Batch and Dose of Angiogenic Factor

After the VEGF implants, neovessels appear and grow progressively from day 3 to day 6, and stabilize thereafter. When comparing the extent of neovascularization, FGF-2 appears more efficient since its effects are both on endothelial cells and on stromal cells which are activated to produce VEGF [35, 37].

Variability among growth factors in inducing angiogenesis has been found considering different angiogenic factors, different providers and batch of preparation. Usually the dose of VEGF or FGF-2 able to give positive angiogenic response varies in the range 200–400 ng/pellet using R&D Systems as source.

6.2 Perspectives

The essential role of VEGF in corneal angiogenesis was shown by the inhibition of neovascularization after stromal implantation of anti-VEGF blocking antibody or peptides in rat and rabbit corneal models [7].

New medical and surgical treatments, including angiostatic steroids, non-steroidal inflammatory agents, anti- VEGF agents, argon laser photocoagulation, and photodynamic therapy, have proven effective in inhibiting corneal neovascularization in animal models [7, 12, 70]. The assessment of antiangiogenic activity of new drugs or strategies can be easily performed in the cornea before its validation in other organs as the retina or tumors. In particular, mice can systemically administered with new antitumor drugs to test the inhibition of corneal vascularization before tumor implantation.

Acknowledgments The work was supported by the Italian Ministry of University (MIUR), the Italian Association for Cancer Research (AIRC) and the Istituto Toscano Tumori (ITT). We thank Dr. Antonio Giachetti for editing the manuscript.

References

1. Folkman J, Shing Y (1992) Angiogenesis. J Biol Chem 267(16):10931–10934
2. Aiello LP, Avery R, Arrigg R et al (1994) Vascular endothelial growth factor in ocular fluid of patients with diabetic retinopathy and other retinal disorders. N Engl J Med 331 (22):1480–1487

3. Ting TD, Oh M, Cox TA et al (2002) Decreased visual acuity associated with cystoid macular edema in neovascular age-related macular degeneration. Arch Ophthalmol 120(6):731–737
4. Sunness JS, Margalit E, Srikumaran D et al (2007) The longterm natural history of geographic atrophy from age-related macular degeneration: enlargement of atrophy and implications for interventional clinical trials. Ophthalmology 114(2):271–277
5. Trieschmann M, Van Kuijk FJ, Alexander R et al (2008) Macular pigment in the human retina: histological evaluation of localization and distribution. Eye (Lond) 22(1):132–137
6. Zhang N, Hoffmeyer GC, Young ES et al (2007) Optical coherence tomography reader agreement in neovascular age-related macular degeneration. Am J Ophthalmol 144(1):37–44
7. Chang JH, Gabison EE, Kato T et al (2001) Corneal neovascularization. Curr Opin Ophthalmol 12:242–249
8. Azar DT (2006) Corneal angiogenic privilege: angiogenic and antiangiogenic factors in corneal avascularity, vasculogenesis, and wound healing. Trans Am Ophthalmol Soc 104:264–302
9. Lee P, Wang CC, Adamis AP (1998) Ocular neovascularization: an epidemiological review. Surv Ophthalmol 43:245–269
10. Williams KA, Esterman AJ, Barlett C et al (2006) How effective is penetrating corneal transplantation? Factors influencing long-term outcome in multivariate analysis. Transplantation 81:896–901
11. Ellenberg D, Azar DT, Hallak JA et al (2010) Novel aspects of corneal angiogenic and lymphangiogenic privilege. Prog Retin Eye Res 29(3):208–248
12. Maddula S, Davis DK, Maddula S et al (2011) Horizons in therapy for corneal angiogenesis. Ophthalmology 118:591–599
13. Ambati BK, Nozaki M, Singh N et al (2006) Corneal avascularity is due to soluble VEGF receptor-1. Nature 443(7114):993–997
14. Mazure MN, Chen EY, Yeh P et al (1996) Oncogenic transformation and hypoxia synergistically act modulate vascular endothelial growth factor expression. Cancer Res 56(15):3436–3440
15. Cursiefen C, Hofmann-Rummelt C, Küchle M et al (2003) Pericyte recruitment in human corneal angiogenesis: an ultrastructural study with clinicopathological correlation. Br J Ophthalmol 87(1):101–106
16. Cursiefen C, Maruyama K, Jackson DG et al (2006) Time course of angiogenesis and lymphangiogenesis after brief corneal inflammation. Cornea 25(4):443–447
17. Bock F, Onderka J, Dietrich T et al (2008) Blockade of VEGFR3-signalling specifically inhibits lymphangiogenesis in inflammatory corneal neovascularization. Graefes Arch Clin Exp Ophthalmol 246:115–119
18. BenEzra D, Griffin BW, Maftzir G et al (1997) Topical formulations of novel angiostatic steroids inhibit rabbit corneal neovascularization. Invest Ophthalmol Vis Sci 38:1954–1962
19. Lipman RM, Epstein RJ, Hendricks RL (1992) Suppression of corneal neovascularization with cyclosporine. Arch Ophthalmol 110:405–407
20. Haynes WL, Proia AD, Klintworth GK (1989) Effect of inhibitors of arachidonic acid metabolism on corneal neovascularization in the rat. Invest Ophthalmol Vis Sci 30:1588–1593
21. Joussen AM, Kruse FE, Völcker HE et al (1999) Topical application of methotrexate for inhibition of corneal angiogenesis. Graefes Arch Clin Exp Ophthalmol 237:920–927
22. Shi W, Gao H, Xie L et al (2006) Sustained intraocular rapamycin delivery effectively prevents high-risk corneal allograft rejection and neovascularization in rabbits. Invest Ophthalmol Vis Sci 47:3339–3344
23. Lepri A, Benelli U, Bernardini N et al (1994) Effect of low molecular weight heparan sulphate on angiogenesis in the rat cornea after chemical cauterization. J Ocul Pharmacol 10:273–280
24. Kruse FE, Joussen AM, Rohrschneider K et al (1998) Thalidomide inhibits corneal angiogenesis induced by vascular endothelial growth factor. Graefes Arch Clin Exp Ophthalmol 236:461–466
25. Dastjerdi MH, Al-Arfaj KM, Nallasamy N et al (2009) Topical bevacizumab in the treatment of corneal neovascularization: results of a prospective, open-label, noncomparative study. Arch Ophthalmol 127:381–389

26. Gerten G (2008) Bevacizumab (Avastin) and argon laser to treat neovascularization in corneal transplant surgery. Cornea 27:1195–1199
27. Auerbach R, Lewis R, Shinners B et al (2003) Angiogenesis assays: a critical overview. Clin Chem 49(1):32–40
28. Gimbrone MA Jr, Leapman SB, Cotran RS et al (1972) Tumor dormancy in vivo by prevention of neovascularization. J Exp Med 136(2):261–276
29. Gimbrone M Jr, Cotran R, Leapman SB et al (1974) Tumor growth and neovascularization: An experimental model using the rabbit cornea. J Natl Cancer Inst 52:413–427
30. Muthukkaruppan V, Auerbach R (1979) Angiogenesis in the mouse cornea. Science 206:1416–1418
31. Polverini PJ, Leibovich SJ (1984) Induction of neovascularization in vivo and endothelial cell proliferation in vitro by tumor associated macrophages. Lab Invest 51:635–642
32. Ziche M, Jones J, Gullino PM (1982) Role of prostaglandinE1 and copper in angiogenesis. J Natl Cancer Inst 69:475–482
33. Ziche M, Morbidelli L, Choudhuri R et al (1997) Nitric oxide-synthase lies downstream of vascular endothelial growth factor but not basic fibroblast growth factor induced angiogenesis. J Clin Invest 99:2625–2634
34. Parenti A, Morbidelli L, Ledda F et al (2001) The bradykinin/B1 receptor promotes angiogenesis by upregulation of endogenous FGF-2 in endothelium via the nitric oxide synthase pathway. FASEB J 15(8):1487–1489
35. Claffey KP, Abrams K, Shih SC et al (2001) Fibroblast growth factor 2 activation of stromal cell vascular endothelial growth factor expression and angiogenesis. Lab Invest 81(1):61–75
36. Xue ML, Thakur A, Willcox M (2002) Macrophage inflammatory protein-2 and vascular endothelial growth factor regulate corneal neovascularization induced by infection with pseudomonas aeruginosa in mice. Immunol Cell Biol 80:323–327
37. Qazi Y, Maddula S, Ambati BK (2009) Mediators of ocular angiogenesis. J Genet 88:495–515
38. Finetti F, Donnini S, Giachetti A et al (2009) Prostaglandin E2 primes the angiogenic switch via a synergic interaction with the fibroblast growth factor-2 pathway. Circ Res 105:657–666
39. Oliveira HB, Sakimoto T, Javier JA et al (2010) VEGF Trap(R1R2) suppresses experimental corneal angiogenesis. Eur J Ophthalmol 20:48–54
40. Amano S, Rohan R, Kuroki M et al (1998) Requirement for vascular endothelial growth factor in wound- and inflammation-related corneal neovascularization. Invest Ophthalmol Vis Sci 39(1):18–22
41. Cursiefen C, Rummelt C, Küchle M (2000) Immunohistochemical localization of vascular endothelial growth factor, transforming growth factor alpha, and transforming growth factor beta1 in human corneas with neovascularization. Cornea 19(4):526–533
42. Philipp W, Speicher L, Humpel C (2000) Expression of vascular endothelial growth factor and its receptors in inflamed and vascularized human corneas. Invest Ophthalmol Vis Sci 41(9):2514–2522
43. Langer R, Folkman J (1976) Polymers for the sustained release of proteins and other macromolecules. Nature 363:797–800
44. Brem H, Folkman J (1975) Inhibition of tumor angiogenesis mediated by cartilage. J Exp Med 141(2):427–439
45. Bard RH, Mydlo JH, Freed SZ (1986) Detection of tumor angiogenesis factor in adenocarcinoma of kidney. Urology 27(5):447–450
46. Gallo O, Masini E, Morbidelli L et al (1998) Role of nitric oxide in angiogenesis and tumor progression in head and neck cancer. J Natl Cancer Inst 90:587–596
47. da Silva BB, da Silva Júnior RG, Borges US et al (2005) Quantification of angiogenesis induced in rabbit cornea by breast carcinoma of women treated with tamoxifen. J Surg Oncol 90(2):77–80
48. Brem SS, Gullino PM, Medina D (1977) Angiogenesis: a marker for neoplastic transformation of mammary papillary hyperplasia. Science 195(4281):880–882
49. Cervenak L, Morbidelli L, Donati D et al (2000) Abolished angiogenicity and tumorigenicity of Burkitt lymphoma by Interleukin-10. Blood 96:2568–2573

50. Woolard J, Wang WY, Bevan HS et al (2004) VEGF165b, an inhibitory vascular endothelial growth factor splice variant: mechanism of action, in vivo effect on angiogenesis and endogenous protein expression. Cancer Res 64(21):7822–7835

51. Marconcini L, Marchio S, Morbidelli L et al (1999) c-fos-induced growth factor/vascular endothelial growth factor D induces angiogenesis in vivo and in vitro. Proc Natl Acad Sci USA 96(17):9671–9676

52. Chen C, Parangi S, Tolentino MT et al (1995) A strategy to discover circulating angiogenesis inhibitors generated by human tumors. Cancer Res 55:4230–4233

53. Voest EE, Kenyon BM, O'Really MS et al (1995) Inhibition of angiogenesis in vivo by interleukin 12. J Natl Cancer Inst 87:581–586

54. Yan H, Qi C, Ling S et al (2010) Lymphatic vessels correlate closely with inflammation index in alkali burned cornea. Curr Eye Res 35(8):685–697

55. Ormerod LD, Abelson MB, Kenyon KR (1989) Standard models of corneal injury using alkali-immersed filter discs. Invest Ophthalmol Vis Sci 30:2148–2153

56. Mahoney JM, Waterbury LD (1985) Drug effects on the neovascularization response to silver nitrate cauterization of the rat cornea. Curr Eye Res 4:531–535

57. Khodadoust AA, Green K (1976) Physiological function of regenerating endothelium. Invest Ophthalmol 15:96–101

58. Ziche M, Morbidelli L, Masini E et al (1994) Nitric oxide mediates angiogenesis in vivo and endothelial cell growth and migration in vitro promoted by substance P. J Clin Invest 94:2036–2044

59. Coman L, Coman OA, Paunescu H et al (2010) VEGF-induced corneal neovascularization in a rabbit experimental model. Rom J Morphol Embryol 51(2):327–336

60. Sharma A, Bettis DJ, Cowden JW et al (2010) Localization of angiotensin converting enzyme in rabbit cornea and its role in controlling corneal angiogenesis in vivo. Mol Vis 16:720–728

61. Dratviman-Storobinsky O, Lubin BC, Hasanreisoglu M et al (2009) Effect of subconjunctival and intraocular bevacizumab injection on angiogenic gene expression levels in a mouse model of corneal neovascularization. Mol Vis 15:2326–2338

62. Taraboletti G, Morbidelli L, Donnini S et al (2000) The heparin binding 25 kDa fragment of thrombospondin-1 promotes angiogenesis and modulates gelatinases and TIMP-2 in endothelial cells. FASEB J 14:1674–1676

63. Morbidelli L, Donnini S, Chillemi F et al (2003) Angiosuppressive and angiostimulatory effects exerted by synthetic partial sequences of endostatin. Clin Cancer Res 9(14):5358–5369

64. Donnini S, Finetti F, Lusini L et al (2006) Divergent effects of quercetin conjugates on angiogenesis. Br J Nutr 95(5):1016–1023

65. Ziche M, Alessandri G, Gullino PM (1989) Gangliosides promote the angiogenic response. Lab Invest 61:629–634

66. Federman JL, Brown GC, Felberg NT et al (1980) Experimental ocular angiogenesis. Am J Ophthalmol 89(2):231–237

67. Presta M, Rusnati M, Belleri M et al (1999) Purine analog 6-methylmercaptopurine ribose inhibits early and late phases of the angiogenesis process. Cancer Res 59(10):2417–2424

68. Ziche M, Morbidelli L (2009) Molecular regulation of tumour angiogenesis by nitric oxide. Eur Cytokine Netw 20(4):164–170

69. Ziche M, Donnini S, Morbidelli L et al (1998) Linomide blocks angiogenesis by breast carcinoma vascular endothelial growth factor transfectants. Br J Cancer 77(7):1123–1129

70. Bock F, Onderka J, Dietrich T et al (2007) Bevacizumab as a potent inhibitor of inflammatory corneal angiogenesis and lymphangiogenesis. Invest Ophthalmol Vis Sci 48(6):2545–2552

Directed In Vivo Angiogenesis Assay (DIVAA) for the Screening of Angiogenesis Modulators

Liliana Guedez and William G. Stetler-Stevenson

Abstract The mechanisms of angiogenesis have been studied in great detail, in part because of the recognition that the formation of new blood vessels constitutes a target for tumor therapy. One of the problems in assessing angiogenic responses in pre-clinical animal models is the difficulty of obtaining reliable and consistent quantifiable measurements of the angiogenic reaction. Here, we describe a technique, named directed-*in vivo*-angiogenesis assay (DIVAA), which consists of the subcutaneous implantation in mice of medical-grade silicone cylinders called Angioreactors that are filled with a small amount (20 μL) of basement membrane preparation pre-mixed with angiogenic factors, anti-angiogenic drugs and/or tumor cells. Angiogenesis is quantified by injecting mice with fluorescence-labeled tracers such as FITC-dextran or lectins prior to removing angioreactors and followed by spectrofluorometric measurement. Anti-angiogenesis effects on the histology, as well as biochemical changes can also be assed in the tissue invading the Angioreactor. The DIVAA represents a significant advancement for *in vivo* angiogenesis assays, greatly decreasing inter-assay variability. DIVAA has a wide applicability as a quantitative assay to determine the potency of agents that stimulate or inhibit angiogenesis.

1 General Considerations

DIVAA is reproducible, and has proved to be accurate for the analysis of effects of inhibitors of angiogenesis [1–3]. The DIVAA was initially designed to test inhibitory effects of compounds or factors on the angiogenesis response induced by a mixture of the angiogenic factors FGF-2 and murine VEGF-A and basement

L. Guedez • W.G. Stetler-Stevenson (✉)
Extracellular Matrix Pathology Section, Radiation Oncology Branch, Center for Cancer Research (CCR), Advanced Technology Center, National Cancer Institute, 8717 Grovemont Circle, Bethesda, MD 20892-4605, USA
e-mail: sstevenW@mail.nih.gov

E. Zudaire and F. Cuttitta (eds.), *The Textbook of Angiogenesis and Lymphangiogenesis: Methods and Applications*, DOI 10.1007/978-94-007-4581-0_19,
© Springer Science+Business Media Dordrecht 2012

membrane Matrigel. Stock and LOT numbers should be kept constant for all these reagents. Phenol-red free Matrigel preparations should be low in growth factors with a protein concentration (12–19 mg/mL), and endotoxin free. Other basement membranes such as Collagen type I from rat-tail should be used at pH = 7 and 2–3% in PBS. One important consideration in using DIVAA is the age and genetic background of mice. Angiogenesis is reduced in young mice while older mice (10–12 weeks) develop robust angiogenic response. Some genetic backgrounds such as C57Bl/6 are less responsive than immune deficient mice. DIVAA can be implanted in both immune deficient and immune competent mice.

The protocol consists in three parts: Preparation of Angioreactors, surgical implantation and removal and analysis of angiogenesis inside the Angioreactors.

2 Materials

1. Mice: Four mice are needed for each test group.
2. Angioreactors: Angioreactors can be created by sectioning 0.125 in. outside diameter surgical grade silicone tubing (New Age Industries, Willow Grove, PA) into 1 cm lengths, and sealing one end of the tubing with silicone adhesive (MED-2000, from NuSil Silicone Technology, Cupertino CA). Rinse angioreactors in 70% ETOH followed by rinse with distilled water and steamed sterilization. Use 2–4 Angioreactors for each mouse.
3. Matrigel (Becton Dickinson, Bedford, MA): Thaw Matrigel overnight at 4°C. For each test group use 300 μL.
4. FGF-2 (Becton Dickinson, MA) final concentration 10 ng/mL, murine VEGF-A final concentration 5 ng/mL (R&D Systems, MN), Heparin (Sigma, MO) stock 10 mg/mL in PBS, use 1 μL for each 1 mL of Matrigel.
5. PBS 1X sterile.
6. Hamilton syringe with blunt needle sterile, kept at 4°C.
7. Anesthesia use should be according to the institute's guidelines.
8. Sterilized Eppendorfs 1.2 mL and racks with lids, petri dishes, and pipette tips.
9. Surgical instruments sterile: Forceps, scissors, surgical staples and applier.
10. Heat pads or heat laps.
11. FITC-dextran 150,000 (Sigma, MO).
12. Black 96 well plates with clear bottom.
13. Fluorescent 96-well plate reader.
14. Normalized formalin (10%).

3 Methods

3.1 Angioreactors

In a tissue culture hood, dispense 1 μL of heparin in an Eppendorf tube and gently mix 1 mL of Matrigel (do not vortex to avoid air bubbles) keep it in wet ice and

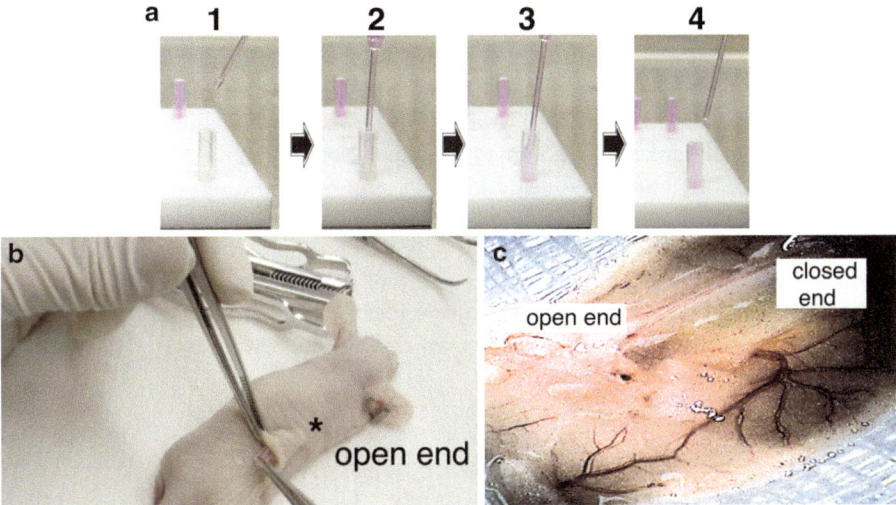

Fig. 1 Scheme of DIVAA Application. Fill Angioreactor to the top with Matrigel solutions for each test group by using a Hamilton syringe or micropipette tip (**a**). Pick it up with sterilized forceps and flip it opened-end down in a sterile marked micro centrifuge tube. Lift skin flap, make a small incision with scissors and implant angioreactor with open-end toward the mouse head (**b**). Skin flap cut around angioreactor 9 days after implantation (**c**)

label it as negative control. In another Eppendorf tube dispense first angiogenic factors (volume ≤ 15 µL) and gently mix in 300 µL of Matrigel-Heparin stock solution per each test group (label positive control). Higher volume of angiogenic factors and Matrigel-Heparin will be prepared accordingly to the number of test compounds for testing. Calculate concentrations of anti-angiogenic compound and in an Eppendorf tube dispense the test drug in PBS (use ≤ 15 µL final volume) and gently mix 300 µL of Matrigel-Heparin containing angiogenic factors, keep all the solutions on wet ice.

Set Angioreactors with the open end upside. Fill Hamilton Syringe or pipette tip with the negative Matrigel and fill the angioreactors as shown in Fig. 1a. Bring the needle tip to the bottom of the Angioreactor and gently fill it to the top. Remove Angioreactor with a forceps and put it upside down to avoid the retraction of the Matrigel in a sterile Eppendorf and label it with the group. Continue filling next group of Angioreactors with the positive control. Rinse syringe in PBS before dispensing the different test solutions in the Angioreactors. Leave the Angioreactors in closed Eppendorfs at room temperature for at least 30 min to allow gelation prior to implantation.

Note: Tumor cells can also be tested on their angiogenic potential by dispensing cells (5,000–10,000 cells/20 µL/Angioreactor) using the negative Matrigel + Heparin mixture as a vehicle.

3.2   Surgical Implantation

Anesthetize mice. In a tissue culture hood, have Angioreactors oriented in the same way in a Petri dish (open side "up"), and have another Petri dish with PBS. Lift flap of skin on dorsal flank side ~0.5 cm above hip socket and make small incision with scissors. Stick blunt scissors inside incision, and open and close to separate skin from underlying tissue. Use forceps to pick up Angioreactor, dip it in the sterile PBS to lubricate, insert it into the incision with the open end facing the head of the mouse (Fig. 1b). Place 2 Angioreactors on either side, then staple each incision shut with 1–2 staples per incision. Leave mice under heating lamp or on heating pad to aid in recovery. Return mice to cages marked with the test group.

3.3   Dissecting Angioreactors and Measurements

Retrieve mice 9–10 days after implantation. For FITC-dextran quantification, 8–10 Angioreactors are needed. For other applications, proceed dissecting Angioreactors as explained below. Prepare FITC-dextran solution 25 mg in 1 mL PBS. Inject 100 μL FITC dextran in the tail vein. After 30 min, euthanize the mice by CO_2 asphyxiation. Make incision down midline on ventral side of mouse; cut out square of skin flap around each Angioreactor (Fig. 1c). Carefully remove all skin from the Angioreactor and very carefully section tissue at the open end of the Angioreactor. Be very careful, as there is a tendency to pull tissue/matrigel out of the Angioreactor. Use a very sharp scalpel, single edge razor blade or scissors. Place the Angioreactor in an Eppendorf containing 300 μL of deionized H_2O with open end down, cut off the Angioreactor closed end, and using small pipette tip push the Matrigel plug out to mix in the water, leave Angioreactor in and label the tube. Cover tubes with foil and freeze overnight or longer (up to 3 months) at − 20°C.

For reading the amount of FITC-dextran inside Angioreactors, thaw Angioreactor mixture in Eppendorfs at room temp in the dark and vortex them. Spin down all Eppendorfs for 2 min at 8,000 RPM. Dispense supernatant (100 μL) of each reactor mix in each well of a black 96-well plate, and determine fluorescence at 485/510 nm. Express results as the relative fluorescence units (RFU), determine anti-angiogenic effects as the reduction of fluorescence in positive control Angioreactors.

3.4   Angioreactor Histological Sections

Angioreactors can alternatively be prepared for histological analysis. Dissected Angioreactors are fixed by cutting the closed-end and immerse them in fixative solution. For formalin, fix the Angioreactors for 1 h, take out the fixed tissue and leave it in formalin for an extra ½ h, transfer it to 70% ETOH. Send tissues

Negative	Positive	Tumor Cells

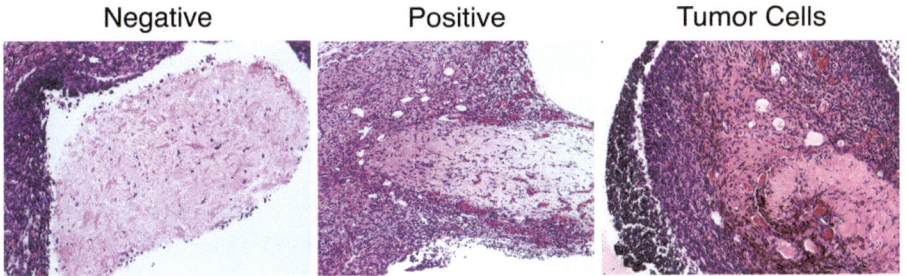

Fig. 2 Histological (H & E) staining of angioreactor Tissues. Negative Matrigel shows few invading cells. Positive Matrigel containing angiogenic factors demonstrates increased cell invasion as well as angiogenic blood vessels. Negative Matrigel mixed with human tumor cells (10,000) demonstrate higher vascularization and cellular invasion as compared with Angioreactors containing Negative and Positive Matrigel

for paraffin embedding and H & E routing staining or immunohistochemistry. Alternatively, the Angioreactors content can be placed in Histogel (Lab Storage Systems, St. Peters, MO) according to manufacturer directions and then processed for formalin fixation or embedded in OCT for frozen sectioning. Figure 2 shows H & E of Angioreactors implanted for 12 days in immunodeficient mice containing Negative Matrigel, Positive Matrigel with angiogenic factors, and Negative Matrigel containing human tumor cells.

Note: Angioreactors can be fixed in 4% paraformaldehyde (PFA) and embedding in OTC freezing medium for frozen sections.

References

1. Guedez L et al (2003) Quantitative assessment of angiogenic responses by the directed in vivo angiogenesis assay. Am J Pathol 162:1431–1439
2. Orgaz JL, Martinez-Poveda B, Fernandez-Garcia NI, Jimenez B (2008) Following up tumour angiogenesis: from the basic laboratory to the clinic. Clin Transl Oncol 10:468–477
3. Napoli C et al (2010) Directed in vivo angiogenesis assay and the study of systemic neoangiogenesis in cancer. Int J Cancer 128:1505–1508

Quantum Dots for Imaging of Angiogenesis

Ashwinkumar Bhirde, Ruijun Xing, Seulki Lee, and Xiaoyuan Chen

Abstract Quantum dots (QDs) are inorganic fluorescent nanoparticles that have found tremendous application in the field of biomedical imaging. QDs have superior photoluminescence properties over traditionally utilized fluorescent probes such as being brighter and more photostable. Their unique advantages make them attractive alternatives to conventional fluorescent dyes and show promise in tumor diagnostics and therapy. In recent years, more and more studies show that QD based nanoparticles with specificity for activated endothelial cells can be used to image ongoing angiogenesis. In this chapter, we will briefly introduce the methods of preparing targeted QDs based probes for angiogenesis imaging.

1 Introduction

QDs are fluorescent semiconductor nanoparticle probes with unique optical properties. They are advancing rapidly into daily lab practice as fluorescent tags for both *in vitro* and *in vivo* studies [1]. The prominent features of QDs include narrow emission bands, broad absorption bands, and size dependent emission that covers most visible and NIR regions [2]. In addition, QDs have high photoluminescence (brightness), high quantum yield, multi-photon absorption cross-section, and exceptional photostability in contrast to traditional dyes like fluorescein. Preparation of QDs has been well established to allow synthesis of particles in the range of 2–10 nm with accurate size control [3]. Therefore, these bright nanoparticles have been used as fluorescent labels for cells such as stem cells, cancer cells and immune cells to facilitate tracking *in vivo* [4]. Furthermore, bioconjugated QDs offer endless

A. Bhirde • R. Xing • S. Lee • X. Chen (✉)
Laboratory of Molecular Imaging and Nanomedicine (LOMIN), National Institute of Biomedical Imaging and Bioengineering (NIBIB), National Institutes of Health (NIH), Bethesda, MD 20892, USA
e-mail: Shawn.Chen@nih.gov

E. Zudaire and F. Cuttitta (eds.), *The Textbook of Angiogenesis and Lymphangiogenesis: Methods and Applications*, DOI 10.1007/978-94-007-4581-0_20, © Springer Science+Business Media Dordrecht 2012

opportunities to integrate multiple ligands and antibodies to construct multifunctional probes for biological imaging and therapeutics [5] in addition to being used as multi-modality imaging probes when combined with magnetic or radioactive materials [6].

Angiogenesis, the growth of new blood vessels from the preexisting vasculature, is essential for normal physiological processes and also plays an important role in tumor growth and progression [7–9]. As previously reported, when a tumor is growing, simple diffusion of oxygen and nutrients from preexisting host microvas-culature becomes insufficient, and therefore leads to generation of a tumor blood supply. This kind of tumor angiogenesis is the main signal for the transition of tumors from a dormant state to a malignant state. Meanwhile, such tumor induced angiogenesis also activate the transcription of certain proangiogenic factors [10] to form new vessels such as vascular endothelial growth factor A (VEGF-A), matrix metalloproteases (MMPs), extracellular matrix (ECM), and integrin $\alpha v \beta 3$.

QDs are mostly made up of binary alloys such as CdTe/CdSe, CdSe/ZnS, InAsxP1-x/InP/ZnSe, CuInSe, or InP/ZnSe [11]. The composition change, in con-junction with size tuning, can yield QDs with emission in the near-infrared (NIR) region (700–900 nm), a spectrum window that has minimal impact from biological material autofluorescence and is suitable for *in vivo* optical imaging. Previously, non-targeted QDs have shown applicability in sentinel lymph node mapping, vasculature imaging, neural imaging and also stem cell tracking [12]. To exploit characteristics of these unique nanocrystals for tumor vasculature-targeted imag-ing, there is a need for them to be effectively, site-specifically and reliably directed to the tumor vasculature [13]. Such target specificity can be achieved by conjug-ating targeting moieties to the surface of QDs [14]. Growth factors, peptides, and other small molecules like peptidomimetics are suitable targeting ligands. A relatively large number of these molecules can be tethered to the surface of a single QD through simple conjugation chemistry. Loading of multiple copies of functional ligands can induce higher receptor binding affinity and possibly, a more desirable targeting efficacy in comparison to an individual ligand due in part to the so called polyvalency effect [15]. Several targeted molecules such as cyclic (arginine-glycine-aspartic acid) (cRGD)/cyclic (asparagine-glycine-arginine) (cNGR)peptide, and VEGF have been widely explored in targeted angiogenesis imaging in the past few years [16].

2 Modification Strategies

Typically, QDs are synthesized in colloidal solutions. A QD usually consists of an inorganic core and an additional organic ligands [17]. Therefore, the main step for biological use is to transfer the hydrophobic QDs into water. Most recently, with the rapid development of specific targeting strategies for inorganic nanoparticles, attaching targeting molecules onto the surface of the QDs has become a new trend for in vivo imaging. Goepferich et al. previously reported that there are several possible methods for the conjugation of biomolecules onto the surface of QDs, especially for targeting agents [18]. Figure 1 exemplifies various biocon-jugation techniques used to develop bioactive QDs, like:

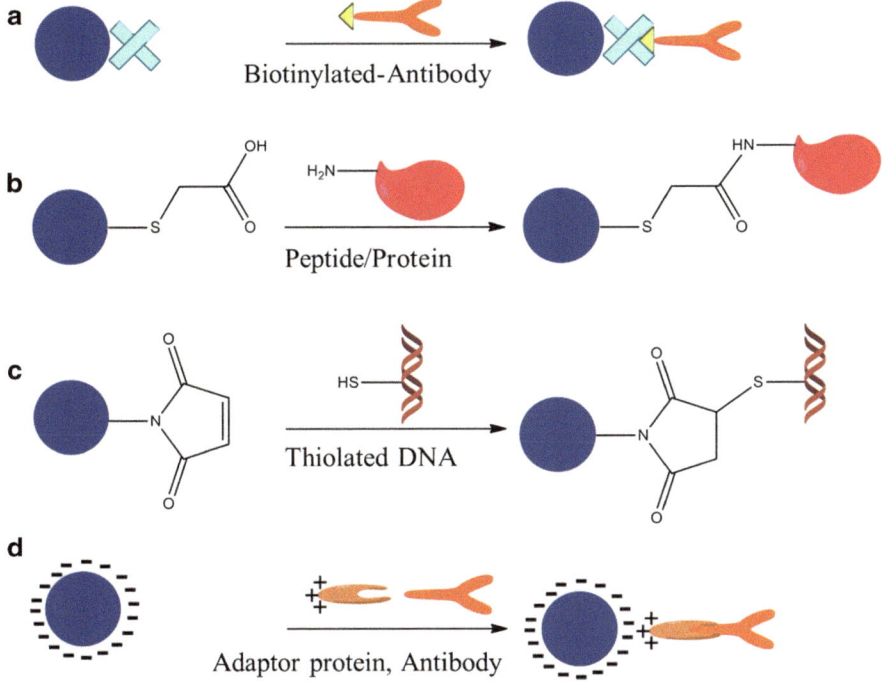

Fig. 1 Examples of various bioconjugation strategies of differently functionalized QDs [18]

1. QDs are easily labeled with streptavidin and can be linked to biotin-tagged biomolecules (Fig. 1a) [19–21].
2. QDs can be coated with mercapto acid via the mercapto group, while the carboxylic acid group can form a stable amide bond with the amine of various biomolecules in the presence of coupling reagents, carbodiimides and N-hydroxysuccinimides (NHS) (Fig. 1b) [22, 23].
3. Maleimide groups of the QDs and the thiol groups of the biomolecules can form stable, covalent bonds (Fig. 1c) [24–26].
4. Electrostatic interactions between the QD-surface and peptides or proteins are also efficient conjugation techniques (Fig. 1d) [27, 28].

3 Targeting Strategies

3.1 RGD Targeted

$\alpha_v\beta_3$-integrin, which is a well-known marker of angiogenic blood vessels, is strongly expressed on activated tumor endothelial cells. However it is only weakly expressed on the endothelial cells of normal blood vessels [29, 30]. Because integrins are an ideal molecule to target for tumor cell imaging, various studies have focused on methods of modifying QDs to enhance their targetability to integrins and their biocompatibility. Previous literature has indicated that RGD peptides are capable of binding with the biomarker, $\alpha_v\beta_3$. Chen group first reported the development of RGD peptide-conjugated QDs for tumor vasculature-targeted imaging [15]. The following are the materials and methods to develop RGD-QD conjugates.

3.1.1 Materials

QDs (Qdot 705 ITK amino (PEG) QDs; Invitrogen)
1,2-Distearoyl-sn-glycero-3-phosphoethanolamine-N-[amino (polyethylene glycol)-2000] (ammonium salt) (DSPE-PEG2000 amine; Avanti Polar Lipids)
Succinimidyl acetylthioacetate (SATA; Sigma)
4-Maleimidobutyric acid N-succinimidyl ester (Sigma)
10 mM sodium borate buffer, pH 8.5 (Sigma)
PBS buffer (Invitrogen)
RGD peptide (c(RGDyK); Peptides International)
N-Succinimidyl S-acetylthioacetate (Pierce Biotechnology Inc.)
Dimethyl sulfoxide (DMSO; ACROS)
High-performance liquid chromatography (HPLC)-grade acetonitrile (Fisher Scientific)
Trifluoroacetic acid (TFA; Fisher Scientific)
Hydroxylamine hydrochloride (Sigma)
Tris (2-carboxyethyl) phosphine hydrochloride (TCEP.HCl; Pierce Biotechnology Inc.)
Sodium hydroxide (Sigma)
Hydrochloric acid (Sigma)
NAP-10 column (GE Healthcare)
U87MG human glioblastoma cells (ATCC)
MCF-7 human breast cancer cells (ATCC)
Low-glucose DMEM (GIBCO)
Minimum essential medium (GIBCO)
Fetal bovine serum (FBS; GIBCO)

3.1.2 Methods

Preparation of QD705-PEG

1. Mix the as-synthesized QD705 (10 nmol) and DSPE-PEG2000 amine (~1.8 μmol) in 200 μL of chloroform. Leave the container open in a fume hood to evaporate the chloroform solvent slowly at room temperature. Pump the dry sample under vacuum for about 4 h to remove chloroform completely and then redisperse the particles in water by gentle sonication.
2. Purify the as-prepared QD705-PEG by centrifugation using Millipore centrifugal filter devices at 4,000 rpm for 20 min to remove any excess amount of DSPE-PEG2000 amine. The final concentration of QD705-PEG in PBS buffer (1×) would be about 10 μM.

Preparation of Thiolated RGD Peptide

1. Mix c(RGDyK) (5 μmol) in 1 mL of borate buffer (pH 8.5) with 100 μL of DMSO solution containing 6 μmol of SATA.
2. Observe the reaction by analytical RP-HPLC (reversed-phase high performance liquid chromatography). In a typical setup, the mobile phase is changed from 95% solvent A (0.1% TFA in water) and 5% solvent B (0.1% TFA in acetonitrile) (0–2 min) to 35% solvent A and 65% solvent B at 32 min. Adjust the flow rate to 1 mL min^{-1}. UV absorbance should be observed at 218 nm. Typically, HPLC retention times (RT) for c(RGDyK) and SATA-c(RGDyK) are 7.9 and 12.1 min, respectively.
3. When the reaction has gone to completion, quench with 100 μL of 2% TFA in water.
4. Freeze-dry the crude mixture and resuspend in 1 mL of water.
5. Add 100 μL of 0.5 M hydroxylamine solution. Adjust the pH to 6.0 (pipette 1–5 μL of the reaction mixture and apply it to pH paper) with 0.5 M sodium hydroxide or hydrochloric acid solution, if needed.
6. When the reaction has gone to completion based on analytical RP-HPLC, purify the RGD-SH by semi-preparative RP-HPLC. The gradient is the same as described above, but instead a semi-preparative HPLC column and a flow rate of 5 mL min^{-1} should be used. The RT for RGD-SH is 10.7 min ($C_{29}H_{43}N_9O_9S$, calculated 693.3, observed 694.5 ([M + H]$^+$). Collect the fractions containing pure RGD-SH and freez-dry).
7. Store RGD-SH under acidic condition (pH 3–4) in water to prevent disulfide formation (Fig. 2).

Synthesis of QD705-RGD and QD705-RAD Bioconjugates

1. Mix the two reagents: QD705-PEG (1 nmol, 125 mL) and 4-maleimidobutyric acid N-succinimidyl ester (1 μmol in 200 μL of borate buffer). Pipette 1–5 μL of the reaction mixture and apply it to a pH paper to make sure that the pH is about 8.5.
2. Incubate the mixture in a 1.5 mL Eppendorf tube at room temperature (20°C) for 1 h with gentle shaking.
3. For purification, equilibrate an NAP-10 column with PBS.
4. Load the reaction mixture onto the NAP-10 column and wait until all the mixture is in the column. Add 2 mL of PBS and collect only the deepest-colored fractions (usually 200–400 μL). A hand-held UV lamp can help visualize the process.
5. Now add 1 μmol of RGD-SH (thiolated RGD peptide, (c(RGDy(ε-acetylthiol) K))) dissolved in minimum volume of PBS to the activated QD705-PEG solution, and adjust the pH to 7.0–7.5 (based on pH paper) with borate buffer if necessary.
6. Again incubate the mixture in a 1.5 mL Eppendorf tube at room temperature for 2 h with gentle shaking.
7. Equilibrate one more NAP-10 column with PBS for purification.

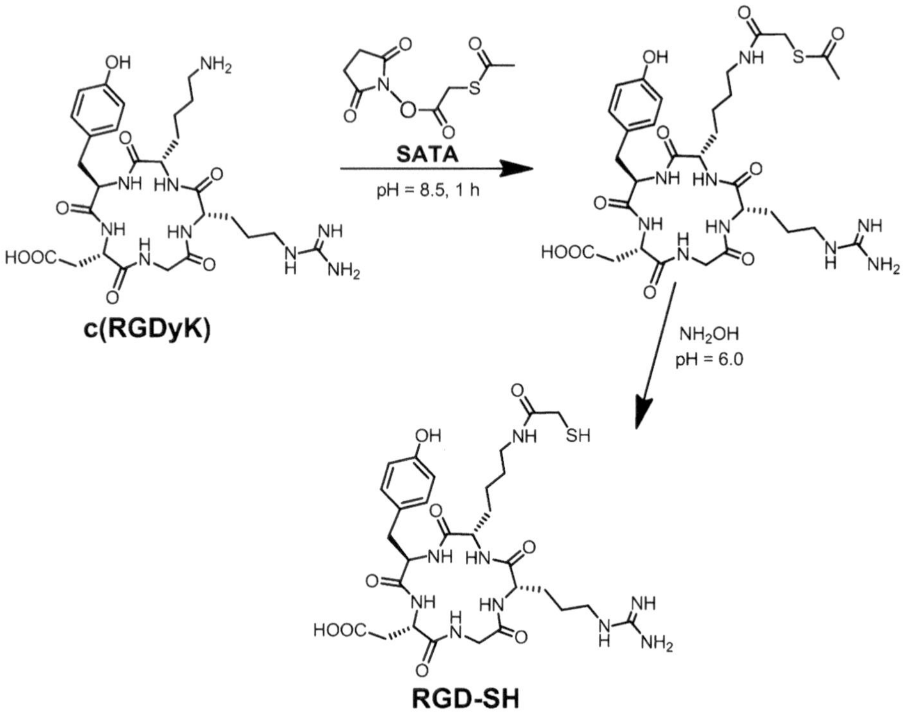

Fig. 2 Converting the amino group in the RGD peptide into a thiol [15]

8. Similar to step 4 described above, load the reaction mixture onto the NAP-10 column and wait until all the mixture is in the column. Add 2 mL of PBS and collect only the deepest-colored fractions (usually 300–500 μL).
9. Store the collected QD705-RGD conjugate in PBS buffer with a concentration of 1 μM fractions at 4°C for later use (Fig. 3).

3.1.3 U87MG Tumor Model

Animals can be used for *in vivo* imaging studies when the U87MG tumor size reaches about 500 mm^3. Typically, 5×10^6 U87MG cells are injected subcutaneously into athymic nude mice. It takes 3–4 weeks for the tumor to reach this size.

3.1.4 *In Vivo* Tumor Imaging with IVIS Imaging System

1. Check and make sure that the U87MG tumor size is about or smaller than 500 mm^3.
2. Set up the IVIS imaging system.

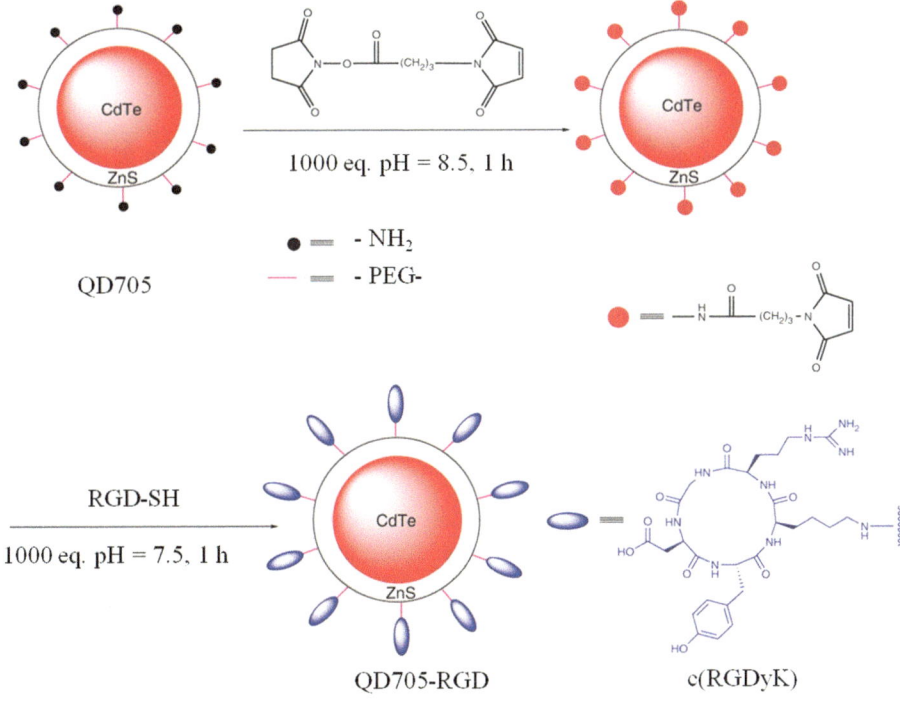

Fig. 3 Synthesis of QD705-RGD [15]

3. Anesthetize the animals using the rodent anesthesia system with isoflurane (2% vol/vol, isoflurane in 0.2 L min^{-1} of O_2 flow).
4. Inject about 200 pmol of QD705-RGD solution in each mouse via tail vein (total volume, 100–200 μL). We recommend injecting some PBS (about 50 μL) afterwards to flush the tail vein.
5. Scan the animal using the IVIS imaging system at serial time points post-injection. We typically choose 10 min, 1 h, 4 h, 12 h and 24 h. Image acquisition time ranges from a few seconds to a few minutes per scan. There is minimal image processing, except adjusting the minimum and maximum value of the color scale using the IVIS system.
6. If needed, harvest the tumor and major organs and image them again using the IVIS imaging system (Fig. 4). *Ex vivo* tissue staining and tissue homogenate fluorescence can also be carried out to validate *in vivo* imaging results.

3.2 VEGF Targeted

Tumor growth depends on angiogenesis. One angiogenesis-related signal pathway is the vascular endothelial growth factor (VEGF)/VEGF receptor (VEGFR) signaling

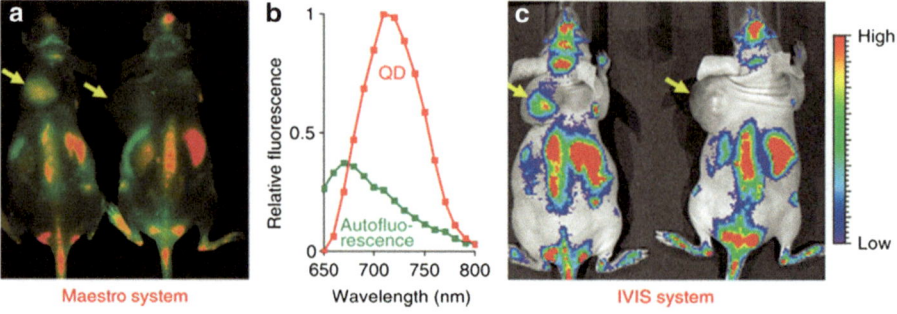

Fig. 4 *In vivo* imaging of U87MG tumor-bearing mice (*yellow arrows*) at 6 h post-injection of 200 pmol of QD705-RGD (*left*) or QD705 (*right*). (**a**) For the Maestro system, the mice autofluorescence is color-coded *green* and the unmixed QD signal is color-coded *red*. (**b**) The pure autofluorescence and QD spectra used for spectral unmixing. (**c**) Image of the same mouse acquired with the IVIS system immediately after being acquired with the Maestro system shown in (**a**) [15]

pathway, which contributes to the formation of new blood vessels [31, 32]. Recently, more studies show that VEGF based targeted agents can be used for tumor imaging and cancer therapeutic targets [33–36]. Previously, the Chen group [37] reported amine-functionalized QD with VEGF protein and macrocyclic chelating agent 1,4,7,10-tetraazacyclododecane-1,4,7,10-tetraacetic acid (DOTA) for VEGFR recognition and ^{64}Cu (t1/2 = 12.7 h) labeling for PET imaging, respectively.

3.2.1 Materials

Tissue culture reagents (Invitrogen)
Rat ant-imouse CD31 primary antibody (BD Bioscience) Cy3- or FITC conjugated goat anti-rat secondary antibody (Jackson ImmunoResearch Laboratories)
Rat anti-mouse VEGFR-2 primary antibody (University of Texas Southwestern Medical Center)
^{64}Cu (the University of Wisconsin-Madison)
All other chemicals were obtained from either Sigma or Fisher Scientific.

3.2.2 Methods

Preparation of DOTA-QD-VEGF-

1. Activate DOTA with EDC and NHS at pH 5.5 for 30 min with a molar ratio of DOTA/EDC/SNHS = 10:5:4.
2. Add the activated DOTA along with a heterobifunctional linker, NHS-MAL, into the QD solution at a pH of 8.5 with 500:500:1 reaction ratio.

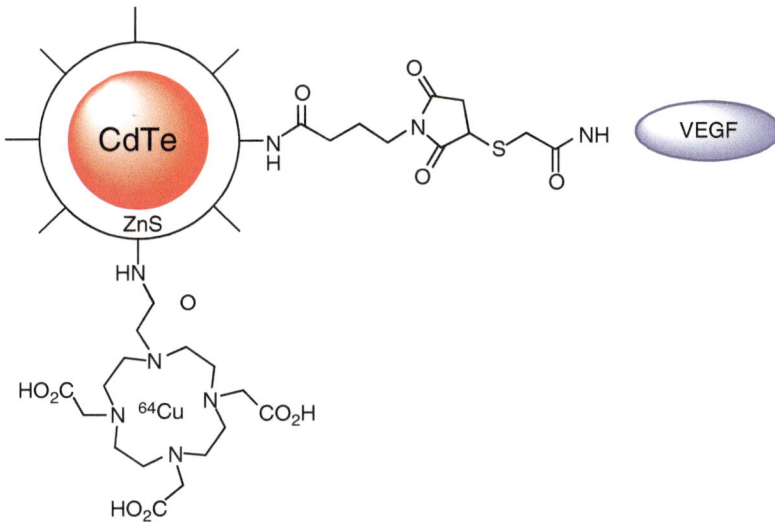

Fig. 5 The Structure of DOTA-QD-VEGF Conjugate. DOTA can chelate ^{64}Cu PET imaging [37]

3. Synthesize thiolated VEGF *in situ* with SATA followed by deprotection with hydroxylamine.
4. Conjugate thiolated VEGF to DOTA–QD–MAL at pH 7.0 with a ratio 10:1. The mixture is incubated for 4 h, and the unreacted materials removed through dialysis (MWCO = 100 kDa). DOTA-QD was also synthesized as a control (the reaction ratio of DOTA/QD was 1,000:1).
5. DOTA–QD–VEGF or DOTA–QD is radiolabeled by the addition of ^{64}Cu in 0.1 N sodium acetate (pH = 6.5) buffer, followed by incubation at 40°C for 45 min.
6. Purify the resulting mixture using a PD-10 column with phosphate-buffered saline (PBS) as the mobile phase (Fig. 5).

3.2.3 U87MG Tumor Model

The U87MG tumor model is established by subcutaneous injection of U87MG cells (5×10^6 in 50 μl of PBS) into the front flank of female athymic nude mice (Harlan). The mice are subjected to imaging when the tumor reaches 200–500 mm^3 (3–4 weeks after inoculation).

3.2.4 *In Vivo* NIRF Imaging

The mice are imaged at multiple time points post injection (p.i.) using the Maestro *in vivo* imaging system (CRI, Woburn, MA, USA; excitation = 575–605 nm, emission = 645 nm long pass) (Fig. 6).

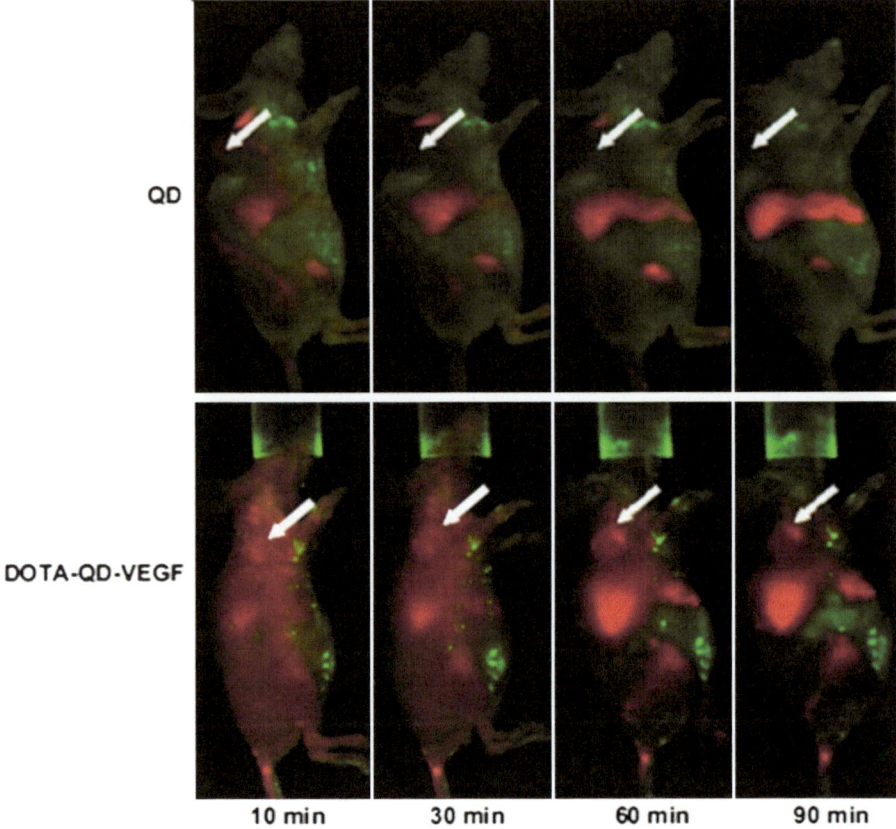

Fig. 6 *In vivo* NIRF imaging of U87MG tumor-bearing mice at 10, 30, 60 and 90 min p.i. of 200 pmol of DOTA-QD-VEGF AND DOTA-QD, respectively (*arrows* indicate the tumor) [37]

References

1. Michalet X, Pinaud FF, Bentolila LA et al (2005) Quantum dots for live cells, in vivo imaging, and diagnostics. Science 307:538–544
2. Chen X, Li ZB, Cai W (2007) Semiconductor quantum dots for in vivo imaging. J Nanosci Nanotechnol 7:2567–2581
3. Su XG, Ma QA (2010) Near-infrared quantum dots: synthesis, functionalization and analytical applications. Analyst 135:1867–1877
4. Nie SM, Gao XH, Yang LL, Petros JA, Marshal FF, Simons JW (2005) In vivo molecular and cellular imaging with quantum dots. Curr Opin Biotechnol 16:63–72
5. Medintz IL, Uyeda HT, Goldman ER, Mattoussi H (2005) Quantum dot bioconjugates for imaging, labelling and sensing. Nat Mater 4:435–446
6. Biju V, Mundayoor S, Omkumar RV, Anas A, Ishikawa M (2010) Bioconjugated quantum dots for cancer research: present status, prospects and remaining issues. Biotechnol Adv 28:199–213
7. Folkman J (1971) Tumor angiogenesis: therapeutic implications. N Engl J Med 285:1182–1186
8. Folkman J (1995) Angiogenesis in cancer, vascular, rheumatoid and other disease. Nat Med 1:27–31

9. Carmeliet P, Jain RK (2000) Angiogenesis in cancer and other diseases. Nature 407:249–257
10. Harris AL (2002) Hypoxia–a key regulatory factor in tumour growth. Nat Rev Cancer 2:38–47
11. Ghasemi Y, Peymani P, Afifi S (2009) Quantum dot: magic nanoparticle for imaging, detection and targeting. Acta Biomed 80:156–165
12. Gao JH, Chen XY, Cheng Z (2010) Near-infrared quantum dots as optical probes for tumor imaging. Curr Top Med Chem 10:1147–1157
13. Medintz IL, Delehanty JB, Mattoussi H (2009) Delivering quantum dots into cells: strategies, progress and remaining issues. Anal Bioanal Chem 393:1091–1105
14. Douroumis D, Obonyo O, Fisher E, Edwards M (2010) Quantum dots synthesis and biological applications as imaging and drug delivery systems. Crit Rev Biotechnol 30:283–301
15. Chen XY, Cai WB (2008) Preparation of peptide-conjugated quantum dots for tumor vasculature-targeted imaging. Nat Protoc 3:89–96
16. Liu ZA, Peng R (2010) Inorganic nanomaterials for tumor angiogenesis imaging. Eur J Nucl Med Mol Imaging 37:S147–S163
17. Zhang Y, Li JB, Wu DD, Miao ZR (2010) Preparation of quantum dot bioconjugates and their applications in bio-imaging. Curr Pharm Biotechnol 11:662–671
18. Goepferich A, Hild WA, Breunig M (2008) Quantum dots – nano-sized probes for the exploration of cellular and intracellular targeting. Eur J Pharm Biopharm 68:153–168
19. Lidke DS, Nagy P, Heintzmann R et al (2004) Quantum dot ligands provide new insights into erbB/HER receptor-mediated signal transduction. Nat Biotechnol 22:198–203
20. Dahan M, Levi S, Luccardini C, Rostaing P, Riveau B, Triller A (2003) Diffusion dynamics of glycine receptors revealed by single-quantum dot tracking. Science 302:442–445
21. Simon SM, Jaiswal JK, Mattoussi H, Mauro JM (2003) Long-term multiple color imaging of live cells using quantum dot bioconjugates. Nat Biotechnol 21:47–51
22. Nie SM, Chan WCW (1998) Quantum dot bioconjugates for ultrasensitive nonisotopic detection. Science 281:2016–2018
23. Ma H, Zhang CY, Nie SM, Ding Y, Jin L, Chen DY (2000) Quantum dot-labeled trichosanthin. Analyst 125:1029–1031
24. Dubertret B, Skourides P, Norris DJ, Noireaux V, Brivanlou AH, Libchaber A (2002) In vivo imaging of quantum dots encapsulated in phospholipid micelles. Science 298:1759–1762
25. Wu XY, Liu HJ, Liu JQ et al (2003) Immunofluorescent labeling of cancer marker Her2 and other cellular targets with semiconductor quantum dots. Nat Biotechnol 21:41–46
26. Srinivasan C, Lee J, Papadimitrakopoulos F, Silbart LK, Zhao M, Burgess DJ (2006) Labeling and intracellular tracking of functionally active plasmid DNA with semiconductor quantum dots. Mol Ther 14:192–201
27. Hanaki K, Momo A, Oku T et al (2003) Semiconductor quantum dot/albumin complex is a long-life and highly photostable endosome marker. Biochem Biophys Res Commun 302:496–501
28. Goldman ER, Anderson GP, Tran PT, Mattoussi H, Charles PT, Mauro JM (2002) Conjugation of luminescent quantum dots with antibodies using an engineered adaptor protein to provide new reagents for fluoroimmunoassays. Anal Chem 74:841–847
29. Griffioen AW, Molema G (2000) Angiogenesis: potentials for pharmacologic intervention in the treatment of cancer, cardiovascular diseases, and chronic inflammation. Pharmacol Rev 52:237–268
30. Sipkins DA, Cheresh DA, Kazemi MR, Nevin LM, Bednarski MD, Li KC (1998) Detection of tumor angiogenesis in vivo by alphaVbeta3-targeted magnetic resonance imaging. Nat Med 4:623–626
31. Chen X, Cai W (2007) Multimodality imaging of vascular endothelial growth factor and vascular endothelial growth factor receptor expression. Front Biosci 12:4267–4279
32. Ferrara N (2002) VEGF and the quest for tumour angiogenesis factors. Nat Rev Cancer 2:795–803
33. Chen X, Wang H, Cai WB et al (2007) A new PET tracer specific for vascular endothelial growth factor receptor 2. Eur J Nucl Med Mol Imaging 34:2001–2010
34. Chen XY, Cai WB, Chen K et al (2006) PET of vascular endothelial growth factor receptor expression. J Nucl Med 47:2048–2056

35. Kim KJ, Li B, Winer J et al (1993) Inhibition of vascular endothelial growth factor-induced angiogenesis suppresses tumour growth in vivo. Nature 362:841–844
36. Ferrara N, Hillan KJ, Novotny W (2005) Bevacizumab (Avastin), a humanized anti-VEGF monoclonal antibody for cancer therapy. Biochem Biophys Res Commun 333:328–335
37. Chen X, Chen K, Li ZB, Wang H, Cai WB (2008) Dual-modality optical and positron emission tomography imaging of vascular endothelial growth factor receptor on tumor vasculature using quantum dots. Eur J Nucl Med Mol Imaging 35:2235–2244

Intravital Imaging of Tumor-Initiated Angiogenesis Using a Dorsal Skin Chamber

Panomwat Amornphimoltham and Roberto Weigert

Abstract Intravital microscopy (IVM) is a powerful tool that has enabled imaging tumor progression and the modifications occurring within the surrounding tissue microenvironment in live animals. Here we describe the installation of an optical window in immunocompromised mice that allows performing longitudinal studies with minimal tissue interference. This approach provides detailed information on the dynamics of the interactions between blood vessels and implanted tumor cells.

Keywords Intravital microscopy • Tumor • Angiogenesis • Dorsal skin chamber

1 Introduction

In order to sustain its proliferative potential, a solid tumor utilizes several strategies to remodel the surrounding tissue microenvironment. Among them is the initiation of new blood vessels from pre-existing ones, referred as tumor-initiated angiogenesis, which is considered as one of the cancer hallmarks [6]. Since tumors require a continuous supply of nutrients and oxygen through the vascular system, the disruption of tumor-initiated vessels has become a primary strategy guiding the development of anti-cancer treatments. Although several anti-angiogenic factors have been already used in clinical trials either as alternative therapy or in parallel with conventional chemo or radiotherapy, the results so far have not been very encouraging [3, 4, 11]. This reflects the fact that the nature of the interactions between the tumor and its surroundings are more complex than anticipated, suggesting that a huge effort has to be devoted toward a better understanding of the molecular

P. Amornphimoltham (✉) • R. Weigert
Intracellular Membrane Trafficking Unit, Oral and Pharyngeal Cancer Branch, National Institute of Dental and Craniofacial Research, National Institutes of Health, 30 Convent Dr. 303A, Bethesda, MD 20892-4340, USA
e-mail: pamornph@nidcr.nih.gov; weigertr@mail.nih.gov

E. Zudaire and F. Cuttitta (eds.), *The Textbook of Angiogenesis and Lymphangiogenesis: Methods and Applications*, DOI 10.1007/978-94-007-4581-0_21, © Springer Science+Business Media Dordrecht 2012

machineries regulating the formation of tumor-initiated blood vessels. To this aim, a very powerful approach comes from the combination of IVM, which enables imaging biological processes at a cellular and subcellular level [2], with the development of implanted optical windows, which grants constant access to the tumor and its microenvironment [1, 2]. Among the optical windows, the dorsal skin chamber (DSC) has been very successful in providing information on the dynamics of blood vessels in rodent tumor xenograft [8], and in assessing the efficacy of potential therapeutic molecules over long periods of time [7]. Earlier versions of the DSC were built in metal, creating an excessive burden on the mice and reducing the time of observation. Here we describe the use of a biocompatible resin-based DSC [10], the procedures utilized to label both tumor cells and blood vessels, and some of the strategies developed to perform long term imaging at different levels of resolution.

2   Materials

2.1   Animals and Housing

1. Athymic nude (*nu/nu*) or SCID mouse, 25–30 g bodyweight (Harlan Laboratories, Frederick, MD).
2. Soft diet for rodent (DietGel® Recovery, Clear H_2O).
3. Sulfadimethoxine 5% oral suspension (Albon; Allivet, Hialeah, FL).
4. All animal studies are carried out according to NIH-approved protocols, in compliance with the Guide for the Care and Use of Laboratory Animals.
5. Due to the fact that the same animal is imaged at least once a day for a period of 10 weeks, all the mice are housed in the satellite animal facility close to the imaging area.
6. Animals are housed individually in ventilated micro-isolator cages and provided with food and water *ad libitum*.

2.2   Surgical Tools and Anesthetics

1. Surgical tools: microspring scissors, fine tip forceps, surgical silk (4–0) with needle, needle holders, stainless steel ball chain with hook and two stainless steel elevating posts (FST, Foster city, CA), fine needle gauge 30 and 1 cm^3. syringe, bulldog serrifine clamps (FST, Foster city, CA)
2. Anestethics: Isofluorane (Forane, 100 ml, Baxter, Deerfield, IL), Ketamine (Ketaved, 100 mg/ml, Fort Dodge Animal Health, Fort Dodge, IA), and Xylazine (Anased, 100 mg/ml Akorn, Decatur, IL) (**see Note 1**)
3. Vaporizer for isoflurane (Isolfuorane V 1.9) connected to a plexiglass restraining tube (Braintree Scientific, Braintree, MA)
4. Electric shaver for animal

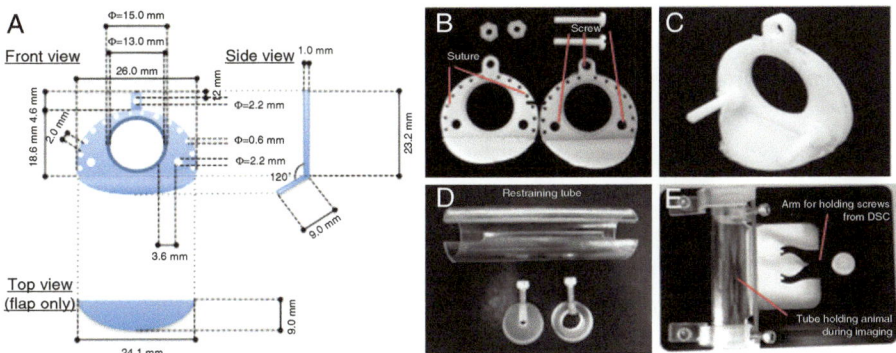

Fig. 1 Mouse dorsal skin chamber and the restraining devices. (a) Schematic drawing of the design and dimension of the DSC. (b) Resin-based DSC and all the assembly parts including resin screws and nuts. (c) The DSC assembled by two complementary frames held together by screws and nuts. (d) Optical clear restraining tube with the closing ring for both tube ends. (e) Restraining tube positioned in the microscope stage. The animal with the DSC is maintained in place with the forked arm by a series of screws

5. Antibiotic solution (Tobramycin 40 mg/ml)
6. Triple ophthalmic antibiotic ointment, (Bausch & Lomb, Tampa, FL)
7. Betadine swab sticks
8. Alcohol swabs
9. Sterile towel drapes
10. Absorption spears (FST, Foster city, CA)
11. Veterinary bonding cement (EMT gel; Trophy, Kansas city, MO)
12. Normal saline and Phosphate buffer saline (PBS)
13. Heat pads

2.3   Dorsal Skin Chamber and Microscope Holder

1. Custom made DSC; polyacetal resin-based (Acetron GP or Duracon®) chambers (weight 1.5 g) (Fig. 1a–c)
2. Acetal pan head screws 0.5 in. with nuts (Small Parts, Inc.) (Fig. 1b, c)
3. Custom made clear acrylic holding tube and mounted stage (Fig. 1d, e)
4. Sterile circular glass microcoverslip 15 mm diameter #1 (EMS, Hatfield, PA)

2.4   Fluorescent Probes and Human Cell Lines Expressing Fluorescently Labeled Proteins

1. Fluorescein isothiocyanate (FITC) or Tetramethyl Rhodamine Isothiocyanate (TRITC)-dextran MW 2000 KD (Invitrogen, Carlsbad, CA)

2. For stable expression of fluorescent proteins in tumor cell lines, we use a lentiviral gene delivery system for expressing fluorescent-tagged protein (i.e. venus, GFP, mCherry) into the tumor cells prior to injection (**see Notes 7 and 8**)

2.5 Microscope

1. IX81 inverted confocal microscope, equipped with a Fluoview-1000 scanning unit (Olympus America, Center Valley, PA) modified for multiphoton microscopy, as described in [9].
2. PlanAPO N × 2 numerical aperture (NA) 0.08 objective (Olympus).
3. XLUMPlanFL 20x NA 0.95 water immersion objective (Olympus).
4. UPlanSAPO 60x NA 1.20 water immersion objective (Olympus).

3 Methods

3.1 Mouse Anesthesia

1. Prepare the anesthetic by mixing 50 μl of xylazine and 50 μl ketamine in 400 μl sterile normal saline. The final working concentration is 10 mg/kg body weight for xylazine and 100 mg/kg body weight for ketamine (**see Note 1**).
2. Weigh the mouse and administer the anesthetic cocktail subcutaneously using a 30 gauge, 1 cm^3 syringe.
3. Rest the mouse on a heated pad until fully anesthetized (5–7 min).
4. Apply the ophthalmic ointment to prevent corneal drying.

3.2 Dorsal Skin Chamber (DSC) Installation

1. If needed, shave the mouse in the mid-scapular region with electrical clippers.
2. Scrub the area where the DSC is going to be installed three times with betadine antiseptic alternated with 70% ethanol swabs.
3. The animal is placed in a prone position (ventral recumbency) and the mid dorsal skin is drawn up into a longitudinal fold using non-serrated forceps. The skin is hold in position by hook chains with posts at both the skin ends. The sterilized frames are aligned with the skin fold sandwiched in between (Fig. 2a). The frames are temporarily held in position with small serrifine clamps. The two lower sterilized screws and nuts are positioned via small incisions made by a sharp sterilized needle. The upper portion of both frames is secured by sutures (sterile 4–0 silk) (**see Note 2**).

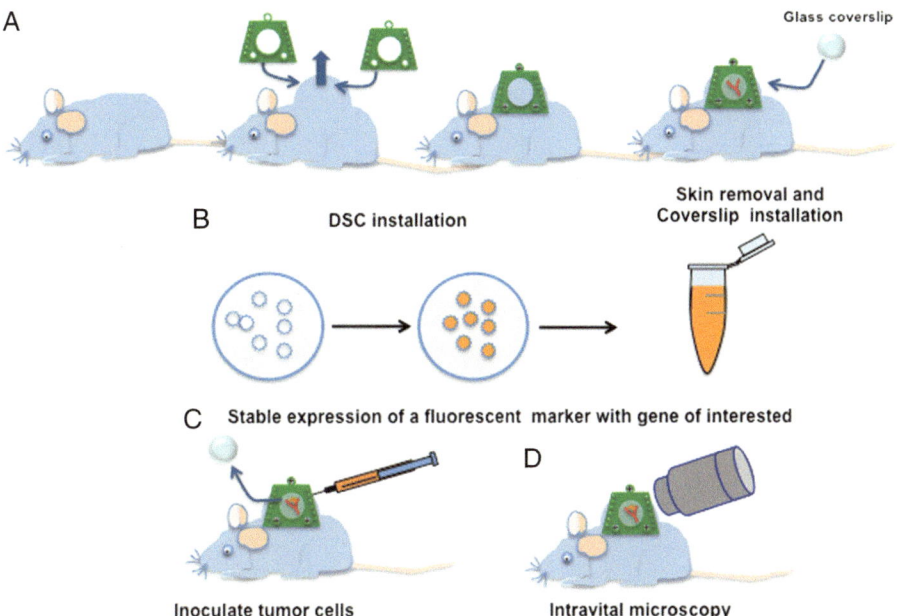

Fig. 2 Installation of the dorsal skin chamber. (**a**) DSC assembly step by step. The DSC is installed by sandwiching the dorsal skin fold between the two frames. The DSC is held in place by screws and sutures. One side of the dorsal skin is removed to reveal the underlying connective tissue. The exposed surface is covered by a round glass coverslip. (**b**) Tumor cell lines are genetically engineered to stably express fluorescent proteins by using lentiviral-mediated gene delivery *in vitro*. (**c**) A few days after the DSC installation, the coverslip is removed and fluorescent tumor cells are injected underneath the connective tissue. (**d**) Once the tumor grows and neo-angiogenesis occurs, the animal is subjected to intravital microscopy

4. The circular observation window is prepared by carefully removing one side of the skin using microsurgical scissors and fine tip forceps. Care must be taken to avoid damaging any major blood vessels. The epidermis is completely removed, exposing the connective tissue and maintaining the fascia intact (**see Note 3**) (Fig. 2a).

5. Antibiotic ointment is applied to the margins of the open skin and the sites where the screws and sutures are placed. Drops of antibiotic solution (40 mg/ml Tobramycin) are applied to the exposed tissue area.

6. The sterile circular glass coverslip is positioned on the exposed connective tissue and fascia and peripherally sealed with veterinary bonding cement (Fig. 2a).

7. Following surgery, the animal is allowed to recover from anesthesia on a heated pad (**see Note 4**), returned to a clean cage and observed for signs of distress or pain (**see Note 5**).

8. After recovering from anesthesia, mice are housed individually in rat cages (**see Note 6**) and antibiotic is administered through the drinking water [Albon, 5% Oral suspension (1:200)]. DietGel is also provided for rapid recovery. The animals are constantly monitored for signs of infection, inflammation, or bleeding.

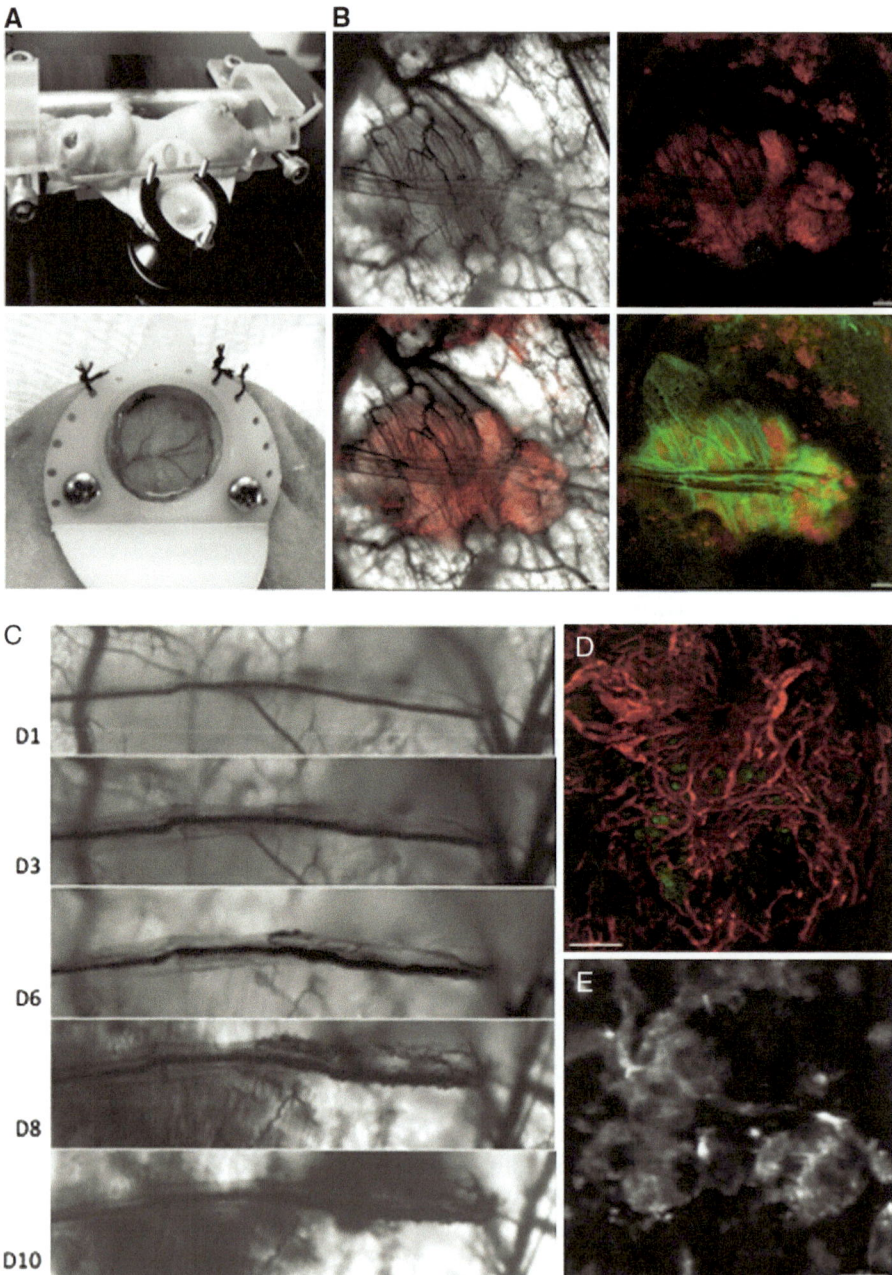

Fig. 3 Imaging tumor angiogenesis in live animal. (a) DSC is shown installed on the nude (*nu/nu*) mouse. The anesthetic mouse was firmly mounted to the holding stage for intravital imaging (in upper panel). Typical view after the DSC installation shown in lower panel. Note the clear, non-bleeding or contamination in the window area. (b) Colon cancer cells (DLD1-tomato in *red*) were inoculated after 7 days. The overall view of tumor mass and blood vessels observed at a low

3.3 Injection of Tumor Cells into DSC

After 2–3 days from the DSC installation, if there is no complication, the animals are ready for the injection of the tumor cells (see Note 7).

1. Grow and maintain tumor cells in growth media in the cell incubator (70–90% confluence).
2. Trypsinize and collect tumor cells, suspend them in media ($2.5–10 \times 10^7$ cells/ml) and centrifuge them ($200 \times g$ for 5 min) (Fig. 2b).
3. For a single injection, resuspend 0.5–2 million cells in 20 µl in serum-free, sterilized cell culture media.
4. The animal is anesthetized using isoflurane and placed in the custom-made holder (Fig. 3) under sterile conditions. The coverslip is removed and the tissue is irrigated with antibiotic in saline solution (40 mg/ml Tobramycin). The excess of fluid is gently blotted with sterilized spear-tip sponges.
5. Cell suspension is injected underneath the connective tissue between the main vessels using a 1 cm^3 syringe equipped with a 30 gauge needle. Antibiotic ointment is applied onto the tissue and a new sterile coverslip is placed and secured (Fig. 2c).

3.4 Mouse Tail Vein Injection

1. Dilate the tail vein by using a heat lamp or immerse the tail for 1–2 min in warm water.
2. Place the mouse into the restraining tube.
3. Clean the tail with 70% ethanol and look for the lateral vein. Stabilize the vein by holding the tail between the thumb and the middle finger while securing the vein with the index.
4. Prepare a 1 ml syringe with a 28-gauge needle containing 100 µl of 20 mg/ml TRITC-dextran or FITC-dextran. Inject the mouse by orienting the needle bevel upward and advancing it gently through the skin, parallel to the vein, half a way from the base of the tail. Easy sliding of the needle into the vein and minimal pressure applied when injecting the solution indicates successful insertions into the vein.

Fig. 3 (continued) magnification (2× lens) using a combination of transmitted light and confocal microscope. The tumor angiogenesis were visualized using confocal microscope by injected 2,000 kD FITC-dextran intravenously. Scale bar– 1,000 µm (**c**) The rapid change of blood vessels approximate to tumor inoculation area depicted as the sequential images captured from day1 to day10 (D1–D10). Noted the new vessels formation and branching out form the existing one. (**d**) Two weeks after the inoculation of leiomyosarcoma (SK-LMS-1) cells expressing venus fluorescent (*green*), the tumor blood vessels were revealed using a 2,000 kD TRITC dextran (*red*) and captured with a 2× objective lens. Scale bar – 100 µm. (**e**) Rab25-venus expressing Hela cells were implanted in the DSC. Subcellular vesicles were visualized using two-photon micros-copy with a 20× objective lens. Excitation wavelength 910 nm. Scale bar – 10 µm

5. Slowly inject the solution into the vein. If too much pressure is required for the injection, carefully withdraw the needle, move closer to the base of the tail and re-attempt the injection.
6. Mount the device to the stage and prepare for imaging.

3.5 Visualization of Tumor-Induced Angiogenesis in DSC by Intravital Two-Photon Microscopy

1. Three to seven days after the injection of the tumor cells (the duration depends on the growth of the tumor and angiogenesis process), the animal is sedated using isoflurane, and anesthetized using ketamine/xylazine as previously described in Sect. 3.1 (**see Note 9**).
2. The animal is placed in the custom-made clear restraining tube in crouching position. A long narrow slit on the tube allows the observation window device to stick out of the tube (Fig. 3a).
3. The observation window is mounted onto a holder attached to a modified microscope stage with the window facing the objective lens of the microscope (Fig. 3a). The window is secured with screws to ensure the elimination of any motion artifact due to the respiration and the heartbeat (Fig. 3a).
4. The mouse's body temperature is maintained at 37–38°C through water circulating or pre-heated pads draped over the restraining tube (**see Note 4**).
5. Since the DSC is optically transparent, the area of the tumor that is going to be imaged is determined by visual inspection (Fig. 3a).
6. Blood vessels and the tumor mass can be imaged initially at a low magnification ($2\times$ lens) using a combination of transmitted light (Fig. 3b) and confocal microscopy, which reveals the fluorescently-labeled tumor cells (Fig. 3b). This approach can be used to follow the progression of the blood vessels (Fig. 3c) and the tumor over time (**see Note 10**).
7. The permeability of the blood vessels can be assessed by injecting fluorescently labeled dyes through tail vein (see Sect. 3.4) and image by either confocal (Fig. 3b) or two-photon microscopy (Fig. 3e) ($20\times$ lens) (**see Note 10**).
8. Cellular and subcellular details of the tumor cells can be imaged by using confocal or two-photon microscopy (Fig. 3e).
9. The acquisition of the data will be completed within 30–45 min. After the imaging session and recovering from anesthesia, the animal will be returned to the cage.
10. These procedures are repeated every day for a maximum of 5 weeks (**see Note 11**).

4   Notes

1. Ketamine induces excellent analgesia, but without sufficient hypnosis and no muscle relaxation. It also induces tachycardia, increases blood pressure and catelepsia. It has been shown to have wide safety margin. Xylazine induces

excellent muscle relaxation and analgesia, accompanied by bradycardia and hyptonia. By using a combination of ketamine-xylazine, the animal sleeps within 5–7 min without excitation. The surgical anesthesia last about 80 min with all the reflexes eliminated. The recovery time is about 100–120 min [5].

2. All the surgical procedures are performed in sterile conditions in a biosafety class II cabinet or downdraft table.

3. All instruments used in the procedures are autoclaved, with the exception of the DSC and the screws, which are sterilized using cold sterilization (soaking in bleach (Clorox®) for 10 min, followed by rinsing several time with distilled water, then 95% ethanol).

4. The body temperature is maintained by placing the mouse on a heated pad covered with a clean surgical drape. Vital signs are monitored throughout the procedure. We found that loss of body temperature during the procedure especially in athymic nude mice, is the main cause of the animal mortality.

5. Animals subjected to this procedure normally resume regular behavior as soon as the anesthesia wears off. If aberrant behavior is observed, veterinary consultation has to be requested and an appropriate treatment implemented.

6. Exception to environmental enrichment for single-housed rodents: Mice must be single housed in rat cages to avoid other mice from damaging the DSC. Single housed mouse is normally provided a paper shack as an environmental enrichment to substitute for not having social housing. Due to the height of the dorsal chamber and the need for vertical clearance of the behaving mouse, paper shacks will not be provided.

7. The lentiviral expression system allows the creation of a replication incompetent, HIV-1-based lentivirus that is used to deliver and express the gene of interest in both dividing and non-dividing cells. Stable integration into the host genome provides long-term expression of the gene that is ideal for long-term xenograft. The major components include: an expression plasmid containing the gene of interest under an EF1 promoter, elements that allow packaging of the construct into virions, a packaging plasmids (psPAX2) and the G glycoprotein gene from Vesicular Stomatitis Virus (VSV-G) as a pseudotyping envelope. This allows the production of a high titer lentiviral vector with a significantly broadened host cell range. The production of lentivirus is carried in 293T cell lines [12].

8. In vitro transductions of the tumor cell lines with lentiviral vectors are considered biosafety level 2 (BSL2), therefore, we follow all NIH published BSL-2 guidelines for handling viral stock and proper waste decontamination. For more information regarding viral agents and Biosafety Level 2 laboratory guidelines and precautions, please refer to the links below: http://oba.od.nih.gov/rdna_rac/rac_guidance_lentivirus.html

9. Since these procedures take 30–45 min, it should rarely be necessary to supplement the anesthesia. However, at any sign of consciousness, the animals will receive a second injection of anesthesia, at half the dose of ketamine of the first injection.

10. The restraining tube is bolted to the stage of the microscope and the DSC is immobilized through a series of screws. This reduces significantly any motion

artifact due to the heartbeat and the respiration enabling the visualization of subcellular details.

11. As a reference point to locate the same imaging area, we use the major blood vessels, whose positions and morphology are relatively constant during the course of our experiments (5 weeks).

12. Tumor-associated blood vessels are particularly leaky. For this reason, we use high molecular weight fluorescent dextrans (2,000 KD) as contrasting reagents that are retained in the vessels for longer period of time.

13. DSCs can be re-utilized in multiple experiments. However, due to the properties of the resin, the sterilization in Clorox and alcohol cannot exceed 10 min.

Acknowledgements This research was supported by the Intramural Research Program of the NIH, National Institute of Dental and Craniofacial Research.

References

1. Alexander S, Koehl GE, Hirschberg M et al (2008) Dynamic imaging of cancer growth and invasion: a modified skin-fold chamber model. Histochem Cell Biol 130:1147–1154
2. Amornphimoltham P, Masedunskas A, Weigert R (2011) Intravital microscopy as a tool to study drug delivery in preclinical studies. Adv Drug Deliv Rev 63:119–128
3. Buzdar AU (2011) Anti-angiogenic therapies in metastatic breast cancer-an unfulfilled dream. Lancet Oncol 12:316–318
4. Carmeliet P, Jain RK (2011) Molecular mechanisms and clinical applications of angiogenesis. Nature 473:298–307
5. Erhardt W, Hebestedt A, Aschenbrenner G et al (1984) A comparative study with various anesthetics in mice (pentobarbitone, ketamine-xylazine, carfentanyl-etomidate). Res Exp Med (Berl) 184:159–169
6. Hanahan D, Weinberg RA (2011) Hallmarks of cancer: the next generation. Cell 144:646–674
7. Jain RK, Munn LL, Fukumura D (2002) Dissecting tumour pathophysiology using intravital microscopy. Nat Rev Cancer 2:266–276
8. Makale M (2008) Chapter 8. Noninvasive imaging of blood vessels. Methods Enzymol 444:175–199
9. Masedunskas A, Weigert R (2008) Internalization of fluorescent dextrans in the submandibular salivary glands of live animals: a study combining intravital two photon microscopy and second harmonic generation. In: Proceedings of SPIE, 6860, 68601V, http://spiedigitallibrary.org/. pp 1605–7422
10. Ushiyama A, Yamada S, Ohkubo C (2004) Microcirculatory parameters measured in subcutaneous tissue of the mouse using a novel dorsal skinfold chamber. Microvasc Res 68:147–152
11. van Kempen LCL, Leenders WPJ (2006) Tumours can adapt to anti-angiogenic therapy depending on the stromal context: lessons from endothelial cell biology. Eur J Cell Biol 85:61–68
12. Zufferey R, Nagy D, Mandel RJ et al (1997) Multiply attenuated lentiviral vector achieves efficient gene delivery in vivo. Nat Biotechnol 15:871–875

Side-View Endomicroscopy for High-Resolution *In Vivo* Imaging of the Gastrointestinal Tract

Pilhan Kim, Euiheon Chung, Rakesh K. Jain, Seok H. Yun, and Dai Fukumura

Abstract Small animal, particularly mouse, models are a useful tool for basic and translational studies of human diseases. However, it has been challenging to optically dissect gastrointestinal tract disease models including colorectal cancer and inflammatory bowel disease. To this end, we have developed a novel side-view endomicroscope that allows for non-invasive, real-time, cellular-level visualization in gastrointestinal tracts of live, anesthetized mice. Here we describe a procedure to visualize both the microvasculature and fluorescently labeled cells in the gastrointestinal mucosa. By obtaining wide-area images at multiple time points in a genetically engineered mouse model of spontaneous colon carcinoma, we can monitor the process of angiogenesis associated with tumor development in real time at the orthotopic site. This powerful new imaging method can provide novel insights into many aspects of gastrointestinal diseases.

P. Kim • S.H. Yun
Wellman Center for Photomedicine, Department of Dermatology,
Harvard Medical School and Massachusetts General Hospital, Boston, MA, USA

Graduate School of Nanoscience and Technology, Korea Advanced Institute of Science
and Technology, Daejeon, Republic of Korea

E. Chung
Edwin L. Steele Laboratory for Tumor Biology, Department of Radiation Oncology,
Harvard Medical School and Massachusetts General Hospital, Boston, MA, USA

Department of Medical System Engineering and School of Mechatronics, Gwangju Institute of
Science and Technology, Gwangju, Republic of Korea

R.K. Jain • D. Fukumura (✉)
Edwin L. Steele Laboratory for Tumor Biology, Department of Radiation Oncology,
Harvard Medical School and Massachusetts General Hospital, Boston, MA, USA
e-mail: dai@steele.mgh.harvard.edu

E. Zudaire and F. Cuttitta (eds.), *The Textbook of Angiogenesis and
Lymphangiogenesis: Methods and Applications*, DOI 10.1007/978-94-007-4581-0_22,
© Springer Science+Business Media Dordrecht 2012

1 Introduction

Cancer is a leading cause of death worldwide [1]. Gastrointestinal (GI) tracts are major sites of tumor incidence and death in humans. Colorectal cancer is 3rd in both incidence and death. Five-year survival is less than 20% in esophageal and gastric cancer. To protect the mucosa from continuous challenge of external substances taken with food, gastrointestinal tracts have a strong epithelial barrier with an unprecedented rate of self-renewal. This "stemness", however, leads to a high probability of malignant transformation. Despite significant progress in under-standing the mechanisms of initiation and progression, mortality rates of GI cancers are still high [1].

Mounting evidence has suggested that the tumor microenvironment plays key roles in tumor development, progression, metastasis and response to treatment. Thus, it is crucial to understand the microenvironment of tumors in order to improve their therapeutic strategies beyond currently available treatment options. A variety of complex interactive processes, such as vascular changes, extracellular matrix modu-lation, and infiltration of inflammatory cells, occur in and around pre-neoplastic and full-malignant lesions. Various genetically engineered mouse models have been developed and are widely used to investigate these processes in GI cancers. However, cellular-level analysis in these spontaneous GI tumor models has been primarily accomplished through *ex vivo* histological examinations. Although these studies have provided significant insight into tumor biology, they miss dynamic information such as functional aspects and time course changes.

Advances in intravital microscopy (IVM) techniques such as laser-scanning confocal and multi-photon microscopies as well as genetic and molecular fluores-cent probes have enabled *in vivo* cellular imaging in various disease models. These technological advances have recently allowed unprecedented insight into the complex cellular process in cancer [11, 19, 26]. In principle, most commercial confocal and multi-photon microscopes, originally designed for excised tissue or *in vitro* observation, can be adapted for *in vivo* imaging of mice after simple modifications [4, 17]. However, a specialized apparatus such as an endoscope is required to visualize GI tracts due to their long cylindrical shapes. Significant efforts have enabled high-resolution non-invasive endoscopy for the GI tract in human. In this setting, endoscopy is used for early diagnosis. Laser-scanning confo-cal endomicroscopy, based on a resonantly vibrating fiber [13] or a fiber bundle [9], has also demonstrated the potential value of *in vivo* cellular-level examination for the early detection of aberrant crypt foci, hyperplasia and dysplasia in the human colon. However, for a mouse, the much smaller size of the GI tracts poses additional technical challenges: miniaturization of optical components and maneuverability. To overcome these hurdles, we used a Graded-index (GRIN) lens in rod shape, which has been widely used as a building block for micro-optics and can be as small as a few hundred micrometers in diameter. A GRIN lens-based miniaturized laser-scanning endomicroscopy has been previously demonstrated to visualize the neural networks in the brain [12, 16] as well as immune cell trafficking in the kidney [5] of a mouse – both in a minimally invasive manner. The GRIN endoscopes used in these studies are small enough to be non-invasively inserted into the GI tract of a mouse. However, their front-view configuration and short working distance requires the GRIN endoscope end-tip to be perpendicular to the intestinal wall. Although such an alignment can be routinely achievable in human the GI tract, the small lumen of the mouse GI tract makes it difficult to maneuver in the same way.

Therefore, we have developed a new endomicroscopy method based on the side-view GRIN endoscope that enables fluorescence cellular imaging in the GI tract of live mice over a wide-area in a non-invasive manner [14]. The side-view design allows the endoscope to be naturally positioned in the lumen, enabling an extensive scanning of colonic mucosa by rotation and pullback manipulation of the probe. A comprehensive map of fluorescently labeled cells and microvasculature of the GI tract can be obtained *in vivo* at multiple time points. Through the longitudinal observation of the colon in a genetically engineered mouse model of colorectal cancer, it is possible to monitor angiogenesis and associated tumor development in the orthotopic site, the colon wall. This method can also be adapted for the study of various cellular processes in other GI tract diseases, also including inflammatory bowel diseases, such as ulcerative colitis and Crohn's disease.

2   Mouse Models of Colorectal Cancer

2.1   Apc *Conditional Knockout Spontaneous Colorectal Mouse Tumor Model*

The vast majority of colorectal cancers in humans are adenocarcinoma arising from premalignant adenomatous polyps initiated by the mutation of *Apc* gene [10, 20]. Notably the *Apc* gene is the key-player initiating adenomatous polyps formation by perturbing *Wnt* signaling through nuclear accumulation of β-catenin [18, 20]. Mutations of the *Apc* gene are detectable in the earliest identifiable lesion, aberrant crypt foci (ACF) [2, 27]. Therefore, *Apc* conditional knockout (*Apc*-cKO) mice in which floxed *Apc* genes can be locally inactivated with adenoviral *Cre* delivery [21] can mimic natural tumorigenesis in which tumor initiating cells in the colon undergo mutation of the *Apc* gene. Here, we used *Apc*-cKO mouse originally developed by Dr. Raju Kucherlapati at the Harvard Medical School [10]. The transduction of adenoviral *Cre* into the colon epithelia can inactivate the *Apc* gene, which mimics somatic *Apc* mutation in human patients with sporadic colorectal cancer.

The *Apc*-cKO mouse model has several distinctive advantages for studying the tumor-microenvironment interaction in early tumorigenesis. Since the tumor develops naturally in colon wall, the microenvironment of the *Apc*-cKO model matches that of the human disease. Thus, it has a much higher potential to predict the true behavior of sporadic spontaneous tumor growth observed in the human patient. Furthermore, although some cancer models do use the orthotopic site for tumor growth, many of these models use the implantation of established malignant cancer cell lines or tissue fragments. As these fully malignant cells already have multiple gene mutations, they may fail to mimic the early stages of cancer development that is initiated from a single-gene mutation in non-transformed cells. In contrast, spontaneous sporadic colorectal tumors initiated by *Apc* gene inactivation can mimic the natural process of early tumorigenesis. Furthermore, as Cre recombinase expression is required for *Apc* gene mutation in *Apc*-cKO mice, the location and timing of tumor development can be controlled. These features enable the *Apc*-cKO model to specifically generate lower/distal colon tumors and thus, it is exceptionally suitable for study with advanced cellular and longitudinal endomicroscopy.

2.2   Procedure to Selectively Inactivate the Apc Gene in Lower/Distal Colon

For the delivery of the adenoviral vector containing the *Cre* recombinase gene into the colon, animals are starved 1 day prior to the procedure. Eight-week old *Apc*-cKO mice are anesthetized and a 3–4 cm midline incision is made along the lower abdomen separating the skin and muscle/peritoneum layer. The abdominal opening is kept open with a wire retractor. Q-tips, immersed in warmed saline, are used to moisten exposed organs during the procedure. The remaining stool inside the colon is pushed to proximal side using an animal feeding needle or by gently massaging the colon with a Q-tip on posterior wall. A 10 g pressure vessel clip (5 × 0.8 mm) is placed proximally around the colon about 2 cm from the anus. A 3 cm long polyethylene tube (0.28 mm inner diameter) attached to a 1 ml syringe is introduced into the colon through the anus until it reaches the proximal clip. The colon is then washed with PBS (~500 μl). A second clip is placed 1 cm distal to the first one and trypsin (150 μl) is administered for 10 min to permeabilize the mucus layer. Then the inner layer of the colon is gently abraded with a small caliber brush three times. These procedures prime colonic surface for transduction. After washing the colon lumen three to four times, the distal side of the colon is clamped (~10 mm from the anus) and 100 μl of adenovirus (more than 10^9 pfu) is infused following incubation for 30 min. After removing both clips, the wound is closed with two layers of sutures. The mouse is kept under a heating lamp until it fully recovers from the anesthesia. At 10–12 weeks after the procedure, several large polyp formations are typically observed. Histological observation of these lesions from H&E-stained slides suggests the formation of adenomas (Fig. 1). Control mice that receive saline do not develop any elevated lesions.

2.3   Materials for Animal Procedures

- Animals: *Apc* conditional knockout (*Apc* cKO) mouse for spontaneous colorectal tumorigenesis model [10]. (Other genetically engineered mouse models for colorectal diseases can also be studied.)

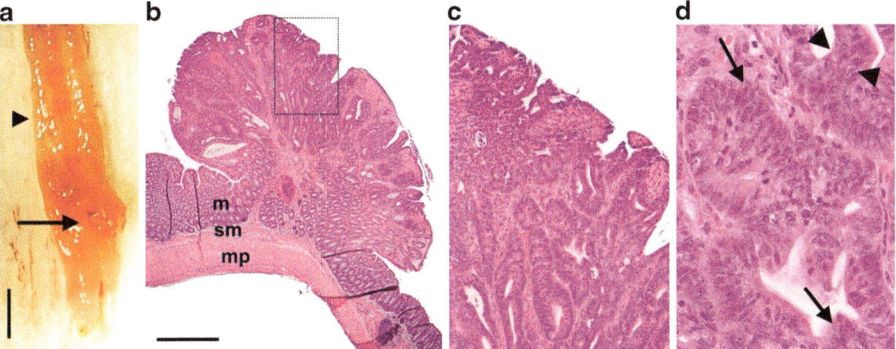

Fig. 1 Histological analysis of colon adenoma in *Apc*-cKO mouse. (**a**) Photo of the dissected colon of *Apc*-cKO mouse showing large polyp (*arrow*). For comparison normal area is marked (*arrowhead*). Scale bar is 5 mm. (**b–d**) Microscopic images of H&E stained colon. (**b**) Lateral view of a large polyp. The irregular arrangement of the colonic pit is shown. Normal mucosa (*m*), submucosa (*sm*) and muscularis propria (*mp*) is visible. Scale bar is 200 μm. (**c**) Magnified image of the area marked by a square in (**b**). Loss of goblet cells, hyperchromasia and occasional dilated glands is observed. (**d**) Loss of polarity of nuclei (*arrowhead*) and stratification of nuclei (*arrow*) is visible

- Analgesic: Buprenorphine (0.1 mg/kg) 20–30 min prior to the surgery, every 8–12 h for 3 days post surgery and thereafter as needed.
- Anesthesia: ketamine/xylazine (100/10 mg/kg) and, if necessary, additional dose (one third of the original dose) for every 30–45 min; or 2 % isoflurane gas anesthesia.
- Heating pump (Gaymar, TP-500, Paragon Medical) and heating pads (TP3E, 3″ by 23″, Paragon Medical) for warming the animal during surgery.
- Eye ointment (Puralube Vet Ointment, Pharmaderm Melville, NY) – a sterile ocular lubricant to keep eyes from drying while under anesthesia.
- Trimmer or hair removal cream to remove abdominal hair prior to surgical procedure
- Surgical instruments: Fechtner conjunctiva forceps (555173f, World Precision Instruments) to prevent perforation of internal organs during the surgical procedure; Wire retractors (500368, World Precision Instruments, 15 mm wire blades and maximum spread 25 mm) to keep the abdomen open during surgery; Surgical suture (5–0 Monocryl, Y490G, Ethicon, Piscataway, NJ) for suturing peritoneal wall; Stainless surgical suture applier (Autoclips™) for closing outer skin; Cautery (disposable) for bleeding control if needed.
- Supplies: Sterile Q-tips and gauzes for absorbing blood; Polyethylene tubing (PE10, i.d. 0.28 mm, Intramedic, Becton Dickinson) for injecting trypsin and adeniviral solution into the colon; HBSS (Hanks' Balanced Salt Solution, Cellgro) for washing colon;
- Colon clipping: Vessel clips (10 g pressure, 15911, World Precision Instruments) for clamping the colon during traipsinization and adenoviral infection (two per animal).

- Colon surface priming: Hyclone Trypsin (0.05%, SH3023601, Fisher scientific); Small caliber brush (use for human GUM brush after chemical sterilization) for mechanical abrasion of colonic surface; Animal feeding needle (9921, Popper and Sons, Inc, Japan) with rubber-tip for washing the colon.
- Ad5CMVCre-eGFP and Ad5CMV-eGFP, Cre recombinase eGFP and control vector, Gene Transfer Vector Core, University of Iowa

3   Side-View Endomicroscopy

3.1   Side-View GRIN Endoscope

The Graded-index (GRIN) lens is a rod shaped optical element in which the refractive index is gradually varied along the radial direction [6]. Light can be focused like a sinusoidal wave along a propagation axis by the GRIN lens. The Simple rod geometry, a flat surface, and a high numerical aperture (NA) in a small diameter make the GRIN lens a preferred component in high-resolution miniature endoscopes [12, 15, 16].

A side-view endoscope is fabricated with a 1-mm-diameter GRIN lens and right-angle micro-prisms (base length = 0.7 mm). A GRIN triplet was built by attaching a coupling GRIN lens (pitch = 0.25, ILW, NA = 0.5) and an imaging GRIN lens (pitch = 0.16, ILW, NA =0.5) to each surface of a relay lens (pitch = 1, SRL, NA = 0.1). A slanted micro-prism with an aluminum coating was attached to the surface of the imaging lens to steer the light to side. The light path in the side-view endoscope is illustrated in Fig. 2a. The assembled side-view endoscope is packaged by a 1.25 mm diameter metal sheath to provide protection from the environment and mechanical strength for *in vivo* live animal imaging (Fig. 2b). At the tip of the side-view endoscope, a metal sheath is ground to expose the micro-prism that provides the optical window for light, which is steered to the side as shown in Fig. 2c, d.

3.2   Imaging System

To accommodate the side-view endoscope, a video-rate laser-scanning confocal microscope system was built [15, 25]. A schematic of the imaging setup is illustrated in Fig. 3a. Three continuous-wave lasers with emission wavelengths of 491 nm, 532 nm and 635 nm are multiplexed by dichroic beam splitters. Laser beams are directed to the raster beam scanner implemented by a rotating polygonal mirror scanner and a galvanometer mirror for fast-axis and slow axis, respectively. The optical system is designed to have a field-of-view (FOV) of $250 \times 250 \ \mu m^2$ using a $40\times$ objective lens (LUCPlanFl, Olympus), which is mounted to an optical axis translator. Three photomultiplier tubes are used as confocal detectors,

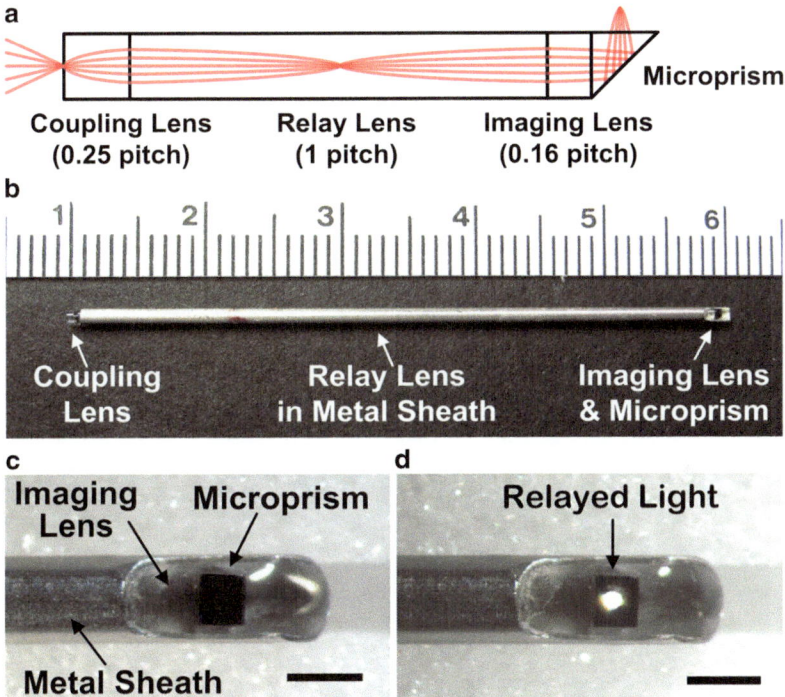

Fig. 2 Side-view endoscope. (**a**) Illustration of the light path inside the side-view endoscope comprised of GRIN lenses and an aluminum-coated *right-angle* micro-prism. (**b**) Photo of a side-view endoscope packaged by metal sheath to provide mechanical strength. Unit of number in ruler is centimeter and the length of side-view endoscope is 5 cm. (**c–d**) Magnified photo of side-view endoscope tip. Exposed micro-prism from the metal sheath (**c**) and the side-directed light scatter (**d**) are shown. Scale bar is 1 mm

and whose output is digitized with a speed of 10^6 samples per seconds by 3-channel frame grabber. Video-rate acquisition of 30 frames per second with 512×512 pixels per frame is accomplished.

The side-view endoscope is integrated to a custom-built XYZ translation/rotation mount and positioned at the focus of the $40\times$ objective lens as shown in Fig. 3b. A mouse platform comprised of a plexiglass plate with flexible silicon heater and XYZ motorized stage for mouse position control is also shown. Figure 3c shows the individual mechanical parts of the side-view endoscope mount and their assembly to achieve accurate XYZ translation and rotation. Individual parts are as following, (1) custom endoscope holder to fix the side-view endoscope, (2) XY translator (Thorlabs, HPT1), (3) pulley, rotation belt and shaft to implement rotation control, and (4) Z translator (Newport, Gothic-Arch). The holder is an aluminum rod with a bore size 1.25 mm that is matched to the outer-diameter of the side-view endoscope. A rubber-tip set screw is used to fix the side-view endoscope tightly. This holder is inserted into the XY translator, which is connected to the rotation shaft by a pulley and rotation belt. The Z translator is used to move the whole integrated system along the optical axis aligned to the objective lens.

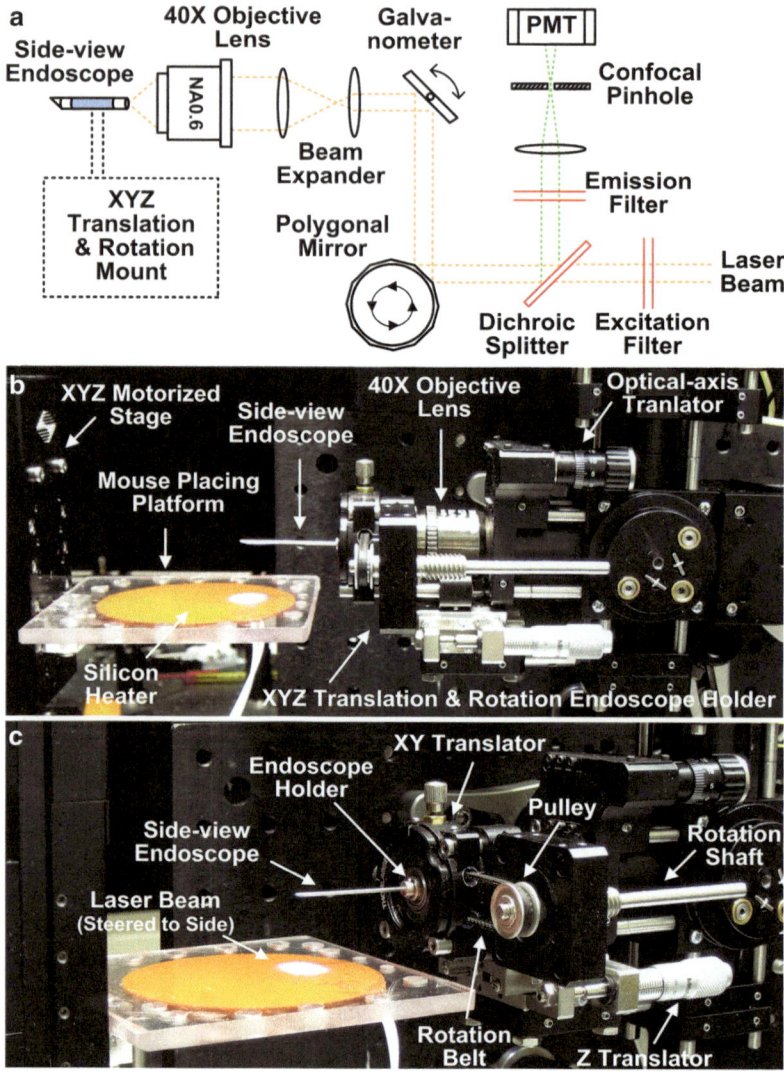

Fig. 3 Side-view endoscopic imaging system. (**a**) Schematic illustration of the confocal microscope system accommodating the side-view endoscope. (**b**) A Photo of the experimental setup showing the side-view endoscope mounted to an XYZ translation and rotation holder that is positioned at the focus of 40× objective lens. The mouse-placing platform attached to an XYZ motorized stage is also shown. (**c**) Photo of individual parts for the side-view endoscope mount and their assembly to implement XYZ translation and rotation

3.3   Side-View Endoscopic Imaging of GI Tract

The side-view endoscope can be noninvasively introduced into the lumen of the gastrointestinal (GI) tract to image the inner-wall, particularly the mucosa. Figure 4a illustrates a representative image of the GI tract wall using the side-view endoscopic.

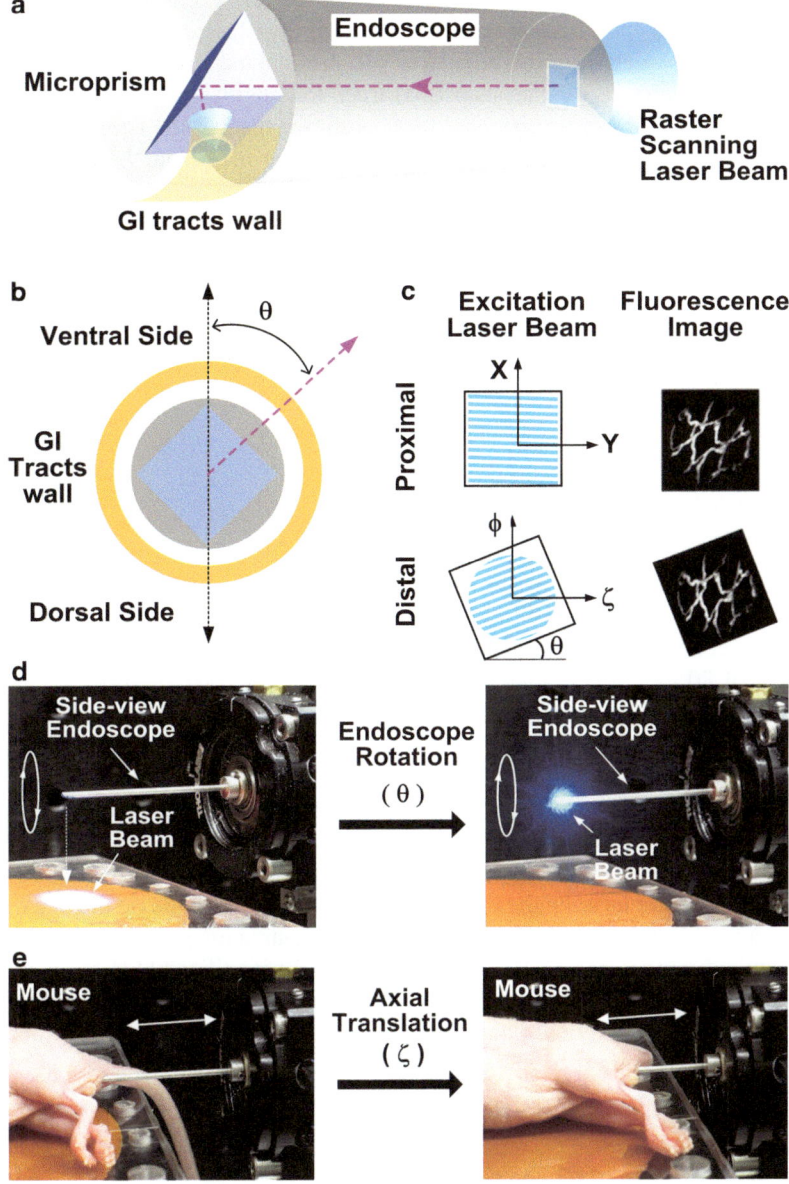

Fig. 4 Side-view endoscopic observation of mouse colon. (**a**) Schematic illustration of side-view endoscopic imaging of the GI wall showing laser beam propagation (*dash line*). (**b**) Schematic illustration of side-view endoscope inside a tubular shaped GI tract. Angular direction of imaging is represented as θ. (**c**) Coordinate relation between the proximal (X-Y) and distal imaging plane (ϕ-ζ). ϕ and ζ represent the circumferential and axial coordinates on the GI wall, respectively. The fluorescence image is a vasculature image obtained in the descending colon *in vivo*. (**d**) Photo of the manipulation of side-view endoscope to rotate (θ). (**e**) Photo of mouse translation along axial axis (ζ) (Figure 4a is modified from [14])

The rater-scanning laser beam is focused by the objective lens and then introduced into the proximal end of endoscope. As shown in Fig. 2a, the laser beam is directed to the side (90 degrees) by reflection off the slanted surface of micro-prism. Once inserted into the tubular shaped GI tract, the angular position of the imaging area on the GI wall can be specified by θ as shown in Fig. 4b. The raster-scanning pattern on the proximal coupling GRIN lens surface in the flat X-Y coordinate is converted to the cylindrical φ-ζ coordinate at the distal imaging site on the GI wall. φ and ζ represent the circumferential and axial axis, respectively. By rotating the shaft shown in Fig. 3c, the angular direction of laser beam emitted from the endoscope can be rotated as shown in Fig. 4d. An axial translation of the image can be accomplished by moving the mouse axially using motorized XYZ stage as shown in Fig. 4e. The side-view endoscope is inserted into the descending colon of the anesthetized mouse through the anus.

3.4   In Vivo *Imaging Procedure for the Mouse Colon*

Prior to colon imaging, the mouse was starved for 24–48 h to avoid the attachment of excrement to the distal end of side-view endoscope, which can severely interfere with the fluorescence signal. The mouse is anesthetized by an intra-peritoneal injection of a mixture of 80 mg/kg ketamine and 10 mg/kg xylazine. The anesthetized mouse is placed on its back or side on the plate with its anus facing the micro-endoscope. The colon is dilated by injecting 0.5 ml of saline water into the colon lumen via the anus using a rubber-tipped needle. A small drop of 2% methylcellulose is applied onto the tip of a side-view endoscope, which serves as a lubricant for smooth insertion of the endoscope into the anus. Once the side-view endoscope is introduced into the lumen of the colon, a rotation and pullback scanning over the entire descending colon is performed. After locating the area of interest, either a single movie file or a multiple of still images can be acquired. To construct a wide-area image mosaic directly from the acquired movie, an image registration algorithm is utilized [14]. As depicted in Fig. 4c, image rotation according to the angular imaging position (θ) is performed to track the φ and ζ axis coordinates of image – this is critical to reconstruct a two dimension image from the rotation scan.

3.5   *Longitudinal Observation of Colorectal Tumorigenesis and Angiogenesis in the Colon of an* Apc-cKO *Mouse*

We inactivated *Apc* in the distal colon of an *Apc*-cKO mouse by administration of adenoviral *Cre* with an invasive procedure depicted in Sect. 2.2. Necropsy, 17 weeks after introduction of Adeno-*Cre*, revealed 1 or 2 adenomatous polyps in the descending colon of all the treated mice (n = 15). In order to visualize the tumor initiation process, we performed *in vivo* imaging with the side-view endoscope every 2 weeks beginning 8 weeks after *Apc* inactivation. Blood vasculature was contrast-enhanced by intravenously injected FITC-dextran (0.5–1 mg) dissolved in PBS. The descending colon (up to 25 mm from the anus) was monitored in order to identify and follow the development of lesions over time.

In the control-injected mice, we observed hexagonal-shaped vasculature indicating a normal crypt structure. In contrast, in the adenoviral Cre-treated mice, several regions of abnormal vasculature where identified. These unique regions had distinct dilated vessel morphology. Figure 5a shows a vascular image obtained at 11 weeks after *Apc* inactivation, revealing a small lesion of ~500 μm in diameter. The same lesion was imaged again at 13 weeks, at which this anomalous area had grown to a size of ~1.5 mm (Fig. 5b). Increased interspacing in the distorted hexagonal vasculature suggests physical manipulation of the surrounding tissue due to the growth of crypt. A strong fluorescent signal in the middle of the lesion indicates accumulation of the fluorescent tracer, most likely due to hampered blood flow. In another mouse, at week 17, we observed a larger lesion, about 4 mm in diameter, exhibiting a more severe vessel dilation, tortuous vasculature and decreased vessel density in its center (Fig. 5c). These findings are typical characteristics of angiogenic vessels associated with tumor development [3, 7]. Magnified images comparing the vasculature in a normal region of the colon and a polyp are shown in Fig. 5d.

We also imaged a modified *Apc*-cKO mouse in which the inactivation of *Apc* is subsequently accompanied by the activation of Green Fluorescent Protein (GFP). This modification enables the long-term tracking and quantification of the growth of *Apc*-inactivated cells in the colonic mucosa. We observed a large polyp expressing GFP 11 weeks after *Apc* inactivation in one mouse (Fig. 6a). The blood vasculature was simultaneously visualized by intravenously injected tetramethylrhodamine (TAMRA)-Dextran. Dilation of blood vessels surrounding the polyp is clearly visible (Fig. 6a). In another mouse, we performed a longitudinal observation at a similar site in the descending colon, identified by both the vasculature and GFP + polyps. Figure 6b shows images obtained at day 10, 12, 14, 28 and 42. We identified multiple groups of GFP + cells from day 10. These micro-nodules showed different behaviors over time. One nodule (arrowhead) clearly grows, whereas others either remains almost unchanged (✻) or vanished (arrow). These results demonstrate the capability of the side-view endoscope to monitor the fate of these cells *in vivo* from the moment of genetic mutation to the formation of large adenomas. In addition to genetic predisposition, the stromal microenvironment is an important factor in the initiation and development of tumors. Side-view endomicroscopy can be utilized to visualize a variety of tumorigenic events, such as vascular changes, matrix modulation and circulating cell infiltration. Therefore we expect it to be a powerful research tool in tumor biology.

3.6 Materials for Endomicroscopy

- Home-built laser scanning confocal microscope system using laser modules of emission wavelength at 488, 532 and 635 nm. For detail, see reference [15, 25].

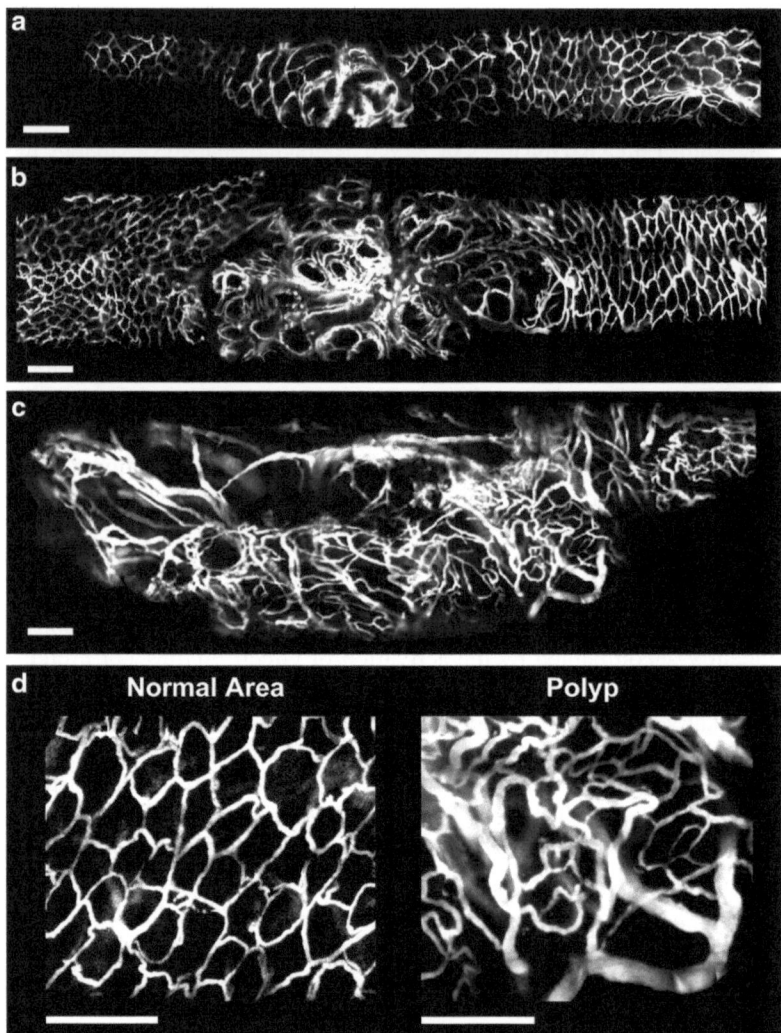

Fig. 5 *In vivo* **fluorescence microangiography during colorectal tumorigenesis in** *Apc***-cKO mouse.** (a) Microvascular image of an *Apc*-cKO mouse 11 weeks after *Apc* inactivation – a small lesion of abnormal vasculature is pictured. (b) A growing legion was observed at the same region of (a) at 13 weeks. Vessel dilation and increased crypt spacing is apparent. (c) Microangiography of a large polyp at week 17. (d) Comparison of the vasculature between a normal area and a polyp at week 17. The scale bars are 200 μm (Figure 5a–c are modified from [14])

- Mouse placing platform: plexiglass plate (100 × 100 × 4 mm) with flexible silicon heater on the bottom, attached by pressure sensitive adhesive (Omega Engineering, SRFR).
- XYZ-axis motorized translator stages (Sutter, MPC-200) holding the mouse platform.

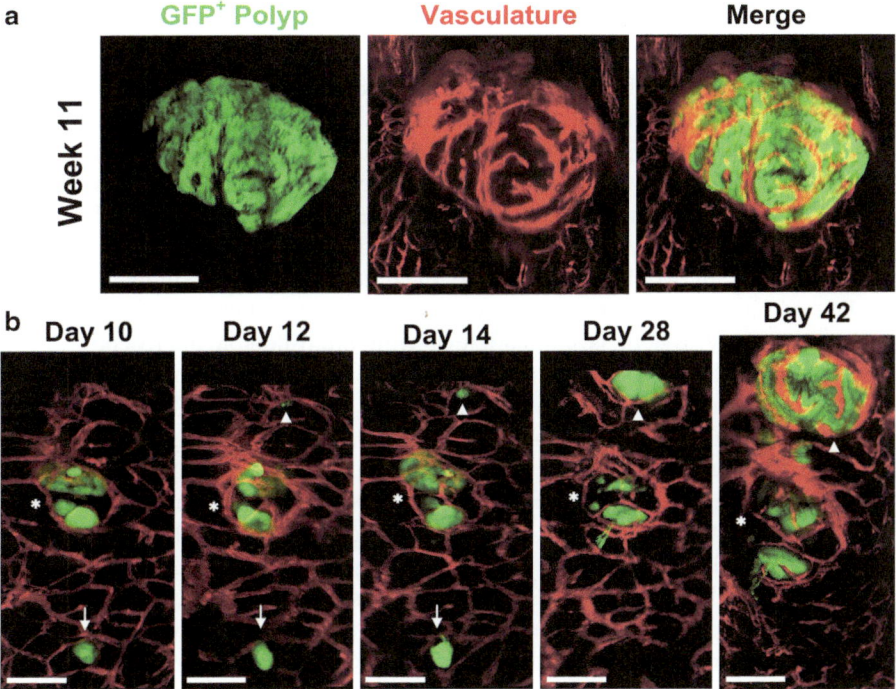

Fig. 6 **Longitudinal *in vivo* imaging of *Apc* inactivated cells and vasculature.** (**a**) Images of a large polyp with GFP + *Apc* inactivated cells (*green*) and vasculature (*red*) obtained 11 weeks after *Apc* inactivation. Blood vessels were contrast-enhanced by intravenously injected TAMRA-dextran. Dilated blood vessels are seen surrounding the polyp. Scale bars are 500 μm. (**b**) Repeated imaging of *Apc* inactivated GFP + cells at the same site on days 10, 12, 14, 28 and 42 after *Apc* inactivation. The images show a GFP + lesion that appears to grow (*arrowhead*), remains unchanged (✱) or disappeared (*arrow*) as well as associated changes in blood vessels visualized by TAMRA-dextran. Scale bars are 200 μm (Figure 6b is modified from [14])

- Side-view endomicroscope composed of a GRIN lens (NSG, Cat No. ILH-100, ILW-100, LRL-100) and micro-prism (POC, Cat No. 8531–601).
- Home-built side-view endomicroscope holder: XYZ-axis translation/360° rotation stage composed by cage-optic XY translator (Thorlabs, HPT1), linear translation stage (Newport, Gothic-Arch), pulley (SDP-SI, A6A51), ball bearing (SDP-SI, 5031) and belt (SDP-SI, A6R51M).
- Fluorescent tracers: Fluorescein Isothiocyanate (FITC) Dextran conjugate, 2,000,000 MW (Sigma Aldrich, FD2000S) and Tetramethylrhodamine (TAMRA) Dextran conjugate, 2,000,000 MW (Invitrogen, D7139), both 1 mg/ml in PBS.
- Other supplies: Pre-warmed distilled water (37°C); Rubber-tip needle (24 G); 2% methylcellulose (Dow, Methocel); Phosphate buffered saline (PBS)

4 Future Perspectives

Intravital microscopy has provided useful insight into angiogenesis and tumor biology [11, 19, 26]. The novel endomicroscopy system described here has significantly extended the application of intravital microscopy – including longitudinal imaging of GI tract. However, several technical challenges still remain. Common to most microscopy techniques currently used, the endomicroscopy method here using confocal microscopy system is surface-weighted. As solid tumors are known to be heterogeneous, it would ultimately be ideal to dynamically study the internal milieu of tumors in order to understand the biology of tumors as a whole. Our endomicroscopy system is compatible a with multi-photon microscopy system [3, 8]. The multi-photon imaging significantly increases the depth of penetration beyond 500 μm, depending on the tissue and tracer used [3, 8]. Other optical methods such as optical coherence tomography (OCT) can image deeper regions and has been applied to an endoscopy system [23]. Doppler-optical frequency domain imaging (OFDI), a second generation OCT system, permits large volume imaging (i.e., $4 \times 5 \times 4$ mm) in a short period of time (~10 min) with high resolution (~10 μm) [24]. These are quite significant advances. However, the major drawback of the OCT/OFDI techniques is lack of specificity, as their image contrast comes from scattering and cannot utilize a fluorescent signal in general. Therefore, the combination of OFDI and multi-photon microscopy would provide complementary information. With more research in this area, we may be able to obtain dynamic images of the whole GI tumor with high spatial resolution and molecular contrast in the near future.

Another challenge comes from the size and bulk of the current device and rigidity of the GRIN-endoscope, which limits their application. This limitation may be overcome with advances in microscope body miniaturization, high speed scanning, and optical fibers. Finally, availability of molecular probes to dissect various biological processes taking place *in vivo* is still limited. Again, development in this area will allow exciting opportunities for unveiling various molecular processes in GI tumors [26]. With the availability of novel fluorophores and computer analysis, we will be able to image multiple events (visualized by distinct colors) simultaneously [22].

Endomicroscopy has high potential to make crucial discoveries in GI tumors which were previously inaccessible by intravital microscopy. Furthermore, with aforementioned improvements in microscopy and molecular probes, it will offer new opportunities for uncovering unexplored aspects of tumor biology in addition to methods for early cancer detection and response to treatment.

Acknowledgments This article is based on the previous article: Kim P, Chung E, Yamashita H et al (2010) *In vivo* wide-area cellular imaging by side-view endomicroscopy. Nat Methods 7:303–305. The work described here was supported by grants from Wellman Center for Photomedicine, Human Frontier Science Program, Tosteson Fellowship, National Research Foundation of Korea, National Institutes of Health, National Science Foundation, and United States Army.

E. Chung is supported by the *Institute of Medical System Engineering* of the GIST, and the Bio & Medical Technology Development Program of the *National Research Foundation* (NRF) funded by the Korean government (MEST) (No. 2011–0019619, 2011–0019632).

References

1. Anonymous (2010) Cancer facts & figures 2010. American Cancer Society, Atlanta
2. Bird RP, Mclellan EA, Bruce WR (1989) Aberrant crypts, putative precancerous lesions, in the study of the role of diet in the aetiology of colon cancer. Cancer Surv 8:189–200
3. Brown EB, Campbell RB, Tsuzuki Y et al (2001) In vivo measurement of gene expression, angiogenesis, and physiological function in tumors using multiphoton laser scanning microscopy. Nat Med 7:864–868
4. Celli S, Bousso P (2007) Intravital two-photon imaging of T-cell priming and tolerance in the lymph node. Methods Mol Biol 380:355–363
5. Fan Z, Spencer JA, Lu Y et al (2010) In vivo tracking of 'color-coded' effector, natural and induced regulatory T cells in the allograft response. Nat Med 16:718–722
6. Gomez-Reino C, Perez MV, Baoby C (2002) Gradient-index optics fundamentals and applications. Springer, Berlin/Heidelberg/New York
7. Hagendoorn J, Tong R, Fukumura D et al (2006) Onset of abnormal blood and lymphatic vessel function and interstitial hypertension in early stages of carcinogenesis. Cancer Res 66:3360–3364
8. Helmchen F, Denk W (2002) New developments in multiphoton microscopy. Curr Opin Neurobiol 12:593–601
9. Hsiung PL, Hardy J, Friedland S et al (2008) Detection of colonic dysplasia in vivo using a targeted heptapeptide and confocal microendoscopy. Nat Med 14:454–458
10. Hung KE, Maricevich MA, Richard LG et al (2010) Development of a mouse model for sporadic and metastatic colon tumors and its use in assessing drug treatment. Proc Natl Acad Sci USA 107:1565–1570
11. Jain RK, Munn LL, Fukumura D (2002) Dissecting tumour pathophysiology using intravital microscopy. Nat Rev Cancer 2:266–276
12. Jung JC, Mehta AD, Aksay E et al (2004) In vivo mammalian brain imaging using one- and two-photon fluorescence microendoscopy. J Neurophysiol 92:3121–3133
13. Kiesslich R, Goetz M, Vieth M et al (2007) Technology insight: confocal laser endoscopy for in vivo diagnosis of colorectal cancer. Nat Clin Pract Oncol 4:480–490
14. Kim P, Chung E, Yamashita H et al (2010) In vivo wide-area cellular imaging by side-view endomicroscopy. Nat Methods 7:303–305
15. Kim P, Puoris'haag M, Cote D et al (2008) In vivo confocal and multiphoton microendoscopy. J Biomed Opt 13:010501
16. Levene MJ, Dombeck DA, Kasischke KA et al (2004) In vivo multiphoton microscopy of deep brain tissue. J Neurophysiol 91:1908–1912
17. Miller MJ, Wei SH, Parker I et al (2002) Two-photon imaging of lymphocyte motility and antigen response in intact lymph node. Science 296:1869–1873
18. Radtke F, Clevers H (2005) Self-renewal and cancer of the gut: two sides of a coin. Science 307:1904–1909
19. Sahai E (2007) Illuminating the metastatic process. Nat Rev Cancer 7:737–749
20. Sansom OJ, Reed KR, Hayes AJ et al (2004) Loss of Apc in vivo immediately perturbs Wnt signaling, differentiation, and migration. Genes Dev 18:1385–1390
21. Shibata H, Toyama K, Shioya H et al (1997) Rapid colorectal adenoma formation initiated by conditional targeting of the Apc gene. Science 278:120–123

22. Stroh M, Zimmer JP, Duda DG et al (2005) Quantum dots spectrally distinguish multiple species within the tumor milieu in vivo. Nat Med 11:678–682
23. Tearney GJ, Brezinski ME, Bouma BE et al (1997) In vivo endoscopic optical biopsy with optical coherence tomography. Science 276:2037–2039
24. Vakoc BJ, Lanning RM, Tyrrell JA et al (2009) Three-dimensional microscopy of the tumor microenvironment in vivo using optical frequency domain imaging. Nat Med 15:1219–1223
25. Veilleux I, Spencer JA, Biss DP et al (2008) In vivo cell tracking with video rate multimodality laser scanning microscopy. IEEE J Selected Topics Quantum Electron 14:10–18
26. Weissleder R, Pittet MJ (2008) Imaging in the era of molecular oncology. Nature 452:580–589
27. Yamada Y, Mori H (2003) Pre-cancerous lesions for colorectal cancers in rodents: a new concept. Carcinogenesis 24:1015–1019

Ex Vivo and In Vivo Assessments of Angiogenesis, Blood Flow and Vasoactive Capability

N.M. Rogers, M. Yao, M.W. Zimmerman, D.D. Roberts, and Jeffrey S. Isenberg

Abstract Cardiovascular disease remains a continuing health threat for much of the world's population. Basic discoveries utilizing cell based assays provide initial insights into signaling pathways and potential molecular targets for developing novel therapeutics. These targets must be validated in more complex vascular cell systems through *in vivo* models of vascular responses to assess their potential efficacy and specificity. We have focused on developing several models that recapitulate angiogenic responses, vascular cell activity and *in vivo* blood flow dynamics to provide such clinically relevant data. Our models have been optimized to be technically straightforward for those not experienced in using composite tissue assays and/or animal work in the hope that these approaches will become more accessible to basic scientists. Our focus on small mammals and in particular mice is based on our belief that studying vascular responses in genetically altered

N.M. Rogers • M. Yao
Vascular Medicine Institute, University of Pittsburgh School of Medicine,
E1226-5A BST, 200 Lothrop Street, Pittsburgh, PA 15261, USA

M.W. Zimmerman
Department of Pharmacology and Chemical Biology,
University of Pittsburgh School of Medicine, Pittsburgh, PA, USA

D.D. Roberts
Laboratory of Pathology, Center for Cancer Research, National Cancer Institute (NCI),
National Institutes of Health (NIH), Bethesda, MD, USA

J.S. Isenberg, M.D., M.P.H., (✉)
Vascular Medicine Institute, University of Pittsburgh School of Medicine,
E1258, BST, 200 Lothrop Street, Pittsburgh, PA 15261, USA

Division of Pulmonary, Allergy and Critical Care Medicine,
University of Pittsburgh School of Medicine, Pittsburgh, PA, USA

Institute for Transfusion Medicine, Pittsburgh, PA, USA

Hemophilia Center of Western Pennsylvania, Pittsburgh, PA, USA
e-mail: jsi5@pitt.edu

E. Zudaire and F. Cuttitta (eds.), *The Textbook of Angiogenesis and Lymphangiogenesis: Methods and Applications*, DOI 10.1007/978-94-007-4581-0_23, © Springer Science+Business Media Dordrecht 2012

animals where target genes are knocked out, knocked in, or conditionally expressed provides powerful insights into the pathogenesis of vascular disease and therapeutic opportunities. Therapeutic strategies developed in mice must also be validated in higher mammals to estimate their potential for treating human disease.

1   A Cellularly Complex Tissue Explant Assay for *Ex Vivo* Angiogenesis

Vascular cells respond to injury by undergoing phenotypic changes mediated by activation of gene signals that stimulate cellular proliferation and migration [1]. The resulting angiogenic response has been assessed *in vitro* by a number of assays using monoculture techniques. However, the *in vivo* angiogenic response is coordinated among a diverse group of cell types including endothelial cells, vascular smooth muscle cells/pericytes, circulating endothelial precursor cells, and various inflammatory cell types [2]. We have optimized a unique *in vitro* assay [3] that captures the cellular complexity of the angiogenic process and simultaneously takes advantage of the power of genetic knockout technology and *in vitro* conditions for maximum control of pharmacology and signal transduction [4, 5]. Aortic ring sections were the first tissue applied to develop a 3-dimensional *ex vivo* angiogenesis assay [6]. However, Weiss *et al.* showed similar vascular outgrowth in a 3D culture using muscle explants [3]. Skeletal muscle has the advantage of being available in larger quantities than aorta, and we find vascular outgrowth from this tissue to be more reproducible and less sensitive than murine aorta to injury during dissection. Our further experience suggests that many tissue types and visceral organ biopsies are amenable to this model and can be miniaturized to approach high-throughput conditions and allow for multiple pharmacologic treatments.

1.1   Method

Supplies (quantities sufficient to make 8 ml gel)

- M199 10x (Gibco) 800 μl
- NaOH 1 M 150–250 μl
- Type I collagen (3 mg/ml) (Originally we used Vitrogen® but this commercial source is no longer available, so we have substituted Purecol® bovine 97% pure type I collagen at 3 mg/ml) 8 ml
- Penicillin/streptomycin 74 μl
- glutamine 100 × 74 μl
- NaHCO₃ 11.8 mg/ml 1 ml
- Sterile pick ups and dissecting scissors (two pair)
- Vascular cell growth medium (Endothelial growth medium, Lonza)

1.1.1 Matrix Composition

In a sterile 15 or 50 ml conical the reagents are combined in the indicated proportions. NaOH (10 M) is used to render the mixture moderately basic (9–10 pH). Add the base by a sterile pipette one drop at a time and swirl the matrix looking for development of purple red color. A drop applied to pH test paper can also help guide this process. The matrix, as it gels in the incubator, will equilibrate with CO_2 to become more acidic, reaching a neutral pH that is ideal for vascular and inflammatory cell activity. If the gel polymerizes at a neutral pH it is rapidly rendered overly acidic upon equilibrating with CO_2, which will suppress any cellular activity. Keep the gel on wet ice until preparing the plate. *On the standard formula one can eliminate the $NaHCO_3$ to increase viscosity of final gel. In this case we then over compensate with 10 N NaOH adding 10–15 μl to obtain a very deep purple color (base over pH 9–10) in the gel. It is always best to have as viscous a gel as possible and adding in extra base compensates for the tendency of CO_2 in incubator to acidify the gel.*

1.1.2 Tissue/Biopsy Harvesting

Age and gender matched animals are humanely euthanized either by cervical dislocation or following inhalation anesthesia and cervical dislocation. Mice should be younger than 4 months of age to ensure robust vascular outgrowth. Under sterile conditions the skin is removed. Using a new set of sterile instruments the target tissue of interest (muscle, cardiac, pulmonary lung parenchyma, trachea, kidney, etc.) is excised and cut sharply into 2–4 mm size biopsy sections.

1.1.3 Plate Set Up

We have miniaturized the assay to be used in the 96-well format. This allows for multiple replicates while conserving expensive treatment reagents such as antibodies or recombinant proteins. Sterile 96-well culture plates such as those from Nunc are ideal for this. Pipette 50–75 μl of gel per well. Cover the plate and place it in an incubator at 37°C for approximately 30 min. Gelation should be complete. Remove plate. Add the tissue biopsy section to each well gently using jewelers forceps. Place the biopsy as close to the middle of the well as possible taking care not to disrupt the lower layer of gel. Then add another 50–75 μl of gel to just cover the tissue fragment, and place the plate again in the incubator for an additional 30 min. After confirming gelation of the second layer, fill each well with 50–75 μl of endothelial growth medium. Alternatively this same technique can be adapted to culture plates with larger well size including standard 6-, 12- and 24-well plates.

Fig. 1 **Angiogenic outgrowth in a tissue explant model.** Microscopic view of a pectoralis major skeletal muscle biopsy explant in type I collagen matrix following 7 days of incubation in endothelial cell standard growth medium (Lonza). The vigorous cell invasion of, and migration through, the matrix can be appreciated. Quantification of the angiogenic response is based on measuring the distance of furthest cell migration from the explant border in 4 equal quadrants

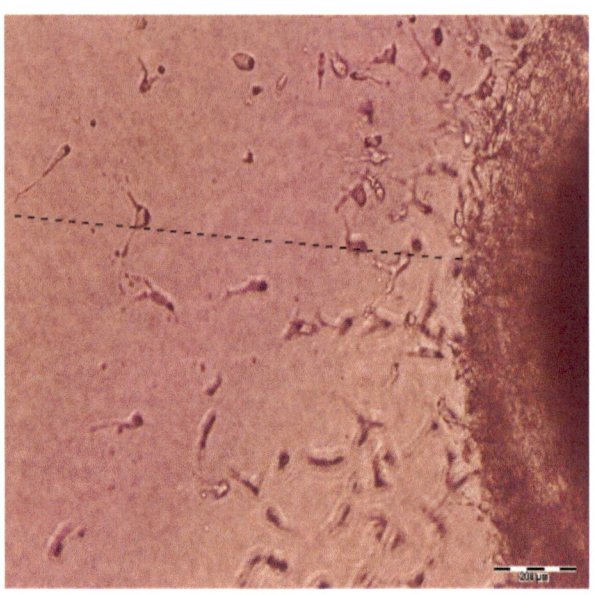

1.2   Evaluation and Data Acquisition

The explants will show an angiogenic response (as defined by cellular migration into and through the matrix) within 24 h (Fig. 1). The plate and each well should be examined daily to acquire a sense of the range of responses. At 72 h assess the distance of farthest cell migration, by measuring with an eye piece grid, the distance from the most outlying cell(s) and the closest tissue border. Standardize the measurements by dividing the area surrounding each explant into 4 roughly equal quadrants, obtain the migration distance for cells in each quadrant and divide by 4 to obtain an average migration distance. A microscope with imaging capabilities can be useful to provide photomicrographs of representative examples of cellular response. Of note, the duration of the angiogenic response may continue for up to 10–14 days, but in most instances begins to resolve within 7 days. Beyond 14 days cellular migration usually ends.

1.3   Isolation of Explant Cellular Response

To conduct functional or biochemical studies upon the cells found in the explant matrix, one can harvest these cells. First, remove the growth medium from above the gel. Then gently remove with fine forceps the biopsy sample. Add 100 ul of sterile basal cell medium (lacking serum) with 0.5% type I collagenase to each well. Incubate for 30 min in a humidified cell culture incubator. This process will result in digestion the matrix gel. Aliquot the resultant into Eppendorf tubes and centrifuged at 800 rpm for 3 min. Aspirate the solution being careful to leave the cell pellet at the bottom of the tube. Suspend the cells in medium in T-25 culture flasks with full growth medium. After surface saturation is achieved (3–6 days) harvest the cells sterilely, incubate in the presence of magnetic antibody coated beads (bearing either CD31 or α-smooth muscle actin) and then use magnetic separation to segregate endothelial cell or vascular smooth muscle cells versus non-vascular cells. Culture selected cell types under cell specific growth medium conditions. This approach will yield a high quality vascular cell monoculture that can be passaged. Alternatively this same technique can used to acquire inflammatory cells by using beads tagged with antibodies specific for a particular cell lineage [7]. Cytospins of cells collected from the digested collagen matrix can also be prepared to aid immunohistologic analysis.

1.4   Strengths of Assay

The tissue explant preparation is a versatile assay, allowing one to study angiogenic responses from unique tissue types and their associated specific vascular beds in a controlled *ex vivo* manner. This latter attribute allows one to also perform classic pharmacology and establish dose response curves for agonists and antagonists of angiogenesis. This assay also allows for the use of tissues from knock out and wild type control mice, providing a powerful genetically pure model to test the role of specific genes in controlling vascular cell responses. The assay does have a certain amount of variability in response between individual wells, hence the need to use high number (4–6 wells/condition) replicates. The explant assay can also be used to study the roles that specific matrix components play in angiogenic responses. These can be added to the type I collagen prior to polymerization. We have also made pure 3D fibrin matrices, although these show less sensitivity to angiogenesis inhibitors due to their innate robust angiogenic promoting activity. One could also construct matrices using alternative collagen types or use proteins from various species (i.e. bovine versus rat).

1.5   A Flexible Platform for Angiogenic Assessment
of Various Composite Tissues

The above composite tissue explant system can be employed to assess angiogenic responses of composite tissues from a range of organs and soft tissues (Fig. 2). We have published results obtained using pectoralis major skeletal muscle biopsies. It is interesting to note that biopsies from chest, back and hind limb skeletal muscle groups from mice have all worked well in this setting, though some differences

Skeletal muscle   Lung
                         parenchyma   Trachea

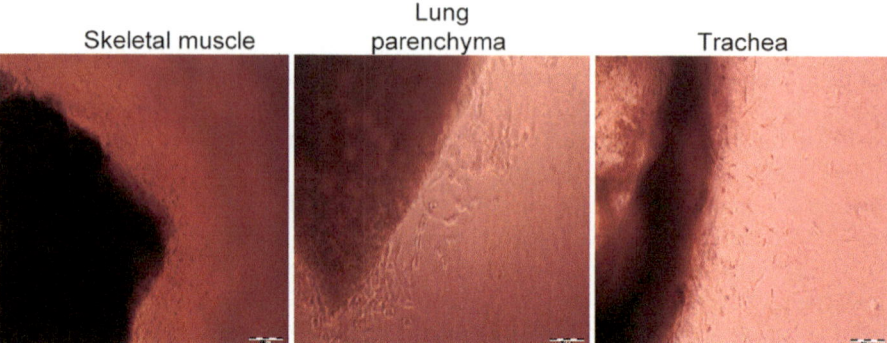

Fig. 2 Angiogenic outgrowth in various tissues demonstrating the wide application of the explant model. Microscopic view of explant angiogenic responses from various tissue types demonstrating the versatility of this technique for studying healing responses in an organ specific manner. Though not pictured, we have also successfully tested the applicability of this technique to uterine and cardiac muscle. Phenotyping of unique cell types obtained from dissolving the matrix after an incubation interval can also be done. Once separated, these cells can be maintained in culture with appropriate support conditions and media. Using tissue biopsies from knockin or knockout mice further enhances the versatility of this model

have been appreciated in explants from different muscle groups, suggesting that one must be consistent upon using the same muscle group in ongoing experiments to make comparisons valid. However, the technique is also effective using many other tissue types. We have found angiogenic responses are reproducible and quantifiable employing lung parenchyma, cardiac ventricular muscle, uterine muscle and trachea. Further, this technique gains significant power when tissue samples from genetically modified animals (lacking or over-expressing a particular protein) are compared to normal control tissues. Alterations in the partial pressure of oxygen with hypoxic or hyperoxic chambers further expands the possible applications of this *ex vivo* platform.

2 *In Vitro* Vascular Smooth Muscle Cell (VSMC) Contraction Assay

VSMCs can be induced to contract in a 3D collagen gel, and the resulting change in gel diameter can be readily quantified. This technique provides a functional means of studying VSMC contraction/relaxation to various activators [8]. It is a very practical approach that is amenable to high throughput since it has been miniaturized to 96-well culture plate platform. Additionally, one can gather mechanistic information by using VSMC harvested from knockout or over-expressing mouse lines.

2.1 Method

2.1.1 Supplies

- As above to prepare type I collagen matrix and explants
- Sterile 96-well culture grade plates
- VSMC freshly harvested and suspended in VSMC growth medium
- VSMC growth medium (Lonza)

2.1.2 Plate Set Up

The fundamentals of the technique are similar to those used to establish the explant angiogenic assays as we have already described in this chapter (see Sect. 1). Type I collagen gel (3 mg/ml) with 10x M199 (Gibco, Grand Island, NY) at a 10:1 ratio of collagen to medium is prepared, pH balanced with NaOH, and seeded with either human VSMC (50,000 cells in 75 μl gel/well) or VSMC harvested from aortic segments from wild type and knockout mice (75,000 cells in 75 μl gel/well) and divided into aliquots onto 96-well plates (Nunc, Denmark). *We have found seeding with higher densities of mouse arterial VSMC maximizes contraction of the collagen disk.* Plates are incubated for 12 h at 37°C and 5% CO_2, allowing for gelation and cell spreading/attachment. The gel disk is then gently released from the well walls with a sterile 2-μl pipette tip, and incubated in serum and additive-deficient growth medium with 0.1% BSA in the presence of the indicated treatment agents for 10 h. Utilization of the 96-well plate format allows for multiple replicates of each treatment condition (n = 5 or more). *Care should be taken in performing matrix disk release from the walls of the well to minimize deformation of the disk.* This preparation allows use of VSMCs from a range of a mammalian species and blood vessels including conduit vessels, such as the aorta, carotid or femoral artery, and resistance vessels, such as mesenteric artery. Alternatively, one can harvest pulmonary arterial VSMCs to study responses of the pulmonary vasculature.

2.2 Evaluation and Data Acquisition

After matrix disk release and incubation for 6–12 h (average 10 h) contraction is determined by measuring the diameter in perpendicular planes (x, y) of each disk, an average obtained, and the surface area is calculated as $(d/2)2\pi$. Experiments are performed in multiple replicates with each condition repeated in from 3 to 6 wells to increase reproducibility and statistical significance. The area of contraction can also be calculated with image analysis approaches using standard soft ware such as Image Pro or Image J. Control wells containing only gel disks without VSMCs are required. The contraction these disks experience is a measure of the elastic properties of a particular matrix preparation, and is subtracted from the contraction values obtained in disks seeded with VSMCs.

2.3   Strengths of Assay

This *in vitro* system allows for real time assessment of VSMC contractility under controlled conditions. Given the total time of the assay, both pharmacologic and gene silencing technologies can be applied to test various questions. Use of VSMCs from knockout mice further enhances the strength of this assay. Additionally, by incubating in conditions of controlled oxygen tension, such as a hypoxia chamber, one can further expand the utility of this assay. The 96-well format is amenable to throughput screening of vasoactive agents to provide a highly appropriate functional read out.

3   *In Vivo* Angiogenic/Tissue Healing Model Systems

Pathobiology of ischemia and ischemia-reperfusion models

Tissue viability requires adequate perfusion that is dependent upon vascular tone, intravascular volume and oxygenation [9]. Interruption of arterial perfusion and blood flow leads to ischemic injury, precipitated by the depletion of adenosine triphosphate and leading to hypoxic cell death [10]. The processes of revascularization, collateralization and angiogenesis restore blood flow to hypoxic and ischemic tissues. In some instances, if this process is too rapid, as in so-called ischemia-reperfusion (I/R) injury a paradoxical worsening of injury occurs through leukocyte infiltration and activation, cytokine and chemokine production, and reactive oxygen species generation [11]. Ischemia and I/R injuries are encountered in multiple clinical scenarios, including myocardial and cerebral revascularization, peripheral vascular disease, and hemodynamic instability due to shock [12]. Ischemia-reperfusion injuries are a significant contributor to failure of solid organ transplantation and microsurgical flaps, both elective and emergent [13, 14]. The development of multiple animal models of ischemia and ischemia-reperfusion allows accurate study of therapeutic approaches with potential clinical applicability [15].

In vivo model systems allow a unique means to translate vascular cell and *in vitro* models with multiple cell types into complex clinically relevant situations. The demands of the model can be tailored to specific needs to address several aspects along a cascade of tissues and vascular stress from models of significant hypoxia and no blood flow to models of limited ischemia and rapid return of blood flow. In each of these below described techniques, a further degree of power can be added by performing the procedure in bone marrow transplant chimers between knockout and wild type mice to study the role of bone marrow-derived and inflammatory and vascular precursor cells versus peripherally recruited vascular endothelial and smooth muscle cells in the process.

3.1 Method

3.1.1 Supplies General to the Following Models

All surgical instruments should be cleaned and autoclaved prior to each procedure and maintained in appropriate surgical quality packaging to protect sterility.

- Scissors
- Toothed forceps
- Blunt iris forceps
- Suture holders
- Atraumatic microaneurysm clamps (the size/force will depend upon the species (i.e. mouse, rat, etc.) and the particular vessel being studied. *As a general rule the clap applying the least amount of pressure needed to obtain complete occlusion is preferred.*)
- Micro-clamp forceps

  Disposables

- Sutures (5–0 nylon)
- Scalpel blades
- 3 ml syringes and fine bore (27 gauge) needles
- Sterile normal saline (0.9%)
- Gauze dressings or plastic wrap

  Operating bench equipment

- Thermostatically regulated heating pad
- Dissecting microscope
- Overhead lamps and infrared heating lamp
- Rectal temperature probe and monitor
- Electric shaver

  Anaesthetic/analgesic agents

- Isoflurane/oxygen (inhalational)
- Ketamine/xylazine (intraperitoneal)
- Thiopentone (intraperitoneal)
- Midazolam to support anaesthesia
- Bupernorphine for post-operative analgesia

There are many choices for general anesthesia of animals undergoing ischemia or I/R injury. All agents provide adequate anesthesia, although the use of inhalational agents allow rapid titration during the procedure and are quite suited to most ischemia models. However in I/R injury, evidence suggests that isoflurane may function to pre-condition and so limit injury induced by I/R [16, 17], though the particular implications this has in the face of a single exposure are not fully known. Nonetheless we have deferred to using i.p. anesthesia with a cocktail of ketamine/xylazine for ischemia-reperfusion studies particularly in studies performed in the liver.

3.1.2   General Points

In all of the following listed models, humane general anesthesia is used to provide pain relief. Post-operative pain relief, if survival based surgery is performed and long term time points studied, should be provided by agents with the least number of effects on platelets, blood pressure and vascular tone. Investigators should work with their veterinary professionals to optimize this. All survival based surgical procedures are also performed under sterile technique, including sterile instruments, gloves, masks, a sterile prep of the site with a topical cleanser such as Betadine®, sterile drapes and suture. Clean technique is acceptable for non-survival surgical models, since infection is not an issue for short time points in animal strains with normal immune systems. Monitoring of core body temperature via rectal probe is also vital to ensure reproducible results particularly if real time assessment of blood flow is planned. Both hypo- and hyperthermia are to be avoided. Heating lamps and warming pads can be employed, but constant monitoring of core temperature is essential.

3.2   Ischemic Tissue Models

3.2.1   Full Thickness Skin Graft

Skin grafts heal through a process of re-vascularization from the wound bed into the existing vascular channels of the dermal layer of the graft [18]. In the 7–14 days this requires, nutritional needs of the graft are met by the process of diffusion [19]. Thus, a full-thickness skin graft does not have an actual circulation for 1–2 weeks post-grafting and experiences a significant amount of hypoxia. It is of note that in the loose skinned mammals, and particularly in the mouse, elevating a true "partial" thickness skin graft is technically challenging [20, 21] and not practical. In the model we describe below we harvest the cutaneous, dermal and sub dermal layers including the so called panniculus carnosus muscle layer. In practical terms, this tissue unit functions and heals comparably to a partial thickness skin graft obtained from human subjects and thus has clinical relevance.

3.2.2   Supplies

Though a list was provided under Sect. 3.1 the requirements for specific model systems will vary.

– General anesthetic protocol of choice. (We have used isoflurane inhalation or ketamine/xylazine in combination. Dosages are published widely but may need slight adjustment dependent of mouse strain as a result of inherent sensitivity to the particular agent found in some null strains compared to wild type.)
– Sterile surgical instruments. (Scissors, pickups, ruler or template for marking the skin graft.)

– Suture material such as 5–0 or 6–0 nylon or Prolene. (Non-absorbable monofilament suture is preferred since it induces less inflammation compared to absorbable or braided sutures.)
– Electric powered sheers and a hair removal agent like Nair®.

3.2.3 Procedure

Animals are provided humane general anesthesia. Fur is removed with electric shears and Nair®. Normal saline can be used then to remove any residual Nair®. This approach is used in all models. (The possible vasoactive properties of Nair® are not considered an issue given the long term assessment points. Use of laser Doppler for real time blood flow analysis in our models requires the skin be hairless.) Skin is then cleaned with Betadine® that is then washed off with sterile saline. A template of 1 cm square dimensions is used to mark the area of surgical excision on the dorsal surface. The size is reflective of limitations in surface area of an average 30 g adult mouse The ideal location is over the dorsal area slightly below the base of the neck to minimize auto-trauma of the graft site. The graft is excised sharply with a fine pair of surgical dissecting scissors functioning better than the scalpel on the loose skin of the mouse. The graft is then sutured in place with four interrupted simple sutures at each corner. Dressings have proven to decrease graft healing rates and do not allow for easy visual graft assessment, and are discouraged. The dorsal location of the graft minimizes/eliminates graft trauma and post-operative contamination.

3.2.4 Read Out

Clinical assessment of graft take can be determined by 3–7 days post grafting [22] and several means as listed, though a combination of approaches is best (Fig. 3):

– Capillary refill test (clinical test that gently applies pressure to tissue, observing blanching, followed by releasing of pressure and observing color flush from tissue vascular refill)
– Pin prick of tissue unit to observe capillary bleeding
– Laser Doppler analysis (see the section below on this technique)
– Determination of tissue necrosis (A clear plastic template is used to trace areas of graft survival versus necrosis, and the ratio of necrosis compared to total tissue unit is determined by weighing respective portions of the traced template)

3.2.5 Applications and Strengths of Model

The full thickness skin graft, by its very nature, experiences a predictable degree of hypoxia and ischemia following grafting, and this slowly resolves as revascularization is completed. Hence the model is a useful tool for studying

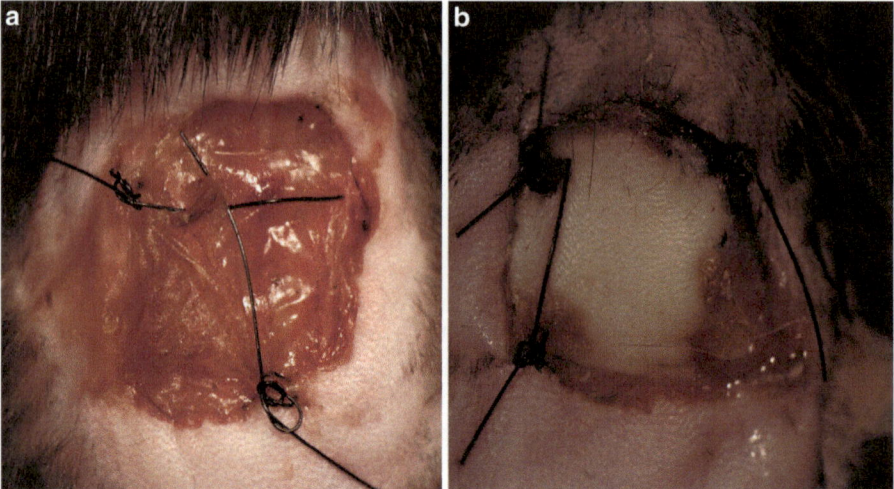

Fig. 3 Full thickness skin grafts in mice demonstrating graft take and graft necrosis. Images of non-take (**a**) and full take (**b**) of murine skin grafts highlight clinically useful differences. The graft that did not take, and hence, did not re-establish perfusion shows necrosis, tissue lysis and reabsorption. The failed graft is collapsed and brittle/desiccated to touch. The graft with near full take has a return of pink coloration and is plump and full

ischemia, angiogenesis and hypoxia. Grafts transplanted between wild type and genetically altered animals allow one to assess the role of particular gene and signaling pathways in the healing process. Additionally, treating the graft, recipient wound area or the entire animals systemically with agents that enhance or retard angiogenic responses and healing increase the model's utility.

3.3 Random Ischemic Flap Model (Fixed Soft Tissue Ischemia)

Ischemia is common to many forms of chronic and acute cardiovascular events [23]. It is also a sequelae of many surgical undertakings. Soft tissue models of fixed ischemia are attractive techniques for studying ischemia and secondary cellular/ tissues responses. Such models are very reproducible and require modest surgical skills to perform.

3.3.1 Procedure

Animals are provided humane general anesthesia. Fur is removed with electric shears and Nair® and skin is then cleaned with Betadine®, which has better antibacterial activity than alcohol and is a preferred skin preparation agent. All flap dimensions are marked according to a template cut to a determined size

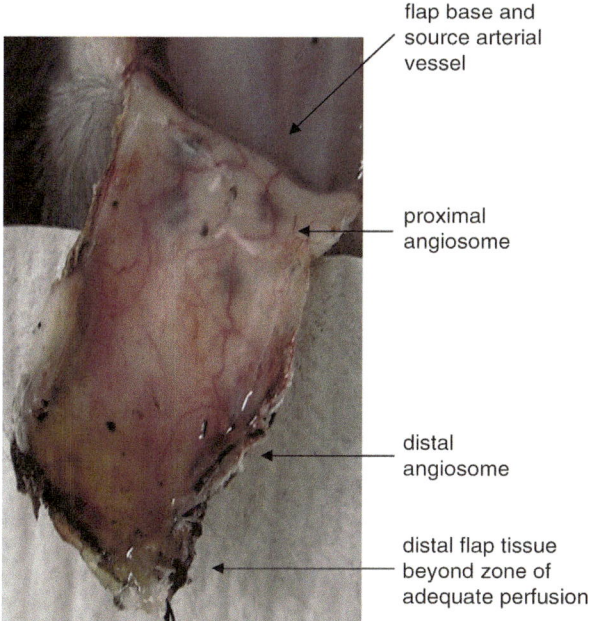

flap base and
source arterial
vessel

proximal
angiosome

distal
angiosome

distal flap tissue
beyond zone of
adequate perfusion

Fig. 4 The surgical and anatomic features of "random" cutaneous flaps. So-called random flaps are thus termed since a defined dominant axial vascular pedicle is lacking. Rather, such flaps have perfusion of a limited nature supplied by one or more minor arterial vessels that only partially traverse the length of the tissue unit. Some degree of perfusion beyond these minor sources of blood flow may also occur through communication and retro-grade flow among neighboring cutaneous vascular territories (angiosomes). The angiosome is an all inclusive perfusion network defined anatomically by a source artery and its vascular field that includes all tissue types including bone, muscle, skin and/or viscera (38). At the level of the skin angiosomes from different source arteries can communicate to provide perfusion under ischemic stress

to ensure consistency of tissue flap size. The long axis of the random flap is orientated parallel to the long axis of the mouse and located on the dorsal surface of the animals. The flap base (the area of the flap from were the tissue blood flow originates may be located in the cephalic or caudal zones of the animals dorsum but we favor the caudal area.

A fundamental rule in designing random soft tissue flaps is to keep the long axis of the flap at least 2 times greater in length than the width of the flap. This ensures predictable ischemia secondary to loss of tissue blood flow and perfusion. Random flaps experience a gradient of ischemia, the most severe being in the region farthest from the flap base (Fig. 4). To induce more severe ischemia one can increase the flap dimensions from a length-to-width ratio of 2:1–3:1 or greater. In our experience in using mice, flap length-to-width ratios of 2:1 through 3:1 induces very robust ischemia and tissue necrosis in wild type control C57BL/6 mice. Under these conditions flaps experience an expected tissue necrosis rate of from 35–55% of total flap area (Fig. 5). This degree of tissue necrosis will allow one

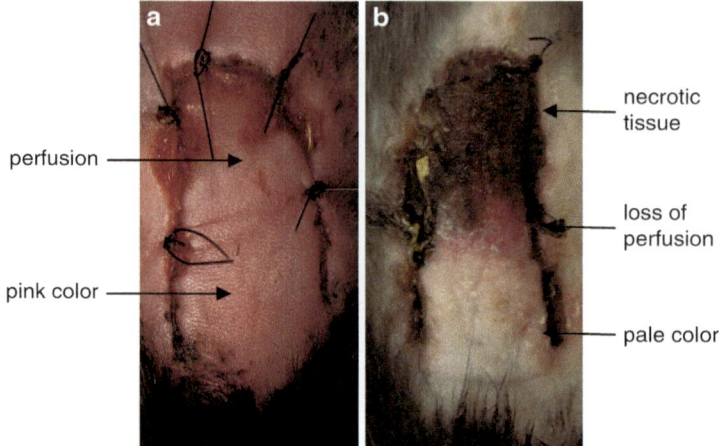

Fig. 5 Clinical correlates of tissue survival in random cutaneous flaps. Presented are representative post-operative images of random cutaneous flaps on the dorsum of mice emphasizing various degrees of perfusion, blood flow and healing. These images were acquired 7 days following surgery. Note the transition zone between sufficient perfusion and inadequate perfusion in flap (**b**). This violaceous coloration is characteristic of underperfused areas in cutaneous tissues. Some of the tissue in this area may go one to necrosis. In contrast observe the well perfused flap (**a**)

to assess effects of therapeutics or gene pathways (knockout mice) in a manner that is statistically significant. Note that random flaps in different mouse strains vary in their resistance to this fixed ischemic injury, and optimal length-to-width ratios should be optimized for a particular strain.

Flaps are sutured to the wound perimeter with simple interrupted nylon suture. *As few sutures as possible is advised since the sutures impose further ischemic stress upon the flap edges. A running suture closure is never recommended.* Location of flaps on the dorsum of animals minimizes auto-trauma and allows for easy daily inspection to assess for complications and healing. Individual housing post-surgery is also recommended to decrease trauma to the flap from cage mates.

We have applied this approach to several species including mice, rats and pigs [[24, 25] and unpublished]. Some variation in vascular networks and perfusion between loose skinned mammals such as mice and rats and dense "adherent" skin mammals such as pigs are appreciated. The degree of perfusion and vascular arcades in the porcine skin and soft tissues is greater than mice and rats and more closely approaches the vascular anatomy of human subjects. Therefore, it is necessary to increase the length-to-width flap ratio to 5:1 or greater in the porcine model to induce predictable distal flap tissue ischemia and necrosis. However, the large surface area of even the inbred miniature pig models allows one to place multiple flaps on the dorsal area of a single animal achieving greater internal control. When working with the porcine model there is a need for professional anesthesia and surgical/post-surgical support from large animal veterinary specialists to perform this work humanely and precisely.

3.3.2 Read Out

As with skin grafts, assessment of flap survival and perfusion can be determined by 3–7 days post-operatively using several means as listed, though a combination of approaches is best:

– Capillary refill test (clinical test that gently applies pressure to the tissue, observing blanching, followed by releasing of pressure and observing the flush of tissue refill).
– Pin prick of tissue unit to observe capillary bleeding.
– Laser Doppler analysis (see section on this technique).
– Tissue oxygen assessment via EPR combined with lithium phthalocyanine implants. This technique is technically demanding, requiring support from experts in the EPR system and data analysis, but provides real time assessment of tissue oxygen levels and the random dorsal soft tissue flap is readily adaptable to this read out as we have published [8]. One can then match healing responses to blood flow, and *in vivo* tissue oxygen levels provide a unique multi-dimensional view of the physiologic process. Though a full description of this approach is beyond the focus of this chapter, we make reference to this approach and our experience with it to highlight how cutting edge technologies such as EPR can be applied to *in vivo* models.
– Determination of tissue necrosis (A clear plastic template is used to trace the area of graft survival versus graft necrosis. The weight ratio of necrosis compared to total tissue is then determined by weighing respective portions of the traced template).
– Tissue can be processed for histology and immunohistochemistry. Further, lysates can be prepared from tissue samples and target protein and mRNA transcript determined by published methods. Additionally other specific markers of vascular mediated signals can be assessed such as the level of reactive oxygen species or cyclic nucleotides (primary agents of vascular health and vasodilation) through analysis of various end products.

3.3.3 Variations on the Basic Model

The random ischemic soft tissue flap model can be modified to address many therapeutic and signaling questions. Flaps can be performed in wild type and genetically altered mice under normoxic and hypoxic conditions. Additionally this model represents an *in vivo* system to test therapeutics that may be delivered either regionally to the flap unit or systemically to the whole animal. The random cutaneous flap model provides analysis of mechanism and therapies appropriate to ischemic stress. Moderate surgical modifications can generate a model that induces ischemia reperfusion, allowing for *in vivo* study of this unique and clinically important pathologic process (see Sect. 3.5).

3.4   Hind Limb Ischemia Model

Skin and soft tissue flaps in small mammals such as mice and rats have limited thickness and potentially can take advantage of diffusion of oxygen from the underlying wound bed to the ischemic tissue, diminishing the potential degree of injury. Though reports of barriers inserted between the flap and the wound surface have claimed this controls wound based diffusion of oxygen and nutrition to the flap, these foreign materials inevitably stimulate inflammation adding additional complications to the model. Composite tissue units such as the hind limb do not demonstrate, in the face of ischemia, similar potential to exploit diffusion. Hind limb ischemia provides a widely used model of severe injury to cutaneous, muscular and osseous structures (with greater metabolic need) to allow more accurate assessment of adaptation of vascular networks or ameliorating treatments. When performed in a genetic model of vasculopathy, such as the ApoE null mouse, the hind limb ischemia model mimics peripheral vascular disease.

3.4.1   Procedure

Hind limb ischemia represents a composite tissue ischemia model. In the mouse the femoral artery provides the majority (>90%) of blood flow into the hind limb. In contrast to people, mice do not have a well developed deep femoral arterial branch. There are several collateral arterial branches identified that come from the posterior pelvis and the acetabulum and provide some degree of flow after femoral artery ligation [26]. Unilateral ligation of the common femoral artery generates critical limb hypoxia and substantial tissue necrosis manifesting as ulceration of the toes and distal limb skin and underlying muscle in wild type C57BL/6 mice. Some groups have described variations of the basic femoral ligation technique including ligation and segmental arterial excision, ligation of the iliac and femoral arteries, and ligation of the artery and cautery of the proximal and distal ends. In our experience, these additional maneuvers are unnecessary to produce a predictable amount of hind limb ischemia, but we have worked primarily with mice in a C57BL/6 background, and as noted above there are strain differences that may require more aggressive surgical techniques.

This procedure is performed on rodents anaesthetized with isoflurane/oxygen titrated to effect. Via a small transverse incision in the inguinal skin crease of the ventral limb, the femoral artery is ligated just as it appears below the inguinal ligament using 5–0 nylon sutures, with the contralateral limb serving as the internal control. The skin incision is closed with interrupted sutures. *There is a slight subtlety in the procedure that is important to take into consideration. At the level of arterial ligation the femoral vein and nerve run adjacent to the artery.* Use of magnification ($2\times$ or $5\times$ loops or a table top operating microscope) helps in avoiding concurrent ligation of these entities. Accidental inclusion of the vein and nerve with the artery sometimes occurs and produces a slightly more dramatic post-operative result in terms of ischemia but does not render the model void. One should be consistent though to either dissect the artery free or do a ligature of all adjacent structures (artery, vein and nerve) as a routine.

3.4.2 Readouts

The assessment of clinical perfusion is useful and relevant with obvious overlap to patients with ischemia and is to be encouraged in pre-clinical models of ischemia but is not a technique with which basic scientists may feel comfortable. We have included a grading scale we use and have taught our staff. Additionally we follow tissue perfusion through assessing skeletal muscle mitochondrial viability, histology, laser Doppler blood flow and, in some cases, blood oxygen level dependent (BOLD) MRI.

- Clinical examination of ischemic injury to be conducted every 24 h: This is a translational skill and once learned will allow the investigator to assess the degree of ischemia and function with animals scored according to a grading scale (0 = normal, 1 + = pale, 2 + = ulceration, 3 + = necrosis) and gait (limping or no limping, but the former is expected if the nerve/vein are also ligated with the artery and hence in this situation gait assessment is removed from the scoring).
- Assessment of hind limb perfusion using laser Doppler imaging (refer to Sect. 4).
- Vascularity index of tissue sections from the vastus medialis muscle: This assay is used to define microscopic alterations in vascularity. Tissue sections from muscle biopsies are reviewed in a blinded fashion. A vessel is defined as a segment traversing 2 branches at $5\times$ magnification. The presence of intravascular red blood cells and wall staining for a VSMC marker, such as α-smooth muscle actin, and an endothelial cell marker, such as CD31, aids in this grading process.
- Evaluation of mitochondrial viability: Mitochondria are very sensitive to oxygen changes and under hypoxia experience loss of metabolic function. Mitochondrial function can be quantified calorimetrically. Tibilais anterior muscle biopsies are weighed and incubated in 3 ml of phosphate buffered saline supplemented with 3-(4,5-dimethylthiazol-2-yl)-2,5 diphenyl tetrazolium solution (Promega, Madison, WI) for 3 h in the dark at 37°C, washed with distilled water and blotted dry. The formazan salt is extracted in 3 ml of 2-propanol for 6 h in the dark at 37°C. Absorbance for 200 μl aliquots is determined using a microplate reader at a wavelength of 425 nm. Muscle samples are dried at 90°C and weighed again, with the absorbance results normalized to dry tissue weight. *A key to this assay is to always select the same muscle group and biopsy the same region of the muscle. It is optimum to take a sample from the midsection of the muscle as the collage tendon bundles at the ends of the muscle (where it forms its boney attachment) will obscure the results.*
Blood oxygen level dependent (BOLD) MRI imaging: *In vivo* analysis of tissue perfusion and blood flow can be achieved through magnetic resonance images acquired by a Bruker Biospin 4.7T scanner. This is a highly precise real time

methodology that assess blood flow at all levels of the hind limb but requires collaborators with specialized expertise in this area. We mention this technique since we have applied it to the hind limb ischemia model in mice [27]. Three days following hind limb arterial occlusion, mice can be anaesthetized using isoflurane to minimize muscular movement and ensure that alterations in oxygenation indicate changes in perfusion as opposed to oxygen consumption, and measurements are taken after body temperature reaches 37°C. Gradient echo-based T1 sequences are used to determine target slice location. T2-weighted gradient echo BOLD image data sets transverse to the femur are acquired for 30 min to monitor temporal changes in blood flow.

3.5   Ischemia-Reperfusion (I/R) Models

The temporary arterial occlusion and rapid restoration of blood flow is termed ischemia-reperfusion. If the occlusion interval is long enough, tissue injury can occur. Though restoration of blood flow would on initial consideration be beneficial, the process can actually initiate additional tissue damage and cell death. I/R injury is clinically relevant in many conditions including myocardial infarct, and subsequent restorative procedures, and stoke. In the realm of visceral organ transplantation some degree of I/R injury occurs in all cases. *In vivo* models of I/R injury are adaptable to various tissue units and visceral organs.

3.5.1   Cutaneous Inferior Epigastric Artery I/R Model

At the level of the cutaneous skin flap, modifications can be made to create a reproducible I/R system. Though the anatomy is similar in both mice and rats we have focused on the model in rats since the target vessels are larger [28]. However, we have also found the model is feasible in mice (Isenberg unpublished). The flap is located on the ventral surface and orientated with the long axis parallel to the long axis of the animal's body. The cutaneous unit is supplied by the inferior epigastric arterial and venous branches that arise from the femoral artery above the inguinal ligament (Fig. 6).

3.5.2   Procedure

The flap should include tissue from both sides of midline and can extend from the level of anterior iliac spine to the mid or upper rib cage. The skin is excised to the level of the deep muscle fascia and easily elevates off the underlying abdominal wall muscles (as with all loose skinned mammals). One can indentify the inferior epigastric vascular bundles from the right and left inguinal areas on the undersurface of the flap. *Care must be taken to not stretch the vascular pedicles.* One pedicle is ligated with suture

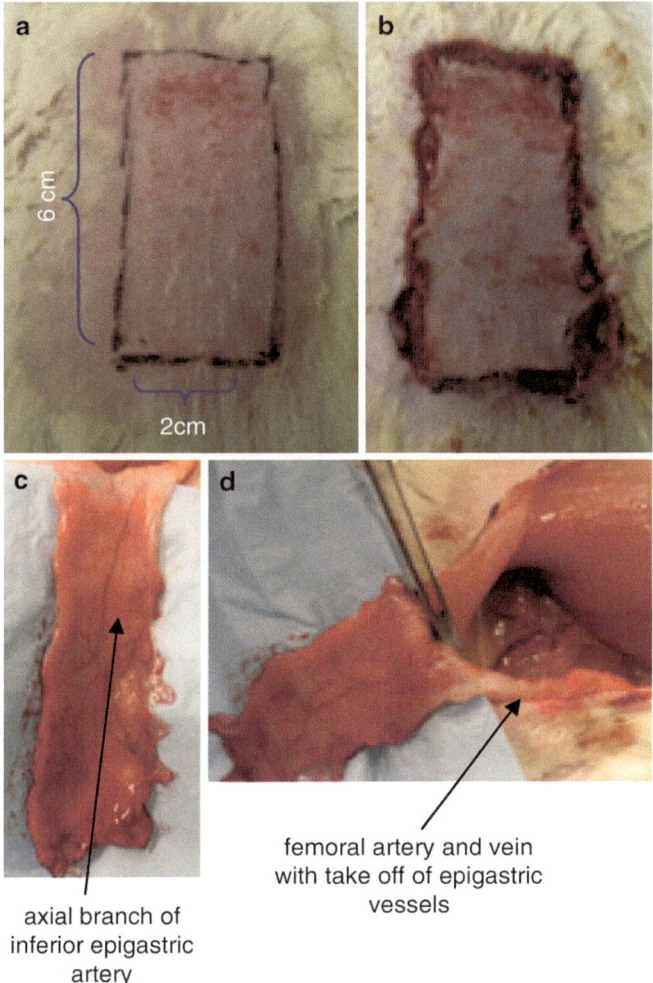

Fig. 6 Cutaneous epigastric artery I/R injury flap. Ventral epigastric 'island' flap in a rat is demonstrated with a length-to-width ratio of 3:1 (**a**). Flap inset is seen in image (**b**). Of note is the location of the vascular pedicles from the right and left side of the animal (**d**). One pedicle is ligated and the remaining is temporarily occluded with a non-crushing microvascular clamp producing an interval of I/R. In contrast to the lack of a major axial vessel in the cutaneous random flap (see Figs. 4 & 5), the epigastric flap is an island flap with a defined axial pedicle highlighted in the image of the flap undersurface (**c**)

and cut. The flap is then sutured back in place with several simple interrupted nylon sutures leaving the closure over the remaining intact epigastric pedicle open. A non-crushing 1 g microvascular clamp is now applied to both the artery and vein. After a defined interval the clamp is removed establishing perfusion in the flap.

3.6   Hind Limb Ischemia I/R Model

The hind limb ischemia model is also readily adapted to an I/R injury model, employing a non crushing microvascular clamp applied to the femoral artery at the level of the inguinal ligament. Alternatively elastic band occlusion at the level of the groin crease has been reported [29]. We have used both methods but find the temporary microvascular clamp affords more control and reproducibility. It is difficult to precisely control the amount of pressure applied by an elastic band encircling the limb at the groin crease leading to variable amounts of I/R.

3.7   Subtotal Liver Warm I/R Model

I/R injury is a major cause of whole organ transplant failure. Attempts have been made to minimize this through blocking I/R injury in transplanted organs. The liver and kidney are widely used target organs in modeling transplant I/R [30, 31]. We have found that both are easily accessed with modest surgical skills. An advantage of the kidney is that the non-I/R challenged organ can serve as an internal control [32], albeit not the same as a sham operated control. A disadvantage of the kidney is that some degree of lethality can occur depending on the severity of the I/R injury, this being especially a problem with any bilateral procedures or in certain strains of mice [33]. Since we have not published yet our experiences with the kidney I/R model we will discuss particulars of the warm liver subtotal I/R model [34].

3.7.1   Procedure

Humane general anesthesia is achieved. In this instance we prefer ketamine/ xylazine in combination. Isoflurane is recognized to cause pre-conditioning and may alter I/R responses. The ventral adnominal and chest hair is removed. A transverse chevron (inverted "v" shaped) incision below the costal margin of the rib cage is made. The hepatic triad is located (under operative magnification) and the branches to the middle and left lobes clamped with non-crushing vascular clamps. The abdominal wall incision is closed with interrupted suture. After the ischemic interval the abdominal incision is re-opened and the vascular clamp removed. The abdominal incision is again closed, the animal allowed to recover for the reperfusion interval prior to euthanasia for final analysis. *There is some variability in both the number of liver lobes and there size between animals. It is useful in launching into this model to do several pilot dissections of the portal triad region and follow the various vascular branching patterns.* Use of a subtotal occlusion model is encouraged since there are less secondary effects on the mesenteric circulation, gastrointestinal tract and other visceral organs.

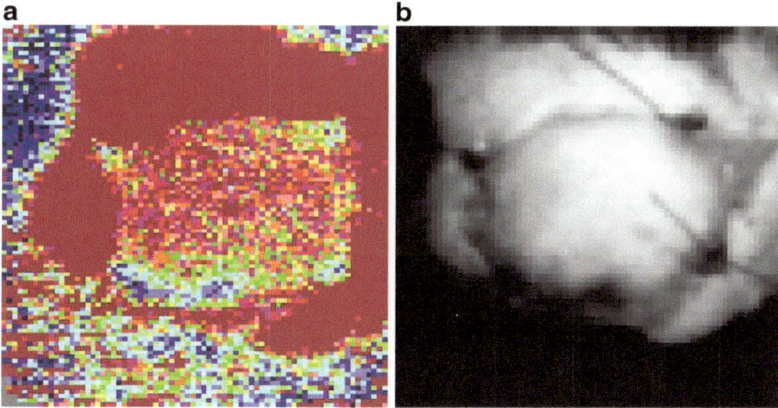

Fig. 7 Laser Doppler blood flow analysis of murine skin and full thickness skin graft. Color laser Doppler (**a**) and photographic image (**b**) of a full thickness skin graft obtained 7 days after surgery. The Doppler image provides a dramatic, easily appreciated and precise quantification of tissue perfusion

3.7.2 Readouts

The readouts discussed in earlier sections are all amenable to I/R with the exception of BOLD MRI. Due to positioning issues of animals in the magnetic field far from the operator, temporary microvascular clamps can not be removed without extracting animals from the magnetic field. Equally important, microvascular clamps will migrate within the magnetic field.

4 Laser Doppler Blood Flow Assessment of Regional Tissue and Visceral Organ Blood Flow

Immediate real-time assessment of blood flow in animals under various stress conditions is a highly useful method of studying vascular and angiogenic responses allowing for the entire internal environment to impact the final result – tissue perfusion. A range of approaches have been developed. Most recently we have favored laser Doppler. This approach is very flexible allowing real-time assessment of blood flow in a range of small mammals from mice to rabbits. It is adaptable to many vascular beds and to visceral organs. With some surgical support it can be used to study intra-abdominal and intra-thoracic organs.

4.1 Applications of Laser Doppler

We and others have found laser Doppler quite versatile and have applied it to a range of angiogenic and blood flow models including:

- skin graft [22] (Fig. 7)
- fixed soft tissue ischemia [27]
- fixed hind limb ischemia [35, 36]
- soft tissue I/R
- hind limb I/R
- warm subtotal liver I/R [34]
- *in vivo* tumor blood flow analysis [37]
- thermal alterations in skin blood flow

In preliminary studies we are adapting the laser Doppler to several other angiogenic and blood flow models targeting specific vascular beds and/or visceral organs. The only limit here is the degree of motion of the particular target tissue. Additional versatility is gained when one adds pharmacologic agents to the study that have known vasoactive or blood pressure effects. These can be added acutely at the time of the laser Doppler studies or chronically to the animal. One relatively straightforward approach is to include such agents as part of the diet in the drinking water (if water soluble and palatable) or chow prior to study.

4.2 Study Design

The laser Doppler allows for immediate analysis of blood flow changes. In any given study the limiting factor is the ability of the animal to tolerate general anesthesia. Our experience with mice is that can tolerate inhalation anesthesia safely for 1–2 h. This limit is especially valid if the animals are breathing spontaneously through a nose cone. If intubated or the airway is controlled via tracheotomy the time window can be extended to 2.5–3 h. However, tracheotomy is only appropriate for terminal experiments. For experiments where blood flow changes are to be studied over time intervals of days or weeks it is critical to re-establish the conditions of the primary study as closely as possible by replicating the degree of general anesthesia and core temperature first employed and also to carefully select the same anatomical region of interest and the prior dimensions of the ROI scanned by the laser. As with all studies a control region similarly sampled should be from a comparable anatomic location and tissue type in the control hind limb.

4.3 Caveats

Several points are critical to obtaining reproducible data with laser Doppler and require emphasis. The control of a constant level of general anesthesia between animals and during a single experiment is critical. This requires close monitoring of the animal's respiration rate if breathing spontaneously. In this instance inhaled isoflurane will allow for rapid and easy adjustment of the level of general anesthesia that is ideal for laser Doppler studies. We have found an average C57BL/6 mouse can tolerate easy 1.5–2 h of inhalation isoflurane general anesthesia. This time interval is sufficient for an assessment of all standard fixed ischemia models and also of adequate length to allow for completion of I/R injury studies of 45 min ischemic interval followed by 60 min of reperfusion. Of note genetically modified mice may demonstrate inherent alterations in sensitivity to isoflurane and other anesthetics and thus doses must be tittered appropriately. The other parameter requiring vigilant monitoring and regulation to insure accurate and reproducible laser Doppler data is core body temperature. We perform all studies with a temperature probe in place. We have found that a warming pad beneath the animal and a warming lap adjacent to the animal provide a flexible system for keeping the core temperature stable. One should not accept any more than 0.1–0.2 degree changes in core temperature through the length of the study. Greater excursion in core temperature will result in significant changes in blood flow in the tissues/organs being studied. Also the laser Doppler study should be performed in an area free from traffic by other researchers since air currents induced from frequent traffic in an area can dramatically alter the core temperature of the animal.

References

1. Folkman J (1995) Angiogenesis in cancer, vascular, rheumatoid and other disease. Nat Med 1:27–31
2. Carmeliet P (2000) Mechanisms of angiogenesis and arteriogenesis. Nat Med 6:389–395
3. Hiraoka N, Allen E, Apel IJ, Gyetko MR, Weiss SJ (1998) Matrix metalloproteinases regulate neovascularization by acting as pericellular fibrinolysins. Cell 95:365–377
4. Isenberg JS, Calzada MJ, Zhou L, Guo N, Lawler J, Wang XQ, Frazier WA, Roberts DD (2005) Endogenous thrombospondin-1 is not necessary for proliferation but is permissive for vascular smooth muscle cell responses to platelet-derived growth factor. Matrix Biol 24:110–123
5. Isenberg JS, Ridnour LA, Perruccio EM, Espey MG, Wink DA, Roberts DD (2005) Thrombospondin-1 inhibits endothelial cell responses to nitric oxide in a cGMP-dependent manner. Proc Natl Acad Sci U S A 102:13141–13146
6. Reed MJ, Karres N, Eyman D, Vernon RB (2007) Culture of murine aortic explants in 3-dimensional extracellular matrix: a novel, miniaturized assay of angiogenesis in vitro. Microvasc Res 73:248–252
7. St Croix B, Rago C, Velculescu V, Traverso G, Romans KE, Montgomery E, Lal A, Riggins GJ, Lengauer C, Vogelstein B, Kinzler KW (2000) Genes expressed in human tumor endothelium. Science 289:1197–1202
8. Isenberg JS, Hyodo F, Matsumoto K, Romeo MJ, Abu-Asab M, Tsokos M, Kuppusamy P, Wink DA, Krishna MC, Roberts DD (2007) Thrombospondin-1 limits ischemic tissue survival by inhibiting nitric oxide-mediated vascular smooth muscle relaxation. Blood 109:1945–1952

9. Hearse DJ (1990) Ischemia, reperfusion, and the determinants of tissue injury. Cardiovasc Drugs Ther 4 Suppl 4:767–776

10. Attanasio S, Snell J (2009) Therapeutic angiogenesis in the management of critical limb ischemia: current concepts and review. Cardiol Rev 17:115–120

11. Granger DN, Hollwarth ME, Parks DA (1986) Ischemia-reperfusion injury: role of oxygen-derived free radicals. Acta Physiol Scand Suppl 548:47–63

12. Kloner RA, Przyklenk K, Whittaker P (1989) Deleterious effects of oxygen radicals in ischemia/reperfusion. Resolved and unresolved issues. Circulation 80:1115–1127

13. Pascher A, Klupp J (2005) Biologics in the treatment of transplant rejection and ischemia/reperfusion injury: new applications for TNFalpha inhibitors? BioDrugs 19:211–231

14. De Santis G, Pinelli M (1994) Microsurgical model of ischemia reperfusion in rat muscle: evidence of free radical formation by spin trapping. Microsurgery 15:655–659

15. Menger MD, Laschke MW, Amon M, Schramm R, Thorlacius H, Rucker M, Vollmar B (2003) Experimental models to study microcirculatory dysfunction in muscle ischemia-reperfusion and osteomyocutaneous flap transfer. Langenbecks Arch Surg 388:281–290

16. Hawaleshka A, Jacobsohn E (1998) Ischaemic preconditioning: mechanisms and potential clinical applications. Can J Anaesth 45:670–682

17. Agnew NM, Pennefather SH, Russell GN (2002) Isoflurane and coronary heart disease. Anaesthesia 57:338–347

18. Smahel J (1977) The healing of skin grafts. Clin Plast Surg 4:409–424

19. Andreassi A, Bilenchi R, Biagioli M, D'Aniello C (2005) Classification and pathophysiology of skin grafts. Clin Dermatol 23:332–337

20. Haramati J, Soppe C, Zuniga MC (2007) A rapid method for skin grafting in mice that greatly enhances graft and recipient survival. Transplantation 84:1364–1367

21. Mayumi H, Nomoto K, Good RA (1988) A surgical technique for experimental free skin grafting in mice. Jpn J Surg 18:548–557

22. Isenberg JS, Pappan LK, Romeo MJ, Abu-Asab M, Tsokos M, Wink DA, Frazier WA, Roberts DD (2008) Blockade of thrombospondin-1-CD47 interactions prevents necrosis of full thickness skin grafts. Ann Surg 247:180–190

23. Das M, Aronow WS, McClung JA, Belkin RN (2006) Increased prevalence of coronary artery disease, silent myocardial ischemia, complex ventricular arrhythmias, atrial fibrillation, left ventricular hypertrophy, mitral annular calcium, and aortic valve calcium in patients with chronic renal insufficiency. Cardiol Rev 14:14–17

24. Isenberg JS, Shiva S, Gladwin MT (2009) Thrombospondin-1-CD47 blockade and exogenous nitrite enhance ischemic tissue survival, blood flow and angiogenesis via coupled NO-cGMP pathway activation. Nitric Oxide 21:52–62

25. Isenberg JS, Romeo MJ, Maxhimer JB, Smedley J, Frazier WA, Roberts DD (2008) Gene silencing of CD47 and antibody ligation of thrombospondin-1 enhance ischemic tissue survival in a porcine model: implications for human disease. Ann Surg 247:860–868

26. Hellingman AA, Bastiaansen AJ, de Vries MR, Seghers L, Lijkwan MA, Lowik CW, Hamming JF, Quax PH (2010) Variations in surgical procedures for hind limb ischaemia mouse models result in differences in collateral formation. Eur J Vasc Endovasc Surg 40:796–803

27. Isenberg JS, Hyodo F, Pappan LK, Abu-Asab M, Tsokos M, Krishna MC, Frazier WA, Roberts DD (2007) Blocking thrombospondin-1/CD47 signaling al

28. Padubidri AN, Browne E Jr (1997) Modification in flap design of the epigastric artery flap in rats–a new experimental flap model. Ann Plast Surg 39:500–504

29. Bonheur JA, Albadawi H, Patton GM, Watkins MT (2004) A noninvasive murine model of hind limb ischemia-reperfusion injury. J Surg Res 116:55–63

30. Que X, Debonera F, Xie J, Furth EE, Aldeguer X, Gelman AE, Olthoff KM (2004) Pattern of ischemia reperfusion injury in a mouse orthotopic liver transplant model. J Surg Res 116:262–268

31. Schneeberger H, Aydemir S, Illner WD, Land W (1997) Nonspecific primary ischemia/reperfusion injury in combination with secondary specific acute rejection-mediated injury of human kidney allografts contributes mainly to development of chronic transplant failure. Transplant Proc 29:948–949

32. Feng L, Xiong Y, Cheng F, Zhang L, Li S, Li Y (2004) Effect of ligustrazine on ischemia-reperfusion injury in murine kidney. Transplant Proc 36:1949–1951

33. Nath KA, Grande JP, Croatt AJ, Frank E, Caplice NM, Hebbel RP, Katusic ZS (2005) Transgenic sickle mice are markedly sensitive to renal ischemia-reperfusion injury. Am J Pathol 166:963–972

34. Isenberg JS, Maxhimer JB, Powers P, Tsokos M, Frazier WA, Roberts DD (2008) Treatment of liver ischemia-reperfusion injury by limiting thrombospondin-1/CD47 signaling. Surgery 144:752–761

35. Isenberg JS, Romeo MJ, Abu-Asab M, Tsokos M, Oldenborg A, Pappan L, Wink DA, Frazier WA, Roberts DD (2007) Increasing survival of ischemic tissue by targeting CD47. Circ Res 100:712–720

36. Isenberg JS, Annis DS, Pendrak ML, Ptaszynska M, Frazier WA, Mosher DF, Roberts DD (2009) Differential interactions of thrombospondin-1, -2, and -4 with CD47 and effects on cGMP signaling and ischemic injury responses. J Biol Chem 284:1116–1125

37. Isenberg JS, Hyodo F, Ridnour LA, Shannon CS, Wink DA, Krishna MC, Roberts DD (2008) Thrombospondin 1 and vasoactive agents indirectly alter tumor blood flow. Neoplasia 10:886–896

Skin Flap Models for Assessment of Angiogenesis

Geraldine M. Mitchell, Zerina Lokmic, Shiba Sinha, and Wayne A. Morrison

Abstract Skin flaps are surgically elevated areas of skin and subcutaneous tissue transferred on their own vascular pedicle from a donor site to a recipient site that is devoid of skin. Skin flaps have been used in clinical reconstructive surgery for over 100 years and remain common procedures to this day. This chapter will describe numerous experimental animal skin flap models that have been developed over the last 40 years, and the various techniques applied to these models to increase their survival including vascular delay, ischemic preconditioning and flap prefabrication. Most of these techniques are now known to significantly increase the blood vessel numbers and blood flow to experimental skin flaps. Skin flap models and applied techniques illustrate various mechanisms by which angiogenesis can be stimulated and manipulated. Additionally, the manipulation of large vascular pedicles associated with flap prefabrication has led to a new and developing area in intrinsic vascularisation of tissue engineered chambers where new tissues and organs can be grown. Basic experimental flap models, techniques to enhance angiogenesis in these flaps and intrinsic vascularisation of tissue engineering chambers will be described in this chapter.

Geraldine M. Mitchell, Zerina Lokmic, and Shiba Sinha contributed equally to this chapter.

G.M. Mitchell, Ph.D. (✉) • W.A. Morrison, MB.BS., MD., FRACS.
O'Brien Institute, University of Melbourne, Department of Surgery at St. Vincent's Hospital, Melbourne, 42 Fitzroy Street, Fitzroy, VIC 3065, Australia

Faculty of Health Sciences, Australian Catholic University, Melbourne, VIC, Australia
e-mail: geraldine.mitchell@svhm.org.au

Z. Lokmic, MSc (Nurs), PhD • S. Sinha, MB.Ch.(Hons), MS.
O'Brien Institute, University of Melbourne, Department of Surgery at St. Vincent's Hospital, Melbourne, 42 Fitzroy Street, Fitzroy, VIC 3065, Australia

E. Zudaire and F. Cuttitta (eds.), *The Textbook of Angiogenesis and Lymphangiogenesis: Methods and Applications*, DOI 10.1007/978-94-007-4581-0_24,
© Springer Science+Business Media Dordrecht 2012

Abbreviations

AAS	(Aminopropyl) triethoxy-silane
AVL	Arterio-venous loop
CAST	Computer assisted stereological toolbox
CT	Computed tomography
CTA	Computed tomographic angiography
DAB	Chromogen diaminobenzidine
DPX	Distyrene, plasticizer, xylene
EC	Endothelial cell
ECM	Extra-cellular matrix
EPC	Endothelial precursor cells
FGF-2	Fibroblast growth factor 2, or basic fibroblast growth factor
HIF	Hypoxia-inducible transcription factor
HRP	Horse radish peroxidase
IP	Ischemic preconditioning
OPS	Orthogonal polarisation spectral
PBS	Phosphate buffered saline
PDGF	Platelet derived growth factor
PPG	Photoplethysmography
RISK	Reperfusion injury salvage kinase
SIE	Superficial inferior epigastric vessels
TUNEL	Terminal deoxynucleotidyl transferase dUTP nick end labeling
TBS	Tris buffered saline
W	Weight
V	Volume
VEGF	Vascular endothelial growth factor

1   Skin Flap Models for Assessment of Angiogenesis

Tissue flaps have been used in reconstructive surgery for centuries. Skin flaps were first described in facial repairs in the sixth century [1]. The reconstructive armamentarium has evolved through increasing understanding of the development and propagation of blood supply within tissues. Skin flaps represent a powerful method of studying *in vivo* angiogenic growth and manipulation and it is largely through extensive animal based studies that we have gained an understanding of the complexities of angiogenesis (sprouting of local endothelial cells from existing capillaries to form neovessels) that have resulted in increasing sophistication of surgical flap design. These *in vivo* skin flap models provide insights into mechanisms of angiogenic stimulation that are not only relevant to the growth of angiogenic tissues in skin flaps, but have far reaching applications within the new fields of tissue engineering and cellular therapy.

A **skin flap** is a unit of tissue involving the layers of skin (epidermis, and dermis) and subcutaneous tissue that can be transferred from a donor site to a recipient site and retains its own blood supply. A **skin graft** is a skin tissue transfer that is not based on a living vascular pedicle and achieves its viability by new autoanastomosis between the existing capillary beds in the graft tissue with capillaries in the recipient site bed, by a process known as inosculation [2], followed by angiogenesis. The O'Brien Institute and other groups have developed and utilised a number of animal skin flap models with the purpose of advancing the development and application of surgical flaps clinically. Our own experience in this field has been primarily in the analysis of the vascular responses in the skin and subcutaneous tissues associated with various skin flap manipulations used in reconstructive surgery.

This chapter provides an overview of the main concepts that have been utilised in order to improve angiogenesis within skin flaps to facilitate their transfer, including vascular delay and ischemic preconditioning. Prefabricated skin flaps and tissue engineering of soft tissue based on a vascular pedicle, represent further evolutions of initial skin flap models and allow *in vivo* assessment of angiogenesis. Methodological guidance will be provided of the standard animal skin flap models and the formation of a prefabricated skin flap, and surgically vascularised tissue engineered constructs.

1.1 Blood Supply Within Human Skin Flaps

Skin comprises a number of layers including the epidermis, dermis and subcutaneous tissue and its blood supply is complex and varies enormously with location. The skin in most mammals serves several functions including thermoregulation and this explains its high density of blood vessels. The skin circulation includes arterio-venous anastomoses in addition to resistance vessels and blood flow is controlled both by sympathetic innervation and local control [3]. Elevating or raising a skin flap involves careful dissection of the skin and underlying subcutaneous tissue whilst preserving the blood supply that will allow this section of tissue to remain viable. Truncal and limb skin blood supply is composed of horizontal and vertical components. The horizontal component includes the vascular plexuses that lie parallel to the skin surface and will distribute blood within a skin flap. The subdermal plexus lies at the junction of the dermis and the subcutaneous fat. The prefascial plexus lies on the deep muscle fascia. The vertical component comprises an artery and two smaller venae comitantes and is referred to as the perforating system of vessels or perforators, as they perforate through the deep fascia from the deeper 'named' arteries and veins. They are the essential well-spring sources of blood supply that distribute blood via their connections (anastomoses) to the horizontal prefascial and subdermal plexuses. Perforators have a specific anatomical distribution throughout the body and each irrigates a territory [4, 5] or angiosome [6]. Each territory is interconnected with its neighbour via their horizontal plexuses, and blood flow may derive from one or other, or both territories according to pressure gradients. Perforators have been extensively characterised in both animals and humans [7, 8].

In the face and head and neck regions where there is no deep fascia blood supply, although ultimately from a defined source is more random and predominantly horizontal in its distribution. Where there are perforators or occasionally where there is a large axial vessel that lies in the horizontal of the flap (axial flap) it is possible to raise flaps by cutting circumferentially around the flap (island flap) provided the vertical perforator or axial vessel remains intact. In this way flaps can be moved greater distances or even transplanted to a distant site by micro-vascular anastomosis.

Alternatively, flaps can be raised based on an intact bridge of skin through which the horizontal blood supply enters. Where the length of a flap relative to its width at the base is less than two to one the flap usually survives in its entirety. This is referred to as a random pattern flap, the horizontal circulation randomly irrigating the elevated tissue. Where elevation encroaches into an adjacent vascular territory, that component of the flap furthest from the base is at risk of necrosis.

1.2   Vascular Delay

Vascular delay or the delay phenomenon is a concept that has been utilised by reconstructive surgeons for centuries [9]. Skin flap reconstruction traditionally involves the partial elevation of a zone of skin and fat (the flap) and transposing or advancing it to cover an adjacent defect. In the process the blood vessel network within the flap is potentially threatened limiting flap size and the distance it can be transferred. To counter this, flap blood supply can be enhanced by partial elevation as a preliminary procedure some days prior to the definitive flap surgery. This "delay" procedure divides blood vessels around the periphery and the resulting ischemia and inflammatory signals stimulate an enhancement of the intrinsic vasculature of the flap.

Milton was the first to demonstrate experimentally in 1965 using a porcine model, superior flap survival using a delayed skin flap technique [10]. Since then there have been a number of experimental studies that have provided potential mechanisms explaining how vascular delay enhances the vascularity of skin flaps [11–14]. These studies have used porcine, rabbit, dog, rat and mouse models and heterogenous study designs and protocols, reviewed by Ghali et al. [9]. Delay procedures have included partial devascularisation of flaps by incision of the skin, partial flap elevation involving the subcutaneous tissues or clipping of a feeding vessel before transposition of the flap. The period of delay that has been used has varied from 2 to 14 days in animal studies.

The underlying mechanism for vascular delay can be divided into early and late stages. When the vessels to a flap are ligated or divided, there is an initial hyperadrenergic state induced due to the division of sympathetic nerves, which causes vasoconstriction affecting the precapillary sphincters resulting in a relative ischemia [15]. Once the flap is elevated there is a resolution of the hyperadrenergic state and this causes vasodilation [12] and increased blood flow. The effect of vasodilation once the flap has been elevated has been demonstrated in numerous subsequent studies to last 2–3 days [13, 16] and in some studies for longer periods, such as 7 days [17]. Anatomical studies demonstrated that the small vascular

interconnections between adjacent territories – 'choke vessels' vasodilate significantly in response to vascular delay [18] at 48–72 h.

Later stages of the vascular delay response results in an increase in the production of arachidonic acid metabolites including an increase in the vasconstrictive
substance thromboxane [19]. The ischemia that occurs in flap tissue after delay also
results in increased neovascularisation within the delayed flap via local increases in
angiogenic growth factors – fibroblast growth factor – 2 (FGF-2), [20] and vascular
endothelial growth factor (VEGF) [21]. Increased levels of these growth factors
lead to the events of angiogenesis- vasodilation, endothelial cell activation, vessel
basement membrane degradation and new capillary sprouting, causing new capillary networks to form, within the flap [9, 22]. In addition it is now believed that bone
marrow derived endothelial precursor cells (EPCs) also contribute new blood vessels
in ischemic flap tissue [9, 23], through a process termed adult vasculogenesis [24].
However, the increased neovascularisation effects of delay may not be evident in the
flap until 7–14 days after the initial delay procedure [9].

1.3   Ischemic Preconditioning

Ischemic preconditioning (IP) is a variant of vascular delay. It is defined as a brief
period of ischemia ("preclamping") followed by tissue reperfusion thereby increasing
the ischemic tolerance for a subsequent longer ischemic period [25]. The initial proof
of principle was demonstrated in the field of cardiovascular surgery by Murry where
repeated brief episodes of coronary artery occlusion followed by reperfusion
improved myocardial muscle survival when the heart was subjected to prolonged
ischemia [26]. The benefit of IP on skin flaps was first demonstrated by Mounsey et al.
in a porcine latissimus dorsi flap model where three cycles of 10 min of ischemia
increased flap survival compared to control flaps when a subsequent 4 h of ischaemia
was applied [27]. There have been numerous studies involving various flap models
(both skin and muscle flaps), utilising different species and flap procedures including
rat muscle flaps [28]; the rat cremaster model [29, 30]; rat musculocutaneous flaps [31]
and murine island skin flaps [32]. There is significant heterogeneity in the protocols
used in IP studies and there is no defined consensus on the optimum protocol for
preconditioning. There is debate regarding the optimal time point of IP application, the
duration of the IP cycle and the total number of cycles to apply. The majority of
protocols described refer to classical IP where the manipulation is applied locally.
Despite its proven efficacy in animal models, classical IP has rarely been used in the
clinical setting. Remote IP protocols have also been described, which aim to reduce
the number of additional procedures and the invasiveness of 'classical' preconditioning described previously. Remote IP involves the non-invasive induction of an
ischemia/reperfusion event (via a tourniquet) in a limb area proximal, or even in the
opposite limb prior to flap elevation. This procedure has been shown in a rat model to
improve skin flap viability [33].

The mechanisms of IP are still not fully understood and have largely been examined in relation to the heart and are thought to involve a number of signalling pathways including the RISK (reperfusion injury salvage kinase) pathway resulting in the activation of prosurvival protein kinases (Akt and Erk 1/2) [34, 35]. Additional pathways may involve transcription factors, that include NF-$_k$B, and STAT 1/3 [36]. Another significant pathway in this response is the hypoxia-inducible transcription factor (HIF) pathway. During hypoxia HIF 1α levels rise in hypoxic cells (under normoxia HIF 1α is degraded) and forms a dimmer with HIF-1ß. This complex binds to the hypoxia response element (HRE), a transcription enhancer [37] that initiates a response potentially involving many genes including those that promote angiogenesis, proliferation, cell survival and erythropoieses [38].

There is a requirement for a standardised and reproducible skin flap ischemic preconditioning model. Many of the models described in the literature have been in large animals where the vascular connections are easier to isolate and manipulate. Smaller animal models have advantages of ease of animal handling, cost and good reproducibility. It is important to note that skin flaps in different animal models [39] and within different strains of the same species [40], and even different skin regions within an animal give diverse responses to different periods of ischemia.

1.4   Prefabricated Flaps

Experimental observations by Erol and Spira determined that it was possible to vascularise a skin graft with a sub-adjacent artery and vein in essence creating a skin flap [41]. The term 'flap prefabrication' was first introduced in 1981 by Shen [42] to create a neo flap by the implantation of a vascular pedicle into tissue, enabling skin territories that may not naturally be perfused by anatomically well defined perforator or axial vessels to be transferred as a flap [42–45]. Prefabricated flaps have been used mostly in head and neck reconstructions where there is often a limitation in local donor tissue and the aesthetic, structural and functional needs are high. The concept can also be used to vascularize a composite of tissue such as skin and cartilage for nose or ear reconstruction. This is referred to as prelamination [46].

The development of flap prefabrication techniques have served as a stepping stone between traditional flaps and vascularised tissue engineering. The experimental observations in early flap prefabrication studies are the foundation for subsequent tissue engineering studies where an arterio-venous loop (AVL) or shunt [41] has been used as the nutrient blood vessel to generate *de novo* vascularised tissue [47–52].

Tissue engineering represents the next evolution in the construction of new tissues but angiogenesis within engineered tissues remains a rate-limiting factor prior to clinical application. Tissue engineering models developed at the O'Brien Institute provide additional models (see later sections) that enable the study of angiogenesis within engineered tissues.

2 Animal Models

There have been a number of well utilized standard animal flap models since the first description in the literature in 1933 describing circulation in tubed skin flaps in a canine model [53]. There are advantages and disadvantages associated with different animal models, which need to be taken into consideration for experimental planning and interpretation of results. Ideally any model of skin flap circulation should match the vascular anatomy of humans as closely as possible, and pig skin approximates human skin most closely [54]. Most skin flap models described are in rats, which are easily obtained and inexpensive. However rat skin, like other loose-skinned animals is supplied by direct cutaneous arteries running parallel to the animal's longitudinal axis with no or few musculocutaneous perforators. In humans, perforators are the main source of blood to the skin [55].

Flap models that are used in the study of angiogenesis need to be reproducible, reliable and have predictable survival patterns. Elevating a skin flap in any model is a significant technical undertaking and a confounding factor will be operator skill at producing reproducible injury patterns. It is recommended that a single surgeon is used for all experiments, who has received appropriate training in raising surgical flaps and has the microsurgical expertise required to form arterio-venous loops and bundles if required. This is an additional cost and it is estimated that approximately 15–20 procedures should be performed during training before competency is achieved with a minimal complication rate. Ideally, models with the capacity for bilateral flap elevation will reduce the number of animals required to achieve statistical significance. Animals will need to be housed and cared for as per local animal ethics guidelines. Measures should be taken to prevent flap auto-cannabilization. It is important to note that the skin flap protocols described here are modified optimal protocols based on the authors' experience.

2.1 McFarlane Flap (Dorsal Rat Skin Flap)

McFarlane introduced one of the first standardised experimental rat skin flap models in 1965 [56]. The original flap consisted of a cranially based 10 cm × 4 cm dorsal skin flap elevated at the level of the deep muscle fascia including the panniculus carnosus, which is a thin layer of striated muscle deep to the subcutaneous tissue. There have been subsequent modifications to this flap, the first of which was by Adamson who altered its base from a cephalic to a caudal position [57]. Numerous experimental studies have described "McFarlane" flaps or "modified McFarlane" flaps [58–60]. The purpose of this dorsal rat flap model was to create a flap with predictable necrosis patterns to provide a scientific basis for the delay phenomenon. It has subsequently been used in ischemia reperfusion injury studies [61, 62]. The McFarlane flap and its modifications are technically simple and therefore can be easily reproduced with minimal training.

2.2 Techique: Elevating a Modified McFarlane Skin Flap

2.2.1 Materials

Sprague-Dawley rats weighing 250–400 g

Anaesthesia: 75 mg/kg Ketamine (Parnell Australia Pty Ltd, Alexandria, New South Wales, Australia), plus 10 mg/kg Ilium Xylzil-20 (Troy Laboratories Pty. Ltd. Glendenning, New South Wales, Australia). Make up 0.75 ml Ketamine plus 0.5 ml xylazil plus 0.75 ml of distilled water for injection. Dose is 0.2 ml/100 g of rat given i.p.

Scalpel blades: No 3 scalpel blade handle with no.15 blades

Skin holding forceps, 4/0 silk or skin clips

Bipolar coagulator, Chlorhexidine 0.5% in 70% ethanol

Analgesia: subcutaneous Carprofen (5 mg/kg of rat in 0.1 ml)

2.2.2 Methods

1. Anesthetize rats as above.
2. Place rats in a prone position on a heating pad.
3. Remove excess fur from dorsal back area with either an electric shaver or depilatory cream applied as per manufacturer's instructions.
4. Prepare skin with Chlorhexidine 0.5% in 70% Ethanol.
5. Palpate external anatomic landmarks: superior margins of the iliac crest, spinous processes and lower scapular angle (Fig. 1a).
6. Mark out a 3 cm × 9 cm flap extending cranially with the edges of the flap approximately 1.5 cm away from the spine and the base at the level of the iliac crests (Fig. 1b).
7. Excise the skin flap (Fig. 1c).
8. Elevate the skin flap ensuring inclusion of the panniculus carnosus, which is a layer of striated muscle deep to the subcutaneous tissue. Inclusion of the panniculus carnosus will have an important effect on skin flap viability.
9. Once intervention as per study protocol is performed suture flap back into position with 3/0 nylon or skin clips (Fig. 1d).
10. Ensure adequate post-operative analgesia and monitoring.
11. Monitor flap survival and harvest the flap at a later time point in a second operation.

2.2.3 Limitations

This model is a pure random pattern flap, which makes it difficult to consistently predict flap survival. The range of necrotic area that has been described is between

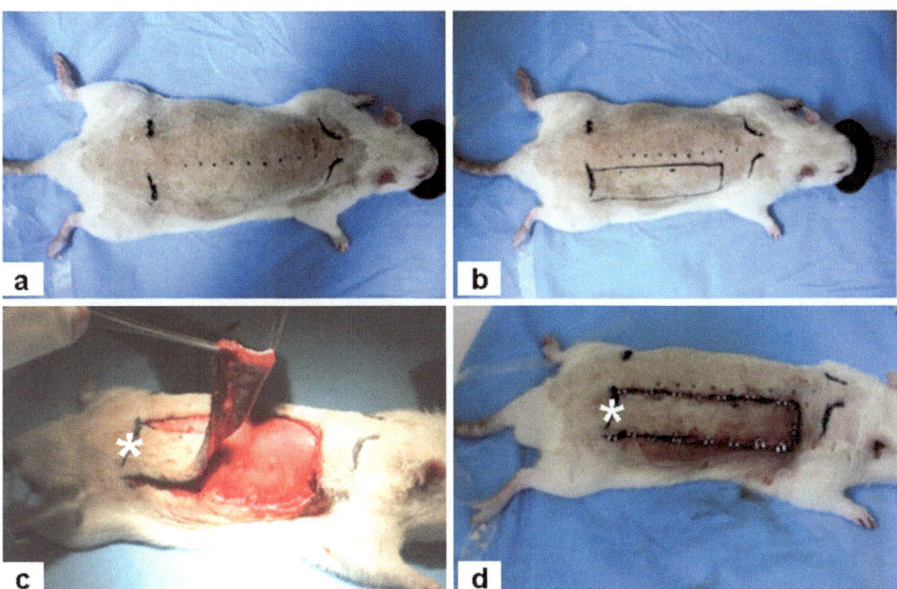

Fig. 1 The McFarlane flap (Modified). (a) On the back of the rat mark out preoperatively the lower tip of the scapula, the spinous process and the iliac crest. (b) Markings of the McFarlane flap boundaries. (c) Elevation of the flap with inclusion of the panniculus carnosus. (d) Insetting of the flap with skin clips. (Note the *white asterisks* in (c) and (d) indicate the intact skin bridge between the flap and the dorsal skin of the rat back)

22% and 52% [55]. A further limitation of the original McFarlane flap is that there are graft bed interactions, which can skew the skin viability profile. This inherent confounding variable makes it difficult to assess the effect of an isolated treatment on a flap. Studies of various designs have demonstrated the favourable effect of placing the flap back on its original bed or by re-exposing the flap to its bed at various time intervals. These studies demonstrated there was increased survival area of the flap with increased exposure time to the vascular bed [63, 64]. The distal end of the McFarlane flap can typically become necrotic. It has been shown that applying pressure can partially salvage this suggesting dependence of the flap on the original bed [65]. Modifications to overcome this confounding variable and thereby create predictable uniform levels of flap survival in a control group so that comparisons can be subsequently made to a treatment group, have included placing silicone, silastic sheeting or other material between the elevated flap and the donor site [63, 64, 66]. Although this is effective in producing uniform percentage survival it can lead to infection and can irritate the rat resulting in wound mutilation. Recently, Kelly has described a blind ended pedicle tube flap based on 3 cm × 9 cm McFarlane flap that extends cranially and is rolled into a tube before being secured to the back [67]. The advantage of the tubed model is that it eliminates nutrient supply from the bed and closes the donor site and the raw surface. Despite their modification there is no significant difference in survival area of this flap compared to the McFarlane flap (15.673 cm^2 +/− SD 3.37 vs 18.904 cm^2 +/− SD 3.79) which suggests there are other hitherto unknown factors influencing survival.

2.3 Dorsal Island or Axial Pattern Flaps

Limitations of the McFarlane random pattern flap include unpredictable necrosis and variations in skin vasculature [68]. With advancing understanding of vascular anatomy including the subcutaneous and angiosome concept, there was a reduced need for random pattern flaps in major reconstructive surgery. Axial pattern flaps have a large vessel that is included in the flap or have a large perforator at the flap base. There are situations in reconstructive surgery where a large defect requires a flap that extends beyond the axial or perforator territory of the flap and may result in necrosis of random parts of the flap. As a consequence there has been continual research to improve flap vascularity by pharmacological and/or surgical manipulation.

Island flaps are extreme versions of axial flaps where the skin is islanded and the whole subdermal plexus supplying the skin is cut, and the pedicle is reduced to a single arterio-venous axis. The majority of experimental studies using axial pattern flaps have been limited to flap survival studies but these flaps can be utilised for evaluating the benefits of ischemic and pharmacologic preconditioning; blood flow assessment techniques, and demonstrating the effects of arterio-venous shunts and the delay procedure on skin flap viability.

Increasing understanding of the dorsal aspect of the rat and mouse enabled new axial pattern and island flaps to be elevated. The rat back blood supply comprises overlap from three angiosomes (lateral thoracic; posterior intercostal and deep circumflex iliac vessels). Murine dorsal flaps can be elevated based on two angiosomes (lateral thoracic and deep circumflex iliac). Syed et al. [69] described the first rat dorsal axial pattern flap model, which these authors subsequently modified in 1997 to include a random portion [70]. Limitations of the modification existed due to the inconsistent vascular anatomy of the posterior intercostal arteries [71]. Yang and Morris attempted to overcome this problem by developing a dorsal island skin flap, which incorporated multiple vascular territories based on the unilateral deep circumflex iliac artery [68].

In an attempt to minimise animal numbers used, paired (bilateral) skin flaps have been designed. Hosnuter et al. described a paired dorsal skin flap model with both axial pattern and random flap characteristics with a comparable flap survival profile to single unilateral flaps [72]. A modification of the paired dorsal rat flap based on the deep circumflex iliac vessel has been suggested by Ohara [73]. Murine dorsal axial skin flaps have also been described. There is a slight variation in the murine dorsal vascular anatomy compared to that of the rat. Murine lateral flaps can be prepared using two angiosomes (lateral thoracic and deep circumflex iliac) on each side of the back [74]. Tatlidede et al. have developed a novel murine island single pedicle flap model, which has demonstrated reproducible injury in IP studies [32].

However the murine model involves a skin flap that is quite small, which makes it impractical for studies that require arterial drug infusion.

2.3.1 Limitations

As long as careful attention is paid to the anatomy, elevating an axial pattern flap in an animal model should result in a reliable flap. It is necessary to use standard anatomical markings so this can take into account animals of different sizes. It is important to note that there appears to be variation in the middle territory of the dorsal rat skin flap with differences in the origin of the cutaneous perforator from the posterior intercostal artery [68, 75]. Axial pattern/island flaps that are based on the deep circumflex iliac artery have a more consistent vasculature. There is a predictable area of distal necrosis of about 30%, which can be taken into consideration when conducting delay or preconditioning studies [68]. In studies assessing pharmacological agents, these can be injected directly into the flap via the iliac branch of the deep circumflex iliac artery entering the abdominal aorta. We have not used the dorsal island flap because of concerns about variability of its vascular source confounding the experimental findings.

2.4 Pedicled and Free Ventral Island Flaps

The pedicled ventral island flap on a rat model was initially described as a unilateral inferiorly based flap by Finseth and Cutting in 1978 [12]. It has since become a reliable flap for a number of interventions aiming to improve flap survival and hence angiogenesis. The flap is based on a reliable neurovascular bundle – the superficial inferior epigastric. It is possible to raise bilateral flaps each based on a single pedicle or as a double width flap based on both pedicles. Ligation of one of the neurovascular bundles creates a random flap on one side, and leaves an axial flap on the contralateral side. The random pattern flap has a standard and reproducible pattern of necrosis. The ligation of arteries and veins in different combinations, have enabled the study of different vascular combinations on ventral island flap viability [76].

Advances in microsurgical technique resulted in the development of free tissue transfers. A free flap is essentially an island flap that is detached from its vascular pedicle and then reattached (using microsurgical anastomotic techniques) to recipient vessels at another site distant from the donor site. Free flap animal models produce experimental conditions analogous to clinical situations allowing studies of flap physiology, design and function, and enable assessment of anastomotic risk on flap viability. Goldwyn reported the first animal free skin flap in a canine model based on the superficial epigastric vessels in 1963 [77]. The ventral island free flap was demonstrated on the rat by Strauch and Murray in 1967 [78].

The ventral flap has been extensively used in delay studies [12]; flap viability studies [79–81] and ischemia reperfusion studies performed by our group [82].

This flap has also been employed in numerous studies investigating pharmacological agents that could be used to promote flap survival [83–88].

A variation of the ventral island flap in rats and mice formed the basis of angiogenic studies conducted at the O'Brien Institute. Brown initially described a ventral island skin flap survival model where tissue survival was measured as an indicator of the development of collaterals around the ligated epigastric artery [89]. Thiele developed this model of bridging angiogenesis further. The epigastric artery was initially cauterized and after a period of 0–21 days for spontaneous angiogenic bridging of the vessel gap, the vessels were re-exposed and a 4 cm × 3 cm ventral epigastric island flap was raised and resutured into position. The percentage survival as assessed 5 days later was a measure of angiogenesis [90]. Kane and Furuta used this skin flap model to demonstrate the influence of nitric oxide synthase 2 (NOS 2) on flap survival through pharmacological and genetic inhibition [91, 92].

2.5   Technique: Elevation of a Ventral (Epigastric) Island Flap

2.5.1   Materials

Sprague-Dawley rats weighing 250–400 g.
Anaesthesia: 75 mg/kg Ketamine (Parnell Australia Pty Ltd, Alexandria, New South Wales, Australia), plus 10 mg/kg Ilium Xylzil-20 (Troy Laboratories Pty. Ltd. Glendenning, New South Wales, Australia). Make up 0.75 ml Ketamine plus 0.5 ml xylazil plus 0.75 ml of distilled water for injection. Dose is 0.2 ml/100 g of rat given i.p.
Scalpel blades: No 3 scalpel blade handle with no.15 blades.
Skin holding forceps, Dissecting scissors.
4/0 Silk or skin clips, Chlorhexidine 0.5% in 70% ethanol.
Operating microscope, Bipolar coagulator.
Analgesia: subcutaneous Carprofen (5 mg/kg of rat in 0.1 ml).

2.5.2   Method

1. Anaesthetize rats as above.
2. Place rat in a supine position on a heating pad positioning the extremities at a 45° angle relative to the trunk and secure with tape (Fig. 2a).
3. Remove excess fur from groin area and abdomen with either an electric shaver or depilatory cream applied as per manufacturer's instructions.
4. Mark out a rectangular flap of required dimension extending from the xiphoid process to the pubis centred in the midline (Fig. 2b).
5. Make an incision in the inguinal region to expose the femoral vessels and superficial inferior epigastric (SIE) vessels arising from it (Fig. 2c).

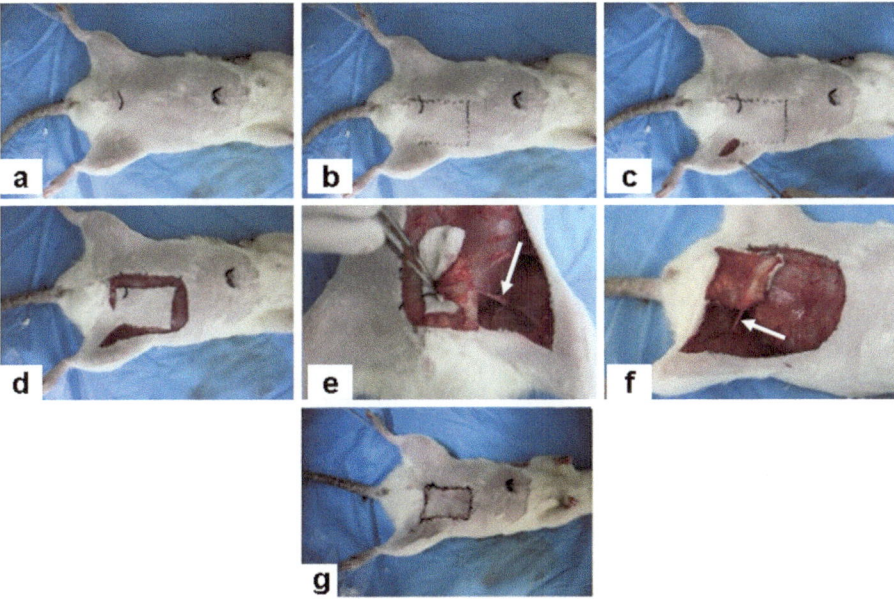

Fig. 2 The ventral epigastric island flap. (a) The rat is placed in supine position. Preoperatively mark out the xiphisternum and symphysis pubis on the rat skin. (**b**) Mark out the ventral island skin flap of required dimension. (**c**) Make a groin incision exposing the femoral vessels. (**d**) Elevate the skin flap extending caudally. (**e** and **f**) Dissect out the superficial inferior epigastric pedicle (*white arrows*). (**g**) Inset the flap into position with 4/0 silk

6. Dissect along the flap markings extending caudally (Fig. 2d). The flap should be carefully elevated with inclusion of the panniculus carnosus. Care should be taken on the lower lateral aspect of the flap not to damage the SIE vessels (Fig. 2e, f). The inguinal fat pad should be raised with the specimen to protect the pedicle from inadvertent injury.
7. Using the operating microscope dissect out the femoral vessels proximally and 2 cm distal to the origin of the SIE vessels.
8. Perform intervention required as per study protocol.
9. Suture flap back into position with 4/0 silk (Fig. 2g).
10. Ensure adequate post-operative analgesia and monitoring.
11. Monitor flap survival, harvest flap and pedicle at a second operation (the timing of which is defined by the study protocol) for histological assessment and/or perform a vascular perfusion.

2.5.3 Limitations

Limitations of the ventral island flap model include controversy over the wound healing in this model. Some have observed the primary flap healing from the edge towards the centre [93] whereas others have shown that new vessels arise from the wound bed thus explaining the early survival of epigastric flaps [94]. This can potentially interfere with the interpretation of various angiogenic augmentation interventions. Inclusion of the inguinal fat pad can potentially overcome this confounding factor [95]. Other factors which can also affect skin flap survival include the dimensions of the skin flap used, inclusion of the panniculus carnosus and the technique of pedicle dissection [87].

2.6   Prefabricated Flaps

Prefabricated flaps which depend on angiogenic connections from implanted blood vessels for their survival represent a technically more challenging skin flap that can be used in animal models. These flaps have furthered the understanding of angiogenesis within flaps. Prefabricated flaps have been described in rat [49, 96, 97]; and rabbit models [50, 98–100]. Different configurations of vascular implantation have been trialled for prefabrication including pre-existing vessels in continuity; arterio-venous shunts or loops (AVL) with or without a vein graft; or ligated arterio venous pedicles. Khouri demonstrated that an AVL can spontaneously generate de novo vascularised tissue when placed in between collagen discs and then placed in a silicone chamber. They also observed that the addition of platelet derived growth factor (PDGF) accentuated tissue formation [101]. The AVL model was further modified by Tanaka et al. who demonstrated that vascularisation occurred in an artificial dermis placed on the loop. The entire construct was stabilised by the addition of a skin graft onto the dermis [102]. This demonstrated that the AVL model has inherent vascularisation properties that can be used in flap prefabrication. Neovascularization from vein grafts alone have been reported experimentally [96, 103] but longer (8–10 cm) vein graft lengths were shown to correlate adversely with survival [99].

2.7   Technique: Prefabricated Skin Flap in a Rat Using an Arterio-Venous Ligated Pedicle

2.7.1   Materials

Sprague-Dawley rats weighing 250–400 g.
Anaesthesia: 75 mg/kg Ketamine (Parnell Australia Pty Ltd, Alexandria, New South Wales, Australia), plus 10 mg/kg Ilium Xylzil-20 (Troy Laboratories Pty. Ltd. Glendenning, New South Wales, Australia). Make up 0.75 ml Ketamine plus 0.5 ml xylazil plus 0.75 ml of distilled water for injection. Dose is 0.2 ml/100 g of rat given i.p.
Scalpel blades: No 3 scalpel blade handle with no.15 blades.
Skin holding forceps.

3/0 Nylon suture, Chlorhexidine 0.5% in 70% ethanol.

Operating microscope, Bipolar coagulator.

Analgesia: subcutaneous Carprofen (5 mg/kg of rat).

2.7.2   Methods

1. Anaesthetize rats as described above.
2. Place rat in a supine position on a heating pad positioning the extremities at a 45° angle relative to the trunk and secure with tape.
3. Remove excess fur from the groin area and abdomen with either an electric shaver or depilatory cream applied as per manufacturer's instructions.
4. Through a longitudinal incision in the groin extending from the medial end of the inguinal ligament down to the inner aspect of the knee expose the femoral vessels.
5. Dissect the femoral vessels free and ligate above the knee with a 4/0 silk suture and mobilize the arterio-venous pedicle proximally (Fig. 3a) to the point of origin of the deep vessels of the leg, which should be preserved.
6. A subdermal pocket should be dissected where the proposed flap will be lifted. The pocket is dissected along the axis of the midline of the proposed flap and the ligated femoral vessels carefully introduced into this pocket above the fascia of the underlying muscle and beneath the panniculus carnosus, the deepest skin layer (Fig. 3a, b).
7. There are several sites using this model where the pedicle can be buried, including lateral to the original site of the femoral vessels in the thigh (Fig. 3a, site 1), or move the pedicle in a cephalad direction, over the inguinal ligament, buried between the fascia over the abdominal muscles and the panniculus carnosis of the lower abdominal skin (Fig. 3a, site 2). The end of the pedicle in this position should reach at least 1 cm cephalad to the inguinal ligament. If a thinner skin flap is required the vascular pedicle may be introduced superficial to the panniculus carnosus.
8. Anchor the end of the pedicle to the undersurface of the overlying skin with a 4/0 nylon stitch.
   Ensure there are no kinks or twists in the vascular pedicle before skin closure. The rat is revived.
9. Administer analgesia.
10. In a second operation again under general anaesthesia, after a defined period of prefabrication as per study protocol the prefabricated flap based on this new pedicle can be elevated.
11. A rectangular flap (of required dimension) should be marked as per the ventral island flap described previously ensuring that the implanted pedicle runs down the midline.
12. Elevate the flap using the original longitudinal incision.
13. Perform intervention as required by study protocol.

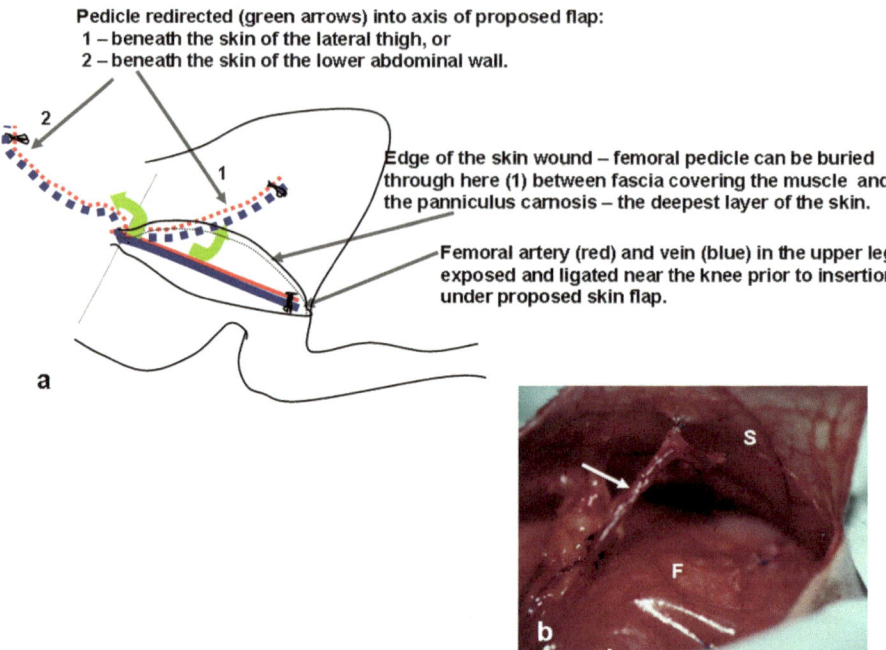

Pedicle redirected (green arrows) into axis of proposed flap:
1 – beneath the skin of the lateral thigh, or
2 – beneath the skin of the lower abdominal wall.

Edge of the skin wound – femoral pedicle can be buried through here (1) between fascia covering the muscle and the panniculus carnosis – the deepest layer of the skin.

Femoral artery (red) and vein (blue) in the upper leg exposed and ligated near the knee prior to insertion under proposed skin flap.

Fig. 3 The prefabricated flap model. (**a**) Diagram illustrating the prefabricated flap surgical method. The femoral artery and vein (*red* and *blue lines*) are isolated and ligated distally near the knee. This vascular pedicle is then buried between the muscle fascia and the overlying skin in either the lateral thigh (position 1) or the lower abdomen (position 2) as indicated by the *dotted red* and *blue lines*. (**b**) Photograph of the isolated and distally ligated femoral vascular pedicle (*white arrow*) which is being buried between the muscle fascia (*F*) and the overlying skin (*S*)

14. Suture flaps back into their original position with 4/0 silk.
15. Monitor flap survival, harvest flap and pedicle at a third operation (the timing of which is defined by the study protocol) for histological assessment and/or perform vascular perfusion.

2.7.3 Limitations

Prefabricated flaps are more technically challenging than previous flap models described, and a confounding factor will be operator skill. There is a risk of venous congestion which can compromise flap viability. This risk increases if there is folding of the pedicle or manipulation prior to transfer [104]. In the second stage of elevating a prefabricated flap, a pedicle which is scarred can endanger the viability of the flap [44]. Furthermore no optimal prefabrication period has been defined and there are conflicting results in different animal models. In early prefabricated flaps in a rabbit model it was demonstrated that 8–10 weeks were needed for maximal development of blood vessels [98] but subsequent studies demonstrated a significantly shorter prefabrication period of 10 days [103]. The key is in ensuring that there is accurate approximation of the vascular pedicle against the underlying skin to enable for early successful elevation and this may be subject to a learning curve. Disadvantages of using an AVL in a prefabricated flap is that it requires microsurgical skill to create an AVL and there is a higher risk of complications including loop thrombosis. For the study of *in vivo* angiogenesis, an arterio venous bundle (pedicle) (described in section 2.7) is simpler to construct, has a vigorous potential for angiogenesis when it is distally ligated, and neovascularisation can be demonstrated angiographically to arise from a large vessel such as the femoral vein [49].

3 Assessment of Skin Flap Survival and Angiogenesis

3.1 Skin Flap Survival

Macroscopically the area of skin flap survival and/or skin flap necrosis can be determined from visual inspection and then surviving and necrotic areas measured as an area. Ideally this should be measured by an independent blinded observer. The necrosed parts of the flaps can be quantitatively measured using an applied grid and point counting techniques or using image analysis techniques [92]. Surviving proportions of the flaps are expressed as a percentage of the whole flap area (surviving flap proportion = surviving flap area/total area × 100).

More generally skin flap survival has been equated with adequate blood supply to the flap which depends on intrinsic vascularization (that is, the flap vascular pedicle or a skin bridge in random pattern flaps). This can be compromised by poor flap design, tight closure or pressure from a haematoma. In addition, various techniques applied to flaps, that is, delay, ischemic preconditioning, and prefabrication using a vascular pedicle may enhance this intrinsic vascularization.

To explore microcirculatory development and extent in skin flaps numerous indirect and direct methods of blood vessel visualization both *in vivo* and *in vitro* have been developed. The following section will describe some of the most commonly used approaches to analyze angiogenesis in skin flap experimental animal models.

3.2 In Vivo *Visualization and Quantification of the Microcirculation*

The most commonly employed method of *in vivo* visualization of skin flap angiogenesis is by intra-vital fluorescence microscopy. Initially to visualize microcirculation in skin, sodium fluorescein [105] and disulphine blue [106] were used. However, these dyes have been largely replaced by more stable fluorochromes such as indocyanine green [107, 108]. Newer fluorescent dyes including fluorescently-conjugated lectins [109, 110] and more recently lipophylic carbocyanine dye DiI (1,1′-dioctadecyl-3,3,3′,3-tertamehyllindocarbocyanine perchlorate) [111] have also been perfused into the flap vascular system enabling enhanced visualization of blood vessels. In addition radio-opaque materials such as barium sulphate (Micropaque) and lead oxide have also been used [112]. By injecting these dyes and radio-opaque materials either via the venous circulation or by cardiac perfusion and visualizing by repetitive intra-vital microscopic analyses, fluorescent angiography or using radiological instruments such as X-rays and computed tomography (CT)-scans, it is possible to visualize the main vessels and their branches, and if using confocal microscopy with fluorescent dyes, the capillary network can be visualized. The assessment of functional capillary density is done by counting the blood vessels visualized either *in vivo* by intra-vital fluorescence microscopy and/or angiography or ex-vivo by processing the skin flap as a whole mount for fluorescent microscopy [113, 114].

The use of many intravascular fluorescent dyes such as sodium fluorescenin and disulphine blue is not suitable for humans as fluorochromes can precipitate unwanted side effects such as nausea, vomiting, itching, flushing and pain [115–117] and possibly phototoxic reactions in the body [118]. Furthermore, the design of the intra vital fluorescence microscope and its limitation to visualizing thin grafts makes this instrument unsuitable for clinical use.

A number of alternative techniques have been developed with various success in measuring skin flap perfusion – these include the laser Doppler flowmeter [119], which quantifies dynamic intrinsic capillary blood flow, but is known to be technically difficult and requires baseline standardization for each measurement; and computed tomographic angiography (CTA) and photoplethysmography (PPG) [120].

Use of orthogonal polarisation spectral (OPS) imaging in assessing microcirculation in both animals and humans has been reported [121]. OPS imaging directly visualizes hemoglobin and hemoglobin carrying structures in the microcirculation using polarized reflected light at a wavelength of 548 nm without the aid of fluorescent dyes. OPS imaging permitted visualization of the microcirculation in both the base and distal part of a single pedicle cutaneous flap in a mouse and permitted quantification of functional capillary density that was directly correlated to functional capillary density estimated by intra-vital fluorescence microscopy [122]. Furthermore, Langer et al. reported that OPS images were superior to intra-vital fluorescent microscopy images due to the absence of any leakage effect seen when fluorescent dyes are used [122]. Olivier WA et al. have demonstrated the high accuracy of OPS when predicting ultimate flap necrosis of random pattern skin flaps in rats [123]. Like intra-vital fluorescence microscopy, OPS can only be used to visualize the surface of the skin flaps. The development of new instruments capable of visualizing microcirculation in thicker tissues and improved resolution is likely to overcome current technical limitations to visualization of microvasculature. In contrast, use of radiography enables visualization of vasculature in thick tissues, however imaging resolution of small capillaries remains an issue which limits the application of radiography when quantifying capillary vascular networks.

In particular, for work performed in animals, conventional angiography is limited to visualizing small blood vessels with diameters >200 μm *in vivo* [124].

To study the microcirculation in flaps, various skin-fold chamber models have also been developed. For example, to assess microcirculation in a bi-pedicle island flap on the mouse dorsum, Ichioka et al. designed an adjustable small skin-fold chamber consisting of two transparent frames held together with polyethylene tube [125]. By visualizing the microcirculation of the island flap through the skin-fold chamber, the authors were able to follow changes in vessel morphology within the skin flap. However, while the model permits morphologic observations, it does not permit accurate quantification or study of the inosculation and sprouting as only a small segment of the flap with previously established microcirculation is seen. Similarly by using the dorsal skin-fold chamber, Lindenblatt et al. transplanted a full-thickness skin graft from the mouse groin to the mouse dorsum so that both the wound bed and the graft could be visualized through the chamber window using intra vital fluorescence microscopy. The visualization was captured by photography and functional capillary density assessed by computer-software assisted image analysis [113].

3.3 Visualisation of the Microcirculation by Space-Occupying Compounds

It is also possible to visualize the microcirculation by perfusing the skin flap whilst in situ with space-occupying materials such as india-ink/gelatine [100], fluorescently conjugated dextran [113, 126–128], low viscosity resins such as methacrylate [129] or Mercox resin [51]. The perfused tissues are then removed and either processed as whole mounts by clearing the tissues in compounds such as xylene, cedar-wood oil or methylsalicylate or alternatively, in the case of infused resins, eroded in weak acid to obtain a cast of the perfused vascular tree. In the case of india ink/gelatine perfusions where the tissue is cleared in xylene, cedar-wood oil or methylsalicylate, whole mount preparations visualizing the entire flap vascular network and pedicle can be prepared, or the flap can be sliced and prepared for paraffin embedding, sectioning and staining to visualize [100] and/or morphometrically count blood vessels within the flap by light microscopy. In contrast, resin casts are best visualized by scanning electron microscopy or confocal fluorescent microscopy [130]. To visualize an india-ink/gelatine perfused microcirculation in rabbit skin flaps, the following protocol has been used in our laboratory.

3.4 Technique: Skin Flap Prefusion to Demonstrate Development of the Flap Microcirculation

The following double labeling technique was established by Hickey et al. in 1998 [100] in a prefabricated flap model similar to that described in '2.7 Technique: A prefabricated skin flap in a rat using an arterio-venous pedicle' (the femoral pedicle) in this Chapter. This method demonstrates both preexisting skin flap vessels and new vascular connections between the transposed vascular pedicle and the overlying skin flap.

3.4.1 Materials

India ink, Monastral blue (Sigma Chemical Co, St. Louis, Mo.), cedar wood oil (Auroma, Australian Botanical Products, Hallum,Victoria, Australia). Normal saline, perfusion cannulae, surgical instruments.

3.4.2 Method

To determine the pattern of vessels arising from the implanted pedicle and to examine the vascular connections between the implanted pedicle and the preexisting dermal vasculature of the flap, either a single-injection or double-injection technique can be used under general anaesthesia for a rat as described in 2.7.

1. For double-injection preparations, at the desired time after vascular implantation and under general anesthesia the original incision (used to isolate the femoral pedicle) is reopened and patency of the pedicle confirmed.
2. The femoral artery in the implanted pedicle is cannulated approximately 3–4 cm proximal to the flap, and the abdominal aorta exposed and cannulated by means of a midline laparotomy.
3. With the pedicle supplying the flap occluded by the cannulae and the lateral margins of the flap unelevated, the abdominal aorta is flushed with normal saline and perfused with 20 ml of India ink. Despite the altered femoral vasculature, perfusion in this manner labels much of the skin of the hind limbs, including through vascular connections with adjacent skin, the preexisting vasculature of the outlined skin flap.
4. To label exclusively the newly formed vessels arising from the implanted pedicle and any connections to preexisting dermal vessels, the artery of the pedicle is flushed with normal saline and perfused with 5 ml of Monastral Blue.
5. Vessels filled with this vivid blue suspension and doubly labeled vessels can be distinguished easily from vessels filled with India ink alone.
6. After perfusion, the pedicles are ligated and divided proximal to the ligature, and the skin flaps are raised as islands, pinned flat on cardboard, and fixed immediately in 10% buffered formalin.
7. Following fixation, the excised flaps and their pedicles are dehydrated in graded ethanol, and cleared in cedar wood oil. Once the surrounding tissues appear clear (no color, that is, transparent) and the blood vessels are clearly visible the flaps can be examined under a dissecting microscope and the vascular pattern photographed.

8. Vertical sections of the flap can also be subsequently embedded in paraffin and prepared for immunohistochemical labeling of blood vessels (see Appendix).

4 The Use of Vascular Pedicles for Angiogenic Tissue Generation and Assessment in Tissue Engineering

The work of Erol and Spira [41, 131] in which isolated vascular pedicles and loops were used to vascularize adjacent skin regions and transplanted skin grafts (leading to the concept of flap prefabrication), also inspired others to use these pedicle models to generate new vasculature unassociated with a skin flap. The severed vascular bundle model of Khouri et al. [101] and Tanaka et al. [49, 102] demonstrated that surgically isolated vascular pedicles could intrinsically and spontaneously generate new, vascularized tissue when sandwiched between sheets of collagenous matrix or when separated from the surrounding tissue within a plastic chamber. These observations led to adaptation of Erol and Spira's arterio-venous loop (AVL) model as the main source of new blood vessel development within tissue engineered constructs [47, 48]. This concept is now known as the intrinsic vascularization model of tissue engineering [132].

In the last decade, as part of a tissue engineering construct vascularization strategy, two animal models have been developed at the O'Brien Institute, in which complete, new arterio-capillary-venous networks are developed by direct sprouting off major vascular pedicles (continuous with the systemic circulation) which are isolated in a 'tissue engineering' chamber *in vivo*. These models enable the formation of a complete new microcirculatory network and supporting tissue suitable for the seeding of stem/progenitor cells and extra-cellular matrices for tissue engineering of specific tissues or organs. The two models are: the mouse flow-through chamber model [133] and the rat AVL chamber model [47, 48, 51]. The growth of these new microcirculatory beds is stimulated by a pro-inflammatory microenvironment, and no doubt the empty chamber space. Commencing with increased vessel permeability that permits endothelial cell (EC) migration and invasion into the surrounding three-dimensional matrix (Matrigel™ in the mouse chamber, spontaneously formed fibrin matrix in the rat chamber), subsequent development into fully functional blood vessels comprised of endothelium, pericytes and smooth muscle cells occurs, and later vessel remodeling and some blood vessel regression also occurs [51].

4.1 The Mouse Flow Through Pedicle Chamber Model

The mouse flow through pedicle chamber model is based on wrapping a longitudinally split silicone tube (termed the chamber) around the epigastric pedicle in the groin which has been cleared of vascular branches and attached connective tissue [133].

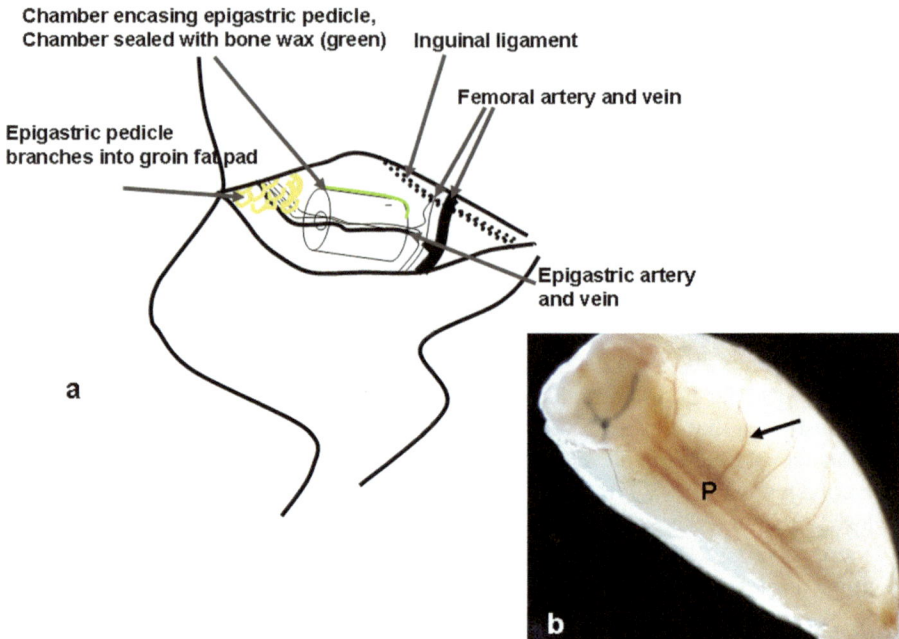

Fig. 4 The mouse flow through pedicle chamber model. (**a**) Diagram illustrating the mouse flow-through pedicle model in the groin of a mouse. The epigastric vascular pedicle branching from the femoral vessels, is cleared of connective tissue and the groin fat pad. A silicone tube (the chamber) is sealed around the epigastric pedicle and contains Matrigel +/− stem or progenitor cells +/− growth factors. (**b**) The mouse chamber construct at harvest at 6 weeks. The vascular pedicle (*P*) is clearly visible on one side of the construct. New vessel branches (*arrow*) have sprouted off the pedicle since the chamber was placed around the pedicle ((**b**) **is Figure 1 e** reprinted from [141], with permission from the American Society of Investigative Pathology)

The longitudinal split and one end of the silicone tube are sealed with dental wax, and the chamber volume is then filled with the desired scaffold such as Matrigel™ with or without stem/progenitor cells. The silicon tube is then sealed with dental wax at the other end (Fig. 4a). Over the next 3–4 weeks new blood vessels sprout off the flow-through pedicle within the chamber and migrate into the scaffold to form a new vascular network (Fig. 4b) able to support the addition of exogenous cells such as: pituitary stem cells [134], thymus tissue [135], pancreatic islets [136–138], liver progenitor cells [139], adipose tissue [140] and support stimulated adipose tissue growth under the influence of growth factors [141]. Different extra-cellular matrices have been used to support the angiogenic growth [142]. The advantages of this model are that genetically modified mice can be used to examine angiogenic development, a chamber can be inserted in each groin, allowing control and experimental chambers in the same mice, and exogenous factors can be easily added to the chamber and sealed off from the surrounding environment. The mouse chamber model can also be created on other vascular pedicles within the mouse (Findlay MW, manuscript in preparation) or in other animals. This is a highly suitable *in vivo* model for assessment of angiogenesis under the influence of various growth factors (or cytokines) as demonstrated in Rophael et al. [141]. Due the relatively "sealed off" nature of this chamber model and the ease of retrieving the chamber, the model can be used as a high through put screening tool for growth factors, or differentiation factors in tissue engineering studies, or anticancer drugs, or other screening applications that are beyond tissue engineering.

4.2   Technique: The Mouse Flow Through Pedicle Chamber Model

[Modified from Cronin et al. [133]].

4.2.1   Materials

Silicon tubing (Dow Corning, North Ryde, New South Wales, Australia) melted bone wax (Ethicon Bone Wax™, Sommerville, NJ, USA), Matrigel™ (BD Biosciences, Mississauga, Ontario, Canada), Bipolar coagulator, microsurgical instruments, sutures (Hygeian Medical Supplies Pty Ltd Singapore), operating microscope (Carl Zeiss Microscopes, Jena, Germany).

Mice: (including knock out or transgenic mice) should be over 8 weeks of age, 20–25 g body weight in a healthy condition.

Anaesthesia and analgesia: Preoperatively mice should receive a subcutaneous injection of the analgesic Rimadyl 5 mg/kg (Pfizer Pty Ltd. West Ryde, New South Wales, Australia).

The operation is performed under general anaesthesia giving chloral hydrate intraperitoneally at 4 mg/g body weight or 0.25 ml of 4% choral hydrate in saline/ 25 g mouse and maintained using isoflurothane 1–2% in oxygen 2 L via a nose cone.

4.2.2   Method

1. Use a depilatory cream to render the groins and upper legs hair free, remove hair with a wet swab. Decontaminate the hairless skin areas with 0.5% chlor-hexidine in 70% alcohol.
2. Place the mouse on its back on a heated pad, under sterile conditions and view under an operating microscope.
3. Using a size 15 scalpel make a transverse groin incision just above the groin fat pad.
4. Using microinstruments – Wescott scissors and a No. 2 jeweller's forceps dissect the superficial epigastric vessels free from the surrounding connective

tissue from their origin at the femoral vessels to their point of entry into the groin fat pad (a distance of approximately 1 cm).

5. Mobilise the entire fat pad from the overlying skin and underlying muscle, creating a space into which the chamber will be introduced.
6. Slit a 6 mm length of silicone tubing (3–4 mm wide) with one cut longitudinally. Encase (wrap around) the first 8 mm of the superficial epigastric vessels (where they are free from the fat pad) with the silicone tube (that is a cylindrical chamber of 6 mm length, 3–4 mm diameter, and volume of approximately 45–50 µl).
7. The tubing is anchored to the subcutaneous tissue with a 10/0 anchoring suture, to prevent the pedicle from being dislodged during subsequent movement.
8. The chamber is sealed at the proximal (femoral) end and along the lateral split with melted bone wax, taking care not to apply the heated wax directly to the epigastric vessels. The wax seal is augmented with two 10–0 nylon micro-sutures, placed at both ends of the lateral split, and the entire chamber is anchored to the underlying muscle (with 10–0 nylon sutures) near the origin of the superficial epigastric vessels, to prevent the pedicle from being dislodged during postoperative mobilization of the animals.
9. The chamber (with the epigastric pedicle intact) is then filled with an appropriate extracellular matrix (Matrigel™) +/− stem cells +/− growth factors.
10. The open end of the chamber is now sealed with bone wax.
11. The entire construct is carefully placed in the groin so that it lies in the dissected space lateral to the femoral vessels (Fig. 4a).
12. The wounds are closed with metal clips.

[Bilateral chambers can be created using both epigastric pedicles, and the chambers can be left *in situ* for an extended period – at least 12 months [143]].

4.2.3 Chamber Harvest

1. Under general anaesthesia (as above) the skin wounds are opened.
2. The epigastric pedicle and chamber are cleared of all connective tissue, taking care to keep the epigastric pedicle entering and exiting the chamber intact.
3. Cut the epigastric pedicle at the distal end close to the chamber and observe and record any bleeding that would indicate pedicle patency.
4. Cut the pedicle at the opposite end and lift the chamber construct (chamber and contents) out of the animal.
5. Remove the bone wax from each end and along the longitudinal split and gently lift out the chamber contents (referred to as 'the construct')- which should, after 3–4 weeks *in situ* be a solid piece of tissue with the pedicle clearly visible running along one side (Fig. 4b).
6. Immediately measure the weight and volume of the construct (see Appendix) prior to fixation – generally in 10% neutral buffered formol saline or 4% paraformaldehyde.

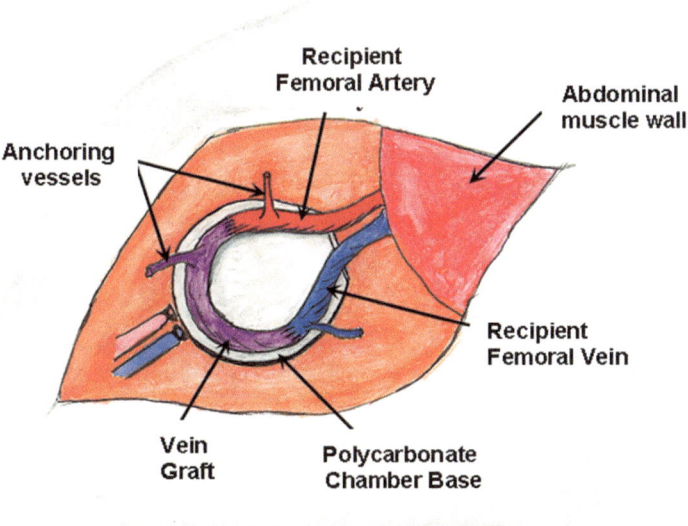

Fig. 5 The rat arterio-venous loop model. Diagramatic representation of the rat arterio-venous loop model, illustrating the vein graft anastomosed to the femoral artery and vein, and the loop positioned on the chamber base. (Illustration by Dr Zerina Lokmic. Modified figure from [155].)

4.3   The Rat Arterio-Venous Loop Chamber Model

The rat *in vivo* AVL model of tissue engineering consists of an AVL, that is, a surgically created diversion of blood flow from an artery directly into a vein, generally with an interposed autologous vein graft harvested from the opposite leg (Fig. 5). The AVL is enclosed in a polycarbonate chamber which is anchored by sutures to the muscle fascia of the rat groin [47].

When the AVL is enclosed into a polycarbonate chamber (Fig. 6), without addition of an exogenous extracellular matrix, it generates intense spontaneous *de novo* vascularized connective tissue that persists for at least 16 weeks [48, 51]. The new vessel development arising from the AVL (Fig. 7a, b) occurs in a wound healing-like environment characterized by AVL vessel hyper-permeability, deposition of an extra-vascular fibrin matrix, transmigration of inflammatory cells from

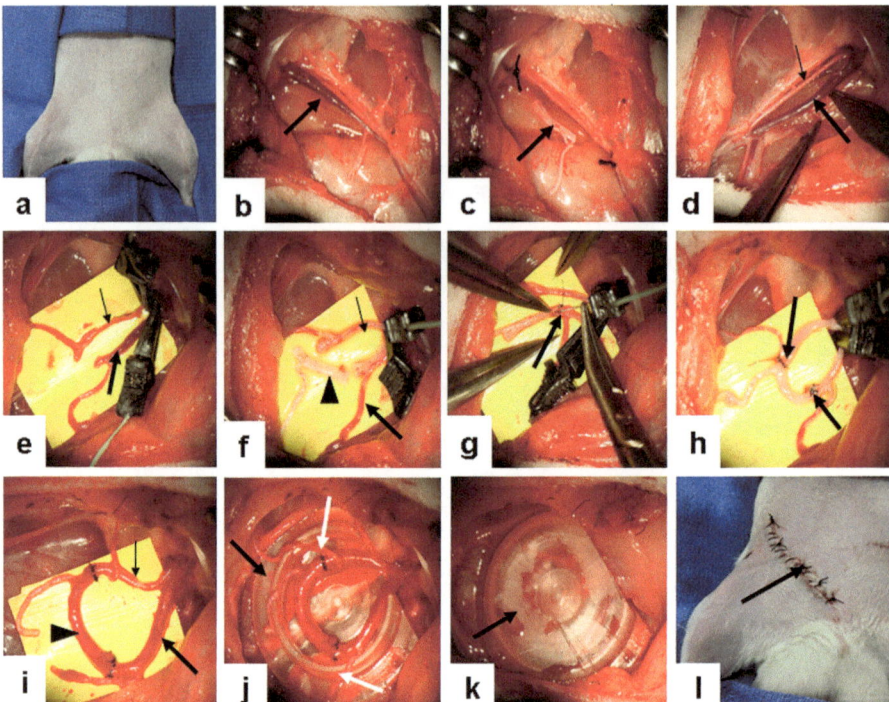

Fig. 6 Surgical creation of the rat arterio-venous loop chamber model. (**a**) The rat is placed in a supine position. Bilateral incisions are then made in the groin of a rat. The left side incision for vein graft harvest and the right side incision for creation of the AV loop and insertion of the chamber. (**b**) The left femoral vein, artery and nerve are exposed and the vein dissected free. *Arrow:* left femoral vein. (**c**) Segment of vein (*arrow*) removed after proximal and distal ligation. Note side branches included with the graft to act as an anchor to stabilize the AV loop in the chamber. (**d**) Right groin opened and femoral vessels (*small arrow*: femoral artery, *large arrow*: femoral vein) exposed and dissected free from each other down to the level of the knee. Note the side branches of both the artery and vein are retained to also anchor the loop. (**e**) The femoral artery (*small arrow*) and vein (*large arrow*) are ligated distally each with their side branches and the proximal vessels clamped with microsurgical clamps. (**f–h: f**) The vein graft (reversed, *triangular arrow* head) is then interposed between the proximal cut ends of the femoral vessels (*small arrow*: artery, *large arrow*: vein) and anastomosed with interrupted 10–0 nylon sutures. Anastomosis to the femoral vein is complete in (**g**) (*single arrow*) and completed to the femoral artery (two anastomoses now complete) in (**h**) (*two arrows*). (**i**) Once the anastomoses are complete the clamps are removed and flow is confirmed in the loop. *Triangular arrowhead*: vein graft in loop, *small arrow*: femoral artery segment of loop, *large arrow*: femoral vein segment of loop. (**j**) The loop is placed on the chamber base plate (*black arrow*) using the side branches as an anchors or handles (*white arrows*) to maintain its position. (**k**) The chamber lid (*arrow*) is then clipped onto the rim of the chamber base locking the branches between the base and lid. (**l**) The skin wounds are closed (*arrow*)

the circulation as well as the migration of blood borne stem and progenitor cells into the fibrin matrix. In a detailed study from 3 to 112 days Lokmic et al. [51] demonstrated that at 3 days a fibrin exudate surrounds the AVL and acts as a scaffold for inflammatory, endothelial and mesenchymal cells, with the first

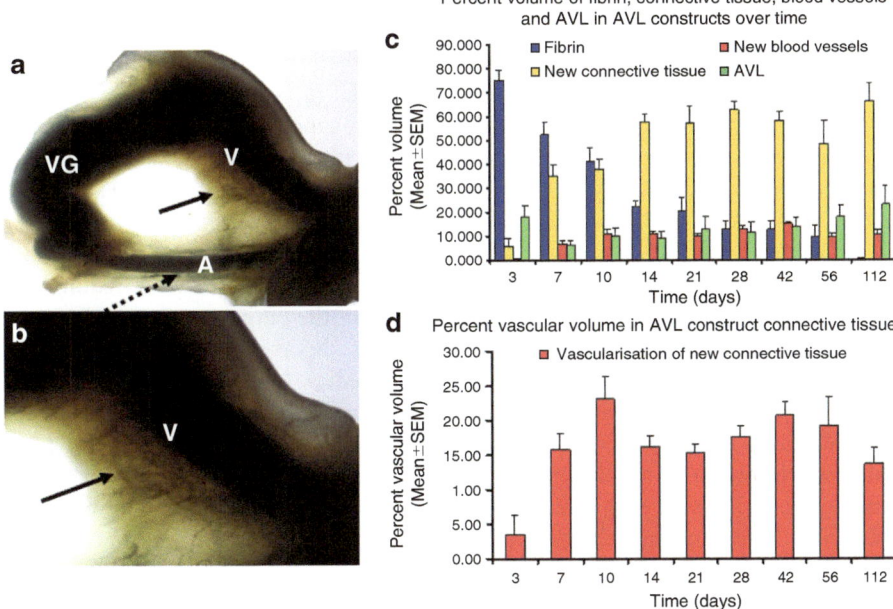

Fig. 7 (**a**) Resin cast of an AVL in a rat chamber at 10 days. *Arrow* indicates capillaries sprouting off the femoral vein (*V*) segment of the loop; *dotted arrow* indicates an arteriole sprouting off the femoral artery (*A*) segment of the loop, *VG*: vein graft segment of the loop. (**b**) Higher power view of (**a**), indicating capillaries (*arrow*) sprouting off the femoral vein (*V*). (**c**) Percent tissue components (fibrin, new connective tissue, new blood vessels and the AV loop) over 112 days of the entire AVL construct (reference compartment), measured by standard morphometric counts as described in the Appendix. (**d**) Percent vascular volume within the new connective tissue (reference compartment) in the AVL over 112 days, measured by morphometric counting. Note peak new blood vessel development is at 10 days ((**a–d**) republished with permission of the FASEB Journal. They are Figures 2I, 2J, 7A and 7B from [51])

capillaries arising from the femoral vein wall of the AVL. A peak of vascular development was observed at day 10 post-implantation (Fig. 7d), an event preceded by maximal cellular proliferation [51]. By day 7, some of the fibrin matrix around the AVL was replaced by newly formed connective tissue and hyper-permeable new capillaries. The first arterioles were observed at 7–10 days post-implantation, whereas definitive new venules were evident at 14–21 days. All vessel types persisted for a minimum of 16 weeks. However, isolated events of vessel remodeling were evident as early as 21 days post implantation and at later time points capillary apoptosis was accompanied by arteriolar apoptosis with pericytes and SMCs positively labeled with TUNEL. Maximal cell apoptosis was observed at 112 days (16 weeks) suggesting that the vessels continue to remodel [51]. The study also clearly showed that whilst the rat AVL model acts as a potent angiogenic source suitable for tissue engineering purposes, it is also a unique environment whereby all processes involved in vessel development, maturation, remodeling and regression can be studied *in vivo*. Furthermore, the model is open to manipulation so that specific growth factors and cytokines can be administered [144] and the effect of different natural, semi-synthetic and synthetic extra-cellular matrix (ECM) scaffolds tested [145–147]. Furthermore, the model supports implantation, survival and growth of tissue flaps [148], and implanted exogenous stem and progenitor cells to generate specific tissues such as cardiac [149, 150] and adipose tissue [151] and skeletal muscle [52]. The AVL chamber model is highly angiogenic, and this angiogenic growth can be visualized three dimensionally with vascular perfusions [51] (Fig. 7a, b) and readily quantified (Fig. 7c, d) [51] (see Appendix).

The disadvantage of this model when studying angiogenesis is the requirement of microsurgical skills to create the AVL, while for the purposes of screening pharmacological agents, growth factors and cytokines, this is also an expensive model. A simpler, more economical (allowing two pedicle-chambers/rat), and almost as efficient model to produce significant angiogenesis is to use an AV pedicle [as described for the prefabricated flap model [49]] rather than an AV loop which involves insertion of a vein graft. The following section describes materials and methods employed to surgically create a rat AVL model. The following Appendix describes how to quantify angiogenesis in the mouse and rat chamber models.

4.4   Technique: The Rat Arterio-Venous Loop Chamber Model

4.4.1   Materials

Sprague-Dawley rats weighing 280–320 g from ARC Western Australia.
Anaesthesia: 75 mg/kg Ketamine (Parnell Australia Pty Ltd, Alexandria, New South Wales, Australia), plus 10 mg/kg Ilium Xylzil-20 (Troy Laboratories Pty. Ltd. Glendenning, New South Wales, Australia). Made up 0.75 ml Ketamine plus 0.5 ml xylazil (Pfizer Pty Ltd. West Ryde, New South Wales, Australia), plus 0.75 ml of distilled water for injection. Dose is 0.2 ml/100 g of rat.
Operating microscope (Carl Zeiss Microscopes, Jena, Germany)
Scalpel, blades: No 4 scalpel blade handle with #15 blades (Swann Morton, Sheffield, England).
Sutures: 6–0 Surgipro suture (Syneture, Norfolk, USA) for suturing the chamber in place and 4/0 suture (Dynek Pty Ltd, Hendon, South Australia, Australia), 10/0 sutures (Hygeian Medical Supplies Pty Ltd Singapore) for AVL anastomoses.
Chlorhexidine/alcohol disinfectant: Chlorhexidine 0.5% in 70% Ethanol
Analgesia: Rimadyl (Carprofen 50 mg/ml, Pfizer Pty Ltd. West Ryde, New South Wales, Australia). Dilute 1 in 10 of sterile water or water for injection. Dose is 5 mg/ kg s/c. The volume injected is 0.1 ml of diluted Carprofen per 100 g of rat given at the time of surgery.

Sterile polycarbonate chamber – base and lid (currently manufactured by Department of Chemical and Biomolecular Engineering, University of Melbourne, Parkville, Victoria, Australia).

4.4.2 Method

Surgical Formation of AVL Constructs

1. Anaesthetize rats (200–250 g) see above, and place in a supine position on a heating pad (Fig. 6a).
2. Remove excess fur from both groin regions using an electric shaver and treat the groin with alcoholic chlorhexidine.
3. To harvest a vein graft, under the operating microscope make a longitudinal incision over the palpable left femoral vessels from the distal femur to the inguinal fat pad to expose the femoral vessels up to the inguinal ligament (Fig. 6b).
4. Dissect the left femoral vein including a length of its superficial inferior epigastric branch free from the surrounding artery and connective tissue and tie with 6.0 silk at both ends. This provides a total length of approximately 1.1 cm between the inguinal ligament and the knee. Ensure that the vessel adventitia is also gently removed
5. Cut the ligated vessel (Fig. 6c) (length of 1.1 cm), irrigate with heparinised saline, and submerge the vein graft into sterile heparinised saline solution.
6. Make a longitudinal incision over the right groin and the underlying inguinal fat pad to expose the right femoral vessels and separate them from the surrounding connective tissue (Fig. 6d).
7. Ligate and divide both the right femoral artery and vein as distal as possible in the thigh, close to the saphenous bifurcation with 6.0 silk. Apply micro clamps proximally in preparation for anastomosis of the intervening vein graft between the proximal stumps of the artery and vein (Fig. 6e).
8. Flush the cut ends with heparinised saline using a 25-gauge needle with silastic tubing on its end, and trim the vessel adventitia.
9. Retrieve the vein graft from heparinised saline and interpose the graft between the right femoral artery and the right femoral vein, with the distal end of the vein graft anastomosed to the artery and the proximal end of the graft to the femoral vein (Figs. 6f–h). Anastomosis is performed using simple interrupted 10.0 nylon sutures. The approximate number of stitches/anastomosis is six-eight.
10. To check the patency of the arterio-venous loop anastomoses, release the micro clamps (venous clamp first) and allow blood flow through the loop (Fig. 6i).
11. Once no leaks are detectable in the loop, place the sterile polycarbonate chamber base under the AVL and suture the base to the inguinal ligament at two points and to the muscle fascia of the groin with 6.0 nylon sutures at three points (Fig. 6j).

12. To anchor the AVL to the chamber base, use the superficial inferior epigastric branch vessels as guy ropes. These vessels will act as the anchoring points ("handles") of the AVL (Fig. 6j).
13. To create the chamber, close the lid over the base containing the AVL so that the "vessel handles" are sandwiched between the opposing edges of the chamber base and the lid and the femoral artery and vein enter and leave the chamber by the entrance/exit window (Fig. 6k).
14. Pull the edges of the inguinal fat and the skin over the chamber and suture the edges with 4/0 silk suture in two layers to create a wound that would heal through primary intention (Fig. 6l). Allow animals to recover from anaesthesia by leaving them on a heated pad until conscious.
15. Provide analgesia as previously described over the following 48 h and monitor animal recovery.

Harvest of the AVL Tissue Construct

1. To harvest the AVL containing chamber, anaesthesise the rat and place in a supine position on a heating pad.
2. Treat the groin with alcoholic chlorhexidine and under the operating microscope reopen the previous incision to expose the chamber and proximal femoral vessels.
3. Carefully dissect away the fibrous capsule that has grown around the chamber since insertion, and dissect the connective tissue associated with the femoral artery and the vein leading to the chamber (approximately 3–5 mm from the entrance to the chamber).
4. To check the patency of the construct (defined as: the contents of the chamber, that is, the AVL and new tissue growth around it), briefly clamp the femoral artery proximal to the chamber and visually assess the blood flow following the release of the clamp.
5. If the construct is patent, doubly ligate each vessel as it enters the chamber and divide the vessels between the ligatures.
6. Open the polycarbonate chamber by placing the tip of blunt forceps at the chamber entry point, and remove the tissue construct away from the chamber base. Measure construct weight and volume as described in Appendix.
7. Euthanise the animals by intraperitoneal injection of Lethobarb (200 mg/kg).

Acknowledgements The authors appreciate the technical assistance of Ms Sue Mc Kay, Ms Liliana Pepe and Ms Anna Deftereos in the Experimental Medical and Surgical Unit at St Vincent's Hospital, Melbourne, We also thank Jason Palmer (O'Brien Institute) for his technical advice.

The O'Brien Institute acknowledges the Victorian State Government's Department of Innovation, Industry and Regional Development's Operational Infrastructure Support Program.

Appendix

Methods for Rat AVL Construct and Mouse Flow Through Pedicle Construct and Skin Flap Tissue Processing, Visualization and Morphometric Assessment of Blood Vessels

Rat AVL Construct or Mouse Flow Through Pedicle Construct Weight Measurements

1. Place an empty 50 ml container on a four decimal balance (Mettler AE 260, DeltaRange®, Mettler-Toledo, Switzerland) and record the initial weight ($W_{initial}$).
2. Blot each construct (rat or mouse) carefully on Whatman filter paper then place the construct in the container on the balance and record the resulting weight (W_{final}).
3. Determine the weight of the AVL construct or mouse chamber construct ($W_{construct}$) by subtracting the initial weight from the final weight recorded: $W_{construct} = W_{final} - W_{initial}$.

Rat AVL Construct or Mouse Flow Through Pedicle Construct Volume Measurements

AVL or mouse construct volume is determined by a method described by Scherle in 1970 [152] in which the water displacement due to construct volume is determined by weighing in saline. To weigh the newly formed AVL construct

1. Tie off the AVL construct artery and vein or mouse chamber construct artery and vein at the point of entry into the chamber (approximately) with a long piece of 6.0 silk thread at harvest.
2. Fill a 50 ml plastic container with 40 ml of isotonic sterile saline (0.9% NaCl) solution, place on a balance and record the initial weight (W_1).
3. Blot the AVL construct or mouse chamber construct on Whatman paper to remove excess fluids from the construct.
4. Attach the AVL or mouse flow through pedicle construct to a silk thread and lower the specimen suspended on the thread into the saline solution so that it is beneath the surface of the liquid but not touching the bottom or the sides of the container.
5. Record the resulting weight (W_2).
6. To calculate the weight of the fluid displaced by the organ (W_O), use the following formula: $W_O = W_2 - W_1$

7. To determine the construct volume (V_0) divide the weight of the fluid displaced (W_0) by the AVL construct or mouse construct by the specific gravity of isotonic saline ($g = 1.0048$): $V_0 = W_0/g$. As the specific gravity of saline is rounded off to $g = 1.0$, then: $V_0 \approx W_0 = W_2 - W_1$.

Slicing the AVL Construct Prior to Tissue Processing

Once weight and volume measurements are complete, the AVL construct should be sliced into 1 mm thick slices. However, if the construct is fragile, usually in the early stages of development (0–14 day construct), it should be fixed for 2 h in the desired fixative (10% neutral buffered formol saline or 4% paraformaldehyde) before attempting to slice. This will ensure that the construct structure does not disintegrate prematurely and therefore affect accurate histomorphometry.

The direction of slicing should be from the top of the construct, adjacent to the chamber lid, to the bottom of the construct, adjacent to the chamber base and from the femoral vessels point of entry into the construct towards the edge of the construct using a razor blade. Slice thickness should be as close as possible to 1–2 mm. This will ensure that cross sections of the AVL are obtained in these slices. After slicing, fixation should take place immediately for approximately 2 days prior to paraffin embedding. If paraffin embedding is delayed tissue should be stored in phosphate buffered saline (PBS) prior to processing. [If immediate fixation prior to slicing is not undertaken, some tissue slices can be removed and placed in liquid nitrogen for subsequent extraction of messenger RNA and total protein prior to fixation of the remaining slices].

Slicing of Skin Flaps

For histology and assessment of angiogenesis skin flaps should be either pinned out flat and fixed [or chemically cleared of all tissues and prepared as a whole mount – see [100]]. Skin flaps that are fixed must be sliced after fixation, at about 4–6 h after the fixative is applied. Flaps should then be cut 'vertically" using a razor blade or scalpel from the epidermis to the fascia, at a slice thickness of 2–4 mm. The skin flap slices are returned to fixative for a minimum of 24 h, and then stored in phosphate buffered saline (PBS) overnight prior to histological processing (see below).

Tissue processing and sectioning

(applies to the AVL construct, mouse flow through pedicle construct and skin flap slices)

The following protocols have been determined by our group to provide optimal conditions that permit clear and consistent tissue staining and successful immunohistochemistry with a strong signal to noise ratio.

1. Post fixation, tissue slices are placed in a tissue processing cassette and stabilized in the cassette with biopsy pads.
2. The processing of tissue into paraffin blocks is done using a tissue embedding processor (HyperCenter XP, Shandon®, Shandon Scientific Ltd, England), where the tissue is dehydrated through graded serial changes of alcohol and embedded into paraffin wax as follows:

   i. 70% Ethanol – 20 min
   ii. 90% Ethanol – 20 min
   iii. 95% Ethanol – 20 min
   iv. 100% Ethanol – 20 min
   v. 100% Ethanol – 30 min
   vi. 100% Ethanol –60 min
   vii. Histolene – 20 min
   viii. Histolene – 20 min
   ix. Histolene – 30 min
   x. Paraffin wax – 20 min under vacuum
   xi. paraffin wax – 60 min under vacuum

3. Following the last step in wax, the tissue is transferred to an embedding stage and embedded into tissue moulds containing melted paraffin wax. Rat chamber slices are placed cut surface down on the base of the mould, all the slices from one construct in one mould, or if large, several slices from a construct are placed in one mould. Skin flap slices are also placed cut surface down on the base of the mould, again several slices may fit into one mould. (Mouse chamber constructs can be left whole, or cut into two halves at this point, either by a single cross section or longitudinal section cut in the mid line, and the two halves embedded cut surface down on the base of the embedding mould).
4. The moulds are then cooled down to 5°C to allow paraffin wax solidification.
5. Section paraffin blocks into 3–5 μm thick serial sections using a microtome (Reichert-Jung, Grale Scientific Australia), float on 3′ (Aminopropyl) triethoxy-silane (AAS) (Sigma®, Sigma-Aldrich, St Louis, USA) coated slides for histology or poly-L-lysine coated slides (Polysine™, Menzel-Glaser®, Germany) for immunohistochemistry and dry overnight in a 37°C oven.

Identification of Blood Vessels in Rat and Mouse Tissue Sections

To identify endothelial lined blood vessels in rat tissue, a two-step indirect immuno-enzymatic method [153] is used to detect B. Simplicifolia lectin by a streptavidin biotin-based immunohistochemical protocol described below. Negative and positive control tissues were included in all immunohistochemical assessments. The negative control consisted of omitting B. Simplicifolia lectin from the reaction. The positive control tissues are rat skin tissues which label with this lectin on blood vessel endothelium. There is no absolutely definitive endothelial marker for rat tissue.

[It should be noted that a number of endothelial labels could be used, none of which are without some problems. The B. Simplicifolia lectin method described here is reliable and labels endothelial cells clearly – however it also labels macrophages and the observer needs also to be able to clearly distinguish small blood vessels from macrophages on morphological grounds. Another antibody which works very well in most rat tissues and labels endothelial cells very clearly is Factor VIII. However in the AVL chamber construct it will also label the fibrin matrix which persists for at least the first 2 weeks. This factor makes blood vessel identification very difficult and it is nearly impossible for the inexperienced to know what is true blood vessel labeling and what is not. We therefore have not included the Factor VIII protocol here. For mouse tissues we have generally used a CD31 antibody. This is a very clear label of endothelial cells, it does not label any inflammatory cells, but it is not specific to vascular endothelial cells as it also labels lymphatic endothelial cells.]

Lectin Iummunolabelling Protocol for Rat Tissue Blood Vessels

1. Dewax 3 μm thick serial tissue sections in histolene, graded ethanol (starting at 100% to 50%) and bring to water.
2. Wash sections in PBS, circle each tissue section with a wax pen. Apply Proteinase K for 5–8 min at room temperature, then wash the sections in three changes of PBS.
3. To block endogenous peroxidise and reduce the background that could be associated with the use of chromogen diaminobenzidine (DAB), blot the sections gently with tissues and treat the sections in 10% hydrogen peroxide (H_2O_2) in 50% methanol for 5 min. Wash sections in three changes of PBS.
4. Apply biotinylated B. Simplicifolia lectin solution (dilution 1:100) for 30 min at room temperature. Incubate the slides in a moist chamber to prevent drying out of the lectin solution. Wash sections in tris buffered saline (TBS) buffer.
5. Apply Streptavidin-HRP for 30 min at room temperature, also in a moist chamber, then wash sections in three changes of 0.05% Tween/TBS buffer.
6. To detect the bound lectin-streptavidin complex, use a chromogen such as DAB for 1–5 min. The sections are then washed in tap water, counterstained in Myer's haematoxylin, dehydrated and cover-slipped with DPX mountant.

CD 31 Immunolabelling Protocol for Mouse Tissue Blood Vessels

1. For CD31 immunolabelling of mouse tissues a rat anti-mouse CD31 antibody is used (BD Pharmingen, San Jose, CA).
2. Paraffin sections are dewaxed and hydrated and have a TBS wash, followed by a peroxidase quench (3% H_2O_2 in 50%methanol/distilled water for 5 min) and another TBS wash.

3. Sections are digested with DAKO proteinase K that is warmed to room temperature (antigen retrieval) for 8 min, washed in TBS, and nonspecific binding blocked using DAKO protein block for 30 min. Blot excess fluid.
4. The primary antibody is applied at 1:100 in DAKO antibody diluent for 1 h, followed by a TBS wash.
5. The secondary antibody (DAKO rabbit anti-rat biotinylated immunoglobulin) is then applied at 1:300 for 30 min followed by a TBS wash.
6. An avidin-biotin-horseradish peroxidase (HRP) detection system (ABC Elite; Vector Laboratories, Burlingame, CA) is then used followed by a TBS wash.
7. Finally diaminobenzidine (DAB) color development for 5 min, is followed by a distilled water wash, and counterstaining with haematoxylin for 1 min.
8. Dehydrate sections in increasing concentrations of alcohol, clear and coverslip with DPX mountant.

Morphometric Analysis of Labelled Blood Vessels

To morphometrically assess the vessel development in the rat AVL model, mouse flow through pedicle model or skin flaps, it is necessary to define targets of interest and reference compartments. In our laboratory, to estimate the proportion and volume of newly developed tissue components, the morphometric technique described by Howard and Reed [154] is used. When assessing rat AVL constructs or mouse flow-through pedicle constructs, the targets of interest include new blood vessels, matrix and new connective tissue and the AVL or flow through pedicle itself. In addition to the target, a reference compartment is also defined, namely a compartment to which a target can be compared to. In AVL constructs, two reference compartments were defined. The first compartment consists of the entire AVL construct comprised of AVL, fibrin matrix, new blood vessels and connective tissue. All points counted in each compartment are added together to give the total reference compartment count. However, to determine the vascularisation of the newly developed connective tissue alone, the second reference compartment is defined as the newly synthesized connective tissue which consists of combined counts for blood vessels and the newly synthesized connective tissue. A number of software tools are available for automated and semi-automated quantification of blood vessels and tissue components. In our laboratory, the Computer Assisted Stereological Toolbox (CAST) system is used (CAST System, Olympus, Denmark).

Method: To count the targets of interest, select a random 3–5 μm tissue section which contains 4–6 slices of AVL construct (1–2 sections of mouse construct) or 2–3 sections of skin flap tissue. Only complete tissue sections should be counted (Note: each of these slices should be approximately 500 to 1,000 μm apart) and labelled with *B. Simplicifolia* lectin (rat tissue) or CD 31 (mouse tissue).

1. Place a section on the automatic stage and outline the area of interest (reference compartment- usually the entire tissue section, or only the new connective tissue area in rat AVL constructs). At this point is it necessary to determine the

percentage of area to be counted so that a representation of all specimens can be achieved. This is done through trial and error and it is best to undertake a small pilot study to determine the percentage area suitable for counting and the number of points suitable for counting. In our case, depending on the model, this percentage value is between 5% and 20% of total area. The software package them randomly and uniformly selects the fields where the counts are to be completed. (Note at least 200 total points per chamber construct should be counted. We generally count between 300 and 800 per chamber construct).

2. To determine vascular volume density, the number of points falling on the blood vessels is divided by the number of points falling on the total reference compartment [51, 141]. The percent vascular volume is then obtained by multiplying that number by 100%. Furthermore, since AVL tissue volume (and mouse chamber construct volume) is available, the percent vascular volume can be multiplied by the organ displacement volume to obtain absolute (total) vascular volume within a construct. This parameter is of particular value when determining the effect of a treatment on the construct development. The final result is represented as a function of time or as the outcome of treatment effect.

References

1. Bhishgratna KKL (1907) Sustrata: an English translation of the Sushruta Samhita (based on original Sanskrit text). Bose, Calcutta
2. Laschke MW, Vollmar B, Menger MD (2009) Inosculation: connecting the life-sustaining pipelines. Tissue Eng B Rev 15:455–465
3. Penington AJ (2010) Local flap reconstruction, 2nd edn. McGraw Hill, Sydney
4. Manchott C (1889) Die Hautarerien des menchlichen Korpers. Vogel, Leipzig
5. Salmon M (1936) Arteres de la Peau. Masson et Cie, Paris
6. Taylor GI, Palmer JH (1987) The vascular territories (angiosomes) of the body: experimental study and clinical applications. Br J Plast Surg 40:113–141
7. Saint-Cyr M, Wong C, Schaverien M et al (2009) The perforasome theory: vascular anatomy and clinical implications. Plast Reconstr Surg 124:1529–1544
8. Taylor GI, Corlett RJ, Dhar SC et al (2011) The anatomical (angiosome) and clinical territories of cutaneous perforating arteries: development of the concept and designing safe flaps. Plast Reconstr Surg 127:1447–1459
9. Ghali S, Butler PE, Tepper OM et al (2007) Vascular delay revisited. Plast Reconstr Surg 119:1735–1744
10. Milton SH (1969) The effects of "delay" on the survival of experimental pedicled skin flaps. Br J Plast Surg 22:244–252
11. Myers MB, Cherry G (1969) Mechanism of the delay phenomenon. Plast Reconstr Surg 44:52–57
12. Finseth F, Cutting C (1978) An experimental neurovascular island skin flap for the study of the delay phenomenon. Plast Reconstr Surg 61:412–420
13. Morris SF, Taylor GI (1995) The time sequence of the delay phenomenon: when is a surgical delay effective? An experimental study. Plast Reconstr Surg 95:526–533
14. Banbury J, Siemionow M, Porvasnik S et al (1999) Muscle flaps' triphasic microcirculatory response to sympathectomy and denervation. Plast Reconstr Surg 104:730–737

15. Pearl RM (1981) A unifying theory of the delay phenomenon-recovery from the hyperadrenergic state. Ann Plast Surg 7:102–112
16. Holzbach T, Neshkova I, Vlaskou D et al (2009) Searching for the right timing of surgical delay: angiogenesis, vascular endothelial growth factor and perfusion changes in a skin-flap model. J Plast Reconstr Aesthet Surg 62:1534–1542
17. Barthe Garcia P, Suarez Nieto C, Rojo Ortega JM (1991) Morphological changes in the vascularization of delayed flaps in rabbits. Br J Plast Surg 44:285–290
18. Callegari PR, Taylor GI, Caddy CM et al (1992) An anatomic review of the delay phenomenon: I. Experimental studies. Plast Reconstr Surg 89:397–407
19. Murphy RC, Lawrence WT, Robson MC et al (1985) Surgical delay and arachidonic acid metabolites: evidence for an inflammatory mechanism: an experimental study in rats. Br J Plast Surg 38:272–277
20. Wong MS, Erdmann D, Sweis R et al (2004) Basic fibroblast growth factor expression following surgical delay of rat transverse rectus abdominis myocutaneous flaps. Plast Reconstr Surg 113:2030–2036
21. Lineaweaver WC, Lei MP, Mustain W et al (2004) Vascular endothelium growth factor, surgical delay, and skin flap survival. Ann Surg 239:866–873
22. Arranz López JL, Suárez Nieto C, Barthe García P et al (1995) Evaluation of angiogenesis in delayed skin flaps using a monoclonal antibody for the vascular endothelium. Br J Plast Surg 48:479–486
23. Tepper OM, Capla JM, Galiano RD et al (2005) Adult vasculogenesis occurs through in situ recruitment, proliferation, and tubulization of circulating bone marrow-derived cells. Blood 105:1068–1077
24. Asahara T, Murohara T, Sullivan A et al (1997) Isolation of putative progenitor endothelial cells for angiogenesis. Science 275:964–967
25. Kuntscher MV, Hartmann B, Germann G (2005) Remote ischemic preconditioning of flaps: a review. Microsurgery 25:346–352
26. Murry CE, Jennings RB, Reimer KA (1986) Preconditioning with ischemia: a delay of lethal cell injury in ischemic myocardium. Circulation 74:1124–1136
27. Mounsey RA, Pang CY, Forrest C (1992) Preconditioning: a new technique for improved muscle flap survival. Otolaryngol Head Neck Surg 107:549–552
28. Carroll CM, Carroll SM, Overgoor ML et al (1997) Acute ischemic preconditioning of skeletal muscle prior to flap elevation augments muscle-flap survival. Plast Reconstr Surg 100:58–65
29. Wang WZ, Anderson G, Firrell JC et al (1998) Ischemic preconditioning versus intermittent reperfusion to improve blood flow to a vascular isolated skeletal muscle flap of rats. J Trauma 45:953–959
30. Adanali G, Ozer K, Siemionow M (2002) Early and late effects of ischemic preconditioning on microcirculation of skeletal muscle flaps. Plast Reconstr Surg 109:1344–1351
31. Zahir KS, Syed SA, Zink JR et al (1998) Ischemic preconditioning improves the survival of skin and myocutaneous flaps in a rat model. Plast Reconstr Surg 102:140–150
32. Tatlidede S, McCormack MC, Eberlin KR et al (2009) A novel murine island skin flap for ischemic preconditioning. J Surg Res 154:112–117
33. Kuntscher MV, Schirmbeck EU, Menke H et al (2002) Ischemic preconditioning by brief extremity ischemia before flap ischemia in a rat model. Plast Reconstr Surg 109:2398–2404
34. Downey JM, Davis AM, Cohen MV (2007) Signaling pathways in ischemic preconditioning. Heart Fail Rev 12:181–188
35. Hausenloy DJ, Yellon DM (2007) Reperfusion injury salvage kinase signalling: taking a RISK for cardioprotection. Heart Fail Rev 12:217–234
36. Bolli R, Li QH, Tang XL et al (2007) The late phase of preconditioning and its natural clinical application–gene therapy. Heart Fail Rev 12:189–199
37. Semenza GL, Wang GL (1992) A nuclear factor induced by hypoxia via de novo protein synthesis binds to the human erythropoietin gene enhancer at a site required for transcriptional activation. Mol Cell Biol 12:5447–5454

38. Bernhardt WM, Warnecke C, Willam C et al (2007) Organ protection by hypoxia and hypoxia-inducible factors. Methods Enzymol 435:221–245
39. Picard-Ami LA Jr, MacKay A, Kerrigan CL (1991) Pathophysiology of ischemic skin flaps: differences in xanthine oxidase levels among rats, pigs, and humans. Plast Reconstr Surg 87:750–755
40. Deune EG, Khouri RK (1995) Rat strain differences in flap tolerance to ischemia. Microsurgery 16:765–767
41. Erol OO, Spria M (1980) New capillary bed formation with a surgically constructed arteriovenous fistula. Plast Reconstr Surg 66:109–115
42. Shen ZY (1981) Vascular implantation into skin flap to form an axial pattern skin flap: experimental study (author's transl). Zhonghua Wai Ke Za Zhi 19:692–695
43. Khouri RK, Upton J, Shaw WW (1992) Principles of flap prefabrication. Clin Plast Surg 19:763–771
44. Morrison WA, Penington AJ, Kumpta SK et al (1997) Clinical applications and technical limitations of prefabricated flaps. Plast Reconstr Surg 99:378–385
45. Pribaz JJ, Fine N, Orgill DP (1999) Flap prefabrication in the head and neck: a 10-year experience. Plast Reconstr Surg 103:808–820
46. Pribaz JJ, Weiss DD, Mulliken JB et al (1999) Prelaminated free flaps reconstruction of complex central facial defects. Plast Reconstr Surg 104:357–365
47. Tanaka Y, Tsutsumi A, Crowe DM et al (2000) Generation of an autologous tissue (matrix) flap by combining an arteriovenous shunt loop with artificial skin in rats: preliminary report. Br J Plast Surg 53:51–57
48. Mian R, Morrison WA, Hurley JV et al (2000) Formation of new tissue from an arteriovenous loop in the absence of added extracellular matrix. Tissue Eng 6:595–603
49. Tananka Y, Sung KC, Tsutsumi A et al (2003) Tissue engineering skin flaps: which vascular carrier, arteriovenous shunt loop or arteriovenous bundle, has more potential for angiogenesis and tissue generation? Plast Reconstr Surg 112:1636–1644
50. Tanaka Y, Sung KC, Fumimoto M et al (2006) Prefabricated engineered skin flap using an arteriovenous vascular bundle as a vascular carrier in rabbits. Plast Reconstr Surg 117:1860–1875
51. Lokmic Z, Stillaert F, Morrison WA et al (2007) An arteriovenous loop in a protected space generates a permanent, highly vascular, tissue-engineered construct. FASEB J 21:511–522
52. Tilkorn DJ, Bedogni A, Keramidaris E et al (2010) Implanted myoblast survival is dependent on the degree of vascularization in a novel delayed implantation/prevascularization tissue engineering model. Tissue Eng A 16:165–178
53. German W, Finesilver EM, Davis JS (1933) Establishment of circulation in tubed skin flaps. Arch Surg 26:27
54. Myers MB, Cherry G (1971) Differences in the delay phenomenon in the rabbit, rat, and pig. Plast Reconstr Surg 47:73–78
55. Donovan WE (1975) Experimental models in skin flap research. In: Grabb WC, Myers MB (eds) Skin flaps, 1st edn. Little Brown and Company, Boston
56. McFarlane RM, Deyoung G, Henry RA (1965) The design of a pedicle flap in the rat to study necrosis and its prevention. Plast Reconstr Surg 35:177–182
57. Adamson JE, Horton CE, Crawford HH et al (1966) The effects of dimethyl sulfoxide on the experimental pedicle flap: a preliminary report. Plast Reconstr Surg 37:105–110
58. Thompson FM, Berakha GJ, Guthrie RH Jr (1977) The effective duration of the delay phenomenon in the rat. Plast Reconstr Surg 60:384–389
59. Conoyer JM, Toomey JM (1979) Dorsal skin flaps in rats as an experimental model. Surg Forum 30:510–511
60. Knox LK, Stewart AG, Hayward PG et al (1994) Nitric oxide synthase inhibitors improve skin flap survival in the rat. Microsurgery 15:708–711
61. Pang Y, Lineaweaver WC, Lei MP et al (2003) Evaluation of the mechanism of vascular endothelial growth factor improvement of ischemic flap survival in rats. Plast Reconstr Surg 112:556–564

62. Aydogan H, Gurlek A, Parlakpinar H et al (2007) Beneficial effects of caffeic acid phenethyl ester (CAPE) on the ischaemia-reperfusion injury in rat skin flaps. J Plast Reconstr Aesthet Surg 60:563–568
63. Angel MF, Kaufman T, Swartz WM et al (1986) Studies on the nature of the flap/bed interaction in rodents–Part I: flap survival under varying conditions. Ann Plast Surg 17:317–322
64. Hammond DC, Brooksher RD, Mann RJ et al (1993) The dorsal skin-flap model in the rat: factors influencing survival. Plast Reconstr Surg 91:316–321
65. Kaufman T, Angel MF, Eichenlaub EH et al (1985) The salutary effects of the bed on the survival of experimental flaps. Ann Plast Surg 14:64–73
66. Jones M, Zhang F, Blain B et al (2001) Influence of recipient-bed isolation on survival rates of skin-flap transfer in rats. J Reconstr Microsurg 17:653–658
67. Kelly CP, Gupta A, Keskin M et al (2010) A new design of a dorsal flap in the rat to study skin necrosis and its prevention. J Plast Reconstr Aesthet Surg 63:1553–1556
68. Yang D, Morris SF (1999) An extended dorsal island skin flap with multiple vascular territories in the rat: a new skin flap model. J Surg Res 87:164–170
69. Syed SA, Tasaki Y, Fujii T et al (1992) A new experimental model: the vascular pedicle cutaneous flap over the dorsal aspect (flank and hip) of the rat. Br J Plast Surg 45:23–25
70. Syed SA, Restifo RJ (1997) An economical axial-pattern flap necrosis model. Plast Reconstr Surg 99:263–264
71. Kayikcioglu A, Akyurek M, Safak T (1998) The importance of vascular territories in designing new experimental skin flaps. Plast Reconstr Surg 101:1155–1157
72. Hosnuter M, Kargi E, Peksoy I et al (2006) An ameliorated skin flap model in rats for experimental research. J Plast Reconstr Aesthet Surg 59:299–303
73. Ohara H, Kishi K, Nakajima T (2008) Rat dorsal paired island skin flaps: a precise model for flap survival evaluation. Keio J Med 57:211–216
74. Taylor GI, Corlett RJ, Caddy CM et al (1992) An anatomic review of the delay phenomenon: II. Clinical applications. Plast Reconstr Surg 89:408–416
75. Ahmed SS, Pierce J, Reid M et al (1997) A new experimental model: the vascular pedicled cutaneous flap over the mid-dorsum of the rat. Ann Plast Surg 39:495–499
76. Adanali G, Seyhan T, Turegun M et al (2002) Effects of different vascular patterns and the delay phenomenon on rat ventral island flap viability. Ann Plast Surg 48:660–664
77. Goldwyn RM, Lamb DL, White WL (1963) An experimental study of large island flaps in dogs. Plast Reconstr Surg 31:528–536
78. Strauch B, Murray DE (1967) Transfer of composite graft with immediate suture anastomosis of its vascular pedicle measuring less than 1 mm in external diameter using microsurgical techniques. Plast Reconstr Surg 40:325–329
79. Angel MF, Mellow CG, Knight KR et al (1989) The effect of time of vascular island skin flap elevation on tolerance to warm ischemia. Ann Plast Surg 22:426–428
80. Angel MF, Mellow CG, Knight KR et al (1989) Prior elevation of vascular island skin flaps: intolerance to ischemia caused by venous obstruction. J Reconstr Microsurg 5:163–165
81. Roberts AP, Cohen JI, Cook TA (1996) The rat ventral island flap: a comparison of the effects of reduction in arterial inflow and venous outflow. Plast Reconstr Surg 97:610–615
82. Angel MF, Mellow CG, Knight KR et al (1990) Secondary ischemia time in rodents: contrasting complete pedicle interruption with venous obstruction. Plast Reconstr Surg 85:789–793
83. Sagi A, Ferder M, Levens D et al (1986) Improved survival of island flaps after prolonged ischemia by perfusion with superoxide dismutase. Plast Reconstr Surg 77:639–644
84. Mellow CG, Knight KR, Angel MF et al (1990) The effect of thromboxane synthetase inhibition on tolerance of skin flaps to secondary ischemia caused by venous obstruction. Plast Reconstr Surg 86:329–334
85. Willemart G, Knight KR, Morrison WA (1998) Dexamethasone treatment prior to reperfusion improves the survival of skin flaps subjected to secondary venous ischaemia. Br J Plast Surg 51:624–628
86. Ueda K, Nozawa M, Nakao M et al (2000) Sulfatide and monoclonal antibodies prevent reperfusion injury in skin flaps. J Surg Res 88:125–129

87. Machens HG, Morgan JR, Berthiaume F et al (2002) Platelet-derived growth factor-AA-mediated functional angiogenesis in the rat epigastric island flap after genetic modification of fibroblasts is ischemia dependent. Surgery 131:393–400

88. Gideroglu K, Yilmaz F, Aksoy F et al (2009) Montelukast protects axial pattern rat skin flaps against ischemia/reperfusion injury. J Surg Res 157:181–186

89. Brown DM, Hong SP, Farrell CL et al (1995) Platelet-derived growth factor BB induces functional vascular anastomoses in vivo. Proc Natl Acad Sci USA 92:5920–5924

90. Theile DR, Kane AJ, Romeo R et al (1998) A model of bridging angiogenesis in the rat. Br J Plast Surg 51:243–249

91. Kane AJ, Barker JE, Mitchell GM et al (2001) Inducible nitric oxide synthase (iNOS) activity promotes ischaemic skin flap survival. Br J Pharmacol 132:1631–1638

92. Furuta S, Vadiveloo P, Romeo-Meeuw R et al (2004) Early inducible nitric oxide synthase 2 (NOS 2) activity enhances ischaemic skin flap survival. Angiogenesis 7:33–43

93. Serafin D, Shearin JC, Georgiade NG (1977) The vascularization of free flaps: a clinical and experimental correlation. Plast Reconstr Surg 60:233–241

94. Nakajima T (1978) How soon do venous drainage channels develop at the periphery of a free flap? A study in rats. Br J Plast Surg 31:300–308

95. Nishikawa H, Manek S, Green CJ (1991) The oblique rat groin flap. Br J Plast Surg 44:295–298

96. Hirase Y, Valauri FA, Buncke HJ (1989) Creation of neovascularised free flaps using vein grafts as pedicles: a preliminary report on experimental models. Br J Plast Surg 42:216–222

97. Falco NA, Pribaz JJ, Eriksson E (1992) Vascularization of skin following implantation of an arteriovenous pedicle: implications in flap prefabrication. Microsurgery 13:249–254

98. Morrison WA, Dvir E, Doi K et al (1990) Prefabrication of thin transferable axial-pattern skin flaps: an experimental study in rabbits. Br J Plast Surg 43:645–654

99. Wilson YT, Kumpta S, Hickey MJ et al (1994) Use of free interpositional vein grafts as pedicles for prefabrication of skin flaps. Microsurgery 15:717–721

100. Hickey MJ, Wilson Y, Hurley JV et al (1998) Mode of vascularization of control and basic fibroblast growth factor-stimulated prefabricated skin flaps. Plast Reconstr Surg 101:1296–1304

101. Khouri RK, Hong SP, Deune EG et al (1994) De novo generation of permanent neovascularized soft tissue appendages by platelet-derived growth factor. J Clin Invest 94:1757–1763

102. Tanaka Y, Tajima S, Tsutsumi A et al (1996) New matrix flap prefabricated by arteriovenous shunting and artificial skin dermis in rats II. Effect of interpositional vein and artery grafts and bFGF on new tissue generation. J Jpn Plast Reconstr Surg 16:679

103. Takato T, Zuker RM, Turley CB (1991) Prefabrication of skin flaps using vein grafts: an experimental study in rabbits. Br J Plast Surg 44:593–598

104. Maitz PK, Pribaz JJ, Hergruter CA (1994) Manipulating prefabricated flaps: an experimental study examining flap viability. Microsurgery 15:624–629

105. Myers MB, Brock D, Cohn I Jr (1971) Prevention of skin slough after radical mastectomy by the use of a vital dye to delineate devascularized skin. Ann Surg 173:920–924

106. Teich-Alasia S (1971) A study of the vascularisation of pedicle flaps using disulphine blue. Br J Plast Surg 24:282–288

107. Eren S, Rubben A, Krein R et al (1995) Assessment of microcirculation of an axial skin flap using indocyanine green fluorescence angiography. Plast Reconstr Surg 96:1636–1649

108. Lamby PL, Gais S et al (2008) Evaluation of the vascular integrity of free flaps based on microcirculation imaging techniques. Clin Hemorheol Microcirc 39:253–263

109. Ezaki T, Baluk P, Thurston G et al (2001) Time course of endothelial cell proliferation and microvascular remodeling in chronic inflammation. Am J Pathol 158:2043–2055

110. Jiao C, Bronner S, Mercer KL et al (2004) Epidermal cells accelerate the restoration of the blood flow in diabetic ischemic limbs. J Cell Sci 117:1055–1063

111. Li Y, Song Y, Zhao L et al (2008) Direct labeling and visualization of blood vessels with lipophilic carbocyanine dye DiI. Nat Protoc 3:1703–1708

112. Bergeron L, Tang M, Morris SF (2006) A review of vascular injection techniques for the study of perforator flaps. Plast Reconstr Surg 117:2050–2057

113. Lindenblatt N, Calcagni M, Contaldo C et al (2008) A new model for studying the revascularization of skin grafts in vivo: the role of angiogenesis. Plast Reconstr Surg 122:1669–1680
114. Lokmic Z, Mitchell GM (2011) Visualisation and stereological assessment of blood and lymphatic vessels. Histol Histopathol 26:781–796
115. Proano E, Perbeck L (1993) Changes in skin blood flow in ischaemic limbs after vascular reconstruction measured by fluorescein flowmetry and laser Doppler flowmetry. Clin Physiol 13:599–609
116. Proano E, Perbeck L (1994) Effect of exposure to heat and intake of ethanol on the skin circulation and temperature in ischaemic limbs. Clin Physiol 14:305–310
117. Proano E, Svensson L, Perbeck L (1997) Correlation between the uptake of sodium fluorescein in the tissue and xenon-133 clearance and laser Doppler fluxmetry in measuring changes in skin circulation. Int J Microcirc Clin Exp 17:22–28
118. Saetzler RK, Jallo J, Lehr HA et al (1997) Intravital fluorescence microscopy: impact of light-induced phototoxicity on adhesion of fluorescently labeled leukocytes. J Histochem Cytochem 45:505–513
119. Sloan GM, Sasaki GH (1985) Noninvasive monitoring of tissue viability. Clin Plast Surg 12:185–195
120. Chubb D, Rozen WM, Ashton MW (2010) Early survival of a compromised fasciocutaneous flap without pedicle revision: monitoring with Photoplethysmography. Microsurgery 30:462–465
121. Groner W, Winkelman JW, Harris AG et al (1999) Orthogonal polarization spectral imaging: a new method for study of the microcirculation. Nat Med 5:1209–1212
122. Langer S, Biberthaler P, Harris AG et al (2001) In vivo monitoring of microvessels in skin flaps: introduction of a novel technique. Microsurgery 21:317–324
123. Olivier WA, Hazen A, Levine JP et al (2003) Reliable assessment of skin flap viability using orthogonal polarization imaging. Plast Reconstr Surg 112:547–555
124. Tokiya R, Umetani K, Imai S et al (2004) Observation of microvasculatures in athymic nude rat transplanted tumor using synchrotron radiation microangiography system. Acad Radiol 11:1039–1046
125. Ichioka S, Minh TC, Shibata M et al (2002) In vivo model for visualizing flap microcirculation of ischemia-reperfusion. Microsurgery 22:304–310
126. Bellhorn MB, Bellhorn RW, Poll DS (1977) Permeability of fluorescein-labelled dextrans in fundus fluorescein angiography of rats and birds. Exp Eye Res 24:595–605
127. Smith LE, Wesolowski E, McLellan A et al (1994) Oxygen-induced retinopathy in the mouse. Invest Ophthalmol Vis Sci 35:101–111
128. Kurozumi K, Hardcastle J, Thakur R et al (2007) Effect of tumor microenvironment modulation on the efficacy of oncolytic virus therapy. J Natl Cancer Inst 99:1768–1781
129. Bonner-Weir S, Orci L (1982) New perspectives on the microvasculature of the islets of Langerhans in the rat. Diabetes 31:883–889
130. Wagner RC, Czymmek K, Hossler FE (2006) Confocal microscopy, computer modeling and quantification of glomerular vascular corrosion casts. Microsc Microanal 12:262–268
131. Erol OO, Spira M (1979) New capillary bed formation with a surgically constructed arteriovenous fistula. Surg Forum 30:530–531
132. Lokmic Z, Mitchell GM (2008) Engineering the microcirculation. Tissue Eng B 14:87–103
133. Cronin KJ, Messina A, Knight KR et al (2004) New murine model of spontaneous autologous tissue engineering, combining an arteriovenous pedicle with matrix materials. Plast Reconstr Surg 113:260–269
134. Lepore DA, Thomas GP, Knight KR et al (2007) Survival and differentiation of pituitary colony-forming cells in vivo. Stem Cells 25:1730–1736
135. Seach N, Mattesich M, Abberton K et al (2010) Vascularized tissue engineering mouse chamber model supports thymopoiesis of ectopic thymus tissue grafts. Tissue Eng C 16:543–551
136. Hussey AJ, Winardi M, Han XL et al (2009) Seeding of pancreatic islets into prevascularized tissue engineering chambers. Tissue Eng A 15:3823–3833

137. Hussey AJ, Winardi M, Wilson J et al (2010) Pancreatic islet transplantation using vascularised chambers containing nerve growth factor ameliorates hyperglycaemia in diabetic mice. Cells Tissues Organs 191:382–393

138. Forster NA, Penington AJ, Hardikar AA et al (2011) A prevascularized tissue engineering chamber supports growth and function of islets and progenitor cells in diabetic mice. Islets 3:271–283

139. Forster N, Palmer JA, Yeoh G et al (2011) Expansion and hepatocytic differentiation of liver progenitor cells in vivo using a vascularized tissue engineering chamber in mice. Tissue Eng C Methods 17:359–366

140. Kelly JL, Findlay MW, Knight KR et al (2006) Contact with existing adipose tissue is inductive for adipogenesis in Matrigel. Tissue Eng 12:2041–2047

141. Rophael JA, Craft RO, Palmer JA et al (2007) Angiogenic growth factor synergism in a murine tissue engineering model of angiogenesis and adipogenesis. Am J Pathol 171:2048–2057

142. Vashi AV, Abberton KM, Thomas GP et al (2006) Adipose tissue engineering based on the controlled release of fibroblast growth factor-2 in a collagen matrix. Tissue Eng 12:3035–3043

143. Findlay MW, Messina A, Thompson EW et al (2009) Long-term persistence of tissue-engineered adipose flaps in a murine model to 1 year: an update. Plast Reconstr Surg 124:1077–1084

144. Simcock JW, Penington AJ, Morrison WA et al (2009) Endothelial precursor cells home to a vascularised tissue engineering chamber by application of the angiogenic chemokine CXCL12. Tissue Eng A 15:655–664

145. Cassell OC, Morrison WA, Messina A et al (2001) The influence of extracellular matrix on the generation of vascularized, engineered, transplantable tissue. Ann N Y Acad Sci 944:429–442

146. Hofer SO, Knight KM, Cooper-White JJ et al (2003) Increasing the volume of vascularized tissue formation in engineered constructs: an experimental study in rats. Plast Reconstr Surg 111:1186–1192

147. Cao Y, Mitchell G, Messina A et al (2006) The influence of architecture on degradation and tissue in growth into three-dimensional poly(lactic-co-glycolic acid) scaffolds in vitro and in vivo. Biomaterials 27:2854–2864

148. Dolderer JH, Thompson EW, Slavin J et al (2011) Long-term stability of adipose tissue generated from a vascularized pedicled fat flap inside a chamber. Plast Reconstr Surg 127:2283–2292

149. Morritt AN, Bortolotto SK, Dilley RJ et al (2007) Cardiac tissue engineering in an *in vivo* vascularized chamber. Circulation 115:353–360

150. Choi YS, Matsuda K, Dusting GJ et al (2010) Engineering cardiac tissue *in vivo* from human adipose-derived stem cells. Biomaterials 31:2236–2242

151. Messina A, Bortolotto SK, Cassell OC et al (2005) Generation of a vascularized organoid using skeletal muscle as the inductive source. FASEB J 19:1570–1572

152. Scherle W (1970) A simple method for volumetry of organs in quantitative stereology. Mikroskopie 26:57–60

153. Hsu SM, Raine L, Fanger H (1981) Use of avidin-biotin-peroxidase complex (ABC) in immunoperoxidase techniques: a comparison between ABC and unlabeled antibody (PAP) procedures. J Histochem Cytochem 29:577–580

154. Howard CV, Reed MG (2005) Unbiased stereology. BIOS Scientific Publishers, Oxford

155. Lokmic Z, Thomas JL, Morrison WA et al (2008) An endogenously deposited fibrin scaffold determines construct size in the surgically created arteriovenous loop chamber model of tissue engineering. J Vasc Surg 48:974–985

Laser Scanning Methodologies for Measuring RBC Velocity, Flux, Hematocrit and Shear Rate in Vascular Networks

Lance L. Munn and Walid S. Kamoun

Abstract There are many normal and pathological processes that alter blood flow in a vascular network including thermal regulation, infarction, wounding and neoplasia. Flow changes caused by formation or loss of vascular connections and modulation of vessel diameters can dramatically affect nutrient and drug delivery, but these changes are poorly understood at the level of individual vessel segments and their connected neighborhood. To address this problem, we developed methodology for quantifying blood flow (velocity, flux and hematocrit) in extended networks at the single capillary level. Our approach relies on deconvolution of signals produced by labeled red blood cells as they move relative to the scanning laser of a confocal or multiphoton microscope, and provides fully-resolved three-dimensional flow profiles within tumor vessels. This methodology has sufficient spatiotemporal resolution for extracting blood velocity profiles in vivo and can be used to detect changes in blood vessel classification based on function.

1 Introduction

The formation and function of blood vessel networks undergoing active angiogenesis are difficult to assess, but critically important in many pathologies. In tumors, for example, blood vessels are heterogeneous in their structure and function, likely due to differential exposure to growth factors in the microenvironment

This chapter is derived from Kamoun et al., Nat Methods 2010. Figures have been reproduced here with permission.

L.L. Munn (✉)
Edwin L. Steele Laboratory, Department of Radiation Oncology, Massachusetts General Hospital and Harvard Medical School, Boston, MA, USA
e-mail: munn@steele.mgh.harvard.edu

W.S. Kamoun
Merrimack Pharmaceuticals, Cambridge MA, USA

E. Zudaire and F. Cuttitta (eds.), *The Textbook of Angiogenesis and Lymphangiogenesis: Methods and Applications*, DOI 10.1007/978-94-007-4581-0_25,
© Springer Science+Business Media Dordrecht 2012

and non-synchronous progression of angiogenesis and maturation [1–4]. Because of technical limitations, most measurements of tumor vessel function are performed on single vessels or measured as bulk parameters over a large network. This limits our ability to detect responses to vascular-targeted drugs, which usually affect specific vessel subpopulations. To determine how individual vessels, or classes of vessels, are involved in tumor growth and response to treatment, tools for studying blood flow in extended vessel networks – at the spatial resolution of single capillaries – are needed.

Many techniques have been used to analyze hemodynamics. Single photon video-rate imaging has been the standard method for measuring blood flow in normal and tumor vessels [1–3]. Coupled with algorithms for automated detection and tracking of fluorescently-labeled red blood cell (RBC) [5] or two slit and four slit cross-correlation methods [6] flow analysis in single vessels is possible. However, single photon techniques have relatively poor spatial resolution in the "z" direction (along the light path), and are less accurate when the vessels do not lie within the x-y plane.

Using multiphoton laser scanning microscopy (MPLSM), it is possible to analyze blood flow in individual vessels, and quantification of blood flow has been demonstrated in cerebral cortical capillaries [7] and tumor blood vessels [8]. In general, these approaches quantify flow velocity within selected blood vessels by scanning along the central axis of the blood vessel at high frequency, relying on contrast between an injected plasma fluorophore and the erythrocytes, which are dark. In an intensity plot of location along the line scan vs. time, streaks are produced with angles proportional to the RBC velocity (Fig. 1a). Axial Line Scanning (ALS) allows quantification of centerline velocity, which can be used to estimate average flow velocity. ALS has good accuracy and sensitivity, especially when implemented with appropriate analysis algorithms [9]. It can measure a wide range of velocities, but does not allow accurate quantification of erythrocyte flux (number of RBCs per second) or hematocrit (fraction of blood volume occupied by RBCs). Also, the technique is tedious and time-intensive, and is therefore not amenable to analysis of entire networks.

To overcome these limitations we developed two MPLSM-based methods for blood flow analysis in vascular networks. The first method, Residence Time Line Scanning (RTLS), allows direct analysis of flow velocity by scanning a line at an arbitrary angle to the vessel (Fig. 1a). The second method, Relative Velocity Field Scanning (RVFS), is a full-field method allowing simultaneous analysis of most of the vessels within a field of view by deconvolving the image distortion produced by cells moving relative to the moving laser scans (Fig. 1b). Here, we describe the operational principles of RTLS and RVFS and then demonstrate the power of the methodologies by first measuring fully-resolved lumenal profiles of velocity, flux, hematocrit and shear rate in tumor vessels undergoing intussusceptive angiogenesis, and second by performing cluster analyses to identify "signatures" of vessels based on location and function.

2   Methods

To perform either RTLS or RVFS, we first prepare the animal model and label erythrocytes, which provide the fluorescence signal:

2.1   Labeling RBCs

RBCs are labeled ex-vivo with a far-red lipophilic fluorescent dye (1,1-dioctadecyl-3,3,3,3-tetramethylindodicarbocyanine perchlorate (DID)) allowing observation deep within the tissue via MPLSM:

1. Collect 0.5–1.4 ml blood via cardiac or retro-orbital puncture from a donor animal of the same strain to be used in the experiment.
2. Separate RBCs from plasma and leukocytes using centrifugation at 500 G for 20 min; remove top (buffy coat) layer.
3. Dilute RBCs 1:100 in PBS (10 ml) and incubate for 20 min with 100 ul 1 mg/ml DID (dissolved in 95% Ethanol).
4. Wash RBCs with PBS and mix with saline, diluting to 50% Hematocrit.

The resulting fraction of fluorescent RBCs can be quantified in vivo by injecting TAMRA-BSA and monitoring RBCs in small capillaries. Non-fluorescent RBCs exclude TAMRA-BSA and appear dark. The ratio of fluorescent to total RBCs should be calculated during each experiment. The labeled RBCs are ideal for long-term studies, with a half-life of approximately 10–14 days in the circulation, comparable to erythrocyte half-life in mice [10].

2.2   Animal Models and Cell Lines

1. To visualize the mouse vasculature, we implant transparent windows over the tissue of the mammary fat pad or brain in 8–10 week old mice [11]. However, the methodology will also work with acute preparations.
2. Small fragments (0.2–0.3 mm diameter) of mammary carcinoma (MCaIV) or glioma (U87 and GL261) are implanted into the mammary fat pad or the left cerebral cortex respectively.
3. To distinguish the tumor from the normal brain tissue, GFP can be stably transfected into cancer cells using a retroviral construct, if desired.
4. All cell lines are maintained in DMEM medium with 10% FBS.
5. All experiments must be approved by the institutional organization responsible for Research Animal Care.

3   Intravital Multiphoton Laser Scanning Microscopic (MPLSM)

Details of our procedures for in vivo multiphoton laser scanning microscopic analysis of glioblastoma and mammary carcinoma vessels have been described previously [11, 12]. The MPLSM consisted of a MilleniaX pumped Tsunami Ti:Sapphire laser (Spectra-Physics). Some important considerations are:

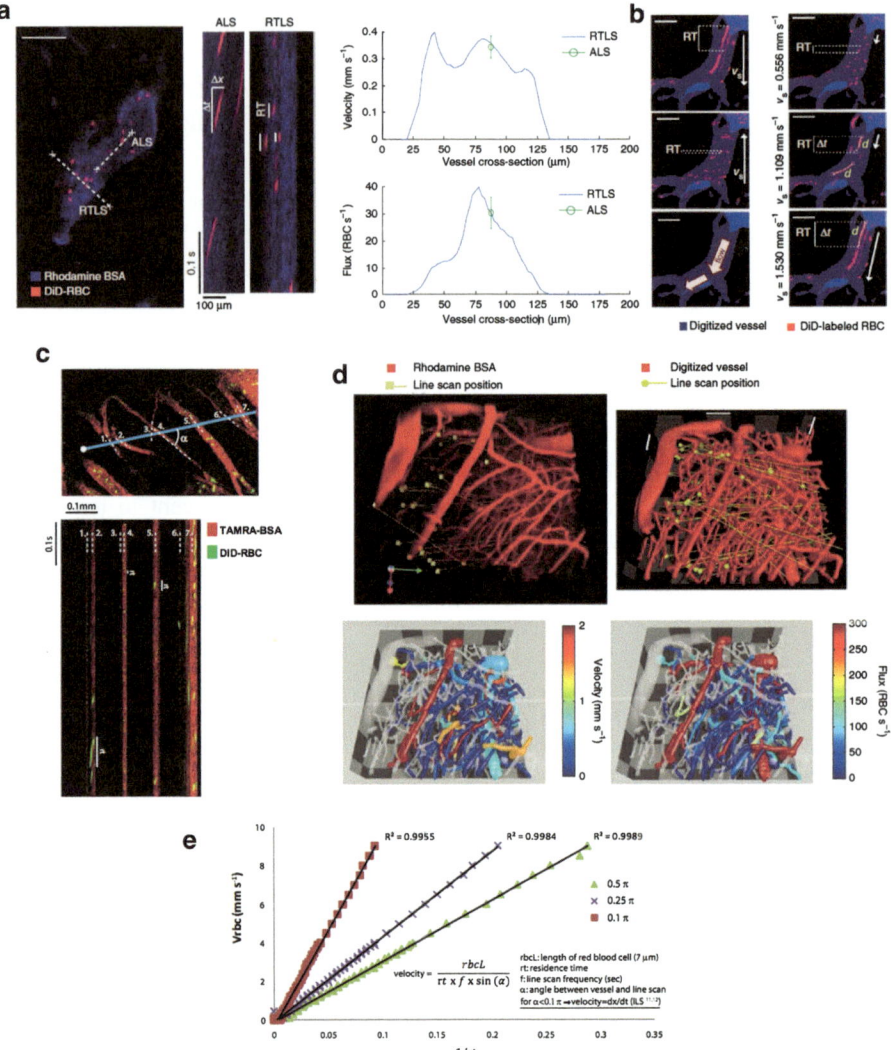

Fig. 1 Description, validation and application of RTLS and RVFS. (**a**) Representative MPLSM angiogram of a tumor blood vessel. x vs t plots were generated by scanning along the centerline of the vessel (ALS) and perpendicular to the vessel (RTLS). Analysis of flow velocity is based on the slope of the RBC signal ($\Delta x/\Delta t$) for ALS and on residence time (rt) for RTLS. RBC velocity and flux measured along the vessel cross-section using RTLS are compared to ALS-based analysis of flow (mean ± s.e.m.). (**b**) A single vessel scanned with a range of scanning velocities in two opposing scanning directions (scanning velocity – Vs). Scanning from top to bottom with velocities from 1.1 to 1.5 mm s^{-1} caused "velocity-matched" red blood cells with higher residence times (RT) and a measurable traveled distance (d), which can be used to measure velocity. (**c**) Representative MPLSM angiogram of a tumor blood vessel from a glioma model imaged through the cranial window visualized by injecting Rhodamine conjugated BSA and DiD labeled red blood cells. Below: x vs t images showing RBC residence times in several vessels analyzed simultaneously.

1. Two photon excitation of the used fluorophores (TAMRA, GFP and DID) can be achieved using 840 nm excitation light.
2. The power at the sample should be 1–5 mW.
3. Our MPLSM microscope consists of an Olympus Fluoview FV300 system customized for multiphoton imaging. Consult the documentation of your specific hardware for proper set-up of laser scan configurations.
4. Vessel angiography should be performed via intravenous injection of 0.1 ml 10 mg/ml TAMRA-BSA (Invitrogen). After injection, acquire 3D stacks of the regions of interest in the GFP (if applicable) and TAMRA channels (dimension: $630 \times 630 \times 250$ μm; Resolution: $2.4 \times 2.4 \times 2.5$ μm/pixel) for the purpose of extracting the tumor and network geometries.
5. If applicable, segment the tumor volume using an appropriate intensity threshold.
6. Trace the structure of the vasculature using a tracing algorithm such as described by Tyrrell et al.[13].

3.1 Residence Time Line Scanning

RTLS is performed by scanning along a single line that intersects the vessel at arbitrary angle (Fig. 1a, c). Repeated scanning along this line generates *x-t* data (fluorescence intensity along the line over time) in which "images" of the fluorescent RBCs are compressed or elongated depending on the residence time of the cells within the scan. The length of each cell signal in the *x-t* domain along the time axis depends on the number of times the laser scans the fluorescent cell as it passes. Thus, we can extract velocities from the line scan frequency, scanning angle and the length of the *x-t* cell images (Fig. 1).

A major advantage of RTLS is that, compared with ALS, it is relatively insensitive to the orientation of the scan line with respect to the vessel. This means that RTLS scans can be randomly placed, or systematically arranged (for example in a grid pattern), removing operator dependence. Furthermore, each scan can intersect several vessels, allowing simultaneous measurement of multiple velocity profiles (Fig. 1c, d). The technique is sufficiently robust to map velocities and fluxes in complex glioma vasculatures with its abundance of small vessels (Fig. 1d).

Fig. 1 (continued) (**d**) 3D MPLSM angiogram of the brain of a tumor-bearing mouse illustrating the position of the line scans performed at several z planes. Comprehensive velocity and flux 3D maps were generated using RTLS. (**e**) Lattice Boltzmann model was used to simulate RBC flow and validate the linear relationship between rt and red blood cell velocity. A single column of red blood cells flowing in a straight channel at a predetermined velocity was modeled. Line scanning at various angles intersecting the vessel from perpendicular ($0.5\ \pi$) to almost parallel to the vessel wall ($0.1\ \pi$) validated the linear relationship between 1/rt and RBC velocity

4   Residence Time Line Scan Procedure

RTLS imaging is performed by scanning a laser line that intersects one or more vessels:

1. Consult your equipment documentation for details of controlling the laser scan and intensity.
2. Line scan length and depth in the tissue can be chosen manually to intersect specific vessels, or specified by an automated procedure. In our system, we generally scan 20–40 iterations of each line at various depths.
3. Using an appropriate computer assisted algorithm, segment the resulting x-t images of fluorescent tracer RBCs based primarily on intensity.
4. Extract the feature "lengths" in the t direction. The velocity of each scanned RBC can be calculated based on the equation in Fig. 1e.
5. Hematocrit is calculated as the fraction of RBC pixels over the total number of pixels within the lane, adjusted for the ratio of labeled to non-labeled RBCs.

5   Shear Rate Calculation

For experiments in which high resolution scans are repeated throughout the depth of a vessel, intra-lumenal shear rate can be calculated:

1. Define "lanes" parallel to the vessel in which the RBCs are traveling—generally a lane is one image pixel wide.
2. For each lane, calculate RBC velocity as the mean velocity of all the RBCs traveling in that lane. Note that each RBC will likely be present in more than one lane– most often two lanes– and therefore can contribute to the average flow, hematocrit and flux of multiple lanes.
3. For the cross-sectional analysis of flow (Fig. 2), shear rate maps are generated by calculating the average shear rate for each pixel at the resolution of 2.5 μm/pixel. Shear rate is equal to the average difference in velocity between the two adjacent pixel lanes, divided by the distance.
4. Wall shear rate is the velocity gradient at the vessel wall calculated as the velocity in the lane closest to the vessel wall divided by the distance (assuming that the velocity at the vessel wall is 0 mm s^{-1}).

6   Relative Velocity Field Scanning

6.1   Theory

RVFS uses conventional laser scanning systems (MPLSM or confocal) to perform full field analysis of flow. Instead of requiring high frame rates, the method takes advantage of the relatively low speed of the laser field scans, which results in

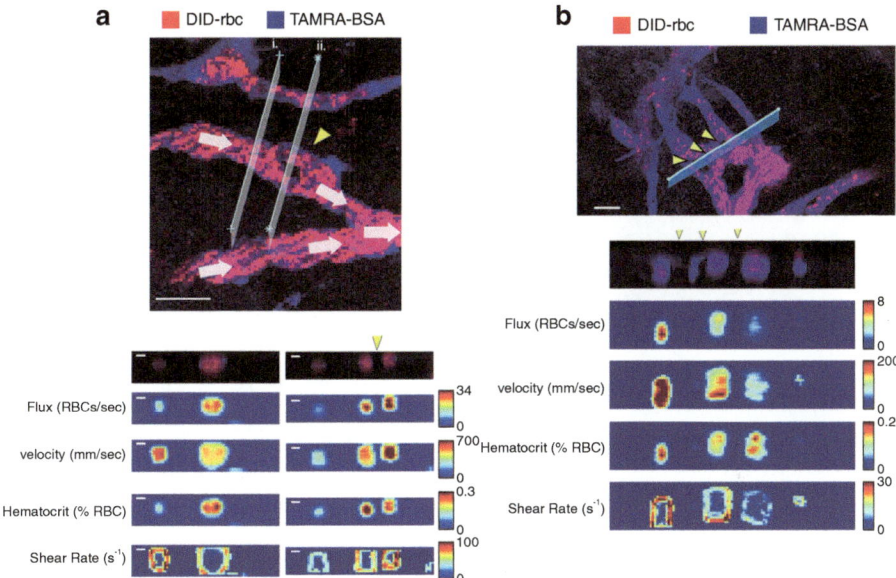

Fig. 2 Analysis of cross-sectional flow profiles within tumor vessels undergoing intussusceptive angiogenesis. Cross-sectional velocity, flux, hematocrit shear rate profiles and raw data maps upstream and at the level of a single (**a**) or multiple (**b**) intussusceptions (*yellow arrow heads*); Scale bars, 50 μm (a), 50 μm (b)

distortion of the fluorescent RBC images. RVFS is based on an analysis of RBC residence times (the length of time a given cell spends in the scan line) and traveled distance, extracted from the distortion of the images of RBCs as they move during scanning. While ALS and RTLS use stationary lines, RVFS uses a moving line scan. Because the sequential lines take a finite amount of time to scan across the field, the resulting *x-y* image contains signals from cells that are distorted, and the distortion depends on the velocity of the cells relative to the progressing scan lines. RVFS analysis is then based on the relationship between RBC distortion and laser scanning velocity. The residence time (RT) is maximum when the scan velocity approximates that of the moving RBCs and is smallest when the RBC is moving in opposite sense to the field scan. To fully deconvolve the velocities and generate reliable data for all vessels in the field, we need to scan in multiple directions and speeds, and collect accurate morphological information of the vessel network (for calculating relative scan angles). Blood velocity can be resolved by comparing the RTs and the scanning velocities for scans in the same direction as the flow and fitting the data to the appropriate function (Figs. 1e and 3 – Equation derived from the Doppler effect). Furthermore, flux in a given vessel can be calculated from the number of RBCs and the scanning velocities.

When RT is maximum, scan velocity is similar to that of the moving RBCs and "velocity -matched" RBC streaks are generated. If the resulting stretched signals are significantly longer than RBC size ($>4 \times 7$ μm) they can be used to directly

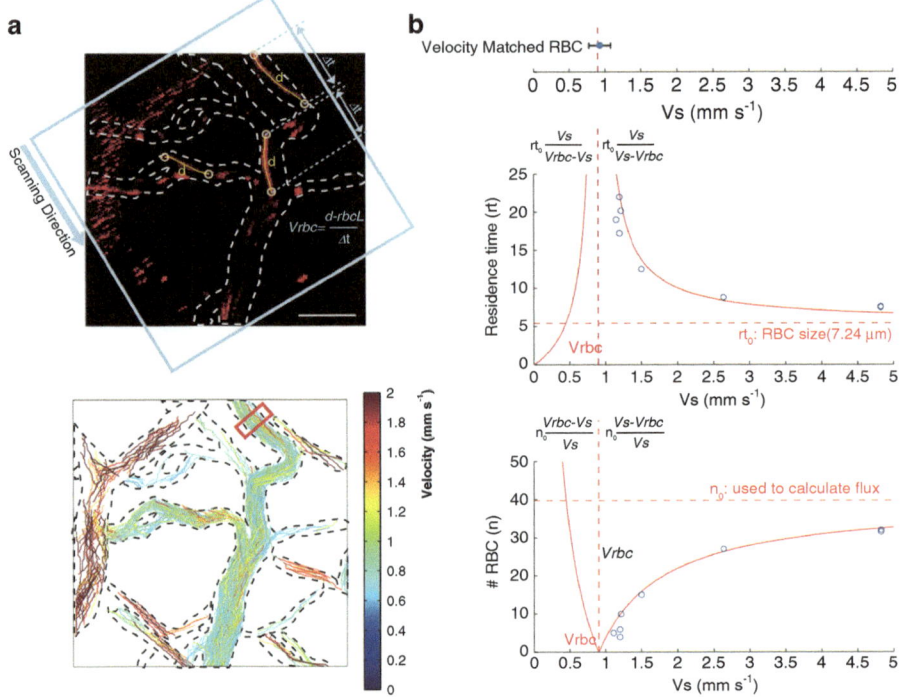

Fig. 3 Relative velocity field scanning (RVFS) method description. (**a**) Representative MPLSM image of brain vasculature imaged through the cranial window visualized by injecting DiD labeled red blood cells. Field was scanned at a 30 angle (*Cyan arrow* and *field*). "Velocity - matched RBCs" (vmRBCs) are shown, and traveled distance (d) and time (Δt) are highlighted (*yellow line* and *cyan arrow*). Velocity map generated by analyzing vmRBC tracks (*color mapped lines*). (**b**) Fitting the number of RBCs and residence times to scanning velocities for scans in the same direction as the flow allows computation of velocity (Vrbc) and number of RBC (n0) used to calculate flux (Equation derived from the Doppler effect). V_{rbc} values computed from the number of RBCs and residence times correlate with V_{rbc} computed from vmRBC. Scale bars, 50 μm

analyze traveled distance and calculate velocity (Figs. 1e and 3). Measurement of flow based on velocity matched RBCs allows averaging of velocity over the imaging time and minimizes the effect of temporal fluctuations. For vessels where no velocity-matched RBCs are generated and RT vs scanning speeds do not follow the doppler equation, it is necessary to increase the range by scanning either faster or slower through the field.

7 Procedure: Relative Velocity Field Scanning

The imaging protocol for RVFS consists of imaging the same field at various scanning velocities.

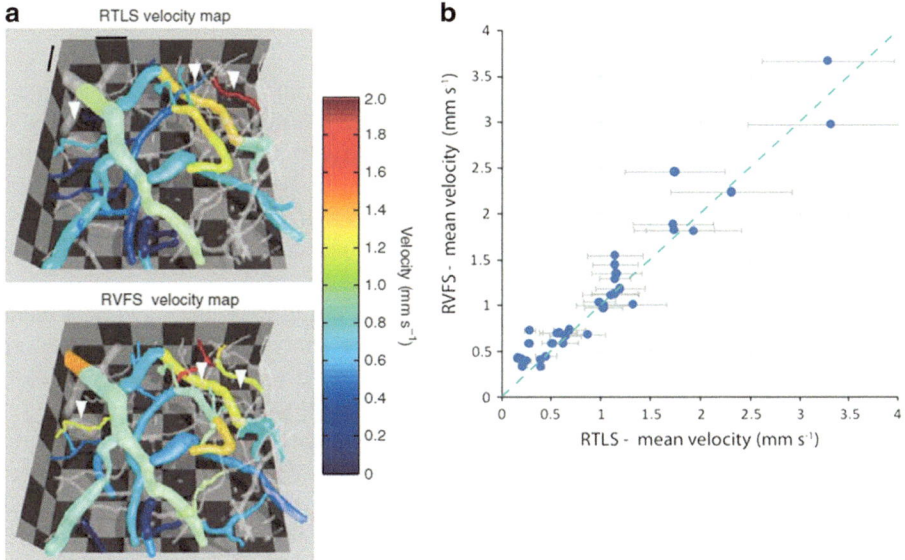

Fig. 4 RTLS & RVFS technique validation. (**a**) 3D Velocity map of a glioma network analyzed by RTLS and RVFS. Vessels in which RVFS velocity measurement differs from RTLS due to low RBC flux (*closed arrowheads*). Scale bars, 100 μm (a), 50 μm (b), 100 μm (c), 100 μm. (**b**) Comparison of RBC velocities measured by RTLS and RVFS shows correspondence between these techniques

1. Using a conventional MPLSM system, perform repetitive 3D imaging of the vessels at variable scanning velocities (1.53, 1.16, 0.78 and 0.39 mm s^{-1}) and at various resolutions (1.3 and 2.5 μm/pixel). Since vessels have various orientations the projected scanning velocity over the vessel axis is different for each vessel and is higher than the vertical velocity.
2. To ensure all vessels are scanned with a significant velocity component along the field scan direction, collect images at four rotation angles 0°, 90°, 180° and 270°.
3. If necessary, register the resulting images using an appropriate algorithm to correct for drift during acquisition.
4. Trace the vessel network and subdivide it into single vessel segments of equal length (50 μm).
5. Within each segment, determine the residence time for segmented RBCs and fit to the scanning velocity adjusted to the vessel angle to extract RBC velocity (Fig. 3b).
6. For RBC tracks representing a traveled distance greater than four times the length of a RBC (28 μm) distance and time can be extracted directly to calculate velocity.

RVFS and RTLS yield similar results when applied to the same vessel segments (Fig. 4). Differences in velocities measured by RVFS and RTLS appear more frequently in small vessels with relatively low RBC flux, where RVFS is less accurate.

8   Applications

By scanning through segments of vasculature undergoing intussusceptive angiogenesis [14], we were able record shear rate profiles around intussusceptive tissue structures. As shown in Fig. 2, there are differences in flow and shear rates between the segments separated by the interstitial tissue structures, and the shear rates vary around the perimeter of each of the segments; this contrasts with vessels not undergoing intussusception, in which the shear rate is uniform around the perimeter.

Another important application of these methodologies is the classification of vessels based on function. Current techniques for analyzing tumor blood vessels mainly rely on vessel structure or morphology to discriminate between vessel subtypes. To demonstrate the ability of our methodology to distinguish and analyze vessel populations, we used RTLS to measure velocities, fluxes and hematocrits of 2,351 vessels from six glioma-bearing animals (Fig. 5). Using green fluorescence protein (GFP) -expressing glioma cells, we were able to record the location of each vessel relative to the tumor mass and classify each as tumor or peritumor; contra-lateral brain hemisphere vessels were also analyzed (Fig. 6a). Considering only the average values, peritumor vessels had higher velocity, flux and wall shear rate compared to tumor or contralateral brain vessels (Fig. 6b). Interestingly, average tumor vessel perfusion parameters (velocity and flux) were not different than the contralateral brain (Fig. 6b). By clustering the vessels into specific phenotypes: Hypoperfused vessels – velocity <0.05 mm s^{-1}, Transport vessels – flux >300 RBC s^{-1}, and, and hemodiluted vessels – hematocrit <0.005, we could analyze the relationship between vessel location and function (Fig. 5a). For example, there is a population of vessels more prevalent in the tumor, characterized by low velocity and flux, high hematocrit and variable diameter. Transport vessels, defined as vessels with high flux, were present in all areas but with the highest fraction in the peritumor area, consistent with peritumoral arteriogenesis. Transport vessels had significantly larger diameters in the tumor and the peritumor compared to the contralateral brain, suggesting extensive dilation of feeding and draining vessels in these regions [4] (Fig. 5a, b). Hemodiluted vessels, defined by their low hematocrit, were present in all compartments but with higher fraction in the tumor and the peritumor areas (Fig. 5b). Low hematocrit vessels within the peritumor area were morphologically normal with diameters similar to normal brain and high RBC velocity. The variation in vessel hematocrit seen in the tumor and peritumor regions is likely due to loss of plasma by leaky vessels, which increases hematocrit. The extravasated plasma, which also increases tumor IFP [15], is eventually reabsorbed by other vessels, diluting their RBCs (Fig. 5c). Thus, RTLS and RVFS provide a framework for distinguishing and classifying vessel populations, providing more sensitive analyses of vascular physiology in and around tumors. This approach should enable focused studies of how vessel subpopulations shift during tumor growth and in response to therapies.

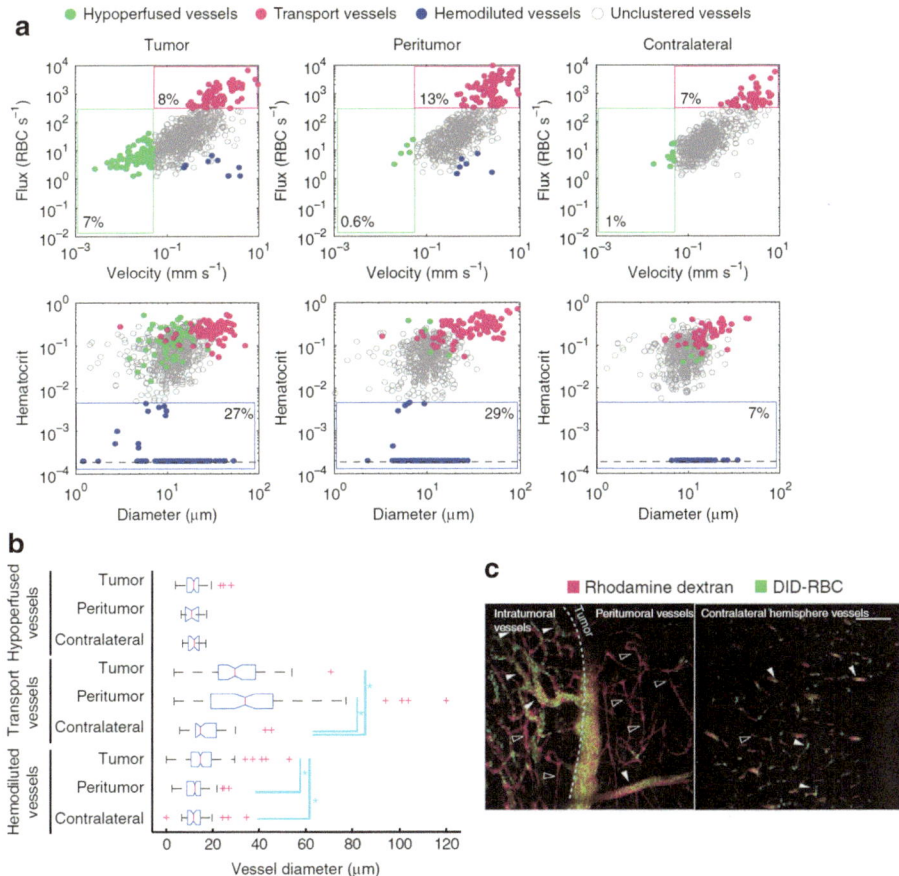

Fig. 5 Multiparametric phenotypic vessel clustering to compare vessels within tumor, peritumor and contralateral brain regions in a glioma model. (a) All the vessels analyzed in U87-bearing animals ($n = 6$) are plotted. Scatter plots of velocity vs flux and diameter vs hematocrit showing the gating applied to analyze three different vessel sub-types (hypoperfused, transport, and hemodiluted). **(b)** Box plots for vessel diameter from clustered vessels. (* $P < 0.05$). **(c)** U87 MPLSM micrographs. Rhodamine BSA was used for the angiographic contrast and fluorescent RBCs (*green*) were injected 1 day prior to imaging. Glioma cells expressed GFP which allowed accurate localization of the tumor and determination of the tumor edge (*light blue dashed line*). Tumor vessels are heterogeneous with high (*closed arrow head*) and low (*open arrow head*) hematocrit vessels. Peritumor vessels are morphologically normal and have low (*open arrow head*) hematocrit when compared to contralateral brain. Scale bars, 100 μm (c)

9   Discussion

Each of these techniques has advantages and disadvantages. RTLS has high sensitivity and allows cross-sectional analysis of flow within vessels, but flow analysis is limited to the location of intersection between the line and the vessel. RVFS allows analysis of entire vessel networks, integrating flow analysis at

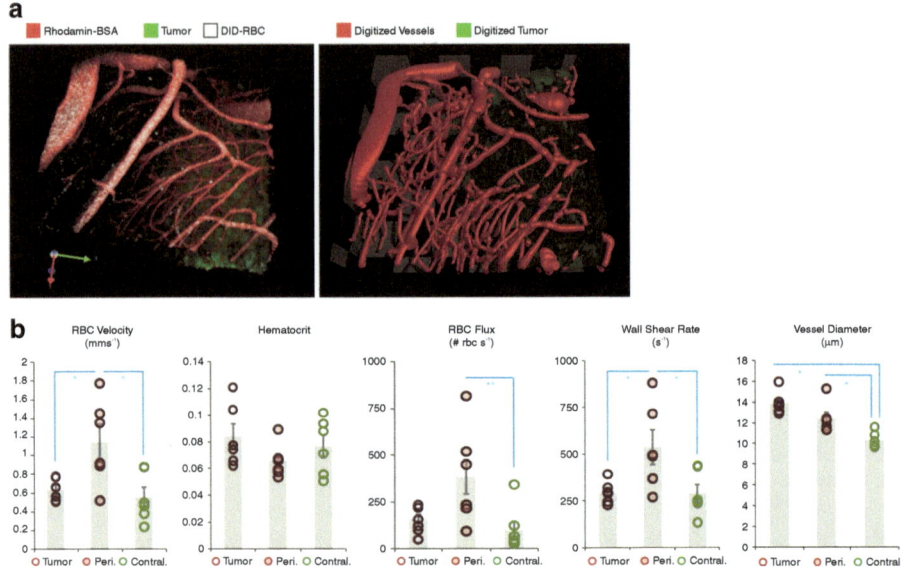

Fig. 6 Analysis of mean flow parameters to compare vessels within tumor, peritumor and contralateral brain regions in a glioma model. (**a**) MPLSM micrographs collected in a glioma model. Glioma cells expressed GFP (GFP-U87), allowing accurate localization of the tumor and determination of the tumor edge. (**b**) mean ± s.e.m. per animal (*circles*) and per experimental group (*bar graph*) of RBC Velocity, hematocrit, RBC flux, wall shear rate and vessel diameter in each region (*P < 0.05, n = 6)

multiple positions along vessels. However, it is slower due to the need for multiple scan angles and speeds, and its accuracy can be affected by low RBC flux. A comparison of ALS, RTLS and RVFS is presented in Table 1.

An important feature of RTLS is its ability to resolve velocities and fluxes spatially within individual vessels. The technique generates profiles of RBC velocity and flux (and therefore hematocrit) across each sampled blood vessel cross section (Fig. 1a, c). Furthermore, when combined with measurements of vessel cross-section assessed using a second fluorophore restricted to the plasma, accurate estimations of wall shear rates and stresses can be extracted.

Such spatially-resolved measurements of wall shear rates have the potential to answer long-standing questions of how blood shear forces exerted on endothelial cells contribute to processes such as atherogenesis and angiogenesis. For example, it is well known that endothelial cells sense shear stress gradients and respond by activating genes related to cell migration, vasoregulation and proliferation [16, 17]. But since there were previously no methods for measuring shear stress gradients in vivo, it is not known to what extent they affect the organization or migration of endothelial cells during angiogenesis. Systematic application of RTLS during the process of angiogenesis may resolve these issues.

Table 1 Comparison of laser scanning velocimetry methodologies

		Axial Line Scan (ALS)	Residence Time Line Scan (RTLS)	Relative Velocity Field Scan (RVFS)
Description	Protocol	Repeatedly scan along a line parallel to the vessel wall (Fig. 1a)	Repeatedly scan along a line intersecting one or multiple vessels (Fig. 1a, c)	Progressively scan a line over an area. Repeat at various scan speeds and angles (Figs. 1b and 3)
	Velocity analysis	Velocity is given by the angle of the signals in the x-t image	Velocity is extracted from the residence times of individual RBCs as they pass the scan line	Velocity is deconvolved from the stretched RBC signal, which is determined by the relative velocity between the field scan and the RBC
	Yield	Single "lane" within the analyzed vessel; resolution determined by resolution of laser scan	Cross-sectional velocity profile of all intersected vessels	Full velocity field within all vessels (within the range of sensitivity)
	Sensitivity	Depends on line scan rate and line length	Depends on line scan rate	Depends on Sampling (number of scan rate and angle variations)
	Application	Monitor velocity with high time resolution in a single vessel with moderate or fast flow	Monitor velocity and shear rate with high time resolution simultaneously in multiple vessels with moderate or slow flow	Measure average flow velocity in all the vessels of a region or volume of interest (within a predefined range of sensitivity)
Limitations	High velocity vessels Arteries	Line scan rates of standard systems[a] limit sensitivity. Solution: increasing line length increases sensitivity	Line scan rates of standard systems[a] limit sensitivity. Solution: analyze residence time at a single pixel instead of a line	Field scan speed and angle specifications have to be optimized specifically to analyze high velocity vessels
	Low velocity vessels	No limitations	No limitations	Field scan speed and angle specifications have to be optimized specifically to analyze low velocity vessels

(continued)

Table 1 (continued)

		Axial Line Scan (ALS)	Residence Time Line Scan (RTLS)	Relative Velocity Field Scan (RVFS)
	3D Vessels	Analyzes flow in the scanned plane. Velocity components perpendicular to the imaged plane are not resolved	No limitations	Analyzes flow in the scanned plane. Velocity components perpendicular to the imaged plane are not resolved
	Yield	Limited to a specific location in the chosen vessel. Biased to operator-selected vessels	Generally limited to a single cross-section of each intersected vessel	Limited to those vessels that have RBC velocity similar to the specified field scan velocity (speed and direction)
	Temporal Resolution	Limited by laser scan frequency (typically 500–1,000 scans/ s)	Limited by laser scan frequency (typically 500 –1,000 scans/ s)	Typical acquisition rates: for X-Y area, 12 s[b]; for X-Y-Z volume: 8 min[c]
	Potential Error	Accuracy of RBC d-t image angle is critical	Accuracy of RBC residence time is critical. RBC preferential orientation can contribute error to the residence time analysis	Accuracy of measurement of the distance travelled by the RBC is critical
Advantages		Higher sensitivity than RTLS for high velocity vessels No need to label tracer RBCs	Measures velocity profile and shear rates Analyzes flow in penetrating and diving vessels Can analyze multiple vessels simultaneously	Measures all vessels in the area of interest

[a]Olympus confocal or multiphoton laser scanning microscope. Olympus FV300 Scanning unit: 256 × 256 pixels (660 × 660 µm) scanned in 0.45 s
[b]Assumes 660 × 660 µm field sampled at six angles with two speeds at each angle
[c]Assumes 660 × 660 × 100 µm field sampled at six angles with two speeds at each angle (40 z-slices)

References

1. Fukumura D, Yuan F, Monsky WL et al (1997) Effect of host microenvironment on the microcirculation of human colon adenocarcinoma. Am J Pathol 151:679–688
2. Brizel DM, Klitzman B, Cook JM et al (1993) A comparison of tumor and normal tissue microvascular hematocrits and red cell fluxes in a rat window chamber model. Int J Radiat Oncol Biol Phys 25:269–276
3. Endrich B, Reinhold HS, Gross JF et al (1979) Tissue perfusion inhomogeneity during early tumor growth in rats. J Natl Cancer Inst 62:387–395
4. Nagy JA, Chang SH, Dvorak AM et al (2009) Why are tumour blood vessels abnormal and why is it important to know? Br J Cancer 100:865–869
5. Kamoun WS, Schmugge SJ, Kraftchick JP et al (2008) Liver microcirculation analysis by red blood cell motion modeling in intravital microscopy images. IEEE Trans Biomed Eng 55:162–170
6. Rosenblum WI, El-Sabban F (1981) Measurement of red cell velocity with a two-slit technique and cross-correlation: use of reflected light, and either regulated dc or unregulated ac power supplies. Microvasc Res 22:225–227
7. Kleinfeld D, Mitra PP, Helmchen F et al (1998) Fluctuations and stimulus-induced changes in blood flow observed in individual capillaries in layers 2 through 4 of rat neocortex. Proc Natl Acad Sci USA 95:15741–15746
8. Brown EB, Campbell RB, Tsuzuki Y et al (2001) In vivo measurement of gene expression, angiogenesis and physiological function in tumors using multiphoton laser scanning microscopy. Nat Med 7:864–868
9. Drew PJ, Blinder P, Cauwenberghs G, Shih AY and Kleinfeld D et al (2009) Rapid determination of particle velocity from space-time images using the Radon transform. J Comput Neuroscience. doi:10.1007/s10827-009-0159-1
10. Lindsey ES, Donaldson GW, Woodruff MF (1966) Erythrocyte survival in normal mice and in mice with autoimmune haemolytic anaemia. Clin Exp Immunol 1:85–98
11. Jain RK, Munn LL, Fukumura D (2002) Dissecting tumour pathophysiology using intravital microscopy. Nat Rev Cancer 2:266–276
12. Winkler F, Kozin SV, Tong RT et al (2004) Kinetics of vascular normalization by VEGFR2 blockade governs brain tumor response to radiation: role of oxygenation, angiopoietin-1, and matrix metalloproteinases. Cancer Cell 6:553–563
13. Tyrrell JA, di Tomaso E, Fuja D et al (2007) Robust 3-D modeling of vasculature imagery using superellipsoids. IEEE Trans Med Imaging 26:223–237
14. Patan S, Tanda S, Roberge S et al (2001) Vascular morphogenesis and remodeling in a human tumor xenograft: blood vessel formation and growth after ovariectomy and tumor implantation. Circ Res 89:732–739
15. Jain RK, Tong RT, Munn LL (2007) Effect of vascular normalization by antiangiogenic therapy on interstitial hypertension, peritumor edema, and lymphatic metastasis: insights from a mathematical model. Cancer Res 67:2729–2735
16. Song J, Munn L (2011) Fluid forces control endothelial sprouting. Proc Natl Acad Sci USA 108:15342–15347
17. Frangos JA, McIntire LV, Eskin SG (1988) Shear stress induced stimulation of mammalian cell metabolism. Biotechnol Bioeng 32:1053–1060

ERRATUM TO

The Murine Hindbrain as a Model to Study the Molecular and Cellular Mechanisms of Angiogenesis in Intact Tissues

Charlotte Maden and Christiana Ruhrberg

DOI 10.1007/978-94-007-4581-0_26

The following reference was inadvertently omitted from the list of references:

Fantin, A.F, Vieira, J.M.V, Plein, A.R., et al., The embryonic mouse hindbrain as a qualitative and quantitative model to study the molecular and cellular mechanisms of angiogenesis. Nature Protocols, doi:10.1038/nprot.2013.015.

The research shown in Fig. 1 was originally published in Blood. Fantin et al. Tissue macrophages act as cellular chaperones for vascular anastomosis downstream of VEGF-mediated endothelial tip cell induction. Blood 2010; 116(5):829–84 © the American Society of Hematology.

E. Zudaire and F. Cuttitta (eds.), *The Textbook of Angiogenesis and Lymphangiogenesis: Methods and Applications*, DOI 10.1007/978-94-007-4581-0_13, © Springer Science+Business Media Dordrecht 2012

Index

E. Zudaire and F. Cuttitta (eds.), *The Textbook of Angiogenesis and*
Lymphangiogenesis: Methods and Applications, DOI 10.1007/978-94-007-4581-0,
© Springer Science+Business Media Dordrecht 2012